A useful reference to units and to physical data is:

Tables of Physical and Chemical Constants, compiled by C.W. Kaye and T.H. Laby. Longman, London and New York (1986)

Basic Units

Quantity	Unit	Abbreviation
Length	meter	m
Mass	kilogram	kg
Time	second	s
Electric current	ampere	A

Composite Units

Quantity	Unit	Abbreviation	Composition
Velocity			$\mathrm{m\,s^{-1}}$
Acceleration			$\mathrm{m\,s^{-2}}$
Force	newton	N	$1\ \mathrm{N} = 1\ \mathrm{kg\,m\,s^{-2}}$
Energy	joule	J	$1\ \mathrm{J} = 1\ \mathrm{N\,m}$
Power	watt	W	$1\ \mathrm{W} = 1\ \mathrm{J\,s^{-1}}$
Magnetic flux density	tesla	T	$1\ \mathrm{T} = 1\ \mathrm{V\,s\,m^{-2}}$

A table of physical data is given inside the back cover.

Elements of Newtonian Mechanics

Springer
Berlin
Heidelberg
New York
Barcelona
Budapest
Hong Kong
London
Milan
Paris
Santa Clara
Singapore
Tokyo

J. M. Knudsen P. G. Hjorth

Elements of Newtonian Mechanics

Including Nonlinear Dynamics

Second Revised and Enlarged Edition
With 324 Figures,
109 Problems with Solutions,
and 92 Worked Examples

Springer

Dr. Jens M. Knudsen

Niels Bohr Institute, Orsted Laboratory, Universitetsparken 5
DK-2100 Copenhagen O, Denmark

Dr. Poul G. Hjorth

Technical University of Denmark, Mathematical Institute, B-303
DK-2800 Lyngby, Denmark

Library of Congress Cataloging-in-Publication Data. Knudsen, J. M. (Jens M.), 1955– Elements of Newtonian mechanics : including nonlinear dynamics / J. M. Knudsen, P. G. Hjorth. – 2nd rev. and enl. ed. p. cm. Includes bibliographical references and index. ISBN 3-540-60841-9 (alk. paper) 1. Mechanics. 2. Mechanics–Problems, exercises, etc. I. Hjorth, P. G. (Poul G.), 1930– . II. Title. QC125.2.K48 1996 531–dc20 96-5697

ISBN 978-3-540-60841-7 ISBN 978-3-642-97673-5 (eBook)
DOI 10.1007/978-3-642-97673-5

Typesetting: Data conversion by Springer-Verlag
SPIN 10768375 56/3144-543210 - Printed on acid-free paper

There is no physical experiment
by which you can determine
and later recover
a single point in astronomical space

Preface to the Second Edition

In the second edition, a number of misprints that appeared in the first edition have been corrected. In addition to this, we have made improvements based on the experience gathered in the use of the first English edition of the book in the introductory course in physics at the University of Copenhagen.

A chapter introducing nonlinear dynamics has been added. The purpose of this chapter is to provide supplementary reading for the students who are interested in this area of active research, where Newtonian mechanics plays an essential role. The students who wish to dig deeper, should consult texts dedicated to the study of nonlinear dynamical systems and chaos. The literature list at the end of this book contains several references for the topic.

The book still contains a one-semester (15 weeks) first university course on Newtonian mechanics. This necessarily introduces some constraints on the choice of topics and the level of mathematical sophistication expected from the reader. If one looks for discussions of technical issues, such as the physics behind various manifestations of friction, or the tensorial nature of the rotation vector, one will look in vain. The book contains what we feel are the essential aspects of Newtonian Mechanics.

It is a pleasure again to thank Springer-Verlag and in particular Dr. H. J. Kölsch and the staff at the Heidelberg office for helpfulness and professional collaboration.

Copenhagen *JMK & PGH*
June 1996

Preface to the First Edition

This book is intended as a textbook for an entry-level university course in Newtonian mechanics for students of physics, astronomy, and the engineering sciences. The material has been used as a first-semester text for first-year undergraduates at the Niels Bohr Institute, which is part of the University of Copenhagen.

Our way of presenting Newtonian mechanics is influenced by the writings of the late Max Born. Also, the *Feynman Lectures on Physics* have been an important source of inspiration. In fact, the idea for the book came when we read Section 16.1 of Volume 1 of the *Feynman Lectures*. Ideas from the well-known *Berkeley Physics Course* may also be traced in the text. All of the books quoted in the literature list have, in one way or another, served as a source for our lectures for undergraduates.

It is assumed that the students already have a rudimentary knowledge of Newtonian mechanics, say at the high-school level. Some background in *vectors* and *elementary calculus* is also required, i.e., the students should know how to add vectors as well as how to differentiate and integrate elementary functions. The Appendix contains the required background for the use of vectors in Newtonian mechanics.

Careful study of the many worked examples will give the student the ability to use the powerful tools of Newtonian mechanics. Furthermore, we emphasize the fundamental problem of motion from the very beginning. This prepares the students for an understanding of Einstein's special and general theory of relativity. The text will demonstrate for the student that the answer to the question "What does it mean that a body moves?" is far from simple.

The authors wish to express their gratitude to many colleagues for discussions and for the encouragement given. Special thanks goes to J. Lyngesen who has read the entire manuscript and provided many valuable suggestions.

It is a pleasure for the authors to thank cand. polyt. E. B. Beran at the Technical University of Denmark, B. Kaluza and Blue Sky Research (makers of Textures) for assistance in LaTeX typesetting, and Springer-Verlag Heidelberg, in particular Dr. H.J. Kölsch, for excellent cooperation in the preparation of the manuscript.

Copenhagen
January 1995

Jens Martin Knudsen
Poul Georg Hjorth

A Note for the Reader

Nearly half of this book consists of examples. These examples form an absolutely essential part of the text and should by no means be skipped. The authors believe that Newton's ideas on the subtle problem of motion can be assimilated only by working with Newton's equations applied to concrete examples.

Listening to lectures is not enough. All processes of learning are somehow connected to active participation, and the learning of physics is no exception. To underline this viewpoint we have, at the beginning of the course, always written on the blackboard, as a kind of motto:

At Home
by Your Desk.

Nearly all the chapters in the book are followed by a set of problems. Very few of these problems are simple "plug-in" exercises. Most problems will demand some independent thinking. If you cannot solve all the problems at first try, do not despair. We have good advice which has worked for many students: study the text, and in particular the examples, one, two, ... many times over. In the end, you will succeed.

For several problems you will need some parameters, e.g., the mass of the Earth, the distance of the Earth from the Sun, etc. The numbers you need can all be found on the inside of the cover of the book. Answers to the problems are found at the back of the book.

Throughout the book SI units have been used. Relevant SI units can be found on the inside of the cover.

A Note for the Instructor

In Copenhagen, less than one quarter of the examples were presented in the lectures. Most of the remaining examples were assigned for recitation sessions.

JMK & PGH

Contents

1. The Foundation of Classical Mechanics

1.1 Principia

It was an important event in the history of physics when Sir Isaac Newton in 1687 published his book *Philosophiae Naturalis Principia Mathematica.*

In this famous work we find a masterly synthesis of the concepts of motion and force. The Newtonian formulation of the laws of motion has, with superior strength and vitality, survived more than 300 years. Although certain fundamental aspects of Newtonian physics have been revised in this century, Newton's principles still find widespread use both in the basic and in the applied sciences. Just one example: the launching of artificial satellites is based directly on these principles.

Newton's work forms the basis upon which theoretical physics as well as modern engineering science rests. We shall here, without going into details of the historical development, discuss some of the most important steps in the evolution of the Newtonian world picture.

1.2 Prerequisites for Newton

There are two main lines of inquiry that lead towards and form the basis for Newton's efforts: one based on motions here on Earth, and one based on the motions of bodies in the heavens. The first line is associated with the name Galileo Galilei (1564–1642); the other with Nicolas Copernicus (1473–1543), Tycho Brahe (1546–1601) and Johannes Kepler (1571–1630). Before Galileo, the "impetus" concept dominated the thinking on motion. This idea is somewhat related to the later concept of momentum, but although it was thought that a body was given a certain impetus when it was thrown, the body was thought to spend its impetus during its flight. When all the initial impetus was spent, the body would stop and – if it was not supported – fall down.

As for the free fall, it had been held since Aristotle (384–322 B.C.) that a heavy object falls faster than a light object. From immediate everyday experience, both the impetus concept and the idea that a heavy object falls faster than a light one, are not unreasonable.

It was Galileo who cleared away these misconceptions and in this way erected one of the pillars on which the work of Newton rests. Let us here state the laws that Galileo formulated:

(1) **The law of inertia:** If a body is left to itself without influence from other bodies it will continue a uniform linear motion if it initially had one, and will remain at rest if it initially was so.
(2) **The laws of free fall:**
   (a) All freely falling objects will, when starting from rest, fall an equal distance in the same time.
   (b) The distance fallen, S, is proportional to the square of the time of fall: $S = \frac{1}{2}gt^2$ where g is a constant (g is the acceleration). According to Galileo the constant acceleration g is *independent of the nature or composition of the falling object.*

What was so revolutionary about Galileo's way of thinking; why is his work today considered as the beginning of modern natural science? There are two basic features which separate Galileo's work from earlier attempts to formulate laws of motion.

First, there is the systematic use of *experiments* to decide what is true and what is false. Before Galileo – and in particular in the Greek tradition – the view was that the laws of nature could be obtained by pure speculation. Galileo showed that experiments are of fundamental importance for our comprehension of the laws of nature.

The other revolutionary feature of Galileo's work was that he took the bold step of extrapolating to a "pure" or *idealized* motion by systematically disregarding features of the motion that need separate analysis. Galileo realized that friction stems from the surroundings and is not a fundamental feature of the motion. In other words, no momentum (or "impetus") is lost in motions where no friction is present – e.g. in the motion of the celestial bodies through interplanetary space no friction is present. In free fall on the Earth we can also – in a first approximation – disregard air resistance, if we focus on small, heavy bodies and let them fall through a short distance so that they do not achieve large velocities. The basis for the other main line of thought leading up to Newton's work was laid down even before the birth of Galileo, by the Polish astronomer Copernicus. In his great work *De Revolutionibus Orbium Coelestium* (1543) he replaced a geocentric model of the world with a heliocentric model.

The idea of a heliocentric system, i.e., a system with the Sun at the center, had been put forward in antiquity, but it had been suppressed by the Ptolemean teaching which was dominant up to the time of Copernicus. In the Ptolemean world model the Earth is at the center of the universe. The Earth is at rest, while the celestial sphere turns once every 24 hours around an axis which intersects the celestial sphere near the North Star. This world model could explain several features of the apparent motion of the Sun, the Moon and the planets. With complicated models of uniform circular motions where circles roll on other circles (producing the so-called epicycles) many features of the motions of celestial bodies could be *quantitatively predicted*.

Copernicus removed the Earth from its central position in the universe. In the world system of Copernicus, the Sun is at the center of the universe. The planets move in circles around the Sun. The planes of the planetary orbits have slightly different positions relative to each other. One of the planets is the Earth, which rotates around its axis once every 24 hours and revolves around the Sun once every year. The Moon on the other hand moves in a circle around the Earth. The stars are suns like our own Sun, and they are assumed to be at rest in the universe. Perhaps Copernicus did not achieve any significant simplification compared to the Ptolemean system, but he had in a decisive way shaken the belief in an immovable Earth, and the axioms of Copernicus show a deep understanding of motion in the planetary system. One can say that Copernicus gave a *qualitative* description of motions in the solar system.

While speculations on the merits of the two world systems were going on, the Danish astronomer Tycho Brahe chose a different path, closely related to that of Galileo. Brahe was of the opinion that the conflict should be resolved *not by philosophical speculations, but through accurate observations* to produce a set of measurements that could form a basis from which to proceed. To this end, he himself built and designed the necessary instruments, and from his observatories on the then Danish island of Hven he measured – through more than 20 years – the positions of planets and fixed stars with an accuracy of about 1 minute of arc. After a controversy with the Danish king, Brahe set up residence in Prague as "Imperial Mathematician". Here he worked briefly with the young Johannes Kepler. They met just at the turn of the century, and when Brahe died nearly two years later, Kepler took over the entire set of observations left by Brahe, and proved himself worthy of this legacy. The confidence Kepler had in the precision of Brahe's measurements was, as we shall see, a decisive factor in his work.

Kepler built on the heliocentric system of Copernicus and decided from the measurements of Brahe to find the orbit (around the Sun) of the planet Mars. He was assuming a circular orbit, but little by little he realized that he had to incorporate an eccentric orbit in which the planet moves in an irregular fashion. In his main work *Astronomica Nova* (1609) Kepler describes how the theory of circular orbits gives rather good agreement with the observations of Brahe. But then suddenly, without any transition, the following crucial statement appears: "Who should have thought it possible? This hypothesis, which agrees so well with the observed oppositions is nevertheless wrong".

What had happened was that Kepler, by considering some of the most accurate of Brahe's observations, had come across a deviation of about 8 minutes of arc from the positions computed on the assumptions of a circular orbit. At this turning point Kepler proves himself worthy of the legacy of Brahe. "On these 8 minutes of arc I shall build a world" was Kepler's comment. Let Kepler speak for himself:

> But for us, who by divine grace has been given an accurate observer like Tycho Brahe, for us it is only fitting to appreciate this divine gift and put it to use... From this point on I shall follow my own ideas toward this goal. For if I had thought that I could ignore these 8 minutes of arc, I would have been patching my hypothesis. But as it is not permissible to ignore them, these 8 minutes of arc show the way to a complete reformation of astronomy; they are the building blocks to a large part of this work.

We see here one of the summits of quantitative science – and we shall see another equally grand example when we describe the work of Newton, towards the end of this chapter. It is on this background one must see the life work of Kepler. By peculiar routes, but governed by an almost infallible intuition, Kepler produced, after having given up the idea of circular orbits, his three famous laws of planetary motion:

(1) The orbit of a planet relative to the Sun lies in a fixed plane containing the Sun, and each planet moves around the Sun in an elliptical orbit with the Sun in one focus.
(2) The radius vector from the Sun to the planet sweeps out equal areas in equal amounts of time.
(3) The square of the period of revolution of a planet is proportional to the third power of the greatest semi axis of the ellipse. If, therefore, T denotes the period of revolution and a the greater semi axis,

$$T^2 = \frac{1}{C}a^3 \quad \text{or} \quad \frac{a^3}{T^2} = C \,, \tag{1.1}$$

where C has the same value for all the planets.

Kepler found the third law in 1618 and published it in *Harmonicus Mundi* (1619). An enormous leap forward had been achieved in the short span of years from 1601 to 1618. In these three elegant laws Kepler summarized all of the enormous amount of data left to him by Brahe. We have here the first *quantitative kinematic* description of the solar system, a description which precisely tells how the planets move, but without pointing to a cause for that motion.

Even though Kepler perhaps has in mind some of the mystic visions which his work also contains, we must acknowledge the justifications of his cry of joy after all these years of work:

> What I dimly perceived 25 years ago before I had discovered the 5 regular solids between the celestial orbits...; what I 16 years ago declared the end goal of my research; what made me spend the best years of my life in studying astronomy, what made me join Tycho

> Brahe and live in Prague – that I have now, by the grace of God... finally uncovered. After having sensed the first inkling of a dawn 18 months ago, daylight 3 months ago, but only a few days ago the full daylight of a wonderful vision – then nothing can stop me now. I revel in joyous bliss. I challenge all mortals with this open confession: I have robbed the golden chalice of the Egyptians for with them to make a tabernacle for my God far from the lands of Egypt. If you forgive me I will rejoice. If you are angry, I can bear it. I have thrown the dice and written the book either for my contemporaries or for posterity. It makes no difference to me. I can wait a hundred years for a reader when God has waited six thousand years for a witness.

Kepler did not have to wait a hundred years for a reader.

1.3 The Masterpiece

Armed with the enormous empirical material which is condensed in the laws of Kepler and Galileo, Isaac Newton (1642–1727) managed to discover his famous laws of motion and the general law of gravitational interaction. In order to continue the historical review in this chapter we postpone a more detailed discussion of the laws of Newton to Chapter 2. We are assuming that the reader already has an elementary knowledge of these laws, so we will here only briefly state Newton's three laws of motion:

(1) **Law of inertia.** A body which is not acted upon by any force will either be at rest or in a state of uniform linear motion.
(2) **Law of acceleration.** The mass m of a body, times its acceleration $\mathbf{a}$ equals the net force $\mathbf{F}$ on the body: $m\mathbf{a} = \mathbf{F}$.
(3) **Law of action and reaction.** If body A acts on body B with a force, then body B will act on body A with an equal but oppositely directed force.

The acceleration $\mathbf{a}$ of a body is a vector, as is the force $\mathbf{F}$. A brief discussion of vectors and vector calculus is provided in the Appendix of this book. Bold face variables are used to designate vector quantities.

Newton arrived at the general law of gravity through an ingenious analysis of Kepler's three laws. This analysis – and its result – is of great importance in the history of the exact sciences. We shall here demonstrate the main features of Newton's method. We restrict ourselves to the simple case of uniform circular motion, so that the physical argumentation will stand out as clearly as possible. But the method can, as we shall see later (Chapter 14), easily be generalized to the more realistic case of an elliptical orbit.

The fundamental law of mechanics connects the force $\mathbf{F}$ acting on a body with the acceleration of that body produced by the force:

$$m\mathbf{a} = \mathbf{F}. \tag{1.2}$$

Newton realized that Kepler's three laws permit a calculation of the acceleration that the planets undergo in their motion around the Sun. *If we assume the validity of (1.2) and compute the acceleration of a planet from Kepler's laws, we can determine the force that acts on the planet.* The determination of the force takes place in a sequence of steps.

(1) **Use of Kepler's 1st law.** Since a circular motion is a special case of elliptical motion (having both foci in the center), the assumption of circular orbits with the Sun at the center is at least not in conflict with Kepler's first law. Incidentally, this assumption gives a "reasonable" approximation to the actual orbits of the planets, which are nearly circular orbits.

(2) **Use of Kepler's 2nd law.** Kepler's second law (the law of areas) demands that the circle is traced out with constant magnitude of the velocity, i.e., the motion must be a uniform circular motion. The analytical expression for the acceleration in a uniform circular motion was first derived by Christian Huygens (1629–1695). In Example 1.2 below, the acceleration in a uniform circular motion is determined by vector analysis. The result is: the acceleration is directed towards the center (the Sun) and has the magnitude

$$a = \frac{v^2}{r}, \tag{1.3}$$

where v is the constant magnitude of the velocity and r is the radius of the orbit.

(3) **Use of Kepler's 3rd law.** In order to use the third law we must introduce the connection between v, r, and the period of revolution, T. We have, for a uniform circular motion,

$$v = \frac{2\pi r}{T}. \tag{1.4}$$

By eliminating v, we get from (1.3) and (1.4) that

$$a = \frac{4\pi^2 r}{T^2}. \tag{1.5}$$

From Kepler's third law we have:

$$\frac{r^3}{T^2} = C, \tag{1.6}$$

where C is the same constant for all planets. Inserting T^2 from (1.6) into (1.5), we find:

$$a = \frac{4\pi^2 C}{r^2}. \tag{1.7}$$

We see that the acceleration of a planet depends only on its distance from the Sun, since the acceleration is inversely proportional to the square of the distance of the planet from the Sun. The constant C does not depend on the mass of the planet since C is the same for all planets. However, C may depend on the mass of the Sun. We shall later demonstrate that (1.7) is valid also for elliptical motions (where r is the time dependent distance between the Sun and planet, see Chapter 14). From Kepler's three empirical laws we have computed the acceleration of a planet as a function of its distance from the Sun.

(4) **Use of Newton's 2nd law.** If we now finally calculate the magnitude of the force on the planet, by inserting the acceleration (1.7) into (1.2) we have, with m the mass of the planet,

$$F = m\frac{4\pi^2 C}{r^2} . \tag{1.8}$$

The force is attractive, i.e., directed towards the Sun, and has a magnitude inversely proportional to the square of the distance to the Sun.

By means of Kepler's laws and Newton's second law we have thus calculated the force acting on a planet. At this point Newton took a decisive step by realizing that the acceleration, as expressed in Equation (1.7) has a very important property in common with the acceleration occurring in free fall near the surface of the Earth. In both cases the acceleration is independent of the mass of the accelerating object.

If we calculate the magnitude of the gravitational force on an object of mass m near the surface of the Earth, we find:

$$F = mg, \tag{1.9}$$

where g is the acceleration of gravity near the surface of the Earth. The force is – just as the force calculated above in (1.8) is – *proportional to the mass of the object.* Starting from this observation, Newton took the bold step of postulating that the two forces – the force pulling an object (say, an apple!) toward the Earth, and the force holding a planet in orbit around the Sun – are of the same physical nature. Both forces are expressions of gravitational attraction. Newton connected the laws for falling bodies here on Earth (found by Galileo) and the laws for the motion of celestial bodies (found by Kepler). The final quantitative test, which is the touchstone for any physical theory, remained. This *quantitative* test is of fundamental importance for the development of Western civilization.

Newton chose to test his theory on the motion of the Moon. Before we retrace the calculations of Newton, there is one problem we must describe more closely: can we use the law of gravity at all, near the surface of the Earth? Obviously, as shown in Figure 1.1, the different parts of the Earth pull in different directions on a given particle A near the surface of the Earth, and it is not clear what the net force on the particle will be.

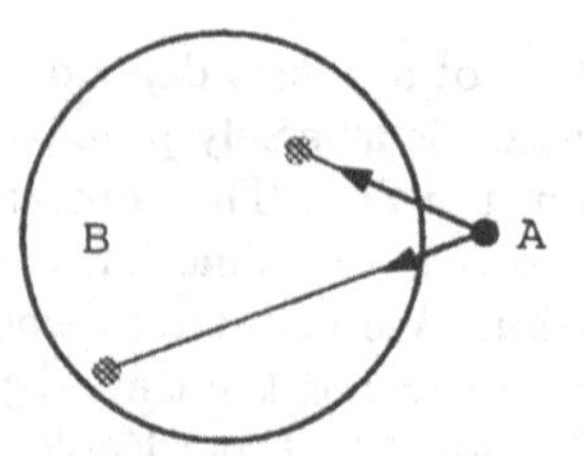

Fig. 1.1. The gravitational interaction between a particle A and a spherically symmetric body B. The body B acts on the particle A as if all of the mass of the body was concentrated in the center

However, in *Principia*, Newton showed the important theorem that *the gravitational force from a body with a spherically symmetric distribution of mass is the same as it would be if the entire mass of the body was concentrated in the center.* We will prove this theorem later in the course (see Section 8.6).

We return now to Newton's calculation. As mentioned, Newton did not test his law of gravity on the motions of the planets around the Sun, but on the motion of the Moon around the Earth. The reason was, that the distance from the Earth to the Moon was known from direct measurements (triangulations). The central body is now the Earth, and the Moon takes the place of a planet. Let the orbital period of the Moon be T , and let r be the radius in the circular orbit of the Moon. The radius of the Earth is denoted ρ.

If the gravitational force on a body near the surface of the Earth has the same origin as the force with which the Earth acts on the Moon, we must have:

$$g = \frac{4\pi^2 C'}{\rho^2} \tag{1.10}$$

(compare with (1.7)). C' has the same value for all objects acted upon by the gravitational force of the Earth. C' depends on the mass of the Earth, just as C in (1.7) depends on the mass of the Sun. We can therefore calculate the constant C' by observing the motion of the Moon. From Kepler's third law,

$$C' = \frac{r^3}{T^2} \, . \tag{1.11}$$

Inserting this into (1.10) we get

$$g = \frac{4\pi^2 r^3}{\rho^2 T^2} \, . \tag{1.12}$$

T is the so-called sidereal month, i.e., the time between two consecutive positions of the Moon when the line connecting the Earth and the Moon has the same direction with respect to the fixed stars.

Note. From the direct measurements the quantities on the right-hand side of (1.12) are known (and were known by Newton).

$$\begin{aligned} T &= 27 \text{ days } 7 \text{ hours } 43 \text{ minutes} \\ r &= 60.1\,\rho \\ \rho &= 6.37 \times 10^6 \text{ m} \end{aligned}$$

Inserting these values into (1.12) we find:

$$g_{\text{calculated}} = 9.8 \text{ m s}^{-2} \,.$$

This means that,

(i) by observing the Moon and analyzing its motion, and
(ii) by assuming gravity to be the common cause, both of the motion of the Moon around the Earth, and of the fall of an apple to the ground,

we have computed the acceleration which gravity imposes on an object near the surface of the Earth.

From direct measurements of the quantity g – for example from experiments with pendulae and watches – we have

$$g_{\text{measured}} = 9.8 \text{ m s}^{-2} \,.$$

This beautiful agreement is rightfully considered a high point in the history of the exact sciences. One can say that these considerations and calculations represent the first piece of theoretical physics, the birth of the method of natural science. This method has changed the conditions of living for man on our planet; both spiritual and material conditions have changed so rapidly in the three centuries since Newton that we can declare that it is as if an entirely new species came into being on our planet in the years 1543–1687.

History tells us that when Newton first performed these calculations, the deviations were so large that he considered his theory to be contrary to the facts and rejected it. Some years later new measurements showed that the astronomers had used a false value for the distance to the Moon. When Newton learned about this, he redid the calculations and found the agreement between the measured and the calculated value.

What remains is to find the final formulation of the law of general gravitational attraction. It is difficult to assert how Newton arrived at this; that feat is a combination of systematic analysis and intuitive insight. A possible route is the following: we saw that the Sun acts on a planet with a force **F** directed toward the Sun and with the magnitude

$$F = m\frac{4\pi^2 C}{r^2} \,, \tag{1.13}$$

where C depends on the mass of the Sun. If a law of general gravitational attraction exists, the planet must act on the Sun with a force $\mathbf{F}'$ of magnitude

$$F' = M\frac{4\pi^2 c}{r^2} , \tag{1.14}$$

where M is the mass of the Sun and c depends on the mass of the planet.

If we assume the validity of Newton's third law (action = reaction), then the force **F** with which the Sun acts on the planet must be equal in magnitude (but oppositely directed) to the force **F**$'$ with which the planet acts on the Sun. We then have

$$m\frac{4\pi^2 C}{r^2} = M\frac{4\pi^2 c}{r^2} , \tag{1.15}$$

or,

$$\frac{C}{M} = \frac{c}{m} , \tag{1.16}$$

Let us define a constant G from the common value of C/M and c/m:

$$\frac{C}{M} = \frac{c}{m} = \frac{G}{4\pi^2} , \tag{1.17}$$

from which

$$4\pi^2 C = GM , \tag{1.18}$$

$$4\pi^2 c = Gm . \tag{1.19}$$

If we insert (1.18) and (1.19) into (1.13) and (1.14), respectively, Newton's law of attraction of mass assumes the following symmetrical form:

$$F = G\frac{mM}{r^2} , \tag{1.20}$$

where G is a universal constant. The numerical value of G depends only on the choice of units. The currently accepted value is

$$G = 6.668 \times 10^{-11}\ \mathrm{N\,m^2\,kg^{-2}} .$$

Newton generalized the above result too: two arbitrary particles attract each other with a force proportional to the mass of each of the two particles and inversely proportional to the square of the distance between the particles. The constant of proportionality is the universal constant G.

For over 200 years this generalization has resisted innumerable tests. Its area of validity is enormous. It is gravity which determines the fall of the apple toward the ground, guides the Moon in its orbit around the Earth, and the Earth and the planets in their orbits around the Sun. Gravity directs the motion of stars in the galaxy, and each galaxy acts with gravitational forces on other galaxies. It is in this general form – that any object in the universe attracts all other objects in the universe – that the gravitational law of Newton is seen as an important step forward. In its original form (1.8), it is derived from the laws of Kepler and is merely a short and precise (although surprisingly simple) way to express these laws. As is well known, Newton's law of gravity was modified in the beginning of this century, by Einstein's general theory of relativity.

Newton discovered his laws through a combination of intuition and systematic analysis of observed facts. These facts are elegantly and precisely expressed in the laws of Galileo and Kepler. One can ponder how the "detached" measurements of planetary positions by Brahe – through the enormous computational effort and unique intuition of Kepler – became the foundation of Newton's laws of motion and in this way was decisive in creating one of the pillars of modern science. Below follow some examples of the use of Newton's laws applied to the study of uniform circular motion in the gravitational field of a spherical distribution of mass.

Example 1.1. The Acceleration of Gravity. Calculate the acceleration of gravity g at the surface of the Earth given the mass of the Earth M and radius ρ. Assume that the Earth has a spherical distribution of mass.

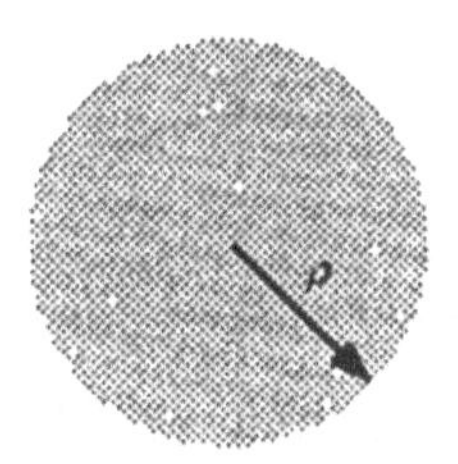

Solution. Consider a particle with mass m placed near the surface of the Earth. The gravitational force $F = |\mathbf{F}|$ on the particle is directed toward the center of the Earth and has the magnitude $F = mg$. We shall now use the fact that the gravitational force outside the Earth is the same as if all the mass of the Earth was collected in the center. For this reason we can write F as:

$$F = G\frac{mM}{\rho^2} .$$

Therefore

$$mg = G\frac{mM}{\rho^2} ,$$

$$g = G\frac{M}{\rho^2} ,$$

By inserting numerical values (from the table on the inside cover of this book), we obtain

$$g = 9.8 \text{ m s}^{-2} .$$

As is well known, **g** is sometimes called the gravitational field strength at the surface of the Earth. The vector **g** is directed toward the center of the Earth. We have:

$$\begin{aligned} \mathbf{g} &= \mathbf{F}/m; \\ g &= |\mathbf{g}| \text{ is therefore equal to the force per mass;} \\ g &= 9.8\ \mathrm{N\,kg^{-1}} = 9.8\ \mathrm{m\,s^{-2}}. \end{aligned}$$

Remember: a force of 1 Newton (1 N) gives to a mass of 1 kg an acceleration of 1 $\mathrm{m\,s^{-2}}$, i.e., 1 N = 1 $\mathrm{kg\,m\,s^{-2}}$.

Question. Compute the gravitational field strength on the surface of the Moon. [Answer: 1.6 $\mathrm{N\,kg^{-1}}$ = 1.6 $\mathrm{m\,s^{-2}}$.] Compute the acceleration of gravity on the surface of the planet Mars. [Answer: 3.7 $\mathrm{N\,kg^{-1}}$ = 3.7 $\mathrm{m\,s^{-2}}$.] △

Example 1.2. Circular Motion

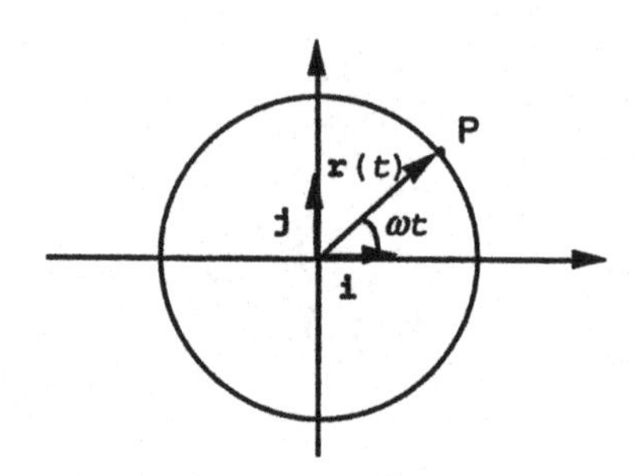

[See the Appendix for a review of vectors.] Let us consider a point P in uniform circular motion. This is described by the radius vector **r** given by:

$$\mathbf{r} = \mathbf{r}(t) = r\cos(\omega t)\mathbf{i} + r\sin(\omega t)\mathbf{j}$$

Here, **i** and **j** are time independent unit vectors, forming an orthogonal coordinate system ($\mathbf{i}\cdot\mathbf{j} = 0, \mathbf{i}^2 = \mathbf{j}^2 = 1$), ω is independent of time and is called the angular velocity. $r = |\mathbf{r}|$.

(1) Find the velocity $\mathbf{v}(t)$ of P and the acceleration $\mathbf{a}(t)$ of P.
(2) Show directly that $\mathbf{v}(t)$ is always perpendicular to $\mathbf{r}(t)$ (i.e., the velocity has no component along the radius). Show that $\mathbf{v}(t)$ is always perpendicular to $\mathbf{a}(t)$.
(3) Show that

$$\mathbf{a}(t) = -\frac{4\pi^2}{T^2}\mathbf{r}(t),$$

where T is the period of revolution (i.e., the acceleration is directed towards the center).

(4) Show that

$$v = |\mathbf{v}| = r\omega,$$

and that

$$a = |\mathbf{a}| = r\omega^2 = \frac{v^2}{r}.$$

Note. Sometimes the word "speed" is used to designate the magnitude of the velocity; i.e., the speed is equal to $|\mathbf{v}|$.

Solution.

(1) $\mathbf{v}(t) = \mathrm{d}\mathbf{r}/\mathrm{d}t = -r\omega\sin(\omega t)\mathbf{i} + r\omega\cos(\omega t)\mathbf{j}$,
$\mathbf{a}(t) = \mathrm{d}^2\mathbf{r}/\mathrm{d}t^2 = -r\omega^2\cos(\omega t)\mathbf{i} - r\omega^2\sin(\omega t)\mathbf{j}$.
(2) $\mathbf{v}(t)\cdot\mathbf{r}(t) = -r^2\omega\sin(\omega t)\cos(\omega t)\mathbf{i}\cdot\mathbf{i} + r^2\omega\cos(\omega t)\sin(\omega t)\mathbf{j}\cdot\mathbf{j} = 0$ since $\mathbf{i}\cdot\mathbf{i} = \mathbf{j}\cdot\mathbf{j} = 1$. In the same manner, $\mathbf{v}(t)\cdot\mathbf{a}(t) = 0$.
(3) $\mathbf{a}(t) = -\omega^2(r\cos(\omega t)\mathbf{i} + r\sin(\omega t)\mathbf{j}) = -(4\pi^2/T^2)\mathbf{r}(t) = -\omega^2\mathbf{r}(t)$ since $\omega T = 2\pi$.
(4) $v = (v_x^2 + v_y^2)^{1/2} = r\omega, \quad a = r\omega^2 = v^2/r$.

Question. A satellite moves in a circular orbit around the Earth. The period of revolution is 98 minutes and the height over the surface of the Earth is 500 km. From this, calculate the mass M of the Earth. [Answer: $M = 5.5 \times 10^{24}$ kg. The accepted value is 5.98×10^{24} kg.] The mass of a planet can be inferred from the motion of its satellites. Using the artificial satellites in orbit around the Earth, we have improved our knowledge of the mass distribution of the Earth. △

Example 1.3. Communication Satellite. A satellite moves in a circular orbit in the equatorial plane of the Earth. The satellite has such a height h above the surface of the Earth that it remains vertically above the same point of the Earth (communication satellite).

(1) Find h.
(2) What is the acceleration of the satellite? With respect to what?

Solution.

(1) The period T is apparently 24 hours. Let the mass of the satellite be m and that of the Earth be M. The distance of the satellite from the center of the Earth is r. The equation of motion for the satellite is Newton's second law with gravity as the force:

$$ma = G\frac{mM}{r^2}.$$

Since

$$a = \frac{4\pi^2 r}{T^2} \,,$$

we find

$$r^3 = \frac{GMT^2}{4\pi^2} \,.$$

Inserting numerical values, one finds:

$$r = 4.22 \times 10^7 \text{ m.}$$

The radius of the Earth is: $\rho = 6.37 \times 10^6$ m. Therefore:

$$h = r - \rho \approx 35.8 \times 10^6 \text{ m} = 35\,800 \text{ km.}$$

Another procedure is to use Kepler's third law directly, together with our knowledge of the orbital period of the Moon and its distance from the Earth:

$$\begin{array}{llll} \text{moon :} & T_\text{m} & = & 27.3 \text{ days,} \\ & r_\text{m} & = & 60.1\ \rho, \qquad \text{where } \rho = \text{radius of the Earth.} \\ \text{satellite:} & T_\text{s} & = & 1 \text{ day,} \\ & r_\text{s} & = & \rho + h. \end{array}$$

We then have

$$\frac{(\rho + h)^3}{(1)^2} = \frac{(60.1\rho)^3}{(27.3)^2} \Rightarrow h = 5.6\rho.$$

(2) The acceleration of the satellite with respect to a coordinate system fixed to the surface of the Earth is of course zero, and the velocity with respect to such a coordinate system also has the constant value zero. The acceleration with respect to a coordinate system with its origin at the center of the Earth and axes fixed in relation to the fixed stars, is:

$$\begin{aligned} a &= \frac{GM}{(\rho + h)^2} = \frac{g}{(1 + h/\rho)^2} = \frac{g}{(6.6)^2} \\ &= 0.23 \text{ m s}^{-2}. \end{aligned}$$

The difficult problems suggested in question 2 of this example will be discussed in the following chapters.

Question. A satellite moves in a circular orbit close to the surface of a planet. The orbital period is T and the mass density of the planet is d. The planet is assumed to have a spherical distribution of mass. Show that dT^2 is a universal constant. △

Example 1.4. Horizontal Throw. When deriving the law of gravitation Newton compared the fall near the Earth (of an apple!) with the fall of the Moon toward the Earth. Newton assumed that there is no essential difference between a fall observed near the surface of the Earth and the motion of the Moon around the Earth. This point was considered carefully by Newton.

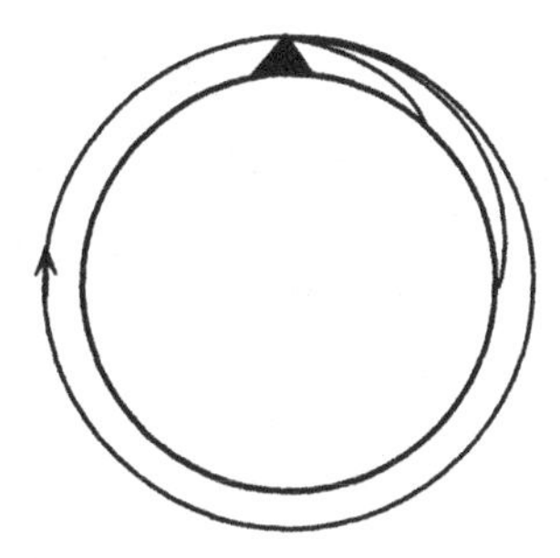

Newton imagined an object thrown with horizontal initial velocity from a tall mountain. With larger and larger initial velocity, the object will reach the Earth further and further from the mountain. See the figure. Newton posed to himself the problem of finding whether it is possible to throw the object with such a large initial velocity that the object although falling under the influence of gravity, never comes closer to the Earth.

Let us assume that the object is thrown with the horizontal velocity v_0. We will investigate what condition v_0 must satisfy for the object never to reach the Earth in spite of the fact that it falls according to the law of free fall, $y = \frac{1}{2}gt^2$, where y is the vertical distance fallen. Consider the figure below.

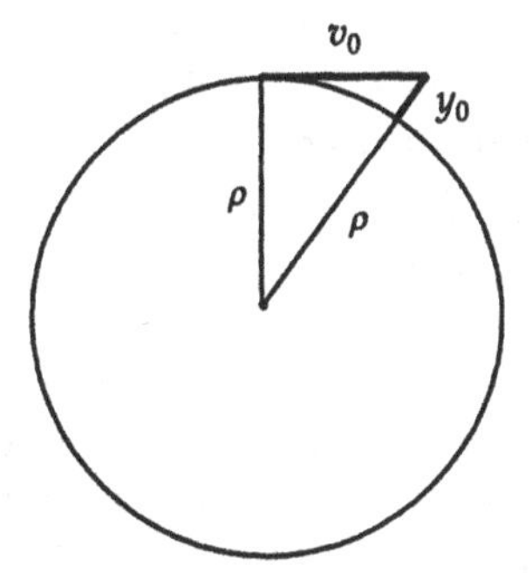

In 1 second the object falls the distance $y_0 = \frac{1}{2}gt^2 = \frac{1}{2}g(1 \text{ second})^2$. The figure is grossly exaggerated; in reality $y_0 \ll \rho$, since ρ is roughly the radius of the Earth, and $y_0 = \frac{1}{2}g(1 \text{ second})^2 \approx 5$ m.

While the object has fallen $y_0 = \frac{1}{2}g(1 \text{ second})^2$, it has moved the distance $v_0 \times (1 \text{ second})$ forward in the inertial motion. From the figure,

$$\rho^2 + (v_0 \times 1)^2 = (\rho + y_0)^2 .$$

Let $(v_0 \times 1) = d$,

$$\rho^2 + d^2 = \rho^2 + y_0^2 + 2\rho y_0 .$$

Using $y_0^2 \ll 2\rho y_0$, we find

$$d^2 = 2\rho y_0 .$$

With $d = (v_0 \times 1)$, and $y_0 = \frac{1}{2}gt^2 = \frac{1}{2}g(1)^2$, we obtain

$$v_0^2 = 2\rho\frac{1}{2}g ,$$

or

$$v_0 = \sqrt{\rho g} .$$

With this value of v_0 as the initial velocity, the body will continue uniformly along a circle. By inserting the numerical values,

$$\rho = 6.4 \times 10^6 \text{ m},$$
$$g = 9.8 \text{ m s}^{-2} ,$$

we get

$$v_0 \approx 8 \times 10^3 \text{ m s}^{-1}.$$

Thus: if we could throw from the top of the highest mountain on Earth a rock with a horizontal initial velocity of about 8 km s^{-1}, the surface of the Earth would "curve away" from the orbit of the rock in such a way that it, although falling toward Earth, would never approach it.

The same result can of course be inferred from direct application of Newton's second law and the law of gravity. The object must under the influence of the gravitational field of the Earth describe a uniform circular motion with radius ρ where $\rho \approx$ radius of the Earth.

Question. Show directly from Newton's second law and the law of gravity that $v_0 = \sqrt{\rho g}$ and calculate the period of revolution. [Answer: $T = 1$ hour 24 min.]

We will conclude this example by calculating the acceleration of the Moon toward the center of the Earth, when the motion of the Moon is considered as a ceaseless fall towards Earth. We shall employ the above results. In the figure, ρ now stands for the radius of the lunar orbit, and we put $\rho = r$.

We let S denote the arc along which the Moon moves during 1 second. S is almost a chord. y_0 denotes the amount the Moon has fallen in 1 second. We have $y_0 \ll S$ (see below). From the results above we infer that

$$y_0 = \frac{S^2}{2r} \quad [\text{note: } S \approx v_0(1 \text{ s})].$$

The distance S through which the Moon moves in 1 second, can be calculated from the values:

radius of lunar orbit:	r	=	3.8×10^8 m,
lunar period:	T	=	2.36×10^6 s.

We find:

$$S = \frac{2\pi r}{T}(1\ \text{second}) \approx 1011\ \text{m}.$$

We can now calculate the distance y_0 through which the Moon falls in one second.

$$y_0 = \frac{(1011)^2}{2 \times 3.8 \times 10^8} = 1.34 \times 10^{-3}\ \text{m}.$$

We see that the condition $y_0 \ll S$ is consistent. From kinematics (i.e. Galileo's law of free fall) we know that

$$y = \frac{1}{2}at^2\ .$$

In our case a is the acceleration in the fall of the Moon. In fact,

$$y = \frac{1}{2}at^2 = \frac{1}{2}a(1\ \text{second})^2\ .$$

We can now determine a:

$$a = 2y_0(1)^{-2} = 2.68 \times 10^{-3}\ \text{m}\,\text{s}^{-2}\ .$$

From Galileo's law of free fall and from an analysis of the motion of the Moon, we have calculated the acceleration in the fall of the Moon. (This acceleration can also be found as: $a \approx 9.8/(60.1)^2 \approx 2.7 \times 10^{-3}\ \text{m}\,\text{s}^{-2}$. Why?).

Question. Calculate directly from the law of gravity the field strength in the gravitational field of the Earth at the position of the Moon, or in other words: calculate the acceleration in the fall of the Moon directly from the law of gravity. [Answer: $a = (GM/r^2) = 2.68 \times 10^{-3}\ \text{N}\,\text{kg}^{-1} = 2.68 \times 10^{-3}\ \text{m}\,\text{s}^{-2}$]

△

Since the Moon falls freely in the gravitational field of the Earth, there is no fundamental difference between the fall of the Moon and the free falls we observe daily at the surface of the Earth. The Moon has simply "been given" an initial velocity perpendicular to the gravitational field of the Earth and of such magnitude that in spite of perpetually falling towards the center of the Earth, the Moon never comes closer to the surface of the Earth.

The contribution of Newton to the understanding of gravitational phenomena must especially be seen in the light of his inferral of the connection between the mutual attraction of celestial bodies and the fall of bodies near the Earth. This connection is by no means trivial. The Aristotelean ideas in their time postulated a fundamental difference between the motion of heavenly bodies and bodies on the Earth.

Example 1.5. The Gravitational Constant. Newton did not know the value of the gravitational constant G. If one assumes an approximately spherical Earth of radius ρ, one gets for the force on a particle of mass m,

$$mg = G\frac{Mm}{\rho^2},$$

where M is the mass of the Earth. Since he knew ρ, Newton could find the product GM, by measuring the acceleration of gravity.

From this he could find the force on a particle at an arbitrary distance r from the center of the Earth:

$$F = G\frac{Mm}{r^2} = G\frac{Mm}{\rho^2}\frac{\rho^2}{r^2} = mg\left(\frac{\rho}{r}\right)^2 .$$

In a famous passage in *Principia* Newton estimates the average density of the Earth: "it is probable that the quantity of the whole matter of the Earth may be five or six times greater than if it consisted of water".

If Newton had used a value of, say, 5.5 $\mathrm{g\,cm^{-3}} = 5.5 \times 10^3\ \mathrm{kg\,m^{-3}}$ for the average density of the Earth, he could – knowing the radius to be $\rho \approx 6371$ km – have calculated G. What value would he have found? [$G = (3g)/(4\pi\rho d)$, where d = density]

From laboratory measurements, H. Cavendish (1731–1810) made an independent determination of G. The currently accepted value of G is:

$$G = 6.67 \times 10^{-11}\ \mathrm{N\,m^2\,kg^{-2}}.$$

Our knowledge of the masses of the objects in the solar system rests entirely on the determination of G. △

Example 1.6. String Force

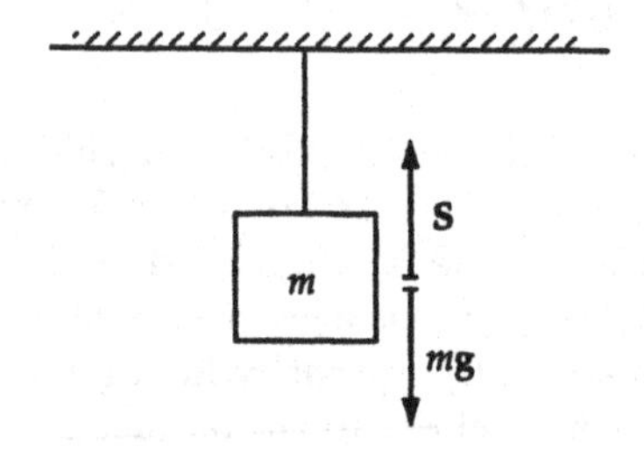

An object of mass m hangs at rest in the laboratory. The object is fastened to the end of a string, the other end of which is fixed to the ceiling of the laboratory. The acceleration relative to the laboratory of the object is nil. From this it follows that the sum of the forces acting on the object is nil. The string force **S** which acts on the object has the direction as shown on the figure, and the magnitude is $|\mathbf{S}| = mg$. This force is a result of a small stretching of the string under the weight of the object. If the prolongation of the string is so small that we can neglect it, the string is called unstretchable.

Question. A boy whirls a string of length 2 m around in an (almost) horizontal plane. At the end of the string a mass $M = 200$ g is fastened. The time of revolution is $T = 0.4$ s. Determine the magnitude F of the string force and compare F to the magnitude of the gravitational force on the mass.

Solution.

$$F = \frac{Mr4\pi^2}{T^2} = \frac{0.2 \times 2 \times 4 \times \pi^2}{0.4^2} \approx 98.7 \text{ N}.$$

$$F_{\text{grav}} = 0.2 \times 9.8 \approx 2 \text{ N}; \quad \text{thus,} \quad \frac{F}{F_{\text{grav}}} \approx 50.$$

△

Example 1.7. Forces and Tension

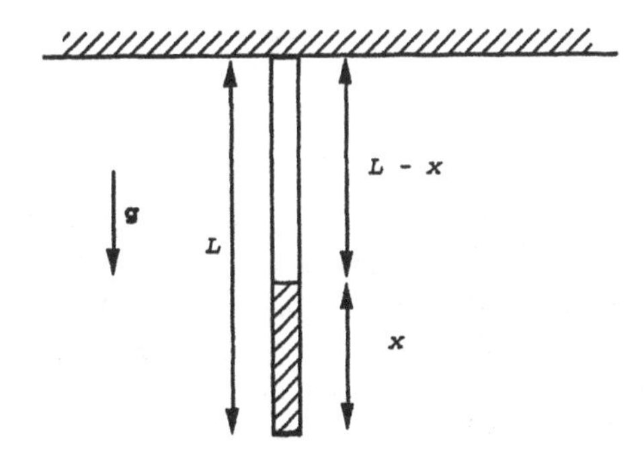

Consider a homogeneous rod of mass M and length L. The rod hangs freely in the gravitational field of the Earth (see the figure). Consider the hatched segment of the rod. This segment has length x. Imagine a cross section dividing the two parts of the rod. The upper part of the rod $(L - x)$ acts on the lower part (x) with contact forces evenly distributed over the cross section of the rod. The tension in the rod is defined as the force per unit area in the cross section. The tension in a particular point of an elastic body is described by a tensor and is treated in the theory of elasticity. The important point here is: at the cross section the segment $L - x$ acts on the hatched segment with a vertical force. This force has the magnitude:

$$F = \frac{Mgx}{L} .$$

(the segment x has acceleration 0 and is being pulled downwards by gravity). At the cross section the segment x acts on the segment $L-x$ with a downward force. Again, this force has the magnitude

$$F = \frac{Mgx}{L} .$$

If our system is a freely hanging string instead of a rod, the quantity

$$T(x) = \frac{Mgx}{L}$$

is often – particularly in theoretical mechanics – called the string tension. $T(x)$ is not a tension, but a force. Usually, confusion is avoided since in (almost) all cases, the quantity in question in problems of mechanics is the force with which a string acts on a body. △

Example 1.8. Dimensional Analysis

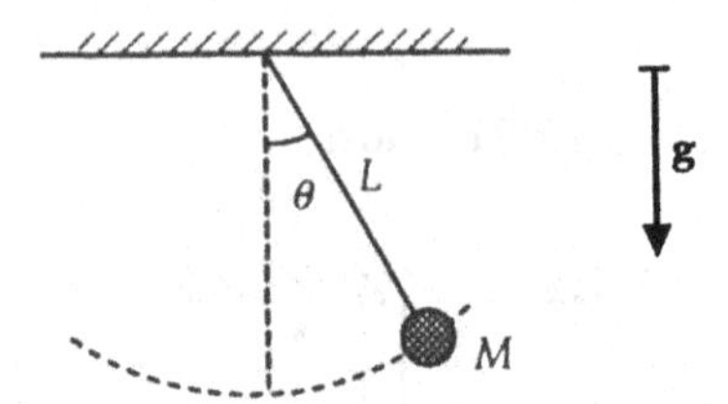

A particle of mass M is fastened to one end of a string. The length of the string is L, and we neglect the mass of the string. This system, which is called the mathematical pendulum, can perform oscillations around the vertical equilibrium position. The acceleration of gravity is **g**. What physical quantities could the period T of the oscillation depend on?

The period might depend on the mass M of the particle, the length L of the string, the magnitude g of the gravitational field, and perhaps on the maximal angle θ of deflection. The angle θ is a dimensionless number (arc length divided by L).

The period T has the dimension of time (s), M has the dimension of mass (kg), L has the dimension of length (m), and g has the dimension of distance per time squared ($\mathrm{m\,s^{-2}}$). We assume that these three quantitites, raised to various powers, enter the formula for the period:

$$T = L^x M^y g^z \ .$$

Or, in terms of units,

$$\begin{aligned} \mathrm{s} &= \mathrm{m}^x (\mathrm{kg})^y \left(\frac{\mathrm{m}}{\mathrm{s}^2}\right)^z \\ &= \mathrm{m}^{x+z} (\mathrm{kg})^y \mathrm{s}^{-2z} \ . \end{aligned}$$

From this:

$$\begin{aligned} x + z &= 0 \ , \\ y &= 0 \ , \\ -2z &= 1 \ , \end{aligned}$$

which leads to $x = 1/2$, $y = 0$, $z = -1/2$. We conclude that

$$T = C\sqrt{\frac{L}{g}}\,,$$

where C is a constant (a pure number) with no dimensions associated. The value of C cannot be determined by this type of analysis.

As we will see in Section 3.4, for small oscillations about the equlibrium position the period is independent of the maximal angle, and C has the value 2π. For large deflections, the value of C depends on the maximal angle of deflection. Since this angle, θ, is a dimensionless number, the general formula above still holds. △

1.4 Concluding Remarks

The main purpose of this book is to describe the problem of motion as this problem was formulated by Newton. We also seek to clarify the content of the modified version of Newtonian mechanics which is normally used today.

In order now to suggest some fundamental problems in the physics of motion we will formulate a few questions, and let them stand unanswered. The questions will be discussed in detail in the following chapters.

(1) Relative to which coordinate system should we measure the velocities and accelerations that appear in Newtonian theory?
(2) Relative to what do the galaxies rotate? The figure following the title page of this book shows a galaxy. A galaxy is a collection of hundreds of billions of stars held together by mutual gravitational forces, just like the planetary system is held together by gravitational forces. Looking at the picture, one can sense the rotation, but is it, as Newton thought, a rotation with respect to "absolute space"? And if so, what is this "absolute space"?
(3) We say that the Earth rotates once every 24 hours. With respect to what does the Earth rotate? Is it relative to the distant stars? Or, is it relative to "absolute space"?

An answer to such – and similar – questions demands a deep understanding of what it really means to say that a body moves. We shall see that in the form in which we usually employ Newtonian mechanics, it is not easy to gain insight into this question.

1.5 Problems

Problem 1.1.

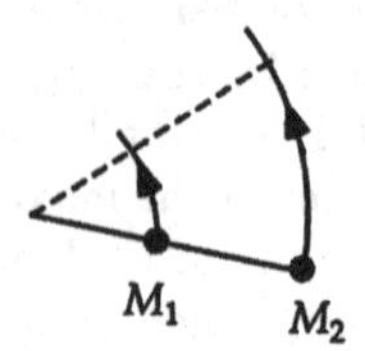

A boy whirls a string of length 2 m around in an (almost) horizontal plane. At the end of the string a mass $M_2 = 0.1$ kg is fastened, and 1 m from the boy's hand another mass $M_1 = 0.1$ kg is fastened. The boy increases the angular velocity ω in the motion of the string. The hand of the boy and the two masses will, when ω is sufficiently large, be almost on a straight line. When the string force reaches the magnitude 327 N, the string will break.

(1) Find the magnitude of the angular velocity when the string breaks, and find where the break occurs.
(2) Is it possible to whirl the two masses in such a way that the two masses on the string are exactly on a horizontal line?

Problem 1.2. The planet Mars has two moons, *Phobos* and *Deimos.*

(1) Phobos has an orbital radius of 9.4×10^3 km and a period of 7 h 39 min. Calculate the mass of Mars from these data.
(2) Assume that the Earth and Mars move in circular orbits around the Sun. Take the radius of the orbit of Mars around the Sun as 1.52 times the orbital radius of the Earth. Calculate the length of the Martian year.

Problem 1.3.

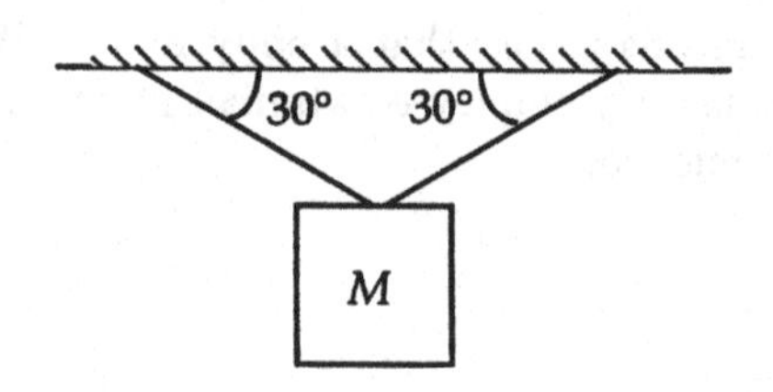

A body of mass M is hanging in equilibrium in the laboratory, as shown on the figure. The two strings have the same length. Determine the string force S in the two strings.

Problem 1.4.

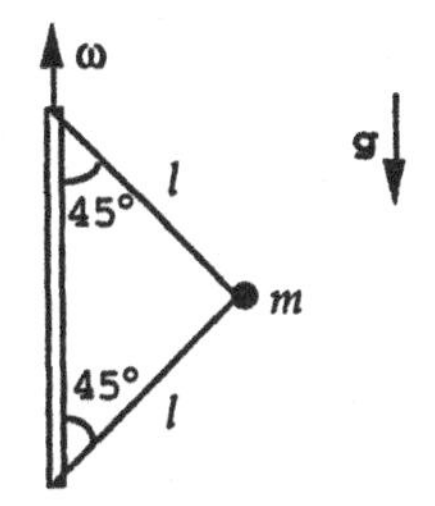

A heavy particle with mass m is attached to a rigid rod by means of two thin, "unstretchable" strings, each of length l. As the strings are unstretchable, one can ignore their stretching during the motion. The entire system rotates with constant angular velocity ω about an axis along the rod, and both strings are taut. The acceleration due to gravity is $\mathbf{g}$.

(1) Calculate the string force in each of the strings.
(2) Find the condition under which the lower string is taut.

Problem 1.5.

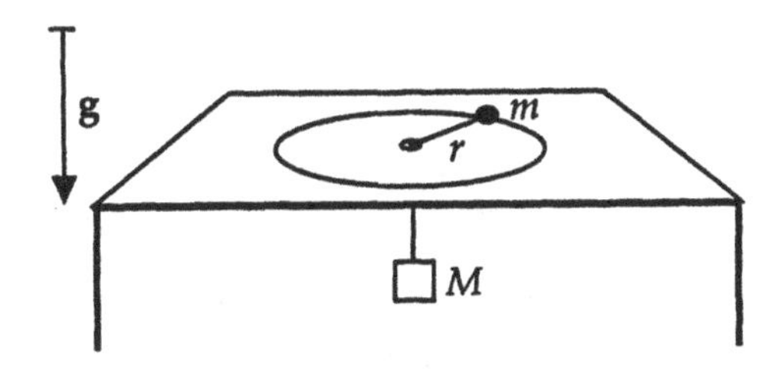

A particle of mass m performs a uniform circular motion on a frictionless table. The particle is attached to a cord that passes through a hole in the table. The other end of the cord is attached to another particle with mass $M > m$. The length of the cord on the table is r. Find the magnitude of the velocity of m such that the system is in equilibrium.

Problem 1.6.

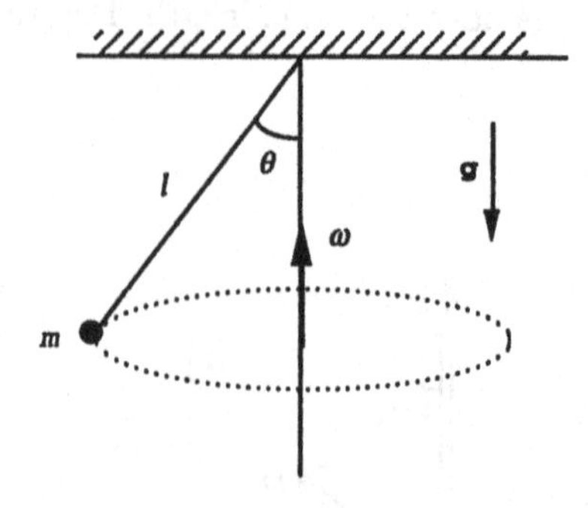

A heavy particle with mass m is tied to one end of a string, which has the length l. The other end of the string is attached to the ceiling of the laboratory. The particle performs a uniform circular motion in a horizontal plane. The angular velocity is ω. The string is at an angle θ to the vertical. This system is called the *conical pendulum*. Ignore the various forms of friction and the mass of the string.

(1) Find the magnitude of the string force S expressed in terms of m, l, and ω.

(2) Express θ as a function of g, l, and ω. Examine the result for $\omega \to \infty$.

For the conical pendulum we have to assume that $\omega > \sqrt{g/l}$. In the derivation of the result of question (1) it is necessary to divide by $\sin\theta$.

(3) State the value of $\sin\theta$ for $\omega = \sqrt{g/l}$.

Problem 1.7.

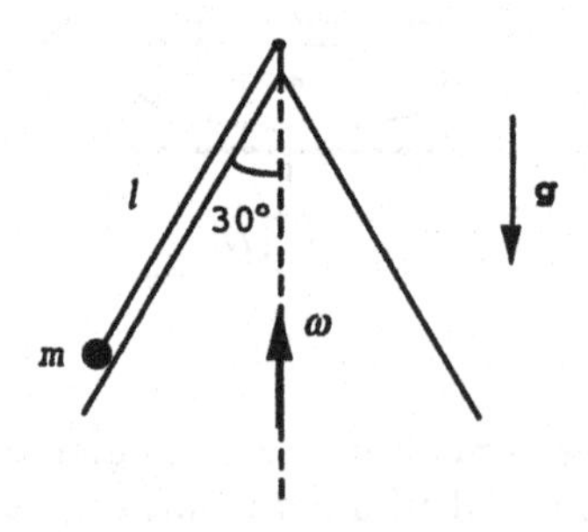

A cone of revolution has sides that form an angle of 30° to the vertical. One end of a thin cord with length l is attached to the apex of the cone. At the other end of the cord a heavy particle of mass m is attached.

The particle performs a uniform circular motion with angular velocity ω in such a way that the particle stays in contact with the cone at all times.

The mass of the string is ignored as well as friction between the particle and the cone, i.e., the reaction force from the cone – acting on the particle – is perpendicular to the surface of the cone.

(1) Find the string force **S** and the force **R** with which the surface of the cone acts on the particle (**R** is the reaction force).
(2) Find the largest value of ω for which the particle can remain in contact with the surface of the cone.

Problem 1.8. If you are not familiar with vectors, we suggest that you read the Appendix carefully. A vector **v** has in a given coordinate system the component $v_y = 2$. The vector **v** forms an angle of 45° with the positive x-axis. The projection of the vector on the xz-plane forms an angle of 60° with the positive z-axis. Find

(1) the components v_x and v_z,
(2) the length $v \equiv |\mathbf{v}|$ of the vector,
(3) the angle between **v** and the positive y-axis,
(4) the angle between **v** and the positive z-axis.

Problem 1.9. Find the angle between any two space diagonals of a cube. Hint: find in a suitable coordinate system the components of the vectors that represent the two body diagonals, and use the dot product (see the Appendix).

Problem 1.10. A particle moves on a straight line with the constant velocity $\mathbf{v} = (1, 1, 2)\ \mathrm{m\,s^{-1}}$. At the time $t = 0$ it passes the origin. A person P is situated at the point (1, 2, 3) m. Find the time at which the particle is nearest P and find the vector **d** from P to the particle at that time (see the Appendix).

Problem 1.11. See the Appendix. In a given coordinate system two vectors **a** and **b** have the following components:

$$\begin{aligned} \mathbf{a} &= (a_x, a_y, a_z), \\ \mathbf{b} &= (b_x, b_y, b_z). \end{aligned}$$

Show that the components of their vector product are given by:

$$\begin{aligned} (\mathbf{a} \times \mathbf{b})_x &= a_y b_z - a_z b_y \ , \\ (\mathbf{a} \times \mathbf{b})_y &= a_z b_x - a_x b_z \ , \\ (\mathbf{a} \times \mathbf{b})_z &= a_x b_y - a_y b_x \ . \end{aligned}$$

(Hint: use the rules for vector products, and $\mathbf{a} = a_x\mathbf{i} + a_y\mathbf{j} + a_z\mathbf{k}$, where $\mathbf{i}, \mathbf{j}$, and $\mathbf{k}$ are unit vectors along the x, y, and z axis, respectively.)

Problem 1.12. At the time $t = 0$ a pendulum of length $r = 0.7$ m is started from rest at an angle of 30° from the vertical.

(1) What is the component of the acceleration along a tangent, called the tangential acceleration a_{T}, and the component of the acceleration along the radius, called the centripetal acceleration a_{N}, at the time $t = 0$?
(2) What is the angular velocity ω and the angular acceleration $\dot{\omega} = \ddot{\theta}$ at $t = 0$?
(3) The mass of the pendulum bob is 0.5 kg. What is the tension in the string (the string force) at $t = 0$?

2. Newton's Five Laws

As mentioned, we will here assume some knowledge of elementary mechanics and the concepts within that theory. As the reader progresses through the book, he or she will find these concepts treated in substantial detail.

2.1 Newton's Laws of Motion

Newton introduced the basic laws of mechanics in an axiomatic form with associated lemmas and definitions. We reproduce the laws here in forms that closely resemble Newton's original version.

Newton's 1st law (the law of inertia):

A body remains in its state of rest or in uniform linear motion as long as no external forces act to change that state.

This law puts the state of rest and the uniform linear motion on an equal footing. No external force is necessary to maintain uniform motion. The motion continues unchanged due to a property of matter we call inertia.

Newton made this law precise by introducing definitions of the concepts *momentum* and *mass*. Momentum is the product of the velocity of a body and the amount of matter the body contains. That amount of matter is called the mass of the body and is a measure of its inertia. Instead of "mass" we shall occasionally use the more precise term "inertial mass". The inertial mass then, is a measure of that property of an object which makes the object resist changes in its state of motion.

We can write

$$\mathbf{p} \equiv m\mathbf{v}\,, \tag{2.1}$$

where m is the inertial mass of the particle, $\mathbf{v}$ its velocity, and $\mathbf{p}$ the momentum. By "$\equiv$" we mean "equal to, by definition". Mathematically, we can express the law of inertia as

$$\text{no external force} \;\Rightarrow\; \mathbf{p} = \text{constant vector.} \tag{2.2}$$

In this way the *law of inertia* coincides with the *law of conservation of momentum for a particle.*

It is by no means simple to grasp the physical content of the law of inertia. We have (almost by definition) no experience with objects that are not subject to some external influence. We can imagine placing an object, not acted upon by any force, in an otherwise empty astronomical space. Can we later by any sort of observation or experiment decide if that object is at rest, or in a state of uniform motion? The answer to – or rather an analysis of – this question is a central theme in this book.

Newton's 2nd law:

The change in the momentum of a body is proportional to the force that acts on the body and takes place in the direction of that external force.

By change one means here the time derivative of the momentum. The law can be expressed as

$$\frac{\mathrm{d}\mathbf{p}}{\mathrm{d}t} = \mathbf{F}, \tag{2.3}$$

where $\mathbf{F}$ is the impressed force. If we assume that the mass of the body is constant, Newton's second law takes the form

$$m\mathbf{a} = \mathbf{F}, \tag{2.4}$$

where $\mathbf{a} \equiv \mathrm{d}\mathbf{v}/\mathrm{d}t \equiv \dot{\mathbf{v}}$ is the acceleration of the body.

In a given (orthogonal) cartesian coordinate system, Newton's second law can be written in component form (see the Appendix):

$$m\frac{\mathrm{d}^2x}{\mathrm{d}t^2} = F_x\,,$$
$$m\frac{\mathrm{d}^2y}{\mathrm{d}t^2} = F_y\,,$$
$$m\frac{\mathrm{d}^2z}{\mathrm{d}t^2} = F_z\,.$$

In the form (2.4) the law states that an impressed force $\mathbf{F}$ *causes* an acceleration $\mathbf{a}$ which is directly proportional to that force; the constant of proportionality $|\mathbf{F}|\,/\,|\mathbf{a}|$ is the inertial mass m of the object.

The whole concept of force has been the subject of much debate since it was introduced by Newton. Let us here note the following: the second law [(2.3) or (2.4)] should *not* be considered as the definition of the concept of force. An essential feature of the law is that the force acting on the particle is supplied by a force law separate from (2.3) or (2.4). One example of such a force law is the law of gravity. Other forms are shown throughout the examples in this chapter.

The second-order differential equation that results when some force law is supplied to the second law is called the *equation of motion* for the particle.

An important property of the concept of a force acting on a given object is that the force has its origin in some other material body.

Thus, Newton introduced the concept of *mechanical force* as the cause of the acceleration of an object. This causal description of the motion of an object constitutes what we call *dynamics.*

In this way a connection is set up between acceleration (a kinematic quantity) and the force (a dynamical quantity). In Chapter 7 we shall see that this coupling of a dynamical and a kinematic quantity was of enormous importance for the understanding of the problem of motion, i.e., for the realization of what it means to say that a given body is in motion.

Newton's 3rd law (the law of action and reaction):

If a given body A acts on another body B with a force, then B will also act on A with a force equal in magnitude but opposite in direction.

This law (postulate) is almost self evident when we consider contact forces in equilibrium. The fact that the law of action and reaction also holds for actions at a distance, is demonstrated for instance by the existence of tides: the Earth keeps the Moon in orbit with a gravitational field, but the Moon acts back on the Earth with a force which (among other things) is the cause of ebb and flood in the oceans.

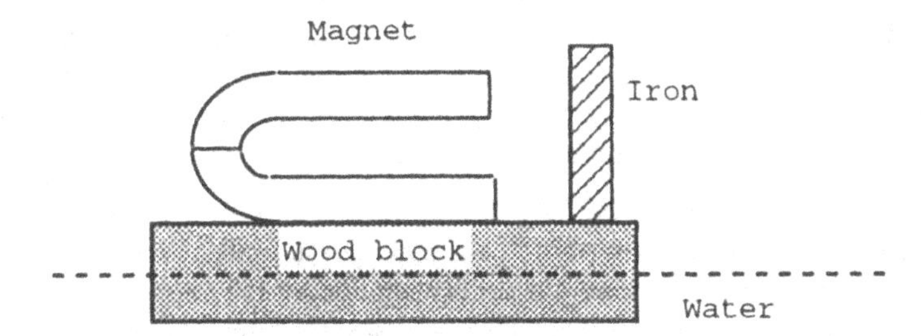

Fig. 2.1. Illustration of Newton's third law

If the law of action and reaction did not hold, one could assemble systems that perpetually increased their velocity through the action of internal forces. Newton tested the law for a special configuration of objects. On a wooden plate he attached a magnet and a piece of iron. The plate floats on a calm surface of water. The magnet acts on the iron with a force directed to the left (see Figure 2.1) and the iron acts on the magnet with a force towards the right. The two forces are apparently equal in magnitude and the system remains at rest.

It is important to realize that Newton's first three laws are without content unless a coordinate system is specified, relative to which velocities and accelerations are to be measured. To the three laws Newton therefore added two postulates. We quote here from the translation into English of Newton's original Latin text [translation by Andrew Motte (1729), the emphasis is ours].

Newton's 4th law (the postulate of absolute time):

> Absolute, true, and mathematical time, of itself, and from its own nature, flows equally *without relation to anything external*, and by another name is called duration: relative, apparent and common time, is some sensible and external (whether accurate or unequable) measure of duration by the means of motion, which is commonly used instead of true time; such as an hour, a day, a month, a year.

Newton's 5th law (the postulate of absolute space):

> Absolute space, in its own nature, *without relation to anything external*, remains always similar and immovable. Relative space is some movable dimension or measure of the absolute spaces; which our senses determine by its position to bodies; and which is commonly taken for immovable space; such is the dimension of a subterranous, an arial, or celestial space, determined by its position in respect of the Earth. Absolute and relative space is the same in figure and magnitude; but they do not always remain numerically the same. For if the Earth, for instance, moves, a space of our air, which relatively and in respect of the Earth remains always the same, will at one time be in one part of the absolute space into which the air passes; at another time it will be in another part of the same, and so, absolutely understood, it will be continually changed.

Newton was well aware of the fact that if his three laws were to have a real content, he had to explicitly point to a coordinate system relative to which all velocities and accelerations ultimately should be referred.

Newton did not choose a material body, say the Sun or the "fixed" stars, as the reference for motion. As the basic system of reference, Newton picked "absolute space" or "astronomical space". Newton has often been criticized for the two postulates above. In particular the phrase "without relation to anything external" has been widely discussed. For, what exists without relation to anything external, cannot be verified by any experiment.

A deep understanding of the postulates of absolute time and absolute space will follow only from *experience with concrete examples* of motion. The following chapters are a means to that end.

In Chapter 7 we will give a more unified discussion of the five fundamental postulates in the Newtonian world picture. By then the reader should understand both the power and the limitation of the Newtonian mechanics. We shall see that Newtonian mechanics – in spite of its enormous success – contains internal weaknesses.

For now we will stick to a presentation of Newtonian mechanics as it is in practical use today. Later we can enter a discussion of what it is all about.

2.2 Integration of the Equation of Motion

Newtonian dynamics is a powerful tool for investigations of Nature. In order that you fully realize this, it is of paramount importance that you actively command essential elements of the theory. That means, you must be able, in an independent manner, to apply the theory to specific examples. In this section we will work out a number of such examples of the integration of Newton's equation of motion. We will consider the motion of a *particle*. By the word "particle" we mean a body whose spatial extensions are unimportant for the problem at hand. We shall furthermore assume that the coordinate systems referred to are reference frames in which the law of inertia holds. Such reference frames are called *inertial frames*. Referred to inertial frames, Newton's laws apply in their simple classical form. How to find inertial frames at all is one of the major themes of this book.

Example 2.1. Constant Force

A particle of mass m moves along the x-axis of a coordinate system under the action of a constant force $K = |\mathbf{K}|$ in the positive x-direction. Analyze the motion; i.e., find the position of the particle as a function of time.

The *equation of motion* is obtained by inserting the force into Newton's second law:

$$m\frac{\mathrm{d}^2x}{\mathrm{d}t^2} = K\,, \tag{2.5}$$

where K is constant. Let us write (2.5) as

$$m\frac{\mathrm{d}}{\mathrm{d}t}\left(\frac{\mathrm{d}x}{\mathrm{d}t}\right) = K\,,$$

or, since the velocity v of the particle by definition is given as $v = \mathrm{d}x/\mathrm{d}t$,

$$m\frac{\mathrm{d}v}{\mathrm{d}t} = K\,. \tag{2.6}$$

The formal integration of equation (2.6),

$$v = \int \frac{dv}{dt} dt = \int \frac{K}{m} dt \,,$$

gives

$$v = \int \frac{K}{m} dt \,.$$

Loosely speaking: we multiply (2.6) on both sides with dt, which gives:

$$m dv = K dt$$

or

$$dv = \frac{K}{m} dt \,. \tag{2.7}$$

Integration of (2.7) gives:

$$v = \int \frac{K}{m} dt \,,$$

or,

$$v(t) = \frac{K}{m} t + c_1 \,, \tag{2.8}$$

where c_1 is a constant of integration. Equation (2.8) can be written in the form

$$\frac{dx}{dt} = \frac{K}{m} t + c_1 \,. \tag{2.9}$$

We multiply both sides with dt:

$$dx = \frac{K}{m} t dt + c_1 dt \,. \tag{2.10}$$

Integration of (2.10) gives

$$x = \int \frac{K}{m} t dt + \int c_1 dt \,,$$

or

$$x = \frac{1}{2} \frac{K}{m} t^2 + c_1 t + c_2 \,, \tag{2.11}$$

where c_2 is a constant of integration.

Note. We have integrated twice. Equation (2.11) thus contains two constants of integration. This is a reflection of the fact that Newton's second law is a second-order differential equation. The integration of the equation of motion (2.5) has not provided a complete solution of the problem from a physical point of view. A complete solution requires specification of x as a function of time. The polynomial (2.11) contains two unknown constants c_1 and c_2. The force K determines only the acceleration of the particle, i.e., x is not completely determined as a function of time by the force alone. In a specific problem the two constants of integration, c_1 and c_2, are determined by the initial conditions: the position and the velocity at the initial time, say, $t = 0$. Part of the strength of differential equations as they are applied to physical problems is that they are insensitive to these arbitrary parts of the motion. Let us assume that at the time $t = 0$ the velocity of the particle is v_0 and the position is x_0. By inserting these initial conditions into (2.8) and (2.11) we find from (2.8) that $c_1 = v_0$, and from (2.11) that $c_2 = x_0$.

The complete solution is therefore:

$$x = \frac{1}{2}\frac{K}{m}t^2 + v_0 t + x_0 \, .$$

If the initial time is $t = t_0$ rather than $t = 0$, we get

$$x = \frac{1}{2}\frac{K}{m}(t - t_0)^2 + v_0(t - t_0) + x_0 \, .$$

This example applies to motions like the free fall (in vacuum) near the surface of the Earth if the total vertical extension is so small that the variation with height of the gravitational force can be disregarded.

Questions.

(1) A locomotive is driving with a velocity of 100 km h^{-1} on a straight railroad track. The mass M of the engine is 10 tons. At the time $t = 0$ the engine begins to brake by blocking the wheels. This causes a frictional force F to appear. We assume F to be constant.
   (a) Write the equation of motion for the train at times $t > 0$.
   (b) Suppose the train comes to a halt after 400 m. Determine the magnitude of F and the time t_1 at which the train stops.

   [Answers: (a): $M\mathrm{d}v/\mathrm{d}t = -F$; (b): note that $x = \frac{1}{2}(-F/M)\,t_1^2 + v_0 t_1$ and $v = 0 = -(F/M)t_1 + v_0 \;\Rightarrow\; F = 9645$ N; $t_1 = 28.8$ s].

(2) A stone is thrown vertically upwards in the gravitational field of the Earth. A point P is at height h above the starting point of the stone. The stone passes P on its way upward, 2 seconds after it was thrown, and 4 seconds after it was thrown the stone passes P on its downward fall. Calculate h and the initial velocity v_0 of the stone. Disregard air resistance. [Answers: $h = 39.2$ m; $v_0 = 29.4$ m s^{-1}.] What is the velocity of the stone when it reaches its point of departure?

△

Example 2.2. The Harmonic Oscillator. A mass m is connected to one end of a spring. The other end of the spring is attached to a wall. The mass slides without friction on a horizontal table. The gravitational force on the mass is therefore balanced by the reaction force from the table. If we compress or stretch the spring, it will act on the mass with a horizontal force. It is possible to construct the spring in such a way that the force acting on the mass is proportional to the displacement of the mass and directed towards

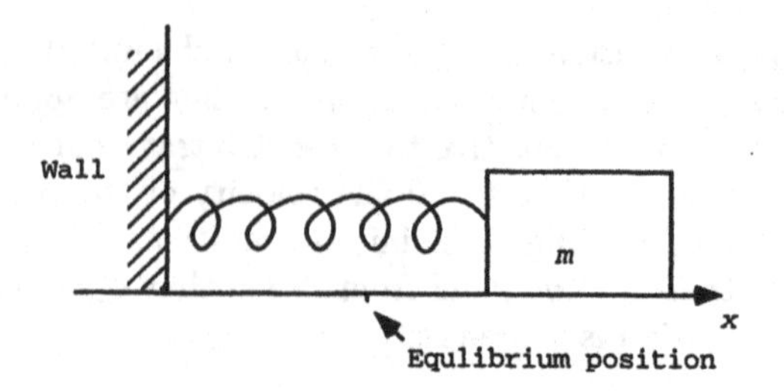

the point of equilibrium. Such a force is often called an elastic force. Let us choose an x-axis with its origin at the equilibrium position of the mass. We can now write an analytical expression for the horizontal force acting on the mass:

$$F_x = -kx \,, \tag{2.12}$$

where k is a constant determined by the physical properties of the spring. The force constant k has units $\mathrm{N\,m^{-1}}$; the minus sign indicates that the force is directed against the displacement, towards the point of equilibrium. The expression (2.12) is known as *Hooke's law.* We neglect the mass of the spring. The equation of motion now follows by inserting (2.12) as the force in Newton's second law:

$$m\frac{\mathrm{d}^2x}{\mathrm{d}t^2} = -kx \,. \tag{2.13}$$

In principle there are two more equations of motion, one for the y-direction and one for the z-direction. However, we do not write these two (trivial) equations, as the problem at hand is, from the point of view of physics, a one-dimensional problem.

(1) What conditions must the constants A, ω, and θ satisfy for

$$x = A\cos(\omega t + \theta) \tag{2.14}$$

to be a solution to (2.13)? (A, ω, and θ are constants, i.e., independent of t).

(2) What is the physical meaning of A?

(3) Find the period T of oscillation, i.e., find the time between two consecutive extremal positions.

(4) Show graphically in an (x, t) diagram the motion of the mass, using the following initial conditions: the mass is pulled to the distance $x = a$ from the equilibrium point and released at $t = 0$ with no initial velocity.

Solution.

(1)

$$m\frac{\mathrm{d}^2x}{\mathrm{d}t^2} = -kx \,.$$

For $x = A\cos(\omega t + \theta)$ we have:

$$\frac{\mathrm{d}x}{\mathrm{d}t} = -A\omega \sin(\omega t + \theta) ,$$

$$\frac{\mathrm{d}^2x}{\mathrm{d}t^2} = -A\omega^2 \cos(\omega t + \theta) .$$

For $x = A\cos(\omega t + \theta)$ to be a solution to (2.13), we must have, for all values of t,

$$m(-A\omega^2)\cos(\omega t + \theta) = -kA\cos(\omega t + \theta) ,$$

or

$$\omega = \sqrt{\frac{k}{m}} . \tag{2.15}$$

Therefore, for ω given by (2.15) and for any value of A and θ, (2.14) will satisfy the equation of motion (2.13). A and θ are two constants of integration (compare with c_1 and c_2 in Example 2.1). The integration constants are determined by the initial conditions. Equation (2.14) is the complete solution of (2.13). Physically this means that with a suitable choice of A and θ, any motion of the mass can be represented by (2.14).

(2) A is the amplitude of the oscillating mass, i.e., the maximal deviation from the point of equilibrium. The angle θ is called the phase angle.

(3) T must satisfy the condition

$$\cos(\omega t + \theta) = \cos[\omega(t + T) + \theta].$$

Since the function cos is periodic with the period 2π, we have

$$\omega T = 2\pi ,$$

or

$$T = \frac{2\pi}{\omega} = 2\pi\sqrt{\frac{m}{k}} . \tag{2.16}$$

Note. The motion described by (2.14) is called a harmonic oscillation and the system is an example of a harmonic oscillator. We have the important result: the period T in a harmonic oscillation is independent of the amplitude. The frequency of the oscillation, i.e., the number of oscillations per second is: $\nu = 1/T$ and consequently $\omega = 2\pi\nu$.

(4) $x(0) = A\cos\theta = a$

$$v(0) = \left(\frac{\mathrm{d}x}{\mathrm{d}t}\right)\bigg|_{t=0} = -A\omega\sin(\theta) = 0 \quad \Rightarrow \theta = 0 , \tag{2.17}$$

i.e.,

$$\begin{aligned} \cos(\theta) &= 1, \\ A &= a, \\ x &= a\cos(\omega t). \end{aligned}$$

The amplitude $A = a$ is here set equal to 1:

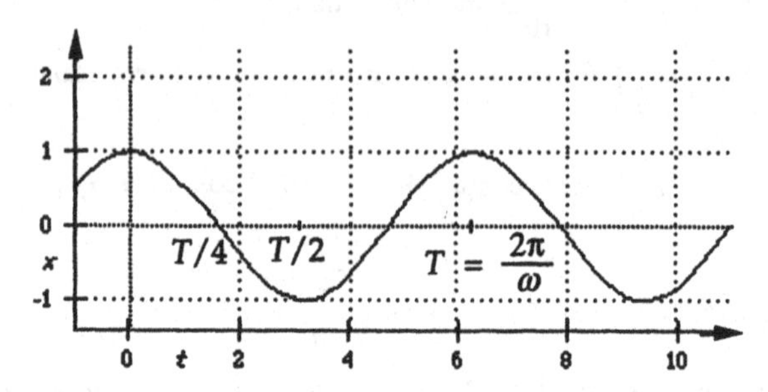

Note that (2.13) is often written as:

$$\frac{\mathrm{d}^2x}{\mathrm{d}t^2} = -\omega^2 x, \tag{2.18}$$

where

$$\omega^2 = \frac{k}{m}. \tag{2.19}$$

Here ω is a characteristic constant for the harmonic oscillator and is called the *cyclic frequency*, the *angular frequency*, or the *eigen frequency*.

Question. Consider a particle with the mass 100 g = 0.1 kg. At time $t = 0$ s, the particle is at the origin ($x = 0$) and the particle has the velocity $v = \mathrm{d}x/\mathrm{d}t = 0.5\ \mathrm{m\,s^{-1}}$. The period of oscillation is $T = 2$ s. Find the force constant k, the phase θ at $t = 0$, and the amplitude A of the motion. [Answer: $k = \pi^2/10\ \mathrm{N\,m^{-1}}$; $\theta = -\pi/2$; $A = 1/2\pi$ m.] △

Example 2.3. Mass on a Spring in the Gravitational Field of Earth

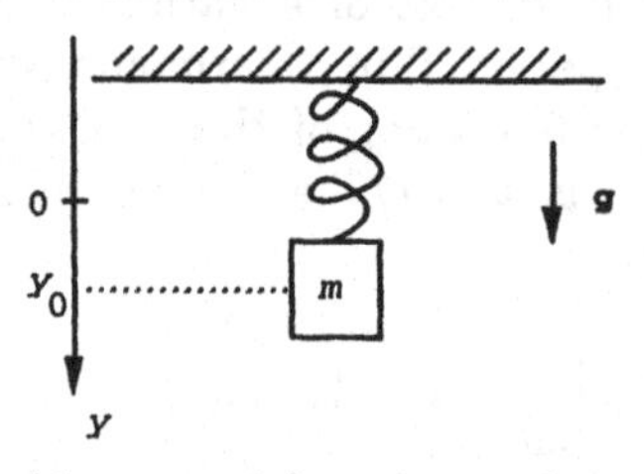

We introduce a vertical (downwards) y-axis with its origin at the extension of the spring when no mass is attached to it. When the mass has the position y, the total force acting on the mass is given by

$$F = -ky + mg\,, \tag{2.20}$$

where k is the spring constant. The equilibrium point of the mass is not $y = 0$, since (2.20) shows that the force acting on the mass at $y = 0$ is not zero, in fact $F(y = 0) = mg$. Let us determine the equilibrium position y_0, i.e., let us determine for which value of y, $F(y) = 0$. From (2.20) we have:

$$0 = -ky_0 + mg, \tag{2.21}$$

$$y_0 = \frac{mg}{k}. \tag{2.22}$$

The equation of motion for the mass as it moves along the vertical y-axis is:

$$m\frac{\mathrm{d}^2y}{\mathrm{d}t^2} = -ky + mg. \tag{2.23}$$

Equation (2.23) can be written in the form

$$\frac{\mathrm{d}^2y}{\mathrm{d}t^2} + \frac{k}{m}(y - y_0) = 0. \tag{2.24}$$

In order to solve this differential equation, we introduce the new variable

$$x = y - y_0. \tag{2.25}$$

From (2.25) we get

$$\frac{\mathrm{d}^2y}{\mathrm{d}t^2} = \frac{\mathrm{d}^2x}{\mathrm{d}t^2}\,. \tag{2.26}$$

By inserting (2.26) and (2.25) into (2.24) we find that the problem has been reduced to that of solving the differential equation

$$\frac{\mathrm{d}^2x}{\mathrm{d}t^2} + \frac{k}{m}x = 0, \tag{2.27}$$

which is identical to (2.13) in example 2.2. The complete solution to (2.27) is again

$$x = A\cos(\omega t + \theta). \tag{2.28}$$

Re-introducing y as variable, we can write

$$y - y_0 = A\cos(\omega t + \theta). \tag{2.29}$$

Equation (2.29) shows that the mass will oscillate around the equilibrium position given by (2.22), with the period

$$T = 2\pi\sqrt{\frac{m}{k}}.$$

Gravity in this case causes just a displacement of the equilibrium position to y_0. This system therefore is equally well suited for the study of harmonic oscillations as that of Example 2.2. △

Example 2.4. Frictional Force: Sphere Falling Through a Liquid. In the previous examples, the force has been dependent "at most" on the position of the particle. We will now consider a case where the force depends on the velocity of the particle. Consider the following physical system:

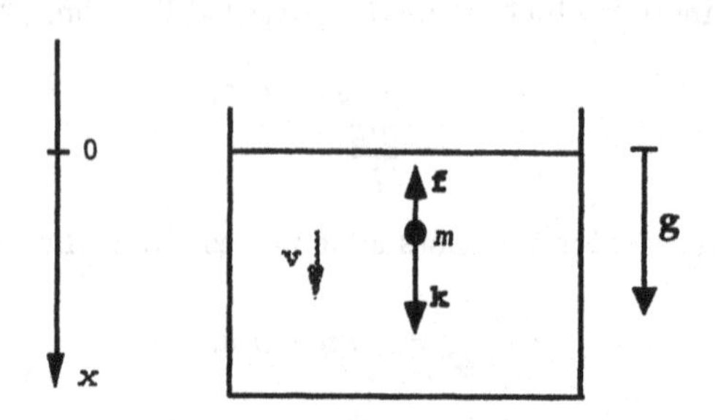

A small heavy ball (say, a lead shot) is sinking through a liquid. At not too large velocities, the frictional force **f** with which the liquid acts on the ball will be proportional to the velocity of the ball. Let us start the ball off in the position $x = 0$ at time $t = 0$ and with velocity $v = 0$. The mass of the ball is m. Note, we neglect the motion of the fluid.

The gravitational force on the ball is in the positive x-direction and has the magnitude mg. Because of the finite volume of the ball there is a buoyant upward force on the ball with magnitude $V\rho g$, where V is the volume of the ball and ρ the density of the liquid. These two forces, gravity and buoyancy, are constant. Let us denote the sum of these by **k** (see the figure). The magnitude of **k** is then

$$|\mathbf{k}| = |m\mathbf{g} - V\rho\mathbf{g}|\,.$$

We now assume that the sphere is also acted upon by a frictional force proportional to the magnitude of the velocity and directed against the motion, that is:

$$\mathbf{f} = -a\mathbf{v},$$

where a is some positive constant.

Find the velocity of the particle as a function of time.

Solution. The *equation of motion* for the particle is

$$m\frac{d\mathbf{v}}{dt} = \mathbf{k} - a\mathbf{v} \tag{2.30}$$

The velocity of the particle has no component perpendicular to the x-axis at $t = 0$, and as all the forces act along the x-axis, the motion will take place along the x-axis. We can thus write (2.30) as

$$m\frac{dv}{dt} = k - av. \tag{2.31}$$

Without integrating (2.31) we note that the velocity is going to increase until the two terms on the right side balance. The velocity will asymptotically approach a *terminal velocity* : $v_t = k/a$. For $v = k/a$, we have $\mathrm{d}v/\mathrm{d}t = 0$.

Let us integrate (2.31):

$$\frac{\mathrm{d}v}{k/a - v} = \frac{a}{m}\mathrm{d}t. \tag{2.32}$$

We multiply both sides of the equation by -1:

$$\frac{\mathrm{d}(-v)}{k/a - v} = -\frac{a}{m}\mathrm{d}t. \tag{2.33}$$

Since k/a is a constant, we can write (2.33) as :

$$\frac{\mathrm{d}(k/a - v)}{k/a - v} = -\frac{a}{m}\mathrm{d}t. \tag{2.34}$$

Integrating both sides of (2.33) results in:

$$\ln\left(\frac{k}{a} - v\right) = -\frac{a}{m}t + c_1 , \tag{2.35}$$

or

$$\frac{k}{a} - v = \exp(c_1)\exp\left(-\frac{at}{m}\right) = c_2\exp\left(-\frac{at}{m}\right), \tag{2.36}$$

where $c_2 \equiv \exp(c_1)$ is an integration constant.

Note. In (2.35) we have taken the logarithm of a quantity which has the dimension of velocity. This is not a proper equation, and we should perhaps write (2.36) directly.

We now use the initial conditions to determine the constant c_2. The relevant initial condition is:

$$v = 0 \quad \text{at} \quad t = 0 ;$$

consequently $c_2 = k/a$. This means that we can write (2.36) as

$$v = \frac{k}{a}\left[1 - \exp\left(-\frac{a}{m}t\right)\right]. \tag{2.37}$$

Note that $v \to k/a$ for $t \to \infty$.

Let us plot v as a function of time.

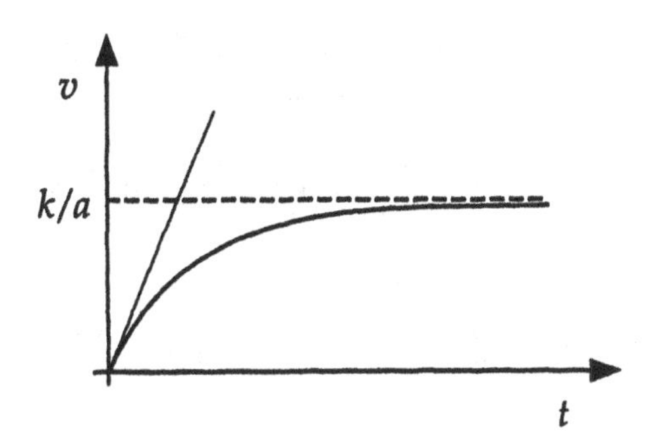

Question. What is the slope of the curve at $t = 0$? [Answer: k/m, use (2.31)].

Find x as a function of time by integrating (2.37):

$$x = \frac{kt}{a} - \frac{km}{a^2}\left[1 - \exp\left(-\frac{at}{m}\right)\right]$$

△

Example 2.5. Frictional Force: Solid Against Solid. When a (dry) object slides across a (dry) surface, the frictional force is to a good approximation independent of the relative velocity. If the object is held against the surface, which reacts by a normal force of magnitude N, we can write the frictional force as:

$$\mathbf{F}_{\text{fr}} = -\mu N \frac{\mathbf{v}}{|\mathbf{v}|}$$

since it is always pointing against the direction of motion. The constant μ is called the *coefficient of friction.* The magnitude of the frictional force is $F_{\text{fr}} = \mu N$.

A wooden block with mass M is projected upwards with an initial velocity v_0 along an inclined plane.

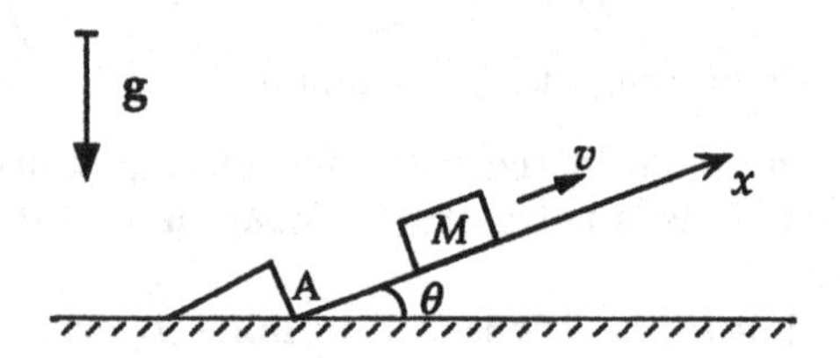

The block starts from the point A. We assume that the block is started in such a way that it moves along a straight line up the inclined plane. During a time T the block moves up the inclined plane, then stops. (It may then move downwards again, see Problem 2.5.)

(1) Find the time T.
(2) Find the distance D which the block moves up the inclined plane before it stops.

Answer. Draw all the forces acting on the body.

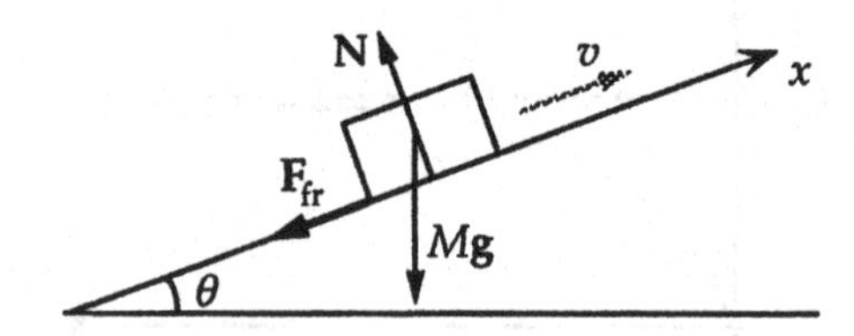

$$N = Mg\cos\theta; \quad F = \mu Mg\cos\theta.$$

The equation of motion is:

$$\begin{aligned} M\frac{\mathrm{d}^2x}{\mathrm{d}t^2} &= \overbrace{-Mg\sin\theta}^{F_x}\ \overbrace{-\mu Mg\cos\theta}^{F_{\mathrm{fr}}}, \\ \frac{\mathrm{d}x}{\mathrm{d}t} &= -g\,(\sin\theta + \mu\cos\theta)\,t + v_0\,, \\ x &= -\frac{1}{2}g\,(\sin\theta + \mu\cos\theta)\,t^2 + v_0 t\,. \end{aligned}$$

(1)

$$T = \frac{v_0}{g\,[\sin(\theta) + \mu\cos(\theta)]}\,.$$

(2)

$$D = \frac{1}{2}\frac{v_0^2}{g\,[\sin(\theta) + \mu\cos(\theta)]}\,.$$

△

Example 2.6. The Atwood Machine

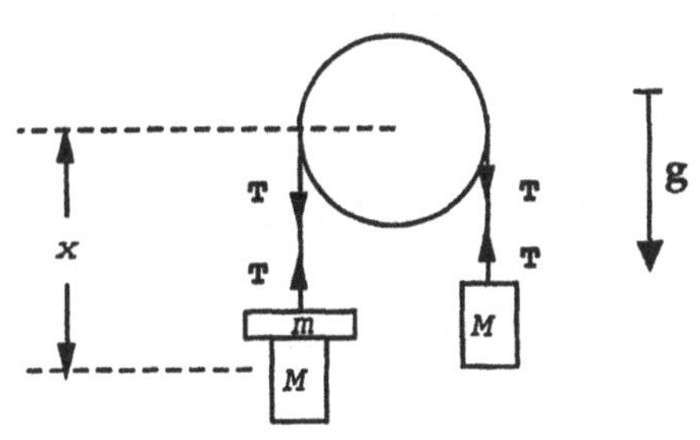

Two equal masses M are fastened to each end of a string that passes over a pulley. On top of one of the masses we place a smaller mass m. The system is then left to move under the influence of gravity. In this discussion of the so-called Atwood machine, we will make the following assumptions:

(a) We shall ignore the mass of the string. This assumption means that the string force is constant along all the free segments of the string. If the string force was not constant along the string, a string segment, assumed to have zero mass, would acquire an infinite acceleration.
(b) We shall neglect the mass of the pulley. This assumption implies that the string force is the same on both sides of the pulley.

The tension $\mathbf{T}$ in the string arises from a small stretching of the string (similar to that of a spring). In a dynamical problem such as the present, where the system moves under the influence of gravity, the string tensions are determined from the equations of motion for the massive objects.

The coordinate x (see the figure) determines the position of the mass $M+m$. The position of the other mass is then determined by the length of the string which we assume unchanged. That means the small change in the length of the string is neglected.

The mass $M+m$ has an acceleration downwards; the other mass has the same acceleration, but upward, since we assume that the string is taut during the entire motion. The acceleration of M is taken as positive when M accelerates upwards. We shall often use the shorthand notation $\dot{x} \equiv \mathrm{d}x/\mathrm{d}t$, $\ddot{x} \equiv \mathrm{d}^2x/\mathrm{d}t^2$.

Let us use this occasion to point to some straightforward rules which one should always keep in mind when applying Newton's second law:

1. Decide to which body you apply Newton's second law.
2. Draw all forces acting upon that body.
3. Write the equation of motion for that body.

By adhering to these rules we get the equation of motion for the mass $M+m$:

$$(M+m)\ddot{x} = (M+m)g - T. \tag{2.38}$$

Similarly, the equation of motion for M is:

$$M\ddot{x} = T - Mg. \tag{2.39}$$

From (2.38) and (2.39) we find:

$$\ddot{x} = \frac{m}{2M+m}g. \tag{2.40}$$

The acceleration $\ddot{x}$ is thus constant, and less than g.

By combining (2.39) and (2.40) one finds:

$$T = Mg\left(1 + \frac{m}{2M+m}\right). \tag{2.41}$$

△

Example 2.7. Force in Harmonic Motion. In general one can say that in Newtonian mechanics there are two types of problems:

(1) We know the forces and determine the motion
(2) We know the motion and determine the force.

We shall now consider an example of case (2). A particle with mass m is moving along the x-axis. The motion is determined by the equation:

$$x = A\sin(\omega t), \tag{2.42}$$

where A and ω are constants.

Find the acceleration $\ddot{x}$ and the force F_x acting on the particle.

Solution.

$$\begin{aligned} \ddot{x} &= -A\omega^2 \sin(\omega t), && (2.43) \\ F_x &= m\ddot{x} = -mA\omega^2 \sin(\omega t), && (2.44) \\ &= -kx, \text{ where } k = m\omega^2 . && (2.45) \end{aligned}$$

We see that true harmonic motion can only be produced by a force of this form, i.e., by a force that is strictly proportional to the distance from the point of equilibrium. On the other hand, harmonic motion is a good *approximation* to many types of oscillations around a point of equilibrium, corresponding to the approximation of a force law $F(x)$ with its tangent at the origin.

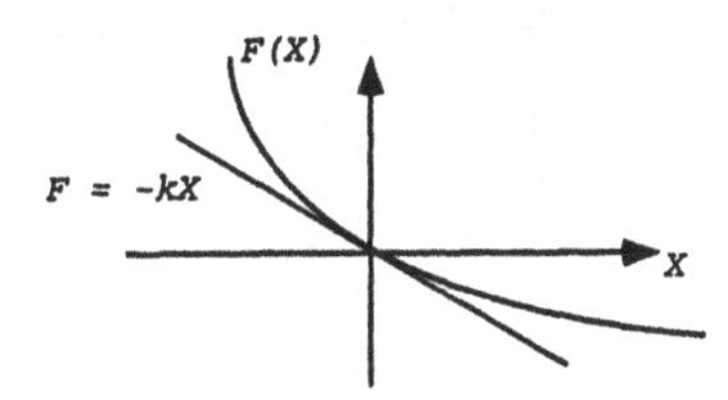

Question. Coin on a plate in harmonic motion.

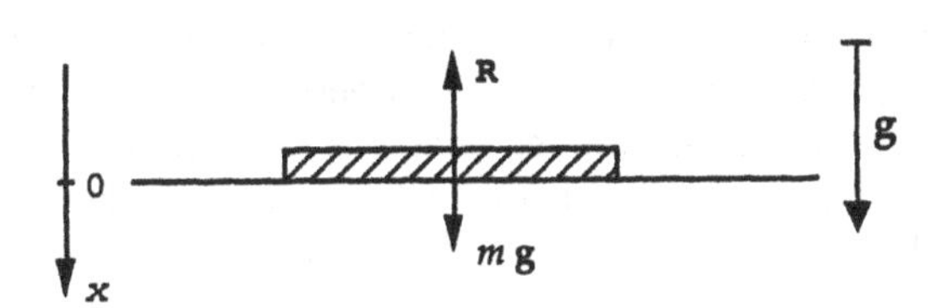

A coin with mass 20 g rests on a plate. The plate is acted on by an external force in such a way that the plate performs a harmonic motion in the vertical direction in the gravitational field of the Earth, described by the equation $x = A\cos(\omega t)$. The amplitude A is 20 cm and the oscillation period is $T = 1$ second.

(1) Find the maximum and minimum values of the force R (the reaction force) with which the plate acts on the coin throughout a period.
(2) Find the smallest value of the period for which the coin will remain in contact with the plate throughout the motion (for fixed amplitude $A = 20$ cm.).

Answers.

(1)

$$\begin{aligned} R_{\text{max}} &= mA\omega^2 + mg = 0.354 \text{ N}; \\ R_{\text{min}} &= mg - mA\omega^2 = 0.0381 \text{ N}. \end{aligned}$$

(2)

$$T_{\min} = 2\pi\sqrt{\frac{A}{g}} = 0.9 \text{ s}.$$

△

Example 2.8. Charged Particle in a Uniform Magnetic Field. We will study the motion of an electrically charged particle in a *magnetic* field which is spatially uniform and constant in time. We denote the mass and the charge of the particle by m and q respectively.

We shall take as an experimentally established fact that the force $\mathbf{F}$ on a charged particle moving with velocity $\mathbf{v}$ in a magnetic field $\mathbf{B}$ is:

$$\mathbf{F} = q(\mathbf{v} \times \mathbf{B}). \tag{2.46}$$

This force is called the *Lorentz Force* after the Dutch physicist H.A. Lorentz (1853–1928).

The *equation of motion* for the particle is thus:

$$m\frac{\mathrm{d}\mathbf{v}}{\mathrm{d}t} = q(\mathbf{v} \times \mathbf{B}). \tag{2.47}$$

For simplicity we shall start by considering the case where the initial velocity $\mathbf{v}_0$ of the particle is perpendicular to the direction of the magnetic field (see figure below – here we have assumed that q is positive and $\mathbf{B}$ is perpendicular to the paper and pointing out of the paper). From the vector equation (2.47) we see that the acceleration of the particle is perpendicular to both the velocity $\mathbf{v}$ and the magnetic field vector $\mathbf{B}$. From this we can deduce that if the initial velocity $\mathbf{v}_0$ is perpendicular to $\mathbf{B}$, the particle will never develop a velocity component parallel to $\mathbf{B}$. The motion will remain in a plane perpendicular to $\mathbf{B}$. Since the acceleration is also perpendicular to $\mathbf{v}$, the vector

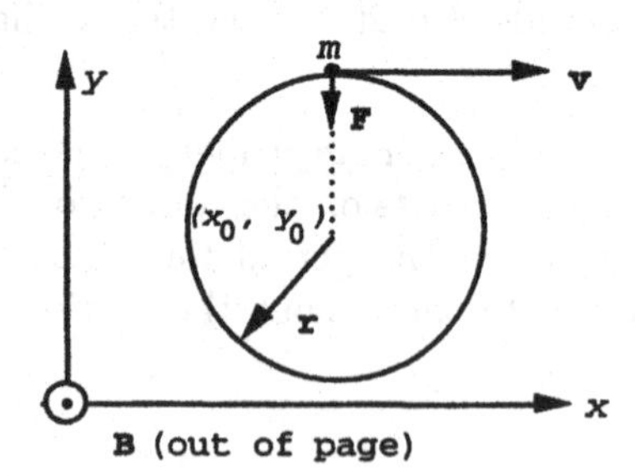

$\mathbf{v}$ can never change magnitude, only direction. From this we conclude that the magnitude of the force and therefore the magnitude of the acceleration is constant. The only motion that can satisfy these conditions is a *uniform*

circular motion (with speed $v_0 = |\mathbf{v}|$). The radius, r_c , in the circular motion is determined by writing the equation of motion in the following way:

$$m\frac{v_0^2}{r_c} = qv_0B, \tag{2.48}$$

where we have used that $\mathbf{v}_0$ is perpendicular to $\mathbf{B}$, i.e.,

$$r_c = \frac{mv_0}{qB}. \tag{2.49}$$

This radius is called the *gyro radius* or the *cyclotron radius* (since it denotes the radius of the particle orbit in a cyclotron accelerator).

The angular velocity, ω_c, is given by

$$\omega_c = \frac{v_0}{r_c} = \frac{qB}{m}. \tag{2.50}$$

We see that *the angular velocity* ω_c – and therefore the number of cycles per second – is *independent* of both the initial velocity and the radius in the orbit, and depends only on the strength of the magnetic field and the ratio of charge to mass of the particle (this also follows from (2.49): since r_c is proportional to v_0, the angular frequency is a constant).

We shall now write a complete solution to the equation of motion. We set up coordinates in such a way that the direction of the $\mathbf{B}$ field is along the z-axis, which gives $\mathbf{B}$ the coordinates $(0, 0, B)$. In the usual way, $\mathbf{i}$, $\mathbf{j}$, and $\mathbf{k}$ denote unit vectors along the x, y, and z axes, respectively. Still assuming $\mathbf{v}$ is perpendicular to $\mathbf{B}$, the vector $\mathbf{v} \times \mathbf{B}$ is given by (see Appendix)

$$\mathbf{v} \times \mathbf{B} = (v_yB, -v_xB, 0). \tag{2.51}$$

We can now write the vector equation (2.46) in terms of three coordinate equations

$$m\frac{dv_x}{dt} = qBv_y , \tag{2.52}$$

$$m\frac{dv_y}{dt} = -qBv_x , \tag{2.53}$$

$$m\frac{dv_z}{dt} = 0. \tag{2.54}$$

From (2.54): the component of $\mathbf{v}$ parallel to the magnetic field is not changed by the field and is therefore constant (zero in our case). Equations (2.52) and (2.53) can be written as

$$\frac{dv_x}{dt} = \frac{qB}{m}v_y , \tag{2.55}$$

$$\frac{dv_y}{dt} = -\frac{qB}{m}v_x . \tag{2.56}$$

or, if we introduce ω_c from (2.50):

$$\frac{dv_x}{dt} = \omega_c v_y , \tag{2.57}$$

$$\frac{dv_y}{dt} = -\omega_c v_x . \tag{2.58}$$

We have here an example of *coupled differential equations*: the change of v_x per unit of time is proportional to v_y and the change of v_y per unit of time is proportional to v_x.

In order to integrate this set of coupled differential equations we proceed in much the same manner as with two algebraic equations of two unknowns. If we differentiate (2.57) with respect to time, and substitute in this the expression for dv_y/dt from (2.58), we eliminate v_y from the equations. Now we have a (second-order) differential equation, in v_x only:

$$\frac{d^2v_x}{dt^2} + \omega_c^2 v_x = 0. \tag{2.59}$$

This equation is recognized as the equation which governs the harmonic oscillator (although here the unknown quantity is the *velocity* component along the x-axis), so the complete solution can be written with constants of integration, v_0 and t_0:

$$v_x(t) = v_0 \cos[\omega_c(t - t_0)] . \tag{2.60}$$

Equation (2.57) now gives

$$v_y(t) = \frac{1}{\omega_c}\frac{dv_x}{dt} = -v_0 \sin[\omega_c(t - t_0)]. \tag{2.61}$$

For simplicity we choose $t_0 = 0$. We can then integrate (2.60) and (2.61) again (using $v_x = dx/dt$ and $v_y = dy/dt$), and get, with the constants of integration x_0 and y_0:

$$x - x_0 = \frac{v_0}{\omega_c}\sin(\omega_c t), \tag{2.62}$$

$$y - y_0 = \frac{v_0}{\omega_c}\cos(\omega_c t), \tag{2.63}$$

displaying a *uniform circular motion* with center (x_0, y_0) and radius

$$r_c = \frac{v_0}{\omega_c} = \frac{mv_0}{qB} . \tag{2.64}$$

The slightly more general case, where initially there is a velocity component v_{0z} parallel to the direction of the magnetic field, can now easily be dealt with. We have already noted that this component of the velocity is not changed by the presence of the magnetic field, so the z component of the motion is a uniform motion:

$$z = v_{0z}t.$$

The resulting orbit of the particle in this case is a helix with axis along the magnetic field and this helix is traced out with a constant magnitude of the velocity:

$$v_0 = |\mathbf{v}_0| = (v_{0x}^2 + v_{0y}^2 + v_{0z}^2)^{1/2} .$$

We can easily find the pitch h of the helix, defined as the z-distance gained through one revolution (i.e., in the time $T_c = 2\pi/\omega_c$)

$$h = v_z T_c = v_z \frac{2\pi}{\omega_c} = \frac{2\pi m v_z}{qB} . \qquad (2.65)$$

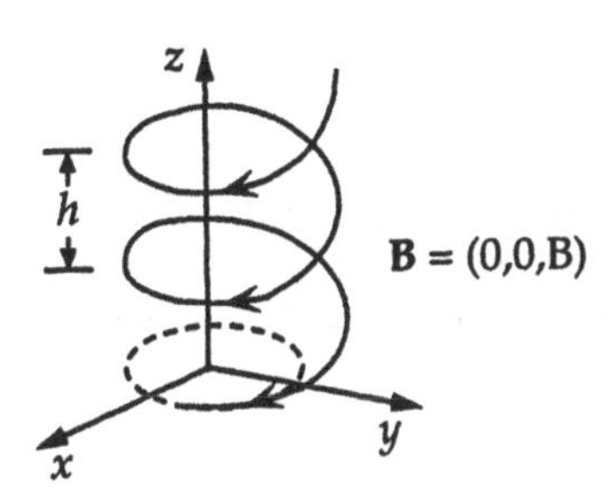

Question.

(1) Find n, the number of cycles per second, for (a) an electron, (b) a proton, both in a uniform magnetic field of strength 1 $\mathrm{Wb\,m^{-2}} = 1\ \mathrm{V\,s\,m^{-2}} = 1\ \mathrm{T} = 1$ Tesla.
(2) Find the cyclotron radius for (a) an electron, (b) a proton, both moving with a velocity of $1 \times 10^6\ \mathrm{m\,s^{-1}}$ in a uniform magnetic field of strength 1 T.

The proton has charge $q = |e| = 1.6 \times 10^{-19}$ C.

[Answers: (1a) $n_e = 2.8 \times 10^{10}\ \mathrm{s^{-1}}$; (1b) $n_p = 1.5 \times 10^7\ \mathrm{s^{-1}}$; (2a) $r_e = 5.7 \times 10^{-6}$ m; (2b) $r_p \approx 10^{-2}$ m = 1 cm.]
We now return to the first figure. ($v_{0z} = 0$). Make a sketch of the motion (use the "right-hand rule") if the particle:

(1) starts with velocity $-\mathbf{v}_0$ and q is positive,
(2) starts with velocity $+\mathbf{v}_0$ and q is negative,
(3) starts with velocity $-\mathbf{v}_0$ and q is negative,
(4) starts with velocity $+\mathbf{v}_0$ and q is negative but $\mathbf{B} \rightarrow -\mathbf{B}$.

Note that if both the magnetic field and the charge change signs, the orbit is unchanged! Consider for comparison a uniform circular motion in a gravitational field:

$$m\frac{v_0^2}{r} = G\frac{Mm}{r^2} . \qquad (2.66)$$

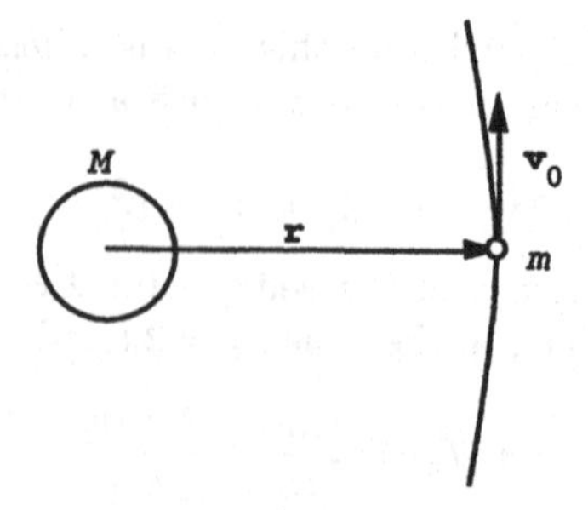

What orbit would the particle describe if we suddenly changed $\mathbf{v}_0$ to $-\mathbf{v}_0$?

△

Example 2.9. Thomson's Experiment. Consider an electron-beam tube where electrons move in crossed electric and magnetic fields. A beam of electrons is accelerated and then passed between two condenser plates.

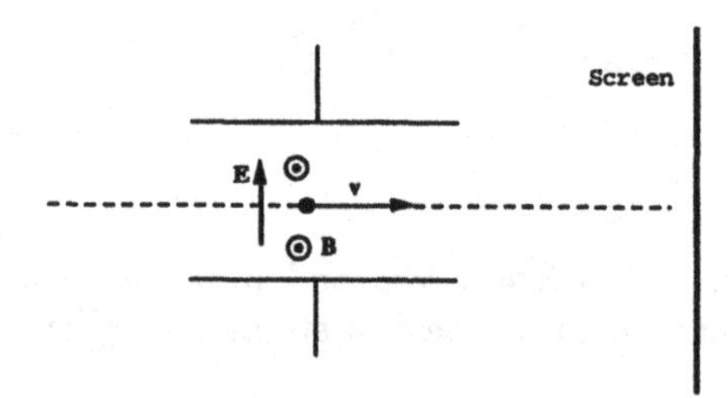

Let the electric field **E** between the condenser plates be parallel to the plane of the paper and let the magnet be placed so that the magnetic field **B** is perpendicular to the paper and pointing out of the paper.

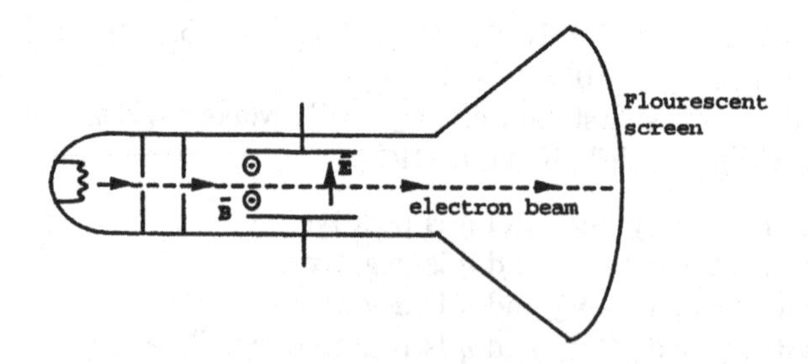

In 1897 Sir J.J. Thomson measured the ratio of the charge q of the electron to the mass m of the electron by studying the motion of electrons in a highly evacuated glass tube (no collisions with air molecules). The sketch above shows a modern version of J.J. Thomson's apparatus. The electron is negatively charged (how can one find out?). Let us write the charge q as $-|e|$. If the electric field has the direction shown in the sketch it will deflect the particle downwards. As the electrons have a velocity and pass through a magnetic field they will be under the influence of the Lorentz force too. If the magnetic field has the direction shown in the sketch, the Lorentz force will deflect the electrons upwards!

The resultant force on an electron is:

$$\mathbf{F} = q\mathbf{E} + q(\mathbf{v} \times \mathbf{B}) = -|e|(\mathbf{E} + \mathbf{v} \times \mathbf{B}).$$

By a suitable choice of strengths of $\mathbf{E}$ and $\mathbf{B}$ the electric and magnetic force on an electron may cancel each other. For, since $\mathbf{E}$, $\mathbf{B}$, and $\mathbf{v}$ are perpendicular to each other the force F will be zero if

$$-|e|E = -|e|vB, \tag{2.67}$$

$$E = vB. \tag{2.68}$$

The ideas behind Thomson's procedure are the following:

(1) The position of the spot with both E and B equal to zero is measured (that means the position of the undeflected beam).
(2) The position of the deflected beam with $B = 0$ and a known value E_0 of the electric field is measured.
(3) A magnetic field is now applied to the beam until the spot again reaches the undeflected position.

Let us call the corresponding value of the magnetic field B_0. Assume that the length of the condensor plate is l and the velocity of the electron is $v = |\mathbf{v}|$. Show that the deflection Δy of an electron in a purely electric field (step 2)

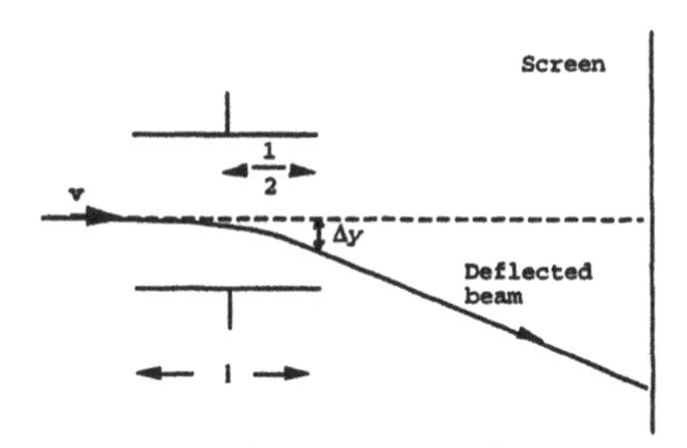

is given by:

$$\Delta y = \frac{|e|E_0 l^2}{2mv^2}, \tag{2.69}$$

where Δy is measured at the far edge of the deflecting plates. Δy is not directly measurable, but it may be calculated easily from the measured displacement of the spot on the fluorescent screen. Thus: Δy, E_0, and l are known. If we can find v we can calculate the ratio $|e|/m$.

The purpose of step 3 is to find v. If the magnetic field is increased until the spot reaches the undeflected position again, we have

$$v = \frac{E_0}{B_0}\,. \tag{2.70}$$

Substituting (2.70) into (2.69) we find

$$\frac{|e|}{m} = \frac{2\Delta y E_0}{B_0^2 l^2}\,. \tag{2.71}$$

Thomson found

$$\frac{|e|}{m} = 1.7 \times 10^{11}\ \mathrm{C\,kg^{-1}}\,.$$

The value quoted today is

$$\frac{|e|}{m} = 1.75888 \times 10^{11}\ \mathrm{C\,kg^{-1}}\,.$$

Questions.

(1) What is the velocity of the beam of electrons when no deflection of the electrons is produced by the simultaneous influence of the electric field

$$E_0 = 3 \times 10^5\ \mathrm{V\,m^{-1}}\,,$$

and the magnetic field

$$B_0 = 10^{-2}\ \mathrm{T},$$

which are both normal to the beam and to each other?

(2) What would be the radius of the orbit of the electron if the electric field was removed?

(3) In the calculation of (2.69) above we have not taken into account the fact that gravity also deflects the electrons downwards. Let us assume that E_0 has the value $E_0 = 3 \times 10^5\ \mathrm{V\,m^{-1}}$ as in the question above and that the velocity of the electrons is the same found above. Let l =1.7 cm. Is it a good approximation to neglect the gravitational force on the electron?

[Answers: (1) $v = 3 \times 10^7\ \mathrm{m\,s^{-1}}$; (2) $r = 1.7$ cm; (3) the deflection due to the electric field is: $\Delta y_E = 8.5$ mm.The deflection due to the gravitational field is:

$$\begin{aligned} \Delta y_{\mathrm{g}} &= 1.6 \times 10^{-15}\ \mathrm{mm}, \\ \frac{\Delta y_{\mathrm{E}}}{\Delta y_{\mathrm{g}}} &= 5.3 \times 10^{15}\,. \end{aligned}$$

We see that we may safely neglect the gravitational force.] △

Example 2.10. Work and Energy in Linear Motion of a Particle. We will in a few examples consider more general methods for integration of the equations of motion.

Note. We shall discuss here the important law of energy conservation for motion in one dimension. When we have gained more experience, this law will be treated in the general case of three dimensions (see Chapter 8).

Consider a particle moving along the x-axis under the influence of some impressed force $\mathbf{F}$. We restrict ourselves to motions along the x-axis, so all quantities are scalars. The equation of motion is

$$m\frac{\mathrm{d}^2x}{\mathrm{d}t^2} = F,$$

or

$$m\frac{\mathrm{d}v}{\mathrm{d}t} = F. \tag{2.72}$$

We multiply (2.72) by $v = \mathrm{d}x/\mathrm{d}t$ on both sides:

$$m\frac{\mathrm{d}v}{\mathrm{d}t}v = F\frac{\mathrm{d}x}{\mathrm{d}t},$$

or

$$mv\,\mathrm{d}v = F\,\mathrm{d}x\,. \tag{2.73}$$

The right-hand side of (2.73) is now written $F\mathrm{d}x = \mathrm{d}W$; the quantity $\mathrm{d}W$ is called the *work done* by the force F over the segment $\mathrm{d}x$.

The left-hand side of (2.73) can be written as $\mathrm{d}\left(\frac{1}{2}mv^2\right)$, where the quantity $\frac{1}{2}mv^2 = T$ is called the *kinetic energy* of the particle. With this, we can write (2.73) as

$$\mathrm{d}\left(\frac{1}{2}mv^2\right) = F\mathrm{d}x\,, \tag{2.74}$$

or

$$\mathrm{d}\,(T) = \mathrm{d}\,(W)\,. \tag{2.75}$$

The change in the kinetic energy of the particle over the segment $\mathrm{d}x$ equals the work done by the force F.

This theorem, which makes a statement about kinetic energy, holds for all forces without exception. It is based on Newton's second law only.

We now restrict ourselves to the special case where the force F depends on position only, $F = F(x)$. (In particular, F is not dependent on time). When the force depends on the position coordinate x only, we can introduce a new quantity, the potential energy $U = U(x)$, defined through

$$\mathrm{d}U(x) \equiv -\mathrm{d}W = -F(x)\mathrm{d}x. \tag{2.76}$$

The change in the potential energy $U(x)$ over the segment $\mathrm{d}x$ is equal to minus the work done by the force F.

In integral form:

$$U(x) = -\int F(x)\mathrm{d}x \tag{2.77}$$

The potential energy, $U(x)$ of a particle in the force field $F(x)$ is through (2.77) determined up to an additive constant; this means for instance that we can choose for which point x_0 we want $U(x_0) = 0$.

With the definitions above, we have

$$\mathrm{d}T = -\mathrm{d}U(x),$$

or, upon integrating,

$$T + U(x) = E. \tag{2.78}$$

The constant E is the value of a quantity called the *total mechanical energy* of the system. The equation (2.78) is called a first integral to the equation of motion since we have – through T – found $\mathrm{d}x/\mathrm{d}t = v$ from the equation of motion containing $\mathrm{d}^2x/\mathrm{d}t^2$. The number E is a constant of integration which can be found from the initial conditions. A force field, where the force only depends on the position coordinate of the particle is called a *conservative force field*, since we have the simple observation:

The total mechanical energy – i.e., the sum of kinetic and potential energy – for a particle moving in a conservative force field, is conserved (constant).

An example of a nonconservative force is a frictional force. Frictional forces depend not on position alone but also on, say, direction of motion. The energy theorem – which we have derived only for one dimensional motion – is of vast importance in physics, and will be discussed in much more detail later. Here we will show, through an example, how the theorem can be used in the integration of the equation of motion. We may write the energy conservation theorem in the form

$$\left(\frac{\mathrm{d}x}{\mathrm{d}t}\right)^2 = \frac{2}{m}\left[E - U(x)\right]. \tag{2.79}$$

△

Example 2.11. Free Fall Towards the Sun from a Great Distance

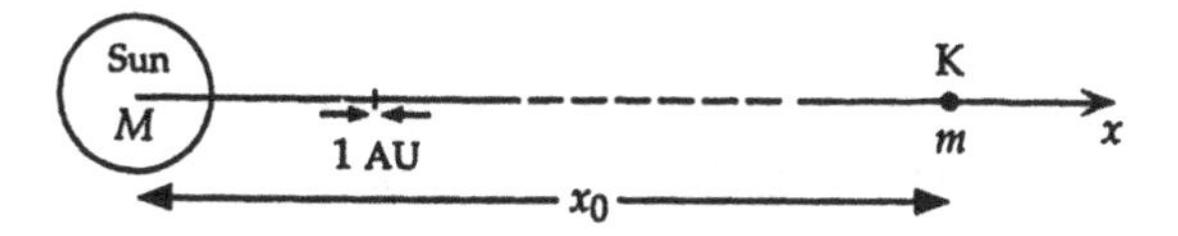

A comet K is found at some time at rest with respect to the Sun. The comet has the distance $x_0 = 50\,000$ AU from the Sun (1 AU = 1 astronomical unit = the average distance of the Earth from the Sun = 1.495×10^{11} m). We assume now that the comet moves only in response to the gravitational pull of the Sun, and we wish to determine the time passed from the comet begins its fall until it is a distance of 1 AU from the Sun. Since the comet has no initial velocity and in particular no initial velocity component perpendicular to the direction toward the Sun, the comet will move on a straight line connecting it to the center of the Sun. The mass of the Sun is considered to be so large that the Sun does not move in response to the (reaction) force from the comet. The mass of the Sun is considered in this sense to be infinite.

The equation of motion for the comet is then

$$m\frac{\mathrm{d}v}{\mathrm{d}t} = -\frac{GMm}{x^2} .$$

We multiply by v and integrate:

$$\begin{aligned} m\frac{\mathrm{d}v}{\mathrm{d}t}v &= -\frac{GMm}{x^2}\frac{\mathrm{d}x}{\mathrm{d}t} , \\ \int_0^v mv\,\mathrm{d}v &= \int_{x_0}^{x} -\frac{GMm}{x^2}\mathrm{d}x . \end{aligned}$$

One finds, since $v(x_0) = 0$,

$$\frac{1}{2}mv^2(x) = \frac{GMm}{x} - \frac{GMm}{x_0} .$$

This result could have been found as a consequence of the energy theorem, since the potential energy of the comet in the gravitational field of the Sun is given by

$$U(x) = -\int -\frac{GMm}{x^2}\mathrm{d}x = -\frac{GMm}{x} + C = -\frac{GMm}{x}$$

The choice of $C = 0$ implies $U(\infty) = 0$.

The energy theorem then implies

$$\frac{1}{2}mv^2(x) - \frac{GMm}{x} = 0 - \frac{GMm}{x_0} .$$

Using this result we have as a first integral to the equation of motion:

$$\frac{dx}{dt} = \pm \left[2GM(\frac{1}{x} - \frac{1}{x_0})\right]^{1/2} .$$

We will use the minus sign of the square root since the dx/dt, with the chosen orientation of the x-axis, is negative. We find for the time t lapsed between x_0 and x:

$$t = -\frac{1}{\sqrt{2GM}} \int_{x_0}^{x} \left(\frac{1}{x'} - \frac{1}{x_0}\right)^{-1/2} dx' .$$

This integral is somewhat cumbersome. To obtain an order of magnitude estimate, we discard the small term $1/x_0$ under the root. Then, after integrating,

$$t = \frac{2}{3}\frac{1}{\sqrt{2GM}} \left(x_0^{3/2} - x^{3/2}\right) .$$

We want the time of fall from $x_0 = 50\,000$ AU to the position $x = 1$ AU. Since $x_0 \gg x$, we get

$$t \approx \frac{2}{3}\frac{1}{\sqrt{2GM}} x_0^{3/2} ,$$

and finally, inserting values, $t \approx 4 \times 10^{13}$ s $\approx 1.3 \times 10^{6}$ years.

Question. Calculate the velocity a particle must have vertically upwards at the surface of the Earth in order just to escape the gravitational field of the Earth. Disregard air resistance and interaction with masses other than the Earth. This velocity is called the escape velocity.

[Answer: 11.2 $\mathrm{km\,s^{-1}}$. Compare this to the velocity of a race car, a supersonic jet and an air molecule!] △

Example 2.12. Momentum Conservation. Consider the equation of motion for a particle:

$$m\frac{d\mathbf{v}}{dt} = \mathbf{F} .$$

Multiply both sides with dt:

$$m d\mathbf{v} = d\mathbf{p} = \mathbf{F} dt,$$

where $d\mathbf{p}$ is an increment of the *momentum* $\mathbf{p} = m\mathbf{v}$ of the particle. In integral form:

$$\mathbf{p}_2 - \mathbf{p}_1 = \Delta\mathbf{p} = \int_{t_1}^{t_2} \mathbf{F} dt ,$$

or, in words:

The integral of the force over time equals the change of the momentum of the particle during that time.

The integral of the force over time is called the *impulse* of the force. We see that the impulse of the force, i.e., $\int F_x \, \mathrm{d}t$, is equal to the change in the momentum. In Example 2.10 we saw that the work of the force, i.e, $\int F_x \mathrm{d}x$, is equal to the change in kinetic energy. These two rules apply to all types of forces since their derivation is based only on Newton's second law.

Consider two interacting particles. The particles may interact through an electric or gravitational field, or they may be in direct contact or interact via a spring attached to each of them. If we assume that the interactions obey Newton's third law, we can derive the important result that *the total momentum of the two particles is conserved.*

The proof is simple. For particle 1 we have

$$\frac{\mathrm{d}\mathbf{p}_1}{\mathrm{d}t} = \mathbf{F}_{12} \, .$$

$\mathbf{F}_{12}$ is the force exerted on particle 1 by particle 2. For particle 2 we have:

$$\frac{\mathrm{d}\mathbf{p}_2}{\mathrm{d}t} = \mathbf{F}_{21} \, .$$

By addition:

$$\frac{\mathrm{d}}{\mathrm{d}t}(\mathbf{p}_1 + \mathbf{p}_2) = \mathbf{F}_{12} + \mathbf{F}_{21} = 0,$$

where we have used Newton's third law. After integration, we find

$$\mathbf{p} \equiv \mathbf{p}_1 + \mathbf{p}_2 = \text{constant vector},$$

i.e., the total momentum of the system is a constant of the motion. The argument may be extended to a system of many interacting particles.

For an isolated (closed) system, i.e., a system where only internal forces act, the total momentum is conserved.

Consider a *collision* between two isolated particles.

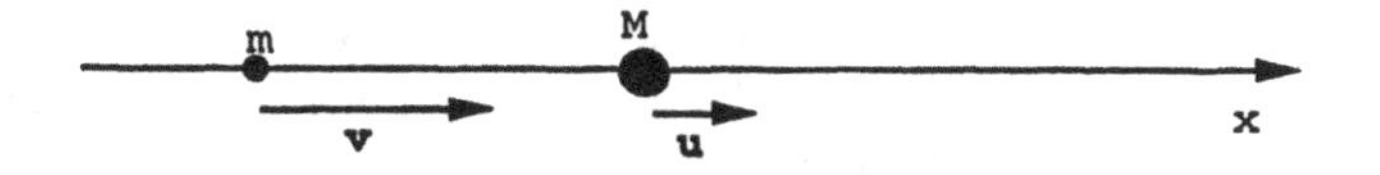

The two particles are constrained to move on a straight line: the x-axis. The two particles collide. Whatever forces act during the collision, the total momentum remains constant – the only assumption about the forces is that they obey Newton's third law. The velocities before the collision are v and u, and after the collision they are v_1 and u_1. We therefore have

$$mv_1 + Mu_1 = mv + Mu \, . \tag{2.80}$$

We call a collision between the two particles *elastic*, if the total kinetic energy is conserved in the collision (equivalently, no heat is produced in the collision; the interaction forces are conservative). For the case of an elastic collision, we have

$$\frac{1}{2}mv_1^2 + \frac{1}{2}Mu_1^2 = \frac{1}{2}mv^2 + \frac{1}{2}Mu^2 . \tag{2.81}$$

The set of equations (2.80) and (2.81) permit a calculation of the velocities v_1 and u_1, i.e., the velocities after the collision. See Problem 2.20. △

Example 2.13. Inelastic Collisions. The law of conservation of momentum for a closed system of particles is truely independent of the forces acting between different parts of the system. By the phrase "a closed system" we mean a system of particles that does not interact with other physical systems.

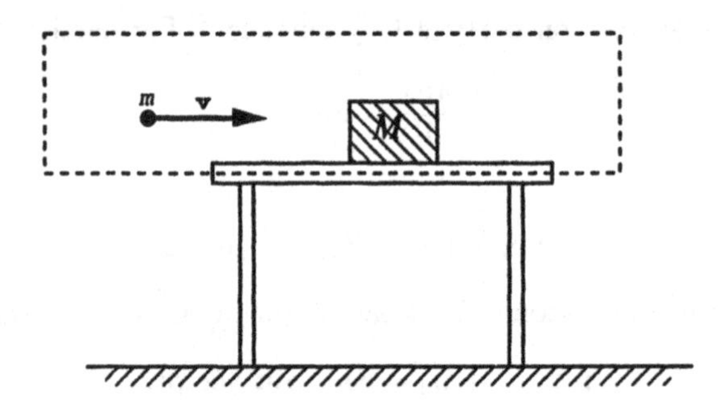

As a simple example of the application of conservation of momentum we consider a bullet (mass m, velocity v) that is shot into a wooden block of mass M resting on a horizontal table. The bullet is stopped by – and embedded into – the block. The system (bullet + block) is not an isolated system. It is under the influence of gravity, the reaction from the table, and air resistance. But if we consider the system for a very short time interval around the time the bullet hits the block, these external forces cannot change the momentum of the system much. If we neglect air resistance and consider the table as frictionless, no external forces are acting on the system (block + bullet) in the horizontal direction. The total horizontal component of the momentum is then the same before, during, and after the collision.

(1) Let m be a gun bullet with mass 5 g and velocity $v = 300\ \mathrm{m\,s^{-1}}$. Let $M = 3$ kg. Find the velocity u of the system $(m+M)$ just after the bullet is stopped.
(2) Let us assume that the gun was fired such that the muzzle was 1 m from the block. While the bullet travels from the muzzle to the block, it is under the influence of gravity. Find the downwards velocity v_y that the bullet gains before it hits the block and find the distance s_y it has fallen below the height of the muzzle.

(3) Just before the collision the system thus has a small vertical momentum mv_y downwards. What happens to that momentum during the collision?

[Answers: (1) $u = 0.50\ \mathrm{m\,s^{-1}}$, (2) $v_y = 3.3\ \mathrm{cm\,s^{-1}}$; $s_y = 0.055$ mm.] Note the order of magnitudes: $v_y = g\Delta t$, $s_y = \frac{1}{2}g(\Delta t)^2$, where Δt is the short time the bullet has fallen. △

Example 2.14. Rocket Propulsion

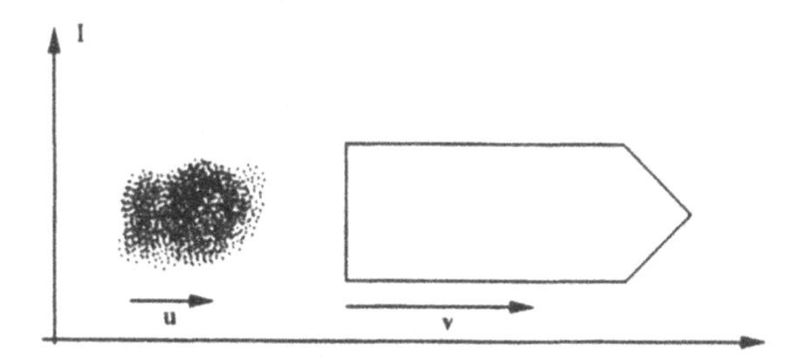

The theorem of conservation of momentum can for certain systems be the key to the integration of the equation of motion.

Let us assume that a rocket moves in force-free space. We describe the motion with respect to an inertial frame **I**. At time $t = 0$, the rocket has velocity 0. Let $m(t)$ be the mass of the hull plus the amount of fuel present at time t. The engine throws out a constant amount of mass per unit time:

$$\mu \equiv -\frac{\mathrm{d}m}{\mathrm{d}t} . \tag{2.82}$$

Note: $\mu > 0$; the mass of the rocket decreases with time! We are assuming that this mass is ejected in the direction of negative x with the constant velocity v_{rel} relative to the rocket. If we denote the velocity of the rocket relative to **I** by $v(t)$, the mass ejected from the rocket at time t will have a velocity

$$u(t) = v(t) - v_{\mathrm{rel}} \tag{2.83}$$

with respect to **I**. All quantities are scalars since the entire motion is one dimensional. Let us consider the hull plus the mass of fuel present at time t. The system is not acted upon by external forces, so we can use the theorem of conservation of momentum. At the time t, the momentum of the system is

$$p(t) = m(t)v(t). \tag{2.84}$$

The total momentum of the same system at time $t + \mathrm{d}t$ is

$$p(t + \mathrm{d}t) = (m + \mathrm{d}m)(v + \mathrm{d}v) + \mu\mathrm{d}t(u + \mathrm{d}u), \tag{2.85}$$

because the amount of material expelled is μdt. Note: in the last term of equation (2.85) we might as well write u instead of $u+du$; the du is multiplied by dt and differentials of second order disappear in the limiting process.

The theorem of conservation of momentum tells us that

$$(m + dm)(v + dv) + \mu dt(u + du) = mv\,. \tag{2.86}$$

Discarding differentials of second order, we find:

$$mdv + vdm + u\mu dt = 0, \tag{2.87}$$

$$mdv + vdm - udm = 0. \tag{2.88}$$

The differential equation for v as a function of m becomes (using (2.83)):

$$m\frac{dv}{dm} + v_{\text{rel}} = 0. \tag{2.89}$$

From (2.89) we obtain:

$$dv = -v_{\text{rel}}\frac{dm}{m}$$

and since $v = 0$ for $m = m(0)$, we get

$$v(t) = v_{\text{rel}} \ln\left[\frac{m(0)}{m(t)}\right]. \tag{2.90}$$

We assume that the rocket starts with velocity zero and that the engine works from $t = 0$ to $t = T$. The final velocity is then:

$$v(T) = v_{\text{rel}} \ln\left[\frac{m(0)}{m(T)}\right].$$

Note.

(1) The final velocity depends logarithmically on the ratio between the initial and final mass.

(2) Given $m(T)$, it is crucial to have a high value of v_{rel}.

Even if the payload is only 1/10 %, i.e., $m(0) = 1000m(T)$, we find $v(T) = v_{\text{rel}} \ln(1000) = 6.9v_{\text{rel}}$; less than ten times the exhaust velocity.

This simple fact, based on the most fundamental mechanical laws shows the difficulty in designing rockets to carry people out of the solar system. Colonizing the galaxy will not be easy! Einstein's special theory of relativity does not change this situation in any fundamental way.

Another numerical example: if the exhaust gasses have a temperature of about 4000°C, the molecules have a velocity of the order of 2000 $\mathrm{m\,s^{-1}}$. If we want a final velocity of the rocket of $v(T) = 8\ \mathrm{km\,s^{-1}}$, we can find $m(T)$ as a ratio of $m(0)$:

$$m(T) = m(0)\exp\left[-\frac{v(T)}{v_{\text{rel}}}\right].$$

or

$$\frac{m(T)}{m(0)} = \exp\left[-\frac{v(T)}{v_{\mathrm{rel}}}\right],$$

$$m(T) = m(0)\,\mathrm{e}^{-4} = m(0) \times 0.018.$$

This corresponds to a payload of about 2 % of the initial mass.

Warning. Consider again the rocket head and the fuel in the rocket head at the time t. The equation of motion at the time t for this total mass is

$$\frac{\mathrm{dp}}{\mathrm{d}t} = 0.$$

From this one might be tempted to conclude

$$\frac{\mathrm{d}}{\mathrm{d}t}(mv) = 0,$$

or

$$m\frac{\mathrm{d}v}{\mathrm{d}t} + v\frac{\mathrm{d}m}{\mathrm{d}t} = 0,$$

or

$$m\frac{\mathrm{d}v}{\mathrm{d}t} = \mu v\,, \quad \mu \equiv -\frac{\mathrm{d}m}{\mathrm{d}t}\,.$$

This is not correct! If the above equations were correct, the acceleration of the rocket head would be independent of the exhaust velocity v_{rel}. The above equations neglect the fact that the part of $m(t)$ which appears in the fuel exhaust also carries a momentum. As shown above:

$$\begin{aligned}
\frac{\mathrm{d}p}{\mathrm{d}t} &= \frac{p(t+\mathrm{d}t) - p(t)}{\mathrm{d}t}\,, \\
p(t) &= m(t)v(t) = mv\,, \\
p(t+\mathrm{d}t) &= (m+\mathrm{d}m)(v+\mathrm{d}v) + \mu\mathrm{d}t(v - v_{\mathrm{rel}}).
\end{aligned}$$

The first term on the right-hand side is the rate of change in the momentum of the rocket head. The second term is the change in momentum of the exhaust fuel per unit time at the instant considered. If an external force of magnitude F acts (e.g. gravity) we find

$$\frac{\mathrm{d}}{\mathrm{d}t}(mv) + \mu(v - v_{\mathrm{rel}}) = F\,.$$

△

Example 2.15. Some Qualitative Remarks on Rocket Propulsion. Let us consider a rocket head containing oxygen plus kerosene. When the oxygen and the kerosene are allowed to interact a very rapid chemical reaction (an explosion) takes place. A lot of potential energy of the electrons inside the molecules of the oxygen and kerosene is converted to kinetic energy of the combustion products.

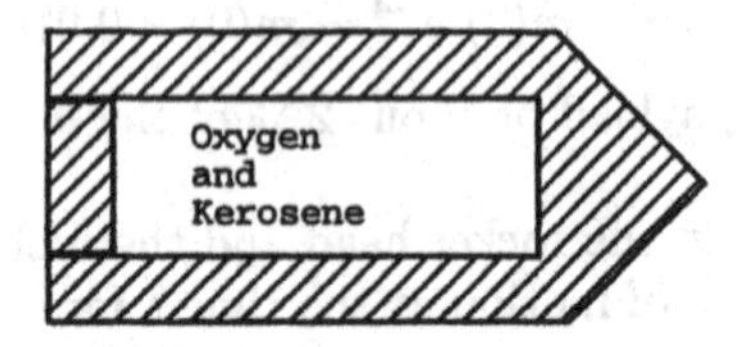

Let us sketch very roughly how a rocket works. After the explosion a lot of kinetic energy is going into the molecules of the combustion products. These molecules move in all directions. On average an equal number moves in each direction. Let us in an extremely rough approximation assume that half the molecules move along straight lines to the left, and the other half along straight lines to the right.

Let us assume that the cap 'A' is still on. We consider the two molecules shown. Let their velocities be v, and their masses m . They are reflected at the end walls.

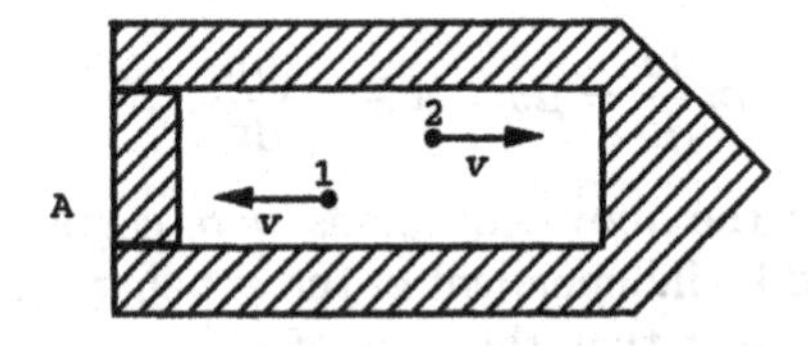

In the reflection from A, molecule 1 gives the momentum $2mv$ to the rocket head to the left (the rocket head has infinite mass compared to one molecule). Molecule 2, in reflection from the end wall, gives a momentum $2mv$ to the rocket head to the right. All in all the rocket head receives no net momentum.

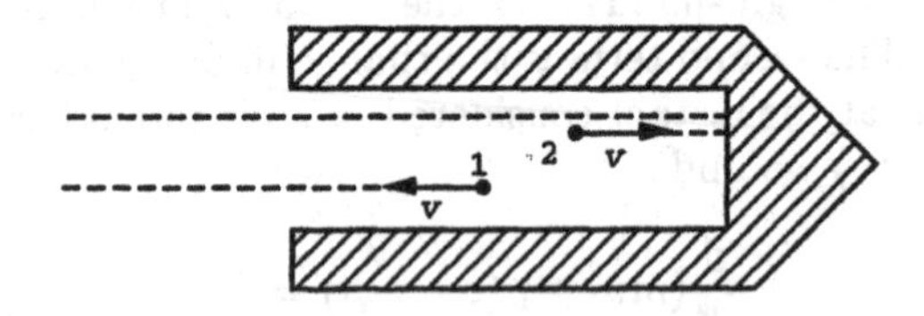

If A is not there, molecule 1 is not reflected but moves out in free space, while molecule 2 gives the momentum $2mv$ to the rocket head and then leaves the rocket head to go out into free space. The rocket head now receives a momentum $2mv$ towards the right.

If a mass $2m$ is expelled to the left with velocity v relative to the rocket head, the rocket head receives a momentum $2mv$ towards the right.

Note. It is not possible to convert all the disorganized thermal energy in the molecules of the combustion products into useful macroscopic kinetic energy of the rocket head. For this to be achieved, all of the molecules should (after the explosion) move in the same direction (first towards the right). This is evidently impossible. With this remark we touch upon one of the most fundamental laws of physics: the second law of thermodynamics. △

Example 2.16. Ball Against a Wall

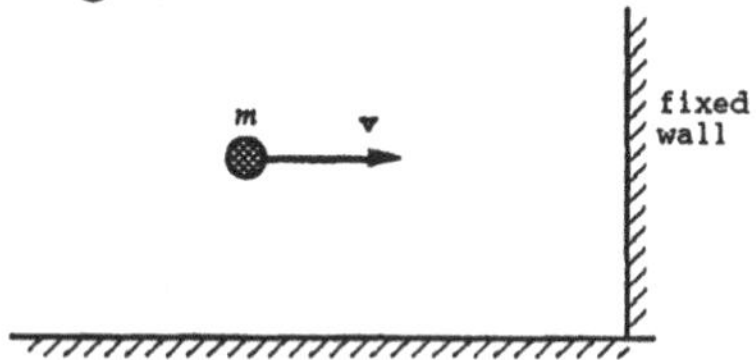

A ball of mass m is thrown horizontally against a fixed wall, so that just before it hits the wall the ball has velocity $\mathbf{v}$. Let us assume that it rebounds with the same speed. The ball is in contact with the wall for a short time Δt.

(1) Let us first consider the ball as our system. Find the average value F of the force by which the wall has acted on the ball during the collision? The value of F should be expressed by m, v, and Δt.
(2) Numerical example: $v = 25\ \mathrm{m\,s^{-1}}$, $m = 100$ g, and $\Delta t = 0.02$ s. Find F. Compare F to the force on a mass of 0.1 kg in the gravitational field of the Earth.
(3) Let us next consider as our system the ball plus the wall, the latter being fixed to the Earth. What about conservation of momentum? It still holds; the Earth picks up the ball's change in momentum. Estimate the change in the velocity of the Earth due to the collision between the wall and the ball.

Solution. The force must be in the direction perpendicular to and outwards from the wall, and the average force can be determined from the equation of motion for the ball:

$$\frac{\mathrm{d}p_x}{\mathrm{d}t} = F_x \,, \tag{2.91}$$

$$\mathrm{d}p_x = F_x \mathrm{d}t \,, \tag{2.92}$$

$$\Delta p_x = \int_1^2 F_x \mathrm{d}t = \langle F_x \rangle \Delta t \,. \tag{2.93}$$

From (2.93) we get

$$\langle F_x \rangle \Delta t = 2mv \ , \tag{2.94}$$

$$\langle F_x \rangle = \frac{2mv}{\Delta t} \ . \tag{2.95}$$

Numerical example: $v = 25\ \mathrm{m\,s^{-1}}$, $m = 100$ g, $\Delta t = 0.02$ s, $\langle F \rangle = 250$ N.

Let us next consider as our system the ball plus the wall, which is fixed to the Earth. What about conservation of momentum? It still holds. The Earth picks up the change in the momentum of the ball.

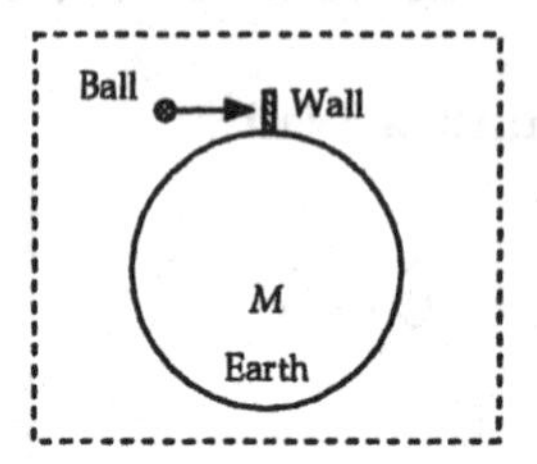

The velocity of the center of mass of the Earth after the collision, measured in the reference frame where the Earth was at rest before the collision, is determined by

$$Mu = 2mv \ , \tag{2.96}$$

$$u = \frac{2mv}{M} \ . \tag{2.97}$$

Inserting numbers,

$$M = 5.98 \times 10^{24} \ \mathrm{kg} \ ,$$

$$u = \frac{2 \times 0.1 \times 25}{5.98 \times 10^{24}} \approx 0.8 \times 10^{-24} \ \mathrm{m\,s^{-1}} \ ,$$

which is a very small velocity indeed.

Question. Equation (2.96) is not exact. Why? △

2.3 Problems

Problem 2.1. Calculate the escape velocity from the planet Mars, and for the (terrestial) Moon. Compare with Example 2.11.

Problem 2.2. A man stands on top of a tall building. From the edge of the building he throws a small heavy particle vertically upwards with an initial velocity of $20\ \mathrm{m\,s^{-1}}$. The particle hits the ground at the foot of the building 5 seconds later. Neglect air resistance.

(1) How tall is the building?
(2) How far above the ground is the highest point of the trajectory of the particle?

Problem 2.3.

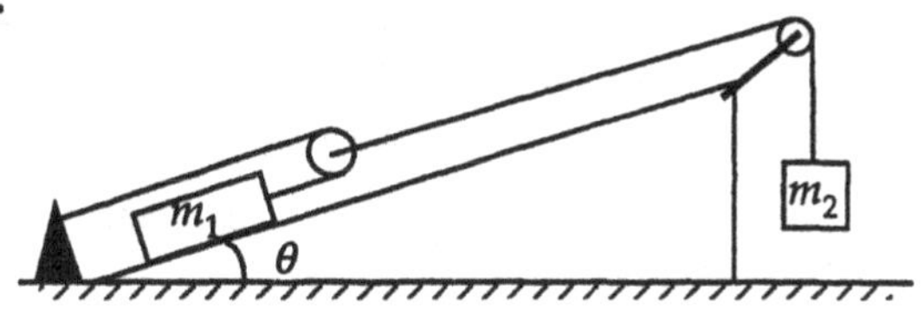

The mass m_1 slides without friction down an inclined plane. $m_1 = 400$ g, $m_2 = 200$ g, and $\theta = 30°$. Neglect the mass of the string and of the pulleys. Find the acceleration of m_2 and determine the string forces.

Problem 2.4.

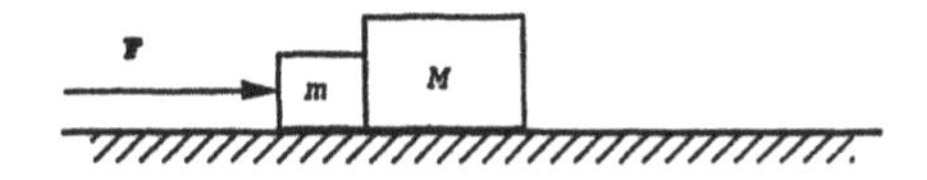

Two blocks, one of mass m and one of mass M, rest on a smooth horizontal table. The blocks are in contact with each other. As shown on the sketch, the system is now acted upon by a horizontal force $\mathbf{F}$, perpendicular to the surface of the block of mass m. Find the force by which the block of mass m acts on the block of mass M.

Problem 2.5.

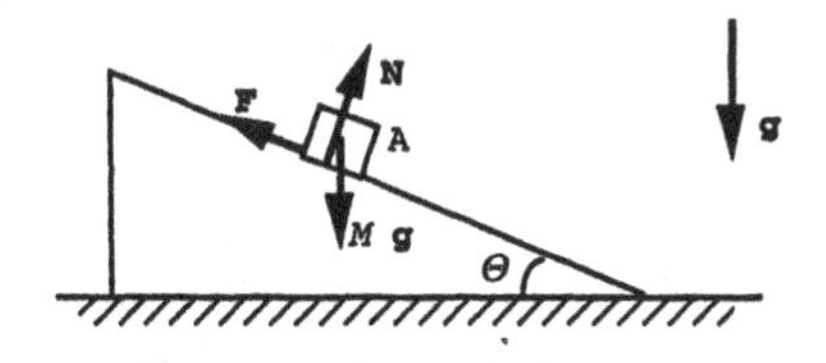

The body A, which has mass M, rests on a fixed inclined plane. The body is acted upon by three forces: $M\mathbf{g}$, $\mathbf{N}$, and $\mathbf{F}$, where $\mathbf{F}$ is the friction force. The friction force is a reaction force. The maximal friction force which the interface between the body A and the inclined plane can supply, is determined by the coefficient of friction μ and the magnitude of the normal force in the following way: $|\mathbf{F}|_{\max} = \mu N$. If the body is in motion down the inclined plane, the friction force assumes this value.

Determine for what values of the angle Θ the body will move down the inclined plane when the coefficient of friction μ is known. (Sometimes it is necessary to distinguish between static friction and friction when the body slides. We shall not discuss this technical point further).

Problem 2.6. A road bend is circular with a radius of $r = 300$ m. The surface is horizontal. The coefficient of friction between car tires and dry asphalt is $\mu = 0.8$. Between car tires and icy road, $\mu = 0.2$.

(1) Determine the maximal speed with which a car can drive through the bend when the road is: (a) dry, and (b) icy.

A gramophone record rotates with an angular velocity corresponding to 45 rpm (revolutions per minute). A small bug crawls slowly outwards in the radial direction. The coefficient of friction between bug and record is $\mu = 0.1$.

(2) Determine how far away from the center the bug will get before it begins to slip.

Problem 2.7.

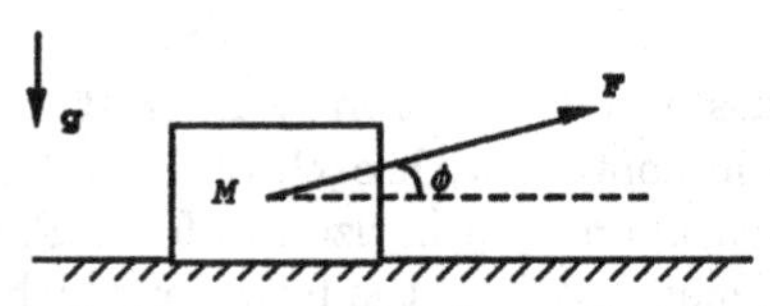

A large block with mass M is going to be pulled with a constant velocity along a horizontal surface. The coefficient of friction between the surface and the block is μ.

(1) Find the angle $\phi = \phi_0$ that the direction of pull shall make with the direction of motion for the pulling force to be as small as possible.
(2) Find this smallest value of F.

Problem 2.8. The mass M is pulled a distance d downwards from the point of equilibrium, and released. Ignore the mass of the string and the mass of the spring. The spring constant is denoted by k.

(1) Determine the maximum value $d = d_{\mathrm{max}}$ for d, if we demand that the string must be stretched during the entire oscillation. $M = 0.1$ kg, $k = 20\ \mathrm{N\,m^{-1}}$.
(2) How far is the spring extended from its unforced position when M is hanging at rest?
(3) What is the magnitude and direction of the acceleration at the lowest position for M when the system is oscillating and $d = d_{\mathrm{max}}$?

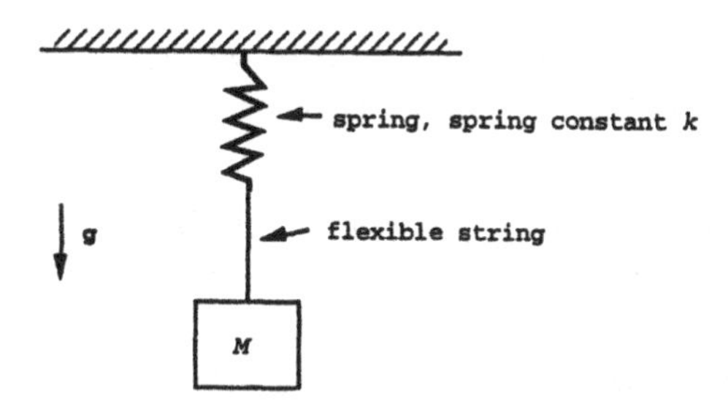

Problem 2.9. A mass M is suspended from two springs attached in series as shown on the sketch. The system is hanging in the gravitational field of the Earth. One spring has the spring constant k_1, the other k_2. We neglect the mass of the springs. Find the period T in the vertical harmonic motion of the mass.

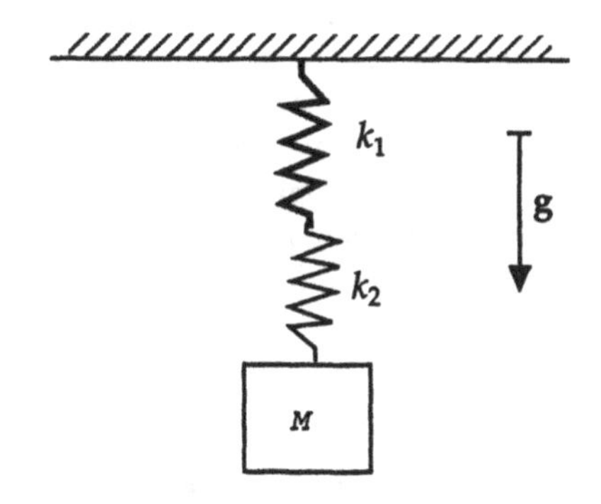

Problem 2.10.

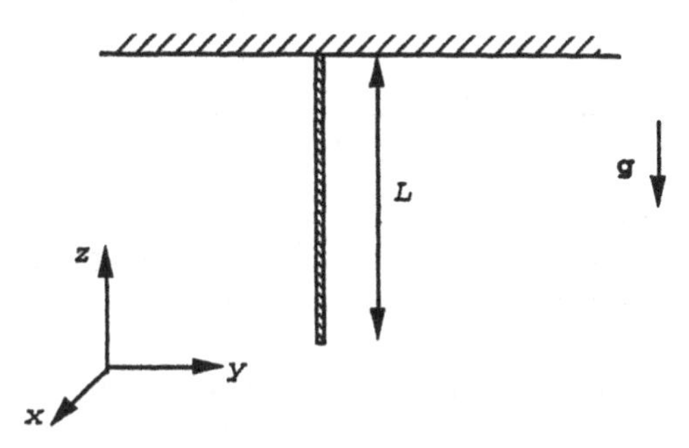

A homogeneous rope with mass M and total length L hangs from the ceiling of the laboratory.

(1) Find the magnitude of the force $F(z)$ by which the top part of the rope acts on the lower segment of the rope at the distance z from the free end of the rope.

Now the same rope is swung in the horizontal plane in such a way that the rope performs a uniform circular motion with constant angular velocity ω. Ignore gravity.

(2) Find the magnitude of the force $F(r)$ by which the inner part of the rope acts on the outer segment of the rope at the distance r from the center of the circular motion.

Problem 2.11.

(1) With what fraction of the escape velocity from the surface of the Earth must a rocket start in order to reach the Moon? The distance between the Earth and the Moon is 60 times the radius of the Earth.
(2) Consider a celestial body belonging to the solar system. Let us assume that this body falls from a large distance towards the Sun. Estimate the magnitude of the largest relative velocity with which the Earth and the body can meet. [Note: an order-of-magnitude estimate is desired, not an exact calculation. When the space probe *Giotto* passed Halley's Comet, the relative velocity of the two bodies was 69 $\mathrm{km\,s^{-1}}$.]

Problem 2.12.

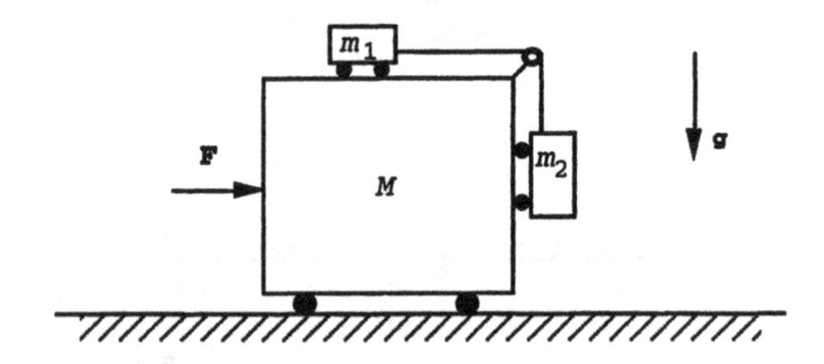

F is a constant horizontal force acting on the system shown in the sketch. **F** is perpendicular to the end surface. The masses m_1 and m_2 can move without friction relative to the mass M which in turn can move without friction relative to the horizontal table.

Determine **F** so that the masses m_1 and m_2 are not moving relative to M. Assume that m_1 and m_2 are initially at rest relative to M.

Problem 2.13. A painter with a mass of 90 kg sits in a chair near the wall of the house he is painting. He holds on to a rope which goes over a pulley and suspends his chair as shown on the figure. He wants to move upwards, and pulls the rope with a force so large that the pressure he exerts on the chair corresponds to the weight of a mass of 50 kg, (i.e., $50\ \mathrm{kg}\times 9.8\ \mathrm{m\,s^{-2}} = 490\ \mathrm{N}$). The mass of the chair itself is 15 kg. Disregard the mass of the pulley.

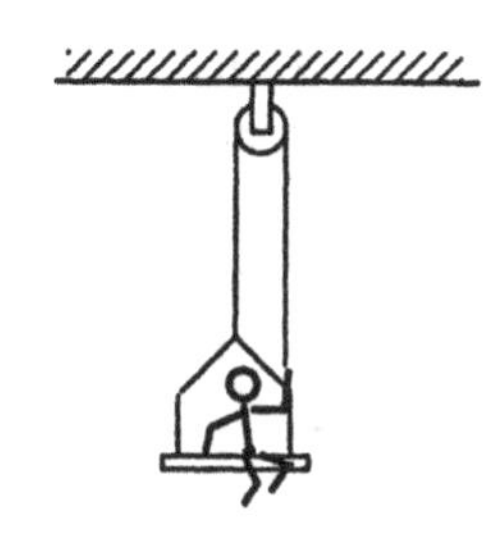

(1) Find the acceleration of painter and chair.
(2) With what force is the painter pulling on the rope (i.e., what is the string force)?

Problem 2.14.

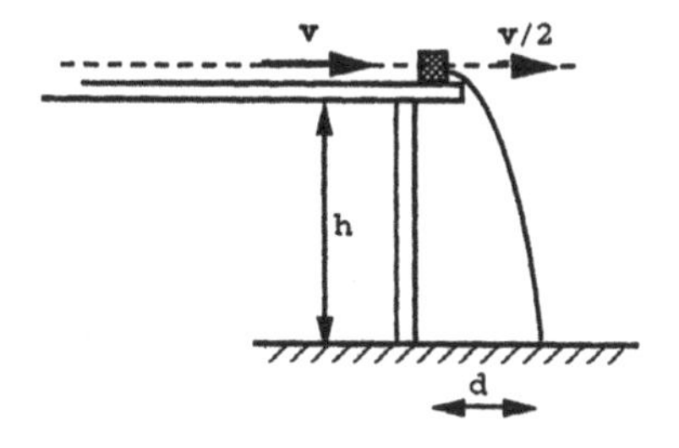

A lead bob has a mass of 1/4 kg and is placed at the edge of a smooth table. A rifle bullet with the mass of 1 g is shot through the bob. The bullet has horizontal velocity $\mathbf{v}$ just before it hits the bob. It moves through the bob in a very short time and exits with velocity $0.5\mathbf{v}$. The table has the height $h = 1$ m . The bob hits the floor a distance $d = 30$ cm from the edge of the table. Find $\mathbf{v}$.

Problem 2.15. A projectile is to be fired in the gravitational field $\mathbf{g}$ near the surface of the Earth. Determine the angle $\theta = \theta_{\max}$ which the aim must form with the horizontal in order that the projectile will get as far away as possible from the spot where it was fired. Ignore air resistance and consider the gravitational field to be homogeneous.

Problem 2.16. A homogeneous, flexible chain of total mass M has the length L. Initially, the chain hangs vertically in the gravitational field in such a way

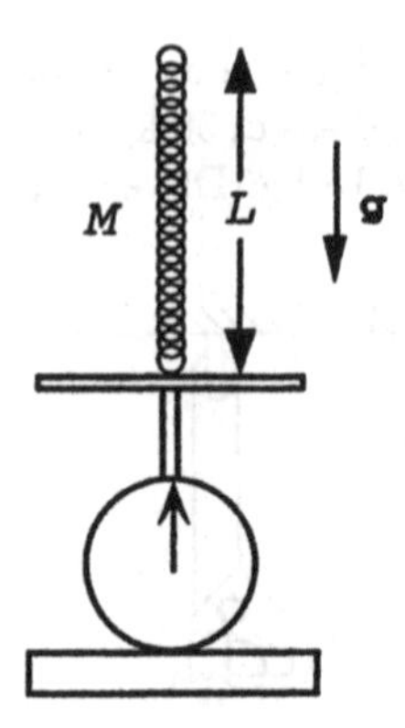

that the lowest point of the chain just touches the horizontal plate of a spring scale (see the figure).

Now the chain is allowed to fall freely in the gravitational field of the Earth. When the chain has fallen the distance S a segment S is resting on the scale.

What will the spring scale show when the chain has fallen the distance S? What will the spring scale show at the instant when the entire chain has reached the scale?

Problem 2.17.

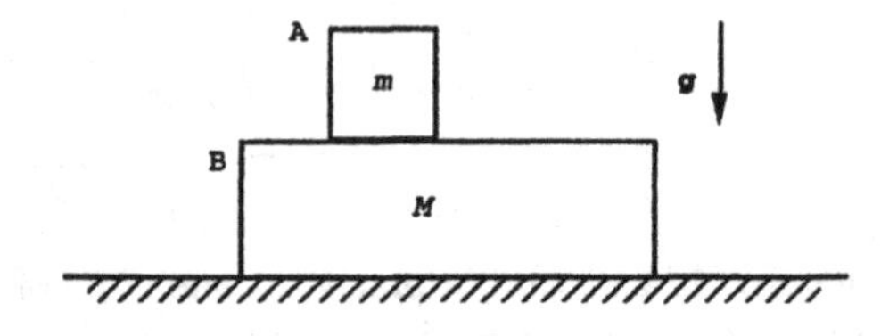

The sketch shows two wooden blocks A and B. A has the mass m and B has the mass M. At the time $t = 0$ B is at rest relative to the table and A is started with a velocity v_0 relative to B. Between the table and B there is no friction. Between A and B there is a coefficient of friction μ. The velocity of A relative to B will decrease and B will start to move along the table.

After a certain time T the block A will be at rest relative to B.

(1) Find the final velocity of the two blocks relative to the table.
(2) Find the time T.
(3) Find the distance D that A has moved relative to B when the velocity of A relative to B has become zero.

Problem 2.18.

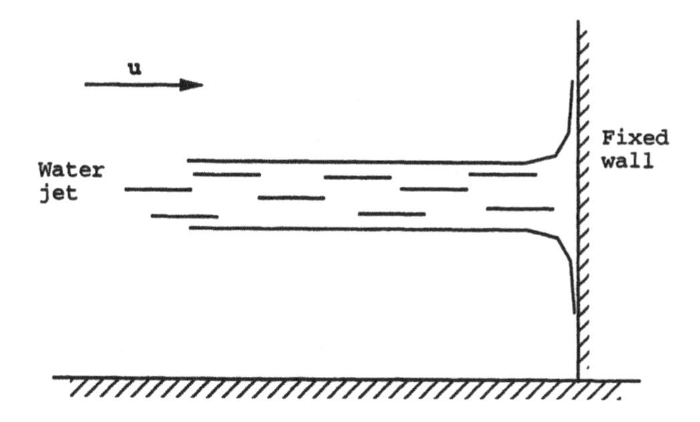

A jet of water from a fire hose impinges horizontally on a wall as indicated in the figure. The speed of the water is u before it strikes the wall (see figure) and it is assumed that the water has no velocity component perpendicular to the wall after it has hit the wall. The cross section of the water jet is a . The density of the water is ρ.

(1) Find the force **F** the water exerts on the wall. The magnitude of **F** should be expressed in terms of ρ, a, and u.
(2) Numerical example: $\rho = 1\ \mathrm{g\,cm^{-3}} = 10^3\ \mathrm{kg\,m^{-3}}$, $a = 25\ \mathrm{cm}^2 = 2.5 \times 10^{-3}\ \mathrm{m}^2$, $u = 60\ \mathrm{m\,s^{-1}}$. Now compute F. Compare the result with the force on a man, of mass 70 kg, in the gravitational field of the Earth.
(3) How far will a water jet from a hose reach before it hits the ground, if it is directed horizontally 1 m above the ground and the muzzle velocity is $u = 60\ \mathrm{m\,s^{-1}}$?
(4) A stream of water impinges on a block B as indicated on the sketch. At a certain time the block has velocity $v < u$ relative to the ground as shown. The velocity of the water before it strikes the block is u relative to the ground. The cross section of the jet is still a and the density ρ. Find the force that the water jet exerts on the block B.

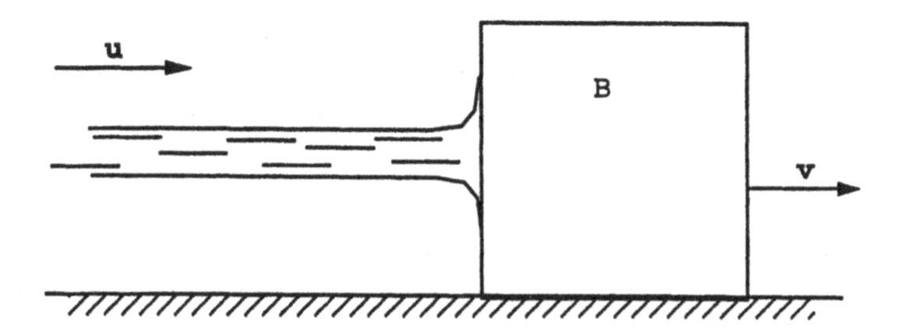

Problem 2.19.

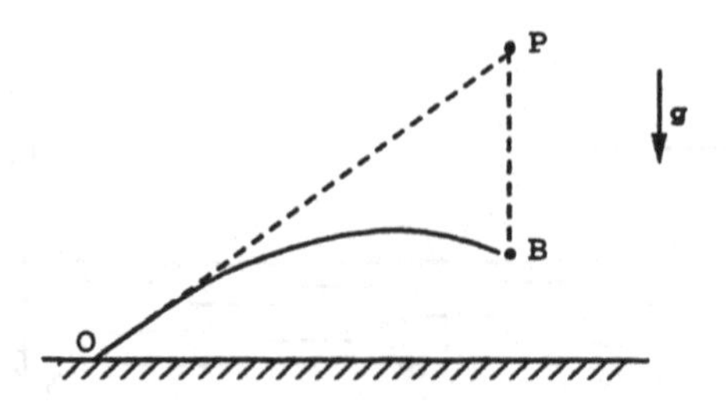

A projectile is shot from a gun located at the origin O, and the projectile is aimed at the point P where an object is located. At the exact moment when the projectile is fired the object at P starts to fall in the gravitational field of the Earth. The projectile and the object will – as indicated – meet at the point B. We neglect air resistance.

Show that this mid-air collision will take place, independently of the value of the muzzle velocity of the projectile, the mass of the projectile and the mass of the object at P. The muzzle velocity of the bullet is assumed to be so large that the bullet will reach as far as the vertical line through P.

Problem 2.20.

A collision is called *central* if the velocities of the particles both before and after the collision lie on the same straight line. A collision is called *elastic* if the sum of the kinetic energies of the colliding bodies is the same before and after the collision. See Example 2.12.

A sphere of mass m and with velocity $\mathbf{v}$ collides in a central and elastic collision with a sphere of mass M initially at rest in the laboratory.

(1) Find the velocities of the two spheres after the collision.
(2) Find the fraction f of the original kinetic energy of m that is transferred to M and the fraction r of the momentum that is transferred. Show that for $M \to \infty$, $f \to 0$.
(3) Show that for $m = M$, $f = 1$.

For $m = M$ we see that all the energy and momentum of the incoming particle is transferred to the particle at rest. From the answer to question 3 it may be understood why fast neutrons from a neutron source are slowed down very little by a lead plate 50 cm thick, while a layer of paraffin nearly stops neutrons completely. In a central collision of the neutrons with a hydrogen nucleus ($M = 1$) in the paraffin the kinetic energy of the neutron ($m = 1$) is transferred to the hydrogen nucleus.

What fraction of the energy of the neutron is transferred to a lead nucleus in a central collision (the atomic mass of lead is about 207)?

Problem 2.21.

(1) A small mass m collides in a central and elastic collision with a large mass M that has velocity V_0 before the collision. What should the velocity v of m be, before the collision, in order that all of the kinetic energy is transferred to M? What happens if $M = m$? If $M < m$?

(2) A moving small sphere of mass m collides elastically (but not necessarily centrally) with another sphere of the same mass, initially at rest. We assume that the two spheres do not start to rotate during the collision. Show that either the two spheres move perpendicular to each other after the collision or the sphere that was moving before the collision is stopped.

[Hint: from the conservation laws it may be shown that $\mathbf{u_1} \cdot \mathbf{u_2} = 0$.]

Problem 2.22. A rocket has an initial total mass (including fuel) of $m(0) = 4 \times 10^3$ kg. The velocity of the exhaust gas is $v_{\mathrm{rel}} = 3.0\ \mathrm{km\,s^{-1}} = 3000\ \mathrm{m\,s^{-1}}$.

(1) Consider the rocket at $t = 0$, i.e., just as it starts. How much mass must be expelled per second (how large must μ be) for the rocket to have an initial acceleration g upwards in the gravitational field of the Earth ($g = 9.8\ \mathrm{m\,s^{-2}}$)?

Let us now consider another rocket of the same design ($m(0) = 4 \times 10^3$ kg, $v_{\mathrm{rel}}= 3.0\ \mathrm{km\,s^{-1}}$). Let us furthermore assume that the rocket engine works in such a way that the mass of the rocket head (hull + fuel) decreases according to the formula $m(t) = m(0)\exp(-bt)$.

(2) Determine the value of b such that the rocket is able to just keep itself hovering in the gravitational field of the Earth.

Problem 2.23.

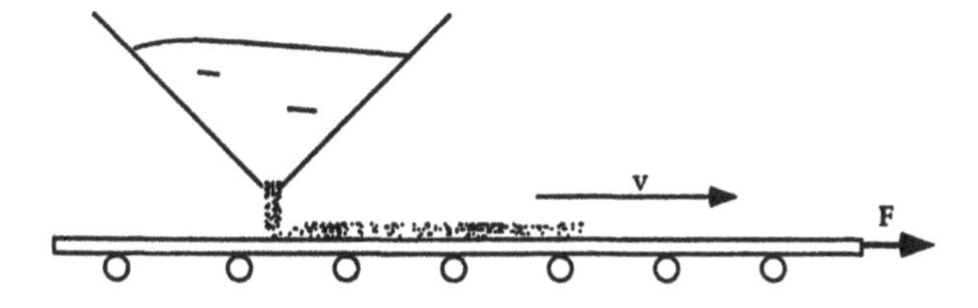

A hopper is at rest relative to the laboratory. Sand drops continuously down onto a conveyor belt below the hopper and comes to rest relative to the conveyor belt.

Assume that the amount of sand dropping down onto the belt per unit time, $\mu = \mathrm{d}m/\mathrm{d}t$, is constant. The velocity of the conveyor belt relative to the laboratory, v, is also constant. Find the force F needed to keep the conveyor belt moving with constant speed v.

3. Gravitational and Inertial Mass

Since the time of Galileo's famous experiments with freely falling bodies it has been known that all material bodies, at the same place on Earth, fall with the same acceleration when air resistance is disregarded. Through many experiments this has become a well documented experimental fact.

After Newton had formulated the dynamics laws for the motion of material bodies, these observations of Galileo took on a deeper significance. In Newtonian mechanics, the force acting on a body is considered to be the *cause* of the acceleration of the body. The mass of the body is a quantitative measure of the fact that the body resists acceleration. The kinematics laws of the free fall, experimentally verified by Galileo, ought to be explainable through the dynamics laws of Newton. Let us briefly recall some relevant features of Newton's theory.

3.1 Gravitational Mass

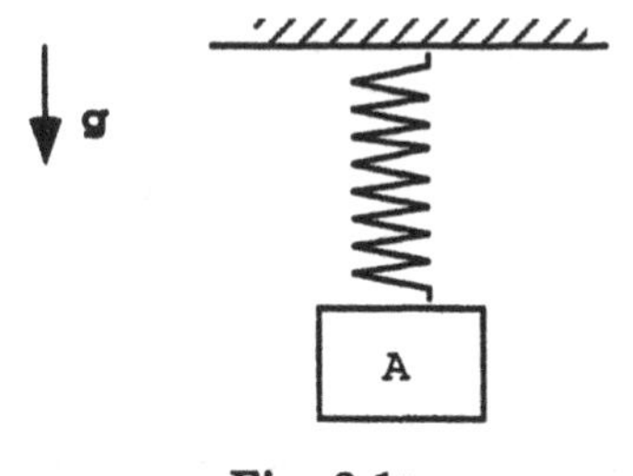

Fig. 3.1:

A mass A hangs from a spring near the surface of the Earth. From Newton's general law of gravity we know that the Earth acts on the mass with a downward force. This force has a direction and magnitude as if all of the mass of the Earth was concentrated in its center. The expression for the magnitude of the force F is:

$$F = G\frac{M_{\mathrm{g}} m_{\mathrm{g}}}{\rho^2} . \tag{3.1}$$

The terms are:

ρ is the radius of the Earth;
M_g is the gravitational mass of the Earth; M_g is a quantitative measure of the property of the Earth which makes it attract – and be attracted by – other material bodies;
m_g is the gravitational mass of the body. Similarly, m_g is a measure of the property of the body causing it to attract and be attracted to other material bodies.

The magnitude of the force F can be measured with the spring scale. Often (3.1) is rewritten as

$$F = m_g \left(\frac{GM_g}{\rho^2} \right) = m_g g \,. \tag{3.2}$$

The more gravitational mass a body has, the bigger its "weight", i.e., the bigger the force by which the Earth acts on the body. It is well known that

$$g = \left(\frac{GM_g}{\rho^2} \right) = 9.8 \text{ N kg}^{-1} = 9.8 \text{ m s}^{-2} \,.$$

3.2 Inertial Mass

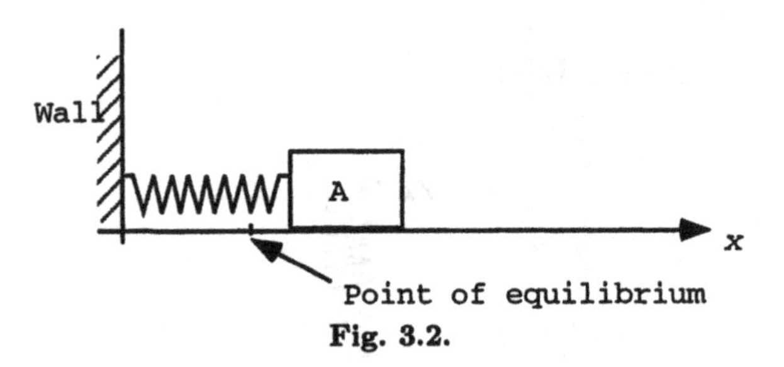

Fig. 3.2.

As Figure 3.2 shows, we now place the body A on a smooth horizontal table. The body A is attached to a spring, the other end of which is fastened to a wall in the laboratory. In this way A can be made to perform horizontal oscillations. We know from Chapter 2 (see Example 2.2), that the equation of motion for the body is

$$m_i \frac{d^2x}{dt^2} = -kx \,. \tag{3.3}$$

Here, the terms are:

k is the spring constant (k has units N m^{-1});

m_i is the inertial mass of the body A. The inertial mass is a quantitative measure of that property of a body which makes the body resist acceleration.

As shown in Chapter 2 the period of oscillation of the harmonic motion is given by $T = 2\pi\sqrt{m_\mathrm{i}/k}$. The greater the inertial mass of the body, the longer the period of oscillation. The gravitational mass does not enter the expression for the period of oscillation. The reaction force from the smooth surface exactly cancels the vertical force by which the Earth attracts the body.

3.3 Proportionality Between Inertial and Gravitational Mass

We now let the body A fall freely in the gravitational field of the Earth. The body starts off with zero initial velocity. See Figure 3.3. We write the

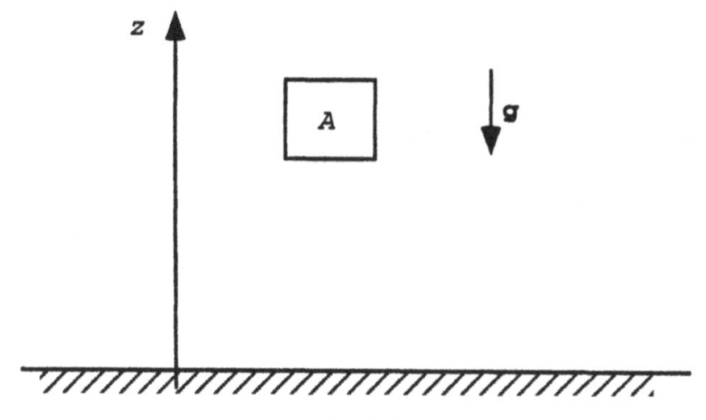

Fig. 3.3.

equation of motion for the body. Note the orientation of the z-axis. We have

$$m_\mathrm{i}\frac{\mathrm{d}^2 z}{\mathrm{d}t^2} = -m_\mathrm{g} g\,, \tag{3.4}$$

$$\frac{\mathrm{d}^2 z}{\mathrm{d}t^2} = -\left(\frac{m_\mathrm{g}}{m_\mathrm{i}}\right) g\,. \tag{3.5}$$

According to Newton all bodies have the following two properties:

(1) they resist acceleration (relative to absolute space);
(2) they attract and are attracted to all other bodies in the universe.

As mentioned, it has been *experimentally* verified with great accuracy that on any given place on the Earth all bodies fall (in vacuum) with the same acceleration . This acceleration is g. In Newtonian mechanics, the observation that all bodies fall with the same acceleration is interpreted as an observation of the equality of inertial and gravitational mass, i.e., $m_i = m_g$.

In other words: that property of a body which conditions the body to resists acceleration is – according to Newtonian mechanics – the same property which causes the body to attract (and to be attracted to) other bodies.

A somewhat more precise expression of the experimental result $m_i = m_g$ is the statement that gravitational and inertial mass are proportional. This statement is more correct for the following reason: Newton's second law expresses proportionality between on one side force, and on the other side mass times acceleration:

$$F = k_1 m_i a\,, \tag{3.6}$$

where k_1 is a constant of proportionality. The general law of gravity can similarly be expressed as

$$F = k_2 \frac{M_g m_g}{r^2}\,, \tag{3.7}$$

where k_2 is a constant. By setting $k_1 = 1$ (dimensionless) and $k_2 = G$ (units for G: $\mathrm{N\,m^2\,kg^{-2}}$), m_i and m_g will get the same dimension. Both kinds of mass should then be measured in kilograms and $m_i = m_g$.

A practical consequence of this remarkable equality (proportionality) between m_i and m_g is that one can compare by a balance measurement, not only the weights but also the masses of two given bodies.

3.4 Newton's Experiment

Newton realized that the proportionality between the gravitational force on a body and the inertial mass of that body, is of fundamental importance for an understanding of the universal law of gravity. Newton carried out experiments with different materials in order to investigate the proportionality between gravitational mass and inertial mass. See Figure 3.4. A body with gravitational mass m_g hangs from a light string with length L as suggested on Figure 3.4. When the body is drawn away from its point of equilibrium, it will oscillate around the vertical, driven by the component of the force of gravity not canceled by the string tension. The tangential acceleration is given by $a_t = L\ddot{\theta}$. The instantaneous value of the tangential force is given by (see Figure 3.4)

$$F_t = -F_g \sin(\theta) \tag{3.8}$$

(the minus indicates the restoring nature of the force). We then find

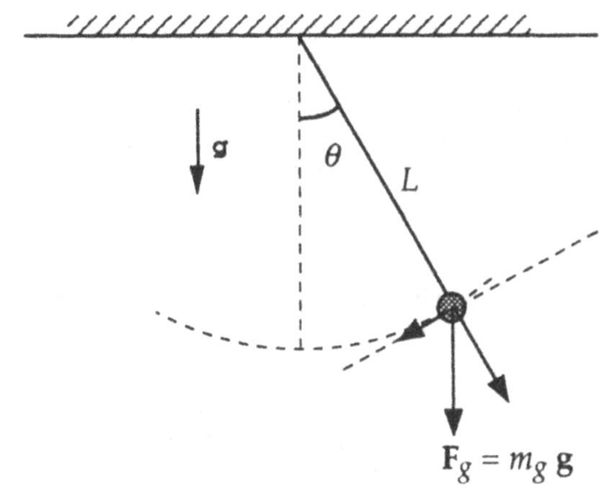

Fig. 3.4.

$$m_\mathrm{i} L\ddot{\theta} = -m_\mathrm{g} g \sin(\theta). \tag{3.9}$$

For small values of θ we use the approximation $\sin(\theta) \approx \theta$. Equation (3.9) now takes the form

$$m_\mathrm{i} L\ddot{\theta} = -m_\mathrm{g} g\theta \tag{3.10}$$

This is a differential equation of the same type as discussed in Example 2.2. We see that for small amplitudes, the mass will perform harmonic oscillations around the equilibrium point. In fact, the period of oscillation can be read directly from (3.10); it is

$$T = 2\pi\sqrt{\frac{L}{g}}\sqrt{\frac{m_\mathrm{i}}{m_\mathrm{g}}} \tag{3.11}$$

All of Newton's observations suggested that T is independent of the magnitude and material quality of the oscillating mass. This experiment convinced Newton that indeed $m_\mathrm{i} = m_\mathrm{g}$. We quote from *Principia:*

> But it has long ago been observed by others, that (allowance being made for the small resistance of the air) all bodies descend through equal spaces in equal times; and, by the help of pendulums, that equality of times may be distinguished to great exactness. I tried the thing in gold, silver, lead, glass, sand, common salt, wood, water, and wheat. I provided two equal wooden boxes. I filled the one with wood, and suspended an equal weight of gold (as exactly as I could) in the center of oscillation of the other. The boxes, hung by equal threads of 11 feet, made a couple of pendulums perfectly equal in weight and figure, and equally exposed to the resistance of the air: and, placing the one by the other, I observed them to play together forwards and backwards for a long while, with equal vibrations. And

> therefore the quantity of matter in the gold was to the quantity of matter in the wood as the actions of the motive force upon the gold to the action of the same upon all the wood; that is, as the weight of the one to the weight of the other. And by these experiments, in bodies of the same weight, I could have discovered a difference of matter less than the thousands part of the whole.

Several researchers have improved the accuracy of this type of measurements. Currently the proportionality between gravitational and inertial mass is known to be valid to $1:10^{10}$.

We have thus seen that by *mechanical experiments* (e.g. by studying pendulum oscillations, masses on springs, collisions of balls, etc.) performed in a gravitational field, it is *not possible to distinguish the two properties of matter: gravitational mass and inertial mass.* These two phenomena, inertia and gravitational attraction which conceptually are so different in Newtonian mechanics, appear to be of the same origin. Inertial mass m_i describes the property of matter which causes it to resist acceleration. The gravitational (heavy) mass m_g describes the property of matter which causes it to attract other bodies.

It is unfortunate in a physical theory to introduce two quantities that are conceptually different, but which cannot be distinguished by any experiment.

Einstein postulated that inertial and gravitational mass are expressions of one and the same property of matter. His general theory of relativity resolves the problem in a very elegant manner; we shall later briefly return to this issue (see, for example, Sections 6.2 and 6.3).

Example 3.1. The Satellite

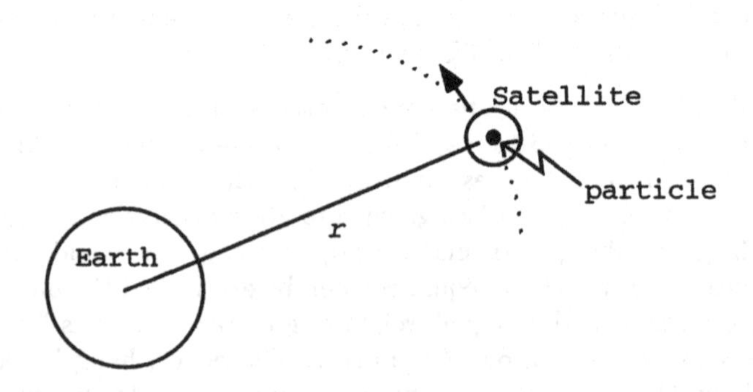

A small spherically symmetrical satellite moves in a circular orbit around the Earth (see the figure). An astronaut inside the satellite carefully places a particle at rest in the center of the satellite. The particle will now maintain that position relative to the satellite while it orbits the Earth. This result is an expression of the exact proportionality between gravitational and inertial mass. Let us consider the situation in more detail.

Mass of the satellite: inertial m_i, gravitational m_g.
Mass of the particle: inertial μ_i, gravitational μ_g.
Gravitational mass of the Earth: M_g.
For the satellite, we have

$$m_i \frac{v^2}{r} = \frac{GM_g}{r^2} m_g \,. \tag{3.12}$$

For the particle:

$$\mu_i \frac{v^2}{r} = \frac{GM_g}{r^2} \mu_g \,. \tag{3.13}$$

Since $m_i = m_g$ and $\mu_i = \mu_g$, we see that both the satellite and the particle fall freely in the gravitational field of the Earth with the acceleration

$$a = \frac{v^2}{r} = \frac{GM_g}{r^2} \,. \tag{3.14}$$

directed towards the center of the Earth. The particle initially had zero velocity relative to the satellite, and the particle and the satellite have the same acceleration. Therefore the particle will never change its position relative to the satellite. We say the particle is weightless relative to the satellite as the satellite falls freely in the gravitational field of the Earth. △

Example 3.2. An Elevator in Free Fall

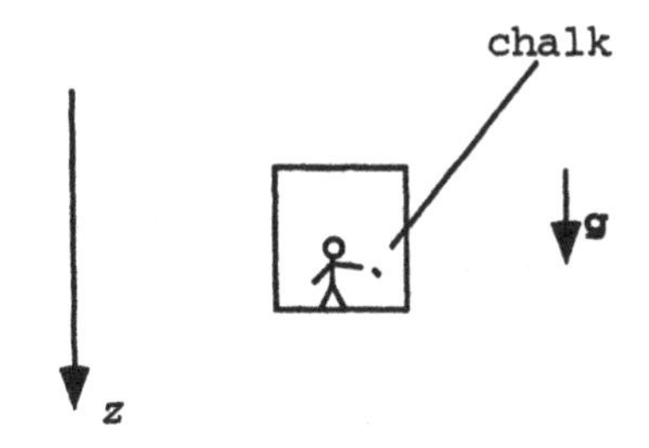

A person stands in an elevator box. At a certain moment two events occur: the cable breaks, and the person releases a piece of chalk. The chalk will now maintain its position relative to the person (and the elevator) because of the equality (proportionality) between gravitational and inertial mass. For an observer on the Earth, the equation of motion for the three bodies – person, chalk and elevator – all reduce to the same equation $d^2z/dt^2 = g$, because the masses are canceled in all the equations. The person and the chalk appear weightless relative to the elevator as the elevator falls freely in the gravitational field of the Earth. △

Example 3.3. Three Balls. Three balls are at rest on a horizontal board.

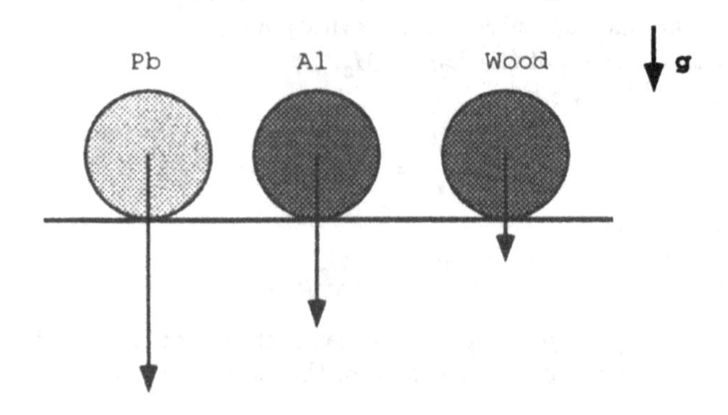

One ball is made of Pb, one of Al, and the third is made of wood. The force with which the Earth acts on each of these balls is different (see the figure). In spite of this, if the board is suddenly removed, the three balls will fall with identical accelerations, remaining flush with each other – because of the universal equality (proportionality) between gravitational and inertial mass.

△

3.5 Problem

Problem 3.1. The Concept of Weight. A man has the mass $m = 70$ kg. He is standing on a spring scale in the gravitational field of the Earth. The weight of the man is defined as the force by which the Earth pulls in the man, i.e.,

$$\text{Weight} = W = mg = 70\ \text{kg} \cdot \times 9.8\ \text{m}\,\text{s}^{-2} = 686\ \text{N}. \tag{3.15}$$

Now, the scale will not show 686 N, but 70 kg. Sometimes in physics one uses a unit kg* for weight. This, however, is not a good idea, because among other things it confuses the concepts of force and mass. But:

(1) How many kg* would the man weigh on the Moon?
(2) How many kg* would the man weigh on Mars?

Let us now assume that the spring scale is placed inside an elevator box near the surface of the Earth.

(3) How many kg* would the scale read if the elevator is accelerating upwards in the gravitational field with an acceleration of $a = 2\ \mathrm{m\,s^{-2}}$?
(4) Same question when the elevator is accelerating downwards with $a = 2\ \mathrm{m\,s^{-2}}$?

4. The Galilei Transformation

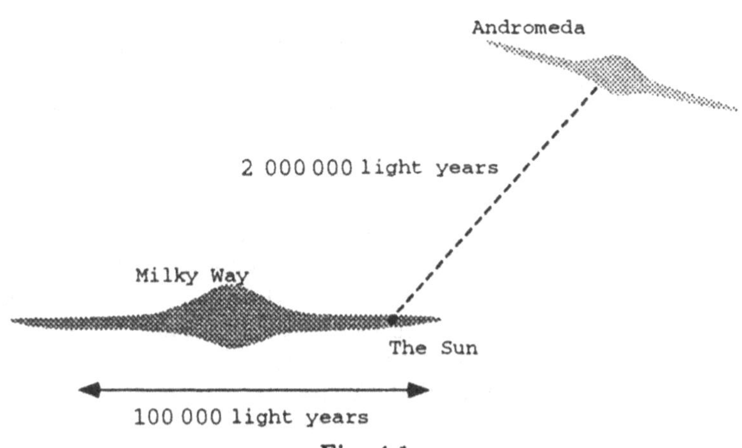

Fig. 4.1.

Our own galaxy is called the Milky Way. The galaxy closest to us is the Andromeda Galaxy. Light from the Andromeda reaching us today initiated out there some 2 000 000 years ago. About this time man was at the beginning of his long journey through evolution.

Let us briefly restate Newtons first three laws of motion:

1. Law of inertia.
2. $m\mathrm{d}\mathbf{v}/\mathrm{d}t = \mathbf{F}$.
3. Law of action and reaction.

As soon as these laws are written we are faced with the following question: *relative to which coordinate system should we measure positions, velocities, and accelerations?* Without an answer to this question the laws loose their meaning.

Should we measure relative to a coordinate system fixed on the surface of the Earth? We know that the Earth revolves once around its axis every 24 hours and orbits the Sun in an approximately circular orbit every year.

The tangential velocity of the Earth in its orbit around the Sun is $v = 3 \times 10^4$ m s^{-1} = 30 km s^{-1}. The radius in that orbit is $r = 1.5 \times 10^{11}$ m. Thus, the Earth is accelerated towards the center of the Sun with an acceleration of magnitude $a = v^2/r = 6 \times 10^{-3}$ m s^{-2}.

A coordinate system at rest relative to the surface of the Earth (say a laboratory) clearly is in a complicated state of motion relative to the Sun. Should we instead measure the kinematical quantities relative to the Sun? The center of the Sun moves in an approximately circular orbit around the center of our galaxy with a velocity of 250 km s^{-1}. The distance of the Sun from the center of the galaxy is about 30 000 light years ($\approx 2.8 \times 10^{20}$ m). The acceleration of the Sun with respect to the center of the galaxy is then $a = v^2/r = 2.2 \times 10^{-10}$ m s^{-2}.

Is it the center of our galaxy or perhaps the center of the Andromeda galaxy which should be taken as the center of our reference frame? Attempts to answer the questions of which coordinate system should be the basis for measurements of positions, velocities, and accelerations will throw light on the basic strengths and weaknesses of the Newtonian world picture, and enable us to understand the profound revision of the problem of motion which was initiated with Einsteins special and general theory of relativity.

Newton answered the question himself by adding to his three basic laws the two postulates: the postulate of absolute time and the postulate of absolute space (see Chapter 2). These two postulates therefore are of crucial importance to the understanding of Newton's mechanics.

We shall, however, later see that Newtonian mechanics – in spite of its enormous success – contains an intrinsic weakness. To perceive this, we shall for a while forget the postulate about absolute space and through the experience we gain in applying the equation of motion, try to find which coordinate systems may be used for a description of the phenomena of motion in Newtonian mechanics.

We begin our discussion with one of the basic tools for comparing measurements taken by observers in uniform motion with respect to each other.

4.1 The Galilei Transformation

By an *inertial frame* we understand a reference frame within which Newton's three laws of motion, in their simple classical form, are valid.

In a certain sense the law of inertia is contained within the second law. When there is no force on a body its acceleration (relative to absolute space!) is zero, from which it follows that the velocity is constant. Newton used the law of inertia to explicitly state that no force is required to maintain uniform motion.

The so-called heliocentric reference frame has its origin in the center of the Sun and coordinate axes that are not rotating relative to the fixed stars, i.e., the three coordinate axis always point toward the same three fixed stars. Experience has shown that the heliocentric reference frame to a high degree of accuracy can be considered an inertial reference frame.

By "experience" we mean in this connection that the predictions based on the assumption that the heliocentric reference frame is inertial, are borne out by experiments and astronomical observations. The fixed stars seem motionless simply because they are so distant. The star next nearest to the Earth is about $4\frac{1}{2}$ light years ($\approx 4 \times 10^{16}$ m) away. Compare this to the distance of the star nearest to the Earth, the Sun (distance 150×10^6 km ≈ 8.3 light minutes).

Every reference frame moving with uniform linear motion relative to the heliocentric frame, is itself an inertial frame. This observation goes hand in hand with the law of inertia, which tells us that there is no fundamental difference between rest and uniform translational motion.

Let us investigate this problem a little further.

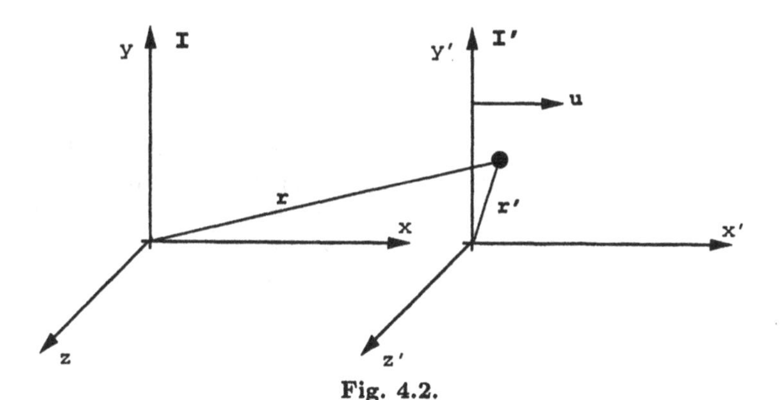

Fig. 4.2.

The Galilei transformation is a prescription for comparing coordinate values and velocities from two inertial frames (see Figure 4.2). We are considering two inertial frames I and I′, where I′ moves with the constant velocity **u** along the x-axis of I. We choose the coordinate systems such that the origins and the axes of the two frames coincide at $t = 0$. You may think of the frame I as the heliocentric reference frame.

Denote by (x', y', z') the coordinate values relative to I′ of a given particle and (x, y, z) are the coordinate values of the same particle relative to I. If the two coordinate systems move relative to each other as in Figure 4.2, the Galilei transformation is given by

$$x' = x - ut\,, \tag{4.1}$$
$$y' = y\,, \tag{4.2}$$
$$z' = z\,, \tag{4.3}$$
$$t' = t\,, \tag{4.4}$$
$$m' = m\,, \tag{4.5}$$
$$q' = q\,, \tag{4.6}$$
$$F' = F\,. \tag{4.7}$$

We have explicitly expressed the relations between time, mass, charge, and force in the two frames. The first three equations can be condensed into

$$\mathbf{r}'(t') = \mathbf{r} - \mathbf{u}t\,. \tag{4.8}$$

Equation (4.8) states the relation between the position vector $\mathbf{r}'$ relative to I$'$, the position vector $\mathbf{r}$ relative to I, and the relative velocity $\mathbf{u}$ of the coordinate systems.

Differentiating (4.8) we get, using $t = t'$,

$$\frac{\mathrm{d}\mathbf{r}'}{\mathrm{d}t'} = \frac{\mathrm{d}\mathbf{r}}{\mathrm{d}t}\frac{\mathrm{d}t}{\mathrm{d}t'} - \mathbf{u} = \frac{\mathrm{d}\mathbf{r}}{\mathrm{d}t} - \mathbf{u}\,. \tag{4.9}$$

Introducing $\mathbf{v}' \equiv \mathrm{d}\mathbf{r}'/\mathrm{d}t'$, and $\mathbf{v} \equiv \mathrm{d}\mathbf{r}/\mathrm{d}t$ we get, from (4.9),

$$\mathbf{v} = \mathbf{v}' + \mathbf{u}\,, \tag{4.10}$$

where $\mathbf{v}'$ is the velocity of the particle in I$'$ and $\mathbf{v}$ the velocity in I. Clearly the velocity of a particle is a relative quantity, having different values in different coordinate frames. That a quantity is "relative" means that the quantity has different values when observed from different reference frames. Mass, charge, force, and time intervals are, according to the Galilei transformation, absolute quantities. These quantities have values that are independent of the frame from which they are observed.

We now assume that a force $\mathbf{F}$ is acting on the particle. Then, the equation of motion in I is

$$m\frac{\mathrm{d}^2\mathbf{r}}{\mathrm{d}t^2} = \mathbf{F}\,. \tag{4.11}$$

Going back to (4.9) and differentiating again, we find

$$\frac{\mathrm{d}^2\mathbf{r}'}{\mathrm{d}t'^2} = \frac{\mathrm{d}^2\mathbf{r}}{\mathrm{d}t^2}\,. \tag{4.12}$$

Note: Equation (4.12) is the essential feature of the Galilei transformation. When $\mathbf{u}$ is constant, the acceleration of a particle is the same when observed from the two frames. The acceleration is an absolute quantity according to the Galilei transformation.

Since furthermore $m' = m$ and $\mathbf{F}' = \mathbf{F}$, we conclude that if (4.11) is the equation of motion as seen in I, then the equation of motion in I$'$ is given by

$$m' \frac{\mathrm{d}^2 \mathbf{r}'}{\mathrm{d}t'^2} = \mathbf{F}' \tag{4.13}$$

Comparing (4.11) and (4.13) we see that Newton's second law is the same ("looks the same") in all inertial frames. A somewhat more precise expression is

Newton's fundamental equation of motion is invariant under a Galilei Transformation

This result is called *the relativity principle of mechanics.* The relativity principle of mechanics is often formulated in the following manner: there exist a three-fold infinity of equivalent coordinate frames, all in a state of uniform linear motion with respect to one another, in which the mechanical laws are valid in their simple classical form. Or: the fundamental mechanical laws of motion are the same within all inertial frames.

The physical content of the relativity principle of mechanics is the following: *it is impossible, from mechanical experiments alone performed completely within a given inertial frame, to decide if the frame is "at rest" or in "uniform linear motion"*. By mechanical experiments we mean oscillations of pendulums, collisions between balls, motion of masses connected to springs, etc. The term "mechanical" is used to emphasize that we – for now – will exclude experiments involving electromagnetic phenomena. The mechanical laws of motion are not influenced by uniform translation. One can play tennis on board a ship sailing with uniform velocity. As long as the ship follows a straight course with constant velocity the ball will obey the same laws as it does on land. On land the tennis ball follows the same rules at all seasons without being influenced by the velocity of the Earth through space. Our speed relative to the Sun is 30 $\mathrm{km\,s^{-1}}$ and the speed of the solar system relative to the center of the Milky Way is 250 $\mathrm{km\,s^{-1}}$.

Summing all this up, one can with some justification say that in the form in which Newtonian mechanics is used today, it rests on the following assumptions:

1. The law of inertia is valid in the so-called inertial frames.
2. $m\mathrm{d}\mathbf{v}/\mathrm{d}t = \mathbf{F}$.
3. Action = reaction.
4. Experiments show that the heliocentric reference frame to a high degree of precision is an inertial frame.
5. The Galilei tranformation is valid (the postulate of absolute time is introduced here).

From 4 and 5 we infer that all reference frames moving with a constant velocity relative to the heliocentric reference frame are inertial frames. To this we add: after having written the laws of motion (1, 2, and 3 above)

Newton was faced with the problem of finding a reference frame in which the laws were valid. As one can tell from the postulate of absolute space, Newton decided that a reference frame fixed in some material body (e.g. the Sun) could not be the basic reference frame for the fundamental laws of mechanics. By choosing a reference frame fixed in the Sun, which is, after all, just one of innumerable stars in the universe, one runs the risk of postponing the problem instead of solving it: perhaps it will turn out later that this reference frame is accelerating after all. Instead of a material body as the universal reference point, Newton chose the abstract concept: absolute space.

Similar remarks can be made with regards to the concept of time. If Newton had chosen the rotation period of the Earth (the second!) as the theoretical foundation for the description of time, the law of inertia would have been found not to hold, for there are small irregularities in the rotation of the Earth around its axis. Again we see from the postulate of absolute time that Newton chose an abstract concept. We shall consider the problem further in Chapter 7.

In the following considerations we will stick to the observation firmly rooted in experience, that *to a high degree of approximation the heliocentric reference frame is an inertial frame.*

The concepts of "reference frame" and "coordinate system" are not quite synonymous: if we change the spatial coordinates without reference to time (say, by rotating the coordinate system) we do not consider this as a change of reference frame. A given reference frame contains an infinity of different coordinate systems. A given reference frame contains all the coordinate systems at rest with respect to a given coordinate system.

We will illustrate the physical content of the *relativity principle of mechanics* by describing a simple experiment.

Let us first state as a fact we shall later prove: For motion "over short distances" and in a "short time span" a laboratory at rest on the Earth can be considered to be an inertial system. We shall later state precisely what is meant by "short distances" and "short time spans".

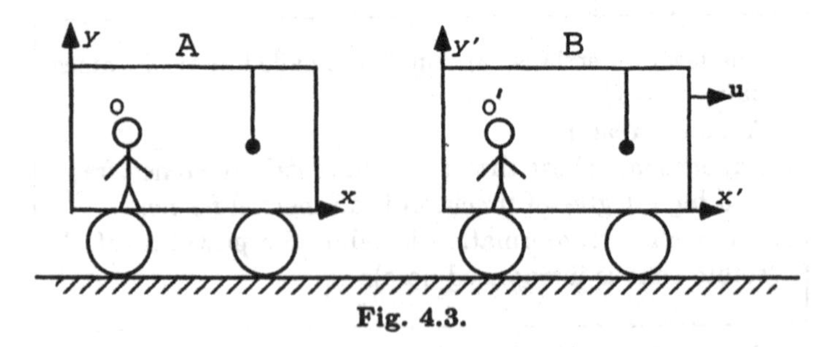

Fig. 4.3.

Consider two railroad cars on a horizontal and straight section of the railroad (see Figure 4.3). A is at rest relative to the ground. B moves with the constant velocity **u** relative to A. The cars are closed, so the observers cannot look out. The observers O and O′ now carry out experiments with identical pendulums inside the cars. Each observer measures the time of oscillation of the pendulum, and its equilibrium position relative to the walls of the cars. They can discuss their observations with each other, over the radio. We disregard all trivial effects like vibrations from the rails, etc. and ask: can the two observers decide from the pendulum experiments alone, who is at "rest" and who is in "motion" with the constant speed u ? In particular, can they determine the magnitude of **u**?

The answer to these questions is no. The mechanical laws are invariant going from one inertial frame to another. The behavior of the pendulae is completely the same in the two cars.

It is an *experimentally* verified fact that by mechanical experiments alone – performed completely within a given inertial system – it is not possible to decide if the coordinate system is "at rest" or it is in uniform motion along a straight line. This experience finds its mathematical expression in the invariance of Newton's second law under a Galilei transformation.

The important point in this connection is the experimental verification and not that the Galilei transformation perhaps can be considered self evident.

Let us illustrate the relativity principle of mechanics by some further remarks. The fundamental law of mechanics is about changes in velocity. To the description of any physical system one can add a constant overall velocity without changing the fundamental behavior of the system. For instance: *by studying how the planets move around the Sun, it is not possible to infer if the solar system as a whole "moves through space"*. According to the Newtonian laws there will be no effect of a translational motion of the Sun through astronomical space. Newton remarks that "The motion of bodies among themselves is the same in space, whether that space is itself at rest relative to the absolute space or moving at a uniform velocity in a straight line". The fundamental mechanical laws contain no reference to an absolute velocity. The laws are independent of uniform translational motion.

There is an important objection to the entire concept of absolute space. We get into trouble if we wish to determine Newton's absolute space by means of mechanical experiments. Suppose we somehow had arrived at the conclusion that a certain reference frame R was at rest with respect to Newton's absolute space. As we have just seen, any frame in uniform linear motion with respect to R can with equal right be claimed to be the one at rest with respect to Newton's absolute space. The outcome of any experiment in the two frames will be exactly the same. Neither frame can be preferred as far as the fundamental laws of mechanics are concerned. A body at rest in one system will, however, be a body in uniform linear motion in the other frame. If an observer in one frame claimed that a particle was at rest in absolute space, an observer in the other frame would claim that the particle was moving with constant velocity along a straight line in absolute space.

The relativity principle of mechanics thus detracts some of the "absoluteness" from Newton's absolute space. Newton was aware of this but, as we shall later see, he had his very good reasons to introduce the absolute space.

4.2 Galileo Speaks

The space and time coordinate transform used in this chapter justly bears the name of the Galilei transformation. Galileo was fully aware of the fact that the mechanical laws are independent of an even translatory motion. We quote from the book by Galileo: *Dialogue Concerning the Two Chief World Systems*:

> Shut yourself up with some friend in the main cabin below decks on some large ship, and have with you there some flies, butterflies, and other small flying animals. Have a large bowl of water with some fish in it; hang up a bottle that empties drop by drop into a wide vessel beneath it. With the ship standing still, observe carefully how the little animals fly with equal speeds to all sides of the cabin. The fish swim indifferently in all directions; the drops fall into the vessel beneath; and, in throwing something to your friend, you need throw it no more strongly in one direction than another, the distances being equal; jumping with your feet together, you pass equal spaces in every direction.
> When you have observed all these things carefully (though there is no doubt that when the ship is standing still everything must happen this way), have the ship proceed with any speed you like, so long as the motion is uniform and not fluctuating this way and that. You will discover not the least change in all the effects named, nor could you tell from any of them whether the ship was moving or standing still. In jumping, you will pass on the floor the same spaces as before, nor will you make larger jumps toward the stern than towards the prow even if the ship is moving quite rapidly, despite the fact that during the time that you are in the air the floor under you will be going in a direction opposite to your jump. In throwing something to your companion, you will need no more force to get it to him whether he is in the direction of the bow or the stern, with yourself situated opposite. The droplets will fall as before into the vessel beneath without dropping towards the stern, although while the drops are in the air the ship runs many spans. The fish in their water will swim toward the front of their bowl with no more effort than toward the back, and will go with equal ease to bait placed anywhere around the edges of the bowl. Finally the butterflies and

the flies will continue their flights indifferently toward every side, nor will it ever happen that they are concentrated toward the stern, as if tired out with keeping up with the course of the ship, from which they will have been separated during long intervals by keeping themselves in the air. And if smoke is made by burning some incense, it will be seen going up in the form of a little cloud, remaining still and moving no more toward one side than the other.
The cause of all these correspondences of effects is the fact that the ship's motion is common to all the things contained in it, and to the air also. That is why I said you should be below decks; for if this took place above in the open air, which would not follow the course of the ship, more or less noticeable differences would be seen in some of the effects noted. No doubt the smoke would fall as much behind as the air itself. The flies likewise, and the butterflies, held back by the air, would be unable to follow the ship's motion if they were separated from it by a perceptible distance. But keeping themselves near it, they would follow it without effort or hindrance; for the ship, being an unbroken structure carries with it a part of the nearby air. For a similar reason we sometimes, when riding horseback, see persistent flies and butterflies following our horses, flying now to one part of their bodies and now to another. But the difference would be small as regards the falling drops, and as to the jumping and throwing it would be quite imperceptible.
Sagredo: Although it did not occur to me to put these observations to the test when I was voyaging, I am sure that they would take place in the way you describe. In confirmation of this I remember having often found myself in my cabin wondering whether the ship was moving or standing still; and sometimes at a whim I have supposed it going one way when its motion was the opposite. Still, I am satisfied so far, and convinced of the worthlessness of all experiments brought forth to prove the negative rather than the affirmative side as to the motion of the Earth.

Example 4.1. Velocity Transformation. Let us consider a particle P moving with constant velocity $\mathbf{v}$ relative to an inertial frame I (see the figure). Let us assume that $\mathbf{v}$ is in the (x, y) plane with an angle θ to the x-axis.

The inertial frame I$'$ moves with the constant velocity $\mathbf{u}$ along the x-axis of I, and the coordinate axis of the two systems are parallel. Find the magnitude of $\mathbf{v}'$ and the angle θ' it forms with the x'-axis.

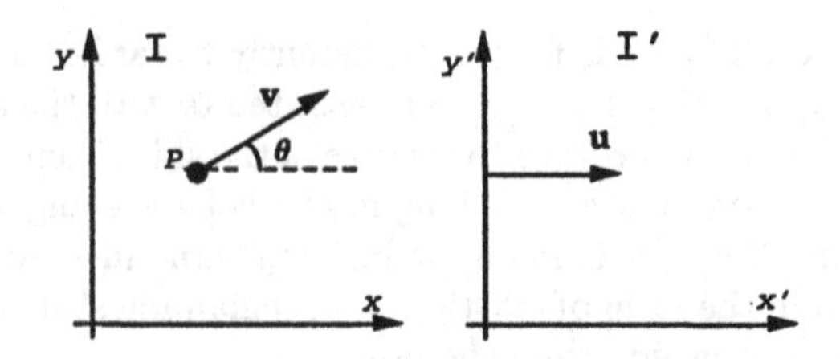

Solution. From the Galilei transformation:

$$v'_x = v_x - u , \tag{4.14}$$

$$v'_y = v_y , \tag{4.15}$$

$$v'_z = v_z . \tag{4.16}$$

In our case:

$$v_x = v\cos(\theta) , \tag{4.17}$$

$$v_y = v\sin(\theta) , \tag{4.18}$$

and

$$v'_x = v'\cos(\theta') , \tag{4.19}$$

$$v'_y = v'\sin(\theta') . \tag{4.20}$$

Thus, (4.14) states that

$$v'\cos(\theta') = v\cos(\theta) - u , \tag{4.21}$$

and (4.15) that

$$v'\sin(\theta') = v\sin(\theta) . \tag{4.22}$$

Dividing (4.22) with (4.21) we find

$$\tan(\theta') = \frac{\sin(\theta)}{\cos(\theta) - u/v} . \tag{4.23}$$

Squaring (4.21) and (4.22), and adding them we find

$$|\mathbf{v}'| \equiv v' = \sqrt{v^2 + u^2 - 2uv\cos(\theta)} . \tag{4.24}$$

These formulas express the vectorial addition of **v** and −**u**. Geometrically we have:

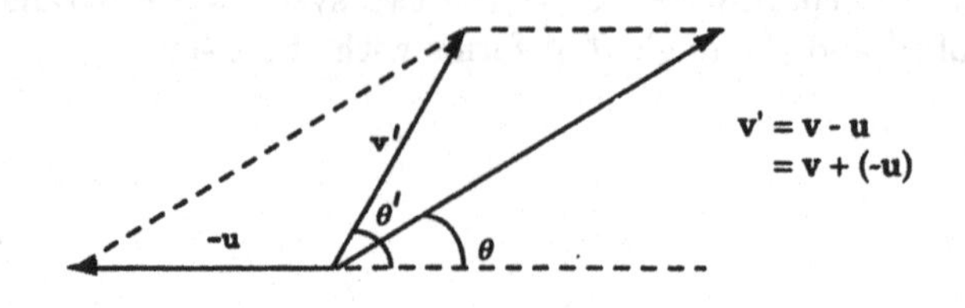

△

4.3 Problems

Problem 4.1. A particle moves along the x-axis of frame I according to the equation $x(t) = 5t^2 + 2t + 4$.

Find the position, velocity and acceleration of the particle, as functions of t in the frame I′. Take I′ to move with the velocity $u = 2\ \mathrm{m\,s^{-1}}$ along the x-axis of I.

Problem 4.2. A spring cannon stands in the chimney of a toy train. The train drives along a straight section with constant velocity **v**. The spring cannon shoots a heavy ball straight upwards from the chimney.

(1) Where does the ball land (disregard air resistance)?
(2) Describe qualitatively the trajectory of the ball in a coordinate system fixed in the laboratory.
(3) Describe qualitatively the trajectory of the ball in a coordinate system fixed on the train.

5. The Motion of the Earth

As a preparation for later chapters we shall now review the description of the motion of the Earth relative to the fixed stars.

Through a few examples we first demonstrate the application of vector calculus to the analysis of the motion of a rigid body. We then apply these results to the motion of the Earth.

5.1 Examples

Example 5.1. Vectors and the Rotation of a Rigid Body

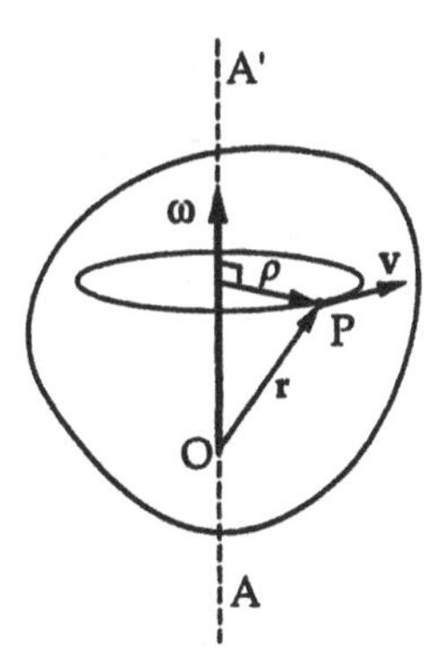

A rigid body rotates with the constant angular velocity ω around a fixed axis A-A$'$. The angular velocity vector $\boldsymbol{\omega}$ is defined as a vector along the rotational axis pointing in the direction where the rotation and $\boldsymbol{\omega}$ defines a right-hand spiral. The magnitude of $\boldsymbol{\omega}$ is $|\boldsymbol{\omega}|$. See the figure.

Let P be some point fixed on the rotating body.

(1) Show that the velocity of P is

$$\mathbf{v} = \boldsymbol{\omega} \times \mathbf{r} . \tag{5.1}$$

where $\mathbf{r}$ is a vector from some origin O on the rotation axis to P.

(2) Show that the acceleration $\mathbf{a}$ of the point P is

$$\mathbf{a} = \boldsymbol{\omega} \times (\boldsymbol{\omega} \times \mathbf{r}) = -\omega^2 \boldsymbol{\rho}, \tag{5.2}$$

where $\boldsymbol{\rho}$ is the projection of $\mathbf{r}$ on a plane perpendicular to $\boldsymbol{\omega}$. See the figure.

Solution. During the rotation, the point P performs a uniform circular motion with the angular velocity ω in a plane perpendicular to $\boldsymbol{\omega}$. The radius in that circular motion is ρ.

(1) The velocity $\mathbf{v}$ of P has the magnitude $|\mathbf{v}| = \omega\rho$ and is in the direction along the tangent to the circle. We obtain a vector in this direction by taking the vectorial product (cross product) $\boldsymbol{\omega} \times \boldsymbol{\rho}$, and the magnitude of this vector is $|\boldsymbol{\omega} \times \boldsymbol{\rho}| = \omega\rho$. Thus:

$$\mathbf{v} = \boldsymbol{\omega} \times \boldsymbol{\rho}. \tag{5.3}$$

Note: when we construct the vector product of two vectors we may translate them to a common origin. The area of the parallelogram spanned by $\boldsymbol{\omega}$ and $\boldsymbol{\rho}$ is the same as the area of the parallelogram spanned by $\boldsymbol{\omega}$ and $\mathbf{r}$ (draw the vectors from a common origin). This means that

$$|\boldsymbol{\omega} \times \mathbf{r}| = |\boldsymbol{\omega} \times \boldsymbol{\rho}| .$$

Since the direction of $\boldsymbol{\omega} \times \mathbf{r}$ and $\boldsymbol{\omega} \times \boldsymbol{\rho}$ is also the same, we conclude that

$$\mathbf{v} = \boldsymbol{\omega} \times \mathbf{r} = \boldsymbol{\omega} \times \boldsymbol{\rho}. \tag{5.4}$$

(2) The acceleration of P has the magnitude $a = \omega^2\rho$ (see Example 1.2) and is directed toward the center of the circle, i.e., toward the axis of rotation. Therefore

$$\mathbf{a} = -\omega^2 \boldsymbol{\rho}. \tag{5.5}$$

Since $\boldsymbol{\omega}$ and $\mathbf{v}$ are mutually perpendicular, we find that the magnitude of $\boldsymbol{\omega} \times \mathbf{v}$ is $|\boldsymbol{\omega} \times \mathbf{v}| = \omega v = \omega^2\rho$. The vector $\boldsymbol{\omega} \times \mathbf{v}$ then points in the direction of $-\boldsymbol{\rho}$. Therefore:

$$\boldsymbol{\omega} \times \mathbf{v} = -\omega^2 \boldsymbol{\rho} = \mathbf{a}. \tag{5.6}$$

Substituting (5.4) into (5.6) we get:

$$\mathbf{a} = -\omega^2 \boldsymbol{\rho} = \boldsymbol{\omega} \times (\boldsymbol{\omega} \times \mathbf{r}). \tag{5.7}$$

Alternatively, (5.7) can be seen as a result of direct differentiation of (5.4):

$$\begin{aligned} \mathbf{a} &= \frac{\mathrm{d}\mathbf{v}}{\mathrm{d}t} = \frac{\mathrm{d}}{\mathrm{d}t}(\boldsymbol{\omega} \times \mathbf{r}) = \boldsymbol{\omega} \times \frac{\mathrm{d}\mathbf{r}}{\mathrm{d}t} \\ &= \boldsymbol{\omega} \times \mathbf{v} = \boldsymbol{\omega} \times (\boldsymbol{\omega} \times \mathbf{r}). \end{aligned}$$

If the rotation of the body is not with constant angular velocity, the acceleration of P is not given by (5.7). But the identity

$$\boldsymbol{\omega} \times (\boldsymbol{\omega} \times \boldsymbol{\rho}) = -\omega^2 \boldsymbol{\rho} \tag{5.8}$$

still holds since it is a purely geometrical formula, valid for any two vectors $\mathbf{r}$ and $\boldsymbol{\omega}$. See the figure below.

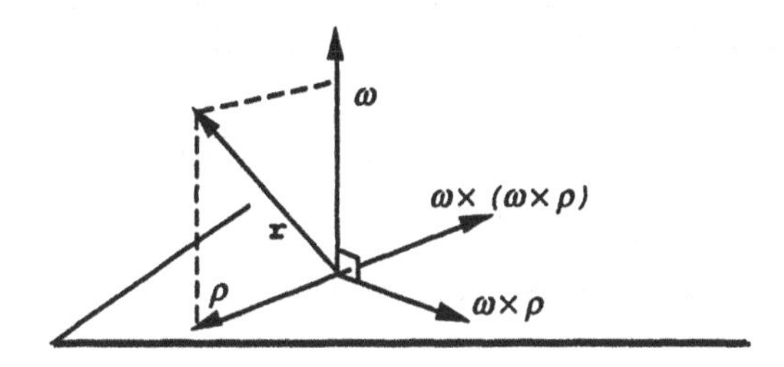

△

Example 5.2. Angular Velocities in the Solar System

(1) Find the angular velocity ω_1 in the daily rotation of the Earth around its axis. The angular velocity is to be taken with respect to the fixed stars, i.e., with respect to the heliocentric reference frame.
(2) Find the angular velocity ω_2 in the annual motion of the Earth around the Sun. ω_2 describes the motion of the center of Earth around the center of the Sun. We take this motion to be a uniform circular motion.
(3) What is the angle between the two rotation vectors ω_1 and ω_2 ?

Solution.

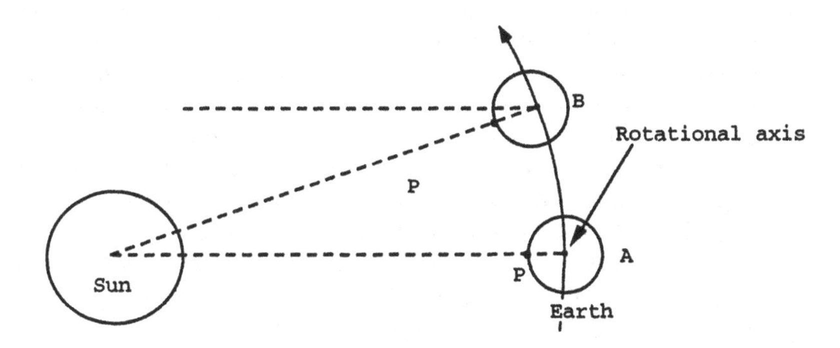

(1) Consider the daily rotation of the Earth. The time T from some point P on the Earth is directed towards the Sun till P again is directly towards the Sun (see the figure) is 24 hours, or 86 400 seconds. During this period,

the Earth has moved relative to the center of the Sun. During the time T, the point P has turned relative to the fixed stars, not only by 2π but by an angle

$$\theta = 2\pi + \frac{2\pi}{365.24} = 2\pi\frac{366.24}{365.24}.$$

(Here we use 1 year = 365.24 solar days, 1 solar day = 24 hours.) The angular velocity of the Earth relative to the fixed stars, i.e., the angular velocity relative to the heliocentric reference frame, is therefore

$$\begin{aligned}\omega_1 &= \frac{\theta}{T} = \frac{2\pi \times 366.24}{86400 \times 365.24}\ \mathrm{s}^{-1} \\ &= 7.29 \times 10^{-5}\ \mathrm{s}^{-1}\end{aligned}$$

A stellar day T_{st} is thus determined by:

$$\omega_1 T_{st} = 2\pi,$$

$$T_{st} = 2\pi/\omega_1 = 23.93 \text{ hours } \approx 23 \text{ hours 56 minutes.}$$

(2) 1 year $\approx 3.16 \times 10^7$ s. From this:

$$\omega_2 \approx \frac{2\pi}{3.16 \times 10^7}\ \mathrm{s}^{-1} \approx 1.99 \times 10^{-7}\ \mathrm{s}^{-1}.$$

(3) About 23.5°.

Questions.

(1) Determine the velocity $\mathbf{v}_2$ and the acceleration $\mathbf{a}_2$ of the center of Earth in the annual motion of the Earth around the Sun; $\mathbf{v}_2$ and $\mathbf{a}_2$ are to be determined relative to the heliocentric reference frame.

(2) Determine the velocity $\mathbf{v}_1$ and the acceleration $\mathbf{a}_1$ of a point P on the surface of Earth at 60° latitude. $\mathbf{v}_1$ and $\mathbf{a}_1$ are to be determined relative to a coordinate system having the *origin in the center of the Earth and axis fixed relative to the fixed stars.* Such a coordinate frame is known as the geocentric reference frame. This frame is not in uniform translation relative to the heliocentric reference frame, and the geocentric reference frame is thus not an inertial system. The *geocentric reference frame* is falling freely in the gravitational field of the Sun.

(3) Consider a coordinate system with its origin at the center of the Earth and axes fixed relative to the Earth. Such a coordinate system, in which the Earth is at rest, is sometimes called the Earth system. Give a qualitative description of how this coordinate system moves relative to the heliocentric reference frame.

Answers.

(1)

$$
\begin{aligned}
|\mathbf{v}_2| &= R\omega_2 = 3.0 \times 10^4 \ \mathrm{m\,s^{-1}} = 30 \ \mathrm{km\,s^{-1}} . \\
|\mathbf{a}_2| &= R\omega_2^2 = 5.9 \times 10^{-3} \ \mathrm{m\,s^{-2}} \ \text{(towards the center of the Sun).}
\end{aligned}
$$

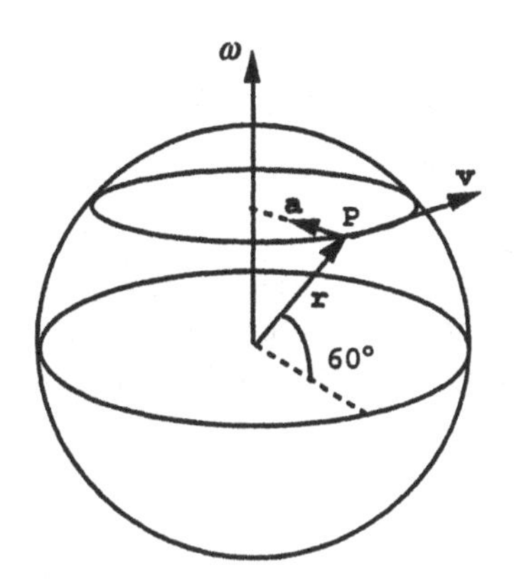

(2)

$$
\begin{aligned}
|\mathbf{v}_1| &= |\boldsymbol{\omega} \times \mathbf{r} \cos(60^\circ)| = 836 \ \mathrm{km/hour} \\
|\mathbf{a}_1| &= |\boldsymbol{\omega} \times \mathbf{v}_1| = 1.7 \times 10^{-2} \ \mathrm{m\,s^{-2}} \quad \text{(direction as on the figure)}
\end{aligned}
$$

Note. We do not sense the velocity $v_1 = 836$ km/hour. This velocity is comparable to that of a jet airliner, inside which we do not sense the velocity either. Our velocity relative to the center of the Sun is $v_2 = 30 \ \mathrm{km\,s^{-1}}$ and our velocity relative to the center of the galaxy is about $250 \ \mathrm{km\,s^{-1}}$. The acceleration in these motions is very small so that at any given moment the motion is almost a uniform translation.

Newton's laws of mechanics are of such a character that according to these laws we do not feel a translational velocity relative to the fixed stars ("the absolute space"?). On the other hand, as we shall see, an acceleration relative to the stars (the heliocentric reference frame) will make itself felt immediately. △

5.2 Problems

Problem 5.1. A rigid body rotates with 240 revolutions per minute around an axis pointing into the first octant as shown on the figure.

The axis forms an angle of 45° with the x-axis and 60° with the y-axis.

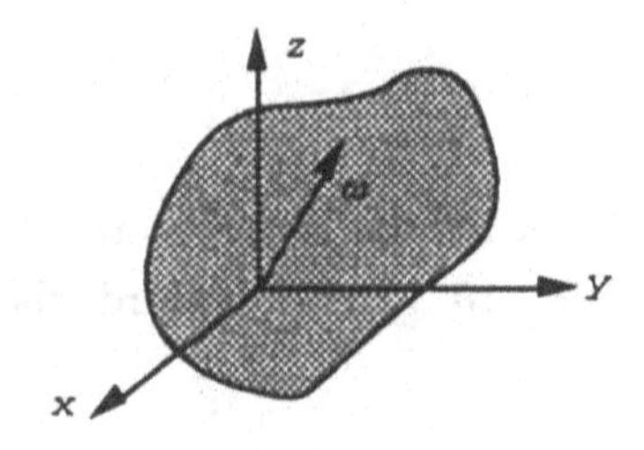

(1) What is the velocity of a point P having the position vector $\mathbf{r} = (0,1,2)$ m?
(2) What is the acceleration of P?

Problem 5.2. Calculate the acceleration of a point on the equator of the Earth. The acceleration should be determined relative to the geocentric reference frame (origin in the center of the Earth, axes fixed relative to the stars).

6. Motion in Accelerated Reference Frames

6.1 Newton's 2nd Law Within Accelerated Reference Frames

This section contains substantial mathematical derivations. The results – not the derivations – are crucial to an insight into fundamental aspects of the Newtonian theory of motion. The main result can be expressed in a simple and elegant manner, and the physical content of this important result is quite transparent (see Equation (6.36)).

Our problem is the following. We wish to find the mathematical equations that describe the motion of a particle in the coordinates of frame S, which moves in an arbitrary way relative to the inertial frames. Expressed differently: we wish to find how Newton's second law transforms from an inertial frame to a frame S whose origin O performs an arbitrary translatory motion relative to I, while at the same time S is rotating around some axis through O. In textbooks on geometry it is shown that the most general motion of a rigid body (a reference frame) can always be resolved into a translation of a point O and a rotation around an axis through O. This result will also be apparent after the reading of this section.

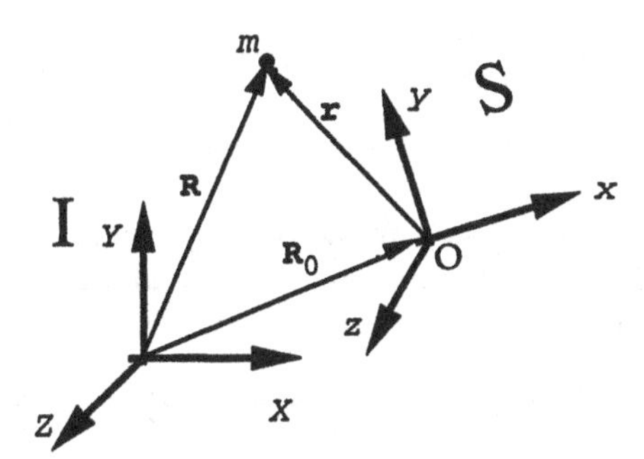

Fig. 6.1. The coordinate frame S moves in an arbitrary way relative to the inertial frame I

Look at Figure 6.1. The reference frame I is an inertial frame. In our terminology at this stage this means simply that I is the heliocentric frame or some frame moving with constant velocity relative to the heliocentric frame. We shall later discuss the peculiar position of the heliocentric reference frame and study in depth why this coordinate frame is a good approximation to an inertial frame.

Consider a particle having the mass m and being influenced by a force $\mathbf{F}$. The equation of motion for the particle relative to I is, as we have seen,

$$m\mathbf{a} = \mathbf{F}.$$

We now seek the equation of motion expressed in the coordinates of the S frame, where S is moving in an arbitrary way relative to I.

One can think of S as an accelerated spaceship that, as we observe it, rotates around its center. We wish to find the laws of motion for a particle as they appear to a passenger of the spaceship. Planet Earth, for instance, is just such a spaceship moving in a circular path and rotating relative to the heliocentric reference frame.

In order to find *the equation of motion as seen from S*, we must find the connection between the kinematic description of the motion as seen from I and as seen from S.

To this end, we denote the vectorial position of the particle relative to I by $\mathbf{R}$ and the vectorial position relative to S by $\mathbf{r}$ (see Figure 6.1).

We begin by finding the connection between the acceleration relative to I, $\mathbf{a} = (\mathrm{d}^2\mathbf{R}/\mathrm{d}t^2)_\mathrm{I}$, and the acceleration relative to S, $\mathbf{a}_r = (\mathrm{d}^2\mathbf{r}/\mathrm{d}t^2)_\mathrm{S}$. We are assuming here, based on Newton's postulate of absolute time, that the time t is common to both observers.

As seen from the inertial frame I the particle is at the position $\mathbf{R} = (X, Y, Z)$ where X, Y and Z are the components of $\mathbf{R}$ along the axes of I. For the velocity $\mathbf{v}_\mathrm{a}$ and the acceleration $\mathbf{a}$ relative to I we have:

$$\mathbf{v}_\mathrm{a} \equiv \left(\frac{\mathrm{d}\mathbf{R}}{\mathrm{d}t}\right)_\mathrm{I} = (\dot{\mathbf{R}})_\mathrm{I} = (\dot{X}, \dot{Y}, \dot{Z}), \tag{6.1}$$

$$\mathbf{a} \equiv \left(\frac{\mathrm{d}^2\mathbf{R}}{\mathrm{d}t^2}\right)_\mathrm{I} = (\ddot{\mathbf{R}})_\mathrm{I} = (\ddot{X}, \ddot{Y}, \ddot{Z}). \tag{6.2}$$

We stress that the differentiations above are performed by an observer at rest with respect to I.

As seen from the coordinate frame S the particle is at the position $\mathbf{r} = (x, y, z)$ where x, y, and z are components of $\mathbf{r}$ along the axes of S. For the velocity $\mathbf{v}_\mathrm{r}$ and acceleration $\mathbf{a}_\mathrm{r}$ in S we have:

$$\mathbf{v}_\mathrm{r} \equiv \left(\frac{\mathrm{d}\mathbf{r}}{\mathrm{d}t}\right)_\mathrm{S} = (\dot{\mathbf{r}})_\mathrm{S} = (\dot{x}, \dot{y}, \dot{z}), \tag{6.3}$$

$$\mathbf{a}_\mathrm{r} \equiv \left(\frac{\mathrm{d}^2\mathbf{r}}{\mathrm{d}t^2}\right)_\mathrm{S} = (\ddot{\mathbf{r}})_\mathrm{S} = (\ddot{x}, \ddot{y}, \ddot{z}). \tag{6.4}$$

Similarly, the differentiations here are performed by an observer at rest with respect to S.

If we denote the three unit vectors along the cartesian axis of S by **i**, **j**, and **k** respectively, we can express position, velocity, and acceleration as:

$$\mathbf{r} = x\mathbf{i} + y\mathbf{j} + z\mathbf{k}\,, \tag{6.5}$$

$$\mathbf{v}_\mathrm{r} = \dot{x}\mathbf{i} + \dot{y}\mathbf{j} + \dot{z}\mathbf{k}\,, \tag{6.6}$$

$$\mathbf{a}_\mathrm{r} = \ddot{x}\mathbf{i} + \ddot{y}\mathbf{j} + \ddot{z}\mathbf{k}\,. \tag{6.7}$$

Here $\mathbf{v}_\mathrm{r}$ and $\mathbf{a}_\mathrm{r}$ are often called the *relative* velocity and the *relative* acceleration, respectively. In contrast, $\mathbf{v}_\mathrm{a}$ (6.1) and $\mathbf{a}$ (6.2) are called the *absolute* velocity and the *absolute* acceleration. This is not necessarily a good terminology. For, what is absolute velocity?

The configuration of the coordinate system S relative to I is given by:

(1) the position of the origin O of S, denoted by its principal vector $\mathbf{R}_0$ in I:
(2) the direction of the three unit vectors **i**, **j**, and **k**. These unit vectors determine the axis of S as they are seen from I.

The position vector **R** for a particle can be written (from I):

$$\mathbf{R} = \mathbf{R}_0 + \mathbf{r}\,, \tag{6.8}$$

or

$$\mathbf{R}(t) = \mathbf{R}_0(t) + x(t)\,\mathbf{i}(t) + y(t)\,\mathbf{j}(t) + z(t)\,\mathbf{k}(t), \tag{6.9}$$

where we have introduced the explicit time dependence (see Figure 6.1). By differentiating (6.9) with respect to time (from I), we find

$$\dot{\mathbf{R}} = (\dot{\mathbf{R}}_0 + x\dot{\mathbf{i}} + y\dot{\mathbf{j}} + z\dot{\mathbf{k}}) + (\dot{x}\mathbf{i} + \dot{y}\mathbf{j} + \dot{z}\mathbf{k}). \tag{6.10}$$

The vector $\dot{\mathbf{R}}$ is the velocity of the particle with respect to I, the velocity we denote $\mathbf{v}_\mathrm{a}$. We have split the terms on the right-hand side into two parts. The terms in the first parenthesis are denoted by the *comoving* velocity $\mathbf{v}_\mathrm{m}$:

$$\mathbf{v}_m = \dot{\mathbf{R}}_0 + x\dot{\mathbf{i}} + y\dot{\mathbf{j}} + z\dot{\mathbf{k}}\,. \tag{6.11}$$

The comoving velocity does not contain $\dot{x}, \dot{y}$, or $\dot{z}$. The vector $\mathbf{v}_\mathrm{m}$ denotes the velocity relative to I that the particle would have if it was fixed in S, i.e., if it was moving only because of the overall motion of S.

The second parenthesis in (6.10) is $(\dot{x}\mathbf{i} + \dot{y}\mathbf{j} + \dot{z}\mathbf{k})$, i.e., the velocity of the particle referred to S, the relative velocity $\mathbf{v}_\mathrm{r}$. Let us write (6.10) as

$$\mathbf{v}_\mathrm{a} = \mathbf{v}_\mathrm{m} + \mathbf{v}_\mathrm{r}\,. \tag{6.12}$$

If you differentiate (6.10) again (be sure to get all terms!), you will find

$$\ddot{\mathbf{R}} = (\ddot{\mathbf{R}}_0 + x\ddot{\mathbf{i}} + y\ddot{\mathbf{j}} + z\ddot{\mathbf{k}}) + (\ddot{x}\mathbf{i} + \ddot{y}\mathbf{j} + \ddot{z}\mathbf{k}) + 2(\dot{x}\dot{\mathbf{i}} + \dot{y}\dot{\mathbf{j}} + \dot{z}\dot{\mathbf{k}}). \tag{6.13}$$

We view the right-hand side of (6.13) as the sum of three terms, as indicated by the parenthesis. Let us introduce a notation: the acceleration of the particle relative to I (the so-called absolute acceleration) is $\ddot{\mathbf{R}} \equiv \mathbf{a}$. The first parenthesis on the right-hand side of (6.13) is called the comoving acceleration $\mathbf{w}_m$:

$$\mathbf{w}_m \equiv (\ddot{\mathbf{R}}_0 + x\ddot{\mathbf{i}} + y\ddot{\mathbf{j}} + z\ddot{\mathbf{k}}). \tag{6.14}$$

The comoving acceleration $\mathbf{w}_m$ is the acceleration the particle would have relative to I if the particle was at rest at its instantaneous position (x, y, z) in S.

The second parenthesis is the acceleration within S, the relative acceleration $\mathbf{a}_r$:

$$\mathbf{a}_r = (\ddot{x}\mathbf{i} + \ddot{y}\mathbf{j} + \ddot{z}\mathbf{k}). \tag{6.15}$$

The third term on the right-hand side of (6.13) is called the Coriolis acceleration (after the French physicist G. Coriolis). We have:

$$\mathbf{w}_{co} \equiv 2(\dot{x}\dot{\mathbf{i}} + \dot{y}\dot{\mathbf{j}} + \dot{z}\dot{\mathbf{k}}). \tag{6.16}$$

The term is complicated because it contains time derivatives of both the relative coordinates (x, y, z) and of the unit vectors **i**, **j**, and **k**. We can now write (6.13) in the compact form:

$$\mathbf{a} = \mathbf{a}_r + \mathbf{w}_m + \mathbf{w}_{co} \,. \tag{6.17}$$

In order to use this fundamental relation between the absolute acceleration **a** and the relative acceleration $\mathbf{a}_r$ it is necessary to derive an expression for the dependence of $\mathbf{w}_m$ and $\mathbf{w}_{co}$ on the motion of the reference frame S relative to the inertial frame I. We will first demonstrate that the comoving velocity of the particle can be split into two velocities, a translation and a rotation about an axis through the origin of S,

$$\mathbf{v}_m = \dot{\mathbf{R}}_0 + \boldsymbol{\omega} \times \mathbf{r}\,, \tag{6.18}$$

and thus determine $\boldsymbol{\omega}$. Next we shall express $\mathbf{w}_m$ and $\mathbf{w}_{co}$ in terms of $\boldsymbol{\omega}$. In order to rewrite the expression (6.11) for the comoving velocity, we will express $\dot{\mathbf{i}}, \dot{\mathbf{j}}$, and $\dot{\mathbf{k}}$ in a more transparent manner. The procedure we will follow is simple.

We will use the fact that S moves as a rigid body relative to I. The mathematical formulation of this fact is that the three unit vectors, **i**, **j**, and **k**, that are fixed in relation to S, remain mutually orthogonal and of unit length no matter how S moves. Only the direction of the vectors relative to I change with time. We therefore have the conditions:

$$\mathbf{i}\cdot\mathbf{i} = \mathbf{j}\cdot\mathbf{j} = \mathbf{k}\cdot\mathbf{k} = 1\,, \tag{6.19}$$

and

$$\mathbf{i}\cdot\mathbf{j} = \mathbf{i}\cdot\mathbf{k} = \mathbf{j}\cdot\mathbf{k} = 0\,. \tag{6.20}$$

By differentiating (6.19) and (6.20), we find that

$$\dot{\mathbf{i}} \cdot \mathbf{i} = \dot{\mathbf{j}} \cdot \mathbf{j} = \dot{\mathbf{k}} \cdot \mathbf{k} = 0\,, \tag{6.21}$$

and

$$\dot{\mathbf{i}} \cdot \mathbf{j} + \mathbf{i} \cdot \dot{\mathbf{j}} = \dot{\mathbf{i}} \cdot \mathbf{k} + \mathbf{i} \cdot \dot{\mathbf{k}} = \dot{\mathbf{j}} \cdot \mathbf{k} + \mathbf{j} \cdot \dot{\mathbf{k}} = 0\,. \tag{6.22}$$

Every vector can be written in terms of three orthogonal vectors. We write the three vectors $\dot{\mathbf{i}}, \dot{\mathbf{j}}$, and $\dot{\mathbf{k}}$ as:

$$\begin{aligned} \dot{\mathbf{i}} &= a_{11}\mathbf{i} + a_{12}\mathbf{j} + a_{13}\mathbf{k}\,, \\ \dot{\mathbf{j}} &= a_{21}\mathbf{i} + a_{22}\mathbf{j} + a_{23}\mathbf{k}\,, \\ \dot{\mathbf{k}} &= a_{31}\mathbf{i} + a_{32}\mathbf{j} + a_{33}\mathbf{k}\,. \end{aligned} \tag{6.23}$$

Note. The coefficients a_{nm} may – as well as the vectors $\mathbf{i}, \mathbf{j}$, and $\mathbf{k}$ – depend on time. Of the nine coefficients a_{nm}, only three are independent. This is a direct consequence of the six equations (6.21) and (6.22). From (6.21) and (6.23):

$$a_{11} = a_{22} = a_{33} = 0\,.$$

From (6.22):

$$a_{12} = -a_{21}\,;\; a_{13} = -a_{31}\,;\; a_{23} = -a_{32}\,.$$

Let us change the notation:

$$\begin{aligned} a_{23} &\equiv \omega_x\,, \\ a_{13} &\equiv -\omega_y\,, \\ a_{12} &\equiv \omega_z\,. \end{aligned} \tag{6.24}$$

We can then write (6.23) as:

$$\begin{aligned} \dot{\mathbf{i}} &= \omega_z\mathbf{j} - \omega_y\mathbf{k}\,, \\ \dot{\mathbf{j}} &= -\omega_z\mathbf{i} + \omega_x\mathbf{k}\,, \\ \dot{\mathbf{k}} &= \omega_y\mathbf{i} - \omega_x\mathbf{j}\,. \end{aligned} \tag{6.25}$$

These expressions can be made more transparent. With this in mind we introduce the vector

$$\boldsymbol{\omega} = \omega_x\mathbf{i} + \omega_y\mathbf{j} + \omega_z\mathbf{k}\,. \tag{6.26}$$

By directly applying the definition of the cross product we find that $\mathbf{i} \times \mathbf{j} = \mathbf{k}$, $\mathbf{k} \times \mathbf{i} = \mathbf{j}$, and $\mathbf{j} \times \mathbf{k} = \mathbf{i}$ (the coordinate system is right handed). By means of these results and using the definition of $\boldsymbol{\omega}$ (Equation (6.26)) we see that, for instance, $\boldsymbol{\omega} \times \mathbf{i} = (\omega_x\mathbf{i} + \omega_y\mathbf{j} + \omega_z\mathbf{k}) \times \mathbf{i} = -\omega_y\mathbf{k} + \omega_z\mathbf{j}$. It is thus easily shown that (6.23) can be written in the compact form:

$$\begin{aligned} \dot{\mathbf{i}} &= \boldsymbol{\omega} \times \mathbf{i}\,, \\ \dot{\mathbf{j}} &= \boldsymbol{\omega} \times \mathbf{j}\,, \\ \dot{\mathbf{k}} &= \boldsymbol{\omega} \times \mathbf{k}\,. \end{aligned} \tag{6.27}$$

By means of the vector $\boldsymbol{\omega}$ we can thus describe how the vectors $\mathbf{i}, \mathbf{j}$, and $\mathbf{k}$ vary in time. Note: (6.27) expresses the fact that the coordinate system S is in rotation. For instance: the rate of change of $\mathbf{i}$, which is $\dot{\mathbf{i}}$, is perpendicular to $\mathbf{i}$ (the cross product). The length of $\mathbf{i}$ does not change, only the direction. The motion of S relative to the inertial frame I can be described by the rate of change of the position vector $\mathbf{R}_0$ to the origin of S, and a specification of the axis of rotation by means of the rotation vector $\boldsymbol{\omega}$ through the origin of S. Compare with the motion of the Earth relative to the heliocentric system, as described in Chapter 5.

By means of (6.27), we can now write (6.11) as:

$$\begin{aligned} \mathbf{v}_{\mathrm{m}} &= \dot{\mathbf{R}}_0 + x\,(\boldsymbol{\omega} \times \mathbf{i}) + y\,(\boldsymbol{\omega} \times \mathbf{j}) + z\,(\boldsymbol{\omega} \times \mathbf{k}) && (6.28) \\ &= \dot{\mathbf{R}}_0 + \boldsymbol{\omega} \times (x\mathbf{i} + y\mathbf{j} + z\mathbf{k}) \\ &= \dot{\mathbf{R}}_0 + \boldsymbol{\omega} \times \mathbf{r}\,. && (6.29) \end{aligned}$$

A particle at rest relative to S has the velocity $\mathbf{v}_{\mathrm{m}}$ (the comoving velocity) relative to I, as indicated by (6.29). This expression for the comoving velocity shows directly that the velocity relative to I of a particle at rest in S can be regarded as a superposition of two velocities: $\dot{\mathbf{R}}_0$ (a translation of the origin of S), and $\boldsymbol{\omega} \times \mathbf{r}$ (a rotation about the axis in the direction of the vector $\boldsymbol{\omega}$ through the origin of S). Note: in general, $\boldsymbol{\omega}$ may depend on time.

The absolute velocity (6.12) of the particle becomes:

$$\mathbf{v}_{\mathrm{a}} = \dot{\mathbf{r}} + \dot{\mathbf{R}}_0 + \boldsymbol{\omega} \times \mathbf{r}\,. \qquad (6.30)$$

Applying (6.27) we can write the Coriolis acceleration (6.16) as

$$\begin{aligned} \mathbf{w}_{\mathrm{co}} &= 2\left(\dot{x}\dot{\mathbf{i}} + \dot{y}\dot{\mathbf{j}} + \dot{z}\dot{\mathbf{k}}\right) \\ &= 2\left(\dot{x}\,(\boldsymbol{\omega} \times \mathbf{i}) + \dot{y}\,(\boldsymbol{\omega} \times \mathbf{j}) + \dot{z}\,(\boldsymbol{\omega} \times \mathbf{k})\right) \\ &= 2(\boldsymbol{\omega} \times \dot{\mathbf{r}}) = 2(\boldsymbol{\omega} \times \mathbf{v}_r)\,. && (6.31) \end{aligned}$$

By differentiation of (6.27) with respect to time, we have that

$$\begin{aligned} \ddot{\mathbf{i}} &= \boldsymbol{\omega} \times \dot{\mathbf{i}} + \dot{\boldsymbol{\omega}} \times \mathbf{i} \\ &= \boldsymbol{\omega} \times (\boldsymbol{\omega} \times \mathbf{i}) + \dot{\boldsymbol{\omega}} \times \mathbf{i}\,, && (6.32) \end{aligned}$$

and similarly for $\ddot{\mathbf{j}}$ and $\ddot{\mathbf{k}}$.

By use of (6.32) the comoving acceleration $\mathbf{w}_{\mathrm{m}}$, specified in equation (6.14), may be written as

$$\mathbf{w}_{\mathrm{m}} = \ddot{\mathbf{R}}_0 + \boldsymbol{\omega} \times (\boldsymbol{\omega} \times \mathbf{r}) + \dot{\boldsymbol{\omega}} \times \mathbf{r}\,, \qquad (6.33)$$

where $\ddot{\mathbf{R}}_0$ is the acceleration vector in I of the origin O of S.

The last term in (6.33) depends on the rate of change of $\boldsymbol{\omega}$. This term will disappear whenever S is rotating around an axis that is fixed in space and with an angular velocity $\omega \equiv |\boldsymbol{\omega}|$ that is constant in time. This happens to be approximately true for the motion of the Earth around its axis. There is actually a small precession around the normal to the ecliptic (i.e., the plane that contains the orbit of Earth) with a period of about 25 800 years.

The second term in (6.33) is called the centripetal acceleration.

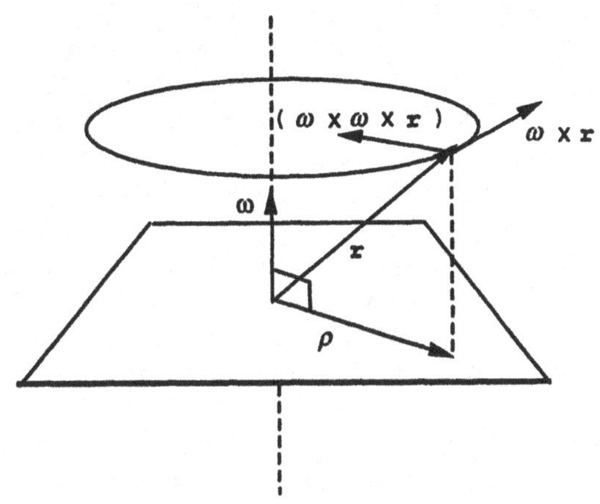

Fig. 6.2. Illustrating the centripetal acceleration in an accelerated (rotating) reference frame. The vector ρ is defined in the text

The centripetal acceleration is often written in the form

$$\boldsymbol{\omega} \times (\boldsymbol{\omega} \times \mathbf{r}) = -\omega^2 \boldsymbol{\rho}, \tag{6.34}$$

where the vector $\boldsymbol{\rho}$ is the projection of the radius vector $\mathbf{r}$ onto a plane perpendicular to $\boldsymbol{\omega}$. The centripetal ("center seeking") acceleration is therefore directed toward the rotation axis, and grows in magnitude as the distance from the axis (see Figure 6.2).

We can now write the complete equation of motion for a particle as seen from the frame S. Recall that the basic equation of motion relative to the frame I is

$$m\mathbf{a} = \mathbf{F}.$$

We now assume that the impressed force $\mathbf{F}$ acting on the particle is the same in S as it is in I. Using the results above, we can write:

$$m(\mathbf{a}_\mathrm{r} + \mathbf{w}_\mathrm{m} + \mathbf{w}_\mathrm{co}) = \mathbf{F}, \tag{6.35}$$

or, writing out the explicit expressions derived for $\mathbf{w}_\mathrm{m}$ and $\mathbf{w}_\mathrm{co}$, we have

$$m\frac{\mathrm{d}^2\mathbf{r}}{\mathrm{d}t^2} = \mathbf{F} - m\ddot{\mathbf{R}}_0 + m\omega^2\boldsymbol{\rho} - 2m(\boldsymbol{\omega} \times \dot{\mathbf{r}}) - m(\dot{\boldsymbol{\omega}} \times \mathbf{r}). \tag{6.36}$$

Here, $\mathbf{r}$, $\dot{\mathbf{r}}$, and $\ddot{\mathbf{r}} \equiv (\mathrm{d}^2\mathbf{r}/\mathrm{d}t^2)$ denote the position, velocity, and acceleration of the particle as seen from the frame S. In other words, (6.36) is the equation of motion within the frame S. $\ddot{\mathbf{R}}_0$ is the translational acceleration of

S, and ω is a vector which describes the rotation of S. The set $\ddot{\mathbf{R}}_0$ and ω completely describe the acceleration of S relative to I. Equation (6.36) is *the basic equation of motion for the particle from within the reference frame S.*

Equation (6.36) shows that we formally can claim that Newton's second law is valid within accelerated frames, if we interpret the last four terms on the right-hand side as forces acting on the particle in addition to **F**. These terms are *fictitious forces.* In Newtonian mechanics, fictitious forces are distinguished from "actual" forces by the fact that the *fictitious forces depend only on the motion ($\ddot{\mathbf{R}}_0$ and ω)of the frame S,* the position (ρ), and the velocity ($\dot{\mathbf{r}}$) of the particle, and on the inertial mass of the particle.

The presence of an actual force **F** can always be related to an interaction between the particle and another material body. For example, if an electrical force acts on the particle (which must then be charged) there is an interaction with other electrically charged particles. If a gravitational force acts, there is an interaction with some body that has gravitational mass. The fictitious forces cannot in any obvious way be related to an interaction with other bodies.

By introducing the fictitious forces, an observer in S can still claim that Newton's first and second laws are valid. Note, however, that the third law, the law of action and reaction, is *not valid for the fictitious forces*; for there is no other body upon which a reaction force can act.

Let us emphasize that the *fictitious forces have important similarities to gravitational forces: like these, the fictitious forces are proportional to the mass of the body.*

In Chapter 7 we will go deeper into the theoretical interpretation of equation (6.36). Here we shall first examine the physical consequences of (6.36) in some special cases.

6.2 The Equivalence Principle of Mechanics

Let us first assume that $\omega = 0$, $\dot{\omega} = 0$, and $\ddot{\mathbf{R}}_0 = (a, 0, 0) \equiv \mathbf{a}$. These assumptions mean that the reference frame S has a constant linear acceleration **a** relative to the inertial frame I (see Figure 6.3).

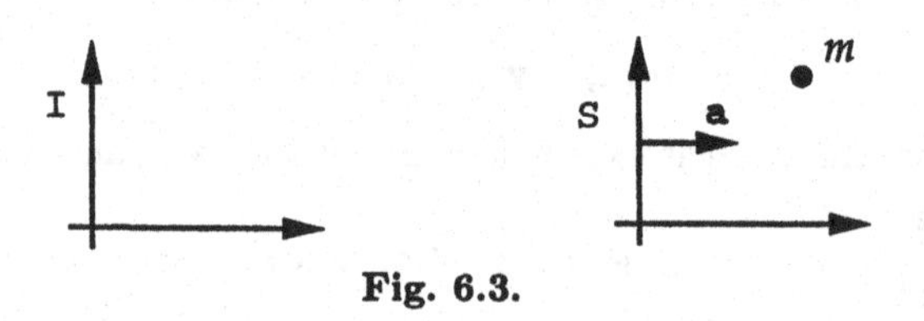

Fig. 6.3.

Our fundamental equation (6.36) now assumes the form:

$$m\frac{\mathrm{d}^2\mathbf{r}}{\mathrm{d}t^2} = \mathbf{F} - m\mathbf{a}\,. \tag{6.37}$$

Equation (6.37) is the equation of motion for the particle, seen from the frame S. The total force $\mathbf{F}'$ on the particle, as evaluated in S, is given by

$$\mathbf{F}' = \mathbf{F} - m\mathbf{a}\,.$$

If $\mathbf{F} = 0$, the total force is given by the fictitious force alone

$$\mathbf{F}' = -m\mathbf{a} = m(-\mathbf{a}).$$

We know that at the surface of the Earth where the gravitational field is nearly homogeneous, the gravitational force on a particle is given by

$$\mathbf{F} = m\mathbf{g}\,,$$

where $\mathbf{g}$ is the acceleration due to gravity.

By comparing the last two equations, we see that the fictitious force acts like a force of gravity, being directly proportional to the mass of the particle. The minus sign reminds us that the fictitious force is directed opposite to the acceleration of the reference frame S.

We collect the result (6.37) and the above considerations in the so-called *equivalence principle of mechanics*, which we can formulate as:

> The motion of a particle, as seen by an observer in a reference frame S which is accelerated with constant acceleration ($\mathbf{a}$) relative to the inertial frames, will be as if a uniform gravitational field ($-\mathbf{a}$) acts in addition to the "natural" forces.

or, simply:

> An acceleration to the right is equivalent to a uniform gravitational field to the left.

The equivalence principle of mechanics – sometimes called the equivalence principle of Einstein, or the strong equivalence principle – has its origin in the proportionality between gravitational and inertial mass. The mass appearing on either side of (6.36) is inertial mass. As the inertial and gravitational masses are equal, the fictitious force may be conceived as a gravitational force (see Chapter 3).

The similarity between gravitational and fictitious forces seems somewhat coincidental in the above considerations. It remained unnoticed for two hundred years. It was to become a major foundation for the general theory of

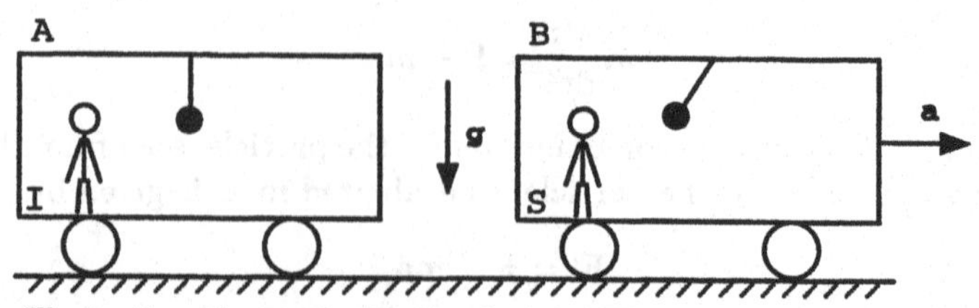

Fig. 6.4. Illustrating the equivalence principle. The railroad car B is accelerated relative to the car A, which represents an inertial frame. The cars are closed, so that the observers cannot look out

relativity put forth by Albert Einstein in 1915. The theory of general relativity is basically a theory of the phenomenon of gravitation.
Look at Figure 6.4. As in Chapter 4, for motion over "short distances" and in "short time spans" a laboratory at rest on the surface of the Earth can be considered to be an inertial frame. A and B are two railroad cars on the surface of the Earth where the gravitational field is uniform. Car A is at rest relative to the ground, standing on a horizontal and straight railroad track. Car B has a constant acceleration relative to the ground. Both cars are closed so that an observer cannot directly look at the outside world. Compare this with Chapter 4.

Equation (6.36) shows that Newton's second law is not invariant under transformations to accelerated reference frames. This is a mathematical expression of the following experimental fact: an observer in the car B, i.e., an observer at rest within the accelerated reference frame S can, by performing mechanical experiments completely within his railroad car, decide that he is accelerated relative to the inertial systems. Specifically he can measure the angle ϕ that a pendulum, hanging from the ceiling and being at rest, forms with the end walls. Through this measurement he can determine the acceleration $\mathbf{a}$.

The pendulum has its equilibrium position parallel to the total gravitational field inside B, which is $\mathbf{g} - \mathbf{a}$. We then have $\tan(\phi) = |\mathbf{a}|/|\mathbf{g}|$. From this experiment alone, the observer in B is unable to tell how much of the acceleration of gravity is caused by a (nearly) homogeneous field of the planet, and how much is due to a linear acceleration relative to an inertial frame. The observer in B can determine the magnitude of the total acceleration field by measuring the time of oscillation of his pendulum, given by

$$T = 2\pi \left[\frac{L}{g_{\text{eff}}}\right]^{1/2} .$$

In particular he would find in the above case

$$T = 2\pi \left(\frac{L}{\sqrt{g^2 + a^2}}\right)^{1/2} .$$

The equivalence principle of mechanics tells us that the observer in B is unable to infer from mechanical experiments alone how much (if any part) of a homogeneous field of gravity is caused by gravitational attraction, and how much is caused by his own reference frame being accelerated relative to the inertial frames.

We saw that the relativity principle of mechanics brought the idea of absolute space into some doubt. It is fair to ask how much existence absolute space has, when no experiment can decide if one is at rest or in uniform motion relative to absolute space. Newton's reply was that the resistance that all bodies has against being accelerated must be interpreted as an action of absolute space. Acceleration is a change of velocity relative to absolute space and for this, says Newton, a force is needed.

Below we shall analyze the equivalence principle of mechanics in further detail.

6.3 The Einstein Box

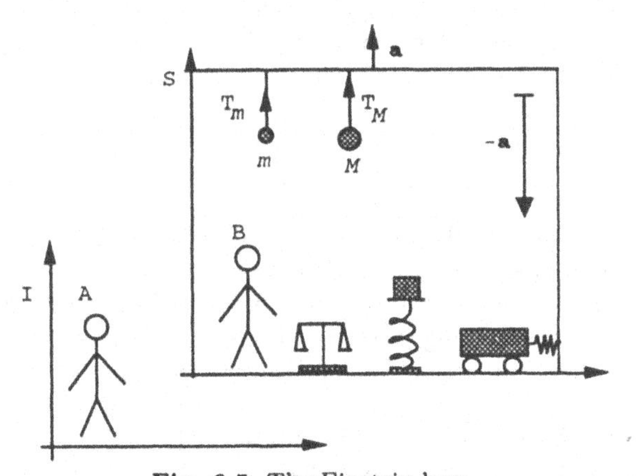

Fig. 6.5. The Einstein box

Einstein considered a closed elevator box S far from all gravitating masses (see Figure 6.5). The box is assumed to have a constant acceleration a relative to the inertial frame I. We do not concern ourselves with how the box acquired this acceleration or how it is maintained; the important feature is that the acceleration has the constant value a. Inside the box, an observer B is present. Two masses, m and M are hanging from equally long strings, as shown on Figure 6.5. B has a balance and a spring scale so he can compare masses and measure forces.

Let us first describe the motion of the masses as seen by an observer A at rest in the inertial frame I, and then as seen by observer B who is at rest in frame S, i.e., at rest in the elevator box.

The observer A in the inertial frame I. The two masses both have an acceleration $\mathbf{a}$. They must consequently be under the influence of some force. *This force must be a "real" force since we are describing their motion from an inertial frame* where no fictitious forces act. The masses are acted on by the string forces $\mathbf{T}_m$ and $\mathbf{T}_M$. The motion is known, so from Newton's law we infer the magnitudes of the string forces:

$$\mathbf{T}_m = m\mathbf{a}, \tag{6.38}$$

$$\mathbf{T}_M = M\mathbf{a}. \tag{6.39}$$

The observer B in the noninertial frame S. By carrying out mechanical experiments (e.g., by measuring the period of a pendulum) B finds that a homogeneous gravitational field exists inside the box. This is a consequence of the equivalence principle of mechanics. B will measure a gravitational field $\mathbf{g}_S = -\mathbf{a}$. The two masses remain at rest relative to B. The conclusion is: both masses are under the influence of a (fictitious) gravitational field, downwards. The force resulting from this (fictitious) field is equal but opposite to the string force. Thus B finds no net force on each mass:

$$\mathbf{T}_m + m(-\mathbf{a}) = 0, \tag{6.40}$$

$$\mathbf{T}_M + M(-\mathbf{a}) = 0. \tag{6.41}$$

Equation (6.38) and (6.40) and Equations (6.39) and (6.41) are formally identical – but the interpretations are profoundly different.

Let us consider a new situation inside the elevator box. At some time $t = 0$ the two strings holding m and M break. What happens? Let us again describe the events as seen and interpreted by the two observers A and B.

The observer A in the inertial frame I. No forces act on the masses m and M when the string tensions disappear. From Newton's second law, the masses will now stop accelerating. The floor of the elevator box, however, will (by assumption) continue to accelerate upwards with acceleration $\mathbf{a}$. The floor (a distance L below when the strings break) catches up with the two masses at a time determined by

$$L = \frac{1}{2}at^2 \quad \text{or} \quad t = \sqrt{\frac{2L}{a}}\,.$$

Note: this result is purely kinematic or, if you like, purely geometric. No gravitational forces are involved in the discussion (for the geometry of the problem, see Figure 6.5 and compare with the Galileo experiment as discussed in Chapter 1.).

The observer B in the accelerated frame S. The two strings of equal length break at time $t = 0$. The two masses have the same height (L) above the box floor, and start falling under the influence of the (fictitious) gravitational field in the elevator. If B orients his coordinate system with the y-axis along the direction of **a**, the fictitious field is in the negative y-direction. The masses follow the equations of motion:

$$m\ddot{y} = -ma\,, \tag{6.42}$$

$$M\ddot{y} = -Ma\,. \tag{6.43}$$

From these two equations, B finds that the two different masses fall with the same acceleration toward the box floor. By integrating these equations, B finds that the masses hit the floor at a time given by

$$t = \sqrt{\frac{2L}{a}}\,.$$

All masses inside the elevator box will, as seen by the observer B, fall with the same acceleration, just as if they fell in a uniform gravitational field. As seen by the inertial observer A, this is entirely obvious!

In the gravitational field of the Earth all masses – regardless of their mass or composition – fall with the same acceleration **g**. As discussed in Chapter 3 this expresses the fact that the inertial and gravitational masses are equal. We can now begin to see that a deep law of nature is hidden in the equivalence of the inertial and the gravitational mass. Gravitation and inertia do not seem to be separate properties of matter, but rather two different aspects of a more fundamental and universal property of space and material particles.

Questions.

(1) What is the total gravitational field at the center of the freely falling satellite and in the freely falling elevator box in Examples 3.1 and 3.2, respectively?

[Answer: zero. The particle in the freely falling satellite and the man in the freely falling elevator box are said to be weightless relative to the satellite and the elevator box respectively.]

(2) On the spring scale in the Einstein box in Figure 6.5 is a 1 kg mass. The scale is calibrated on the surface of Earth. What does the scale show if the acceleration **a** has magnitude 3 m s^{-2} ?

[Answer: 0.306 kg]

We will discuss the equivalence principle of mechanics further in Chapter 7. Let us emphasize here that the perfect equivalence between "acceleration fields" and "gravitation fields" only hold for uniform or, more precisely, homogeneous fields. An easy way to realize this is to consider a box with dimensions comparable to those of the Earth (see Figure 6.6).

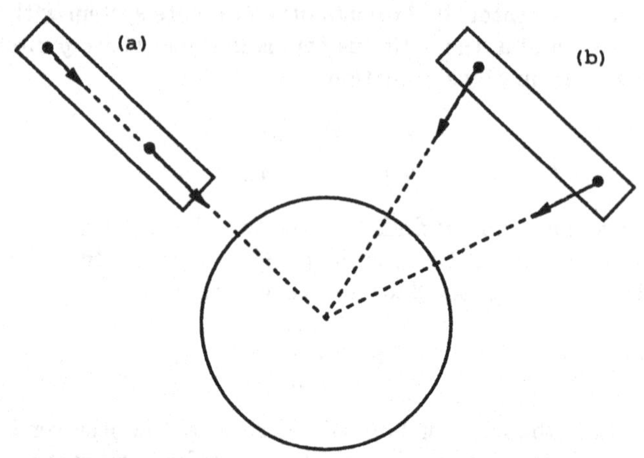

Fig. 6.6. Freely falling elevator boxes. Each elevator box contains two equal masses. The gravitational force from the Earth on each particle is suggested

The gravitational field from the Earth inside each of the freely falling boxes is not homogeneous.

If two particles are released from rest inside the box, they will either (a) separate or (b) approach each other as the box falls freely in the field of the Earth. By observing the behavior of the particles, an observer inside the box can thus, without looking outside, conclude that he falls freely in an inhomogeneous gravitational field. His elevator box can not be used as an inertial frame.

We shall return to a detailed study of this problem in Section 6.5 (tidal forces).

Example 6.1. Balloon in Accelerated Frame

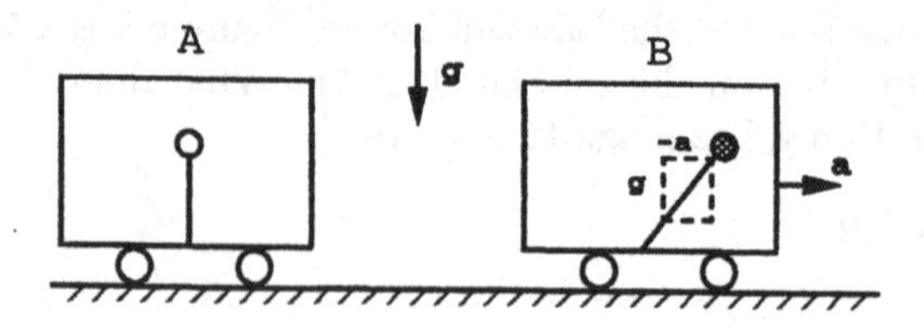

Toy balloons are tied to the floors of the two railroad cars A and B (see the figure). The car A is at rest (or in uniform motion relative to an inertial frame). The balloon inside A is parallel to the end walls. The car B has the acceleration **a** relative to A. Find the equilibrium direction of the string to the balloon in B.

Solution. The string of a balloon is always parallel to the local acceleration of gravity, $\mathbf{g} - \mathbf{a}$, in B. The string will form an angle θ with the floor where θ can be found from $\tan(\theta) = |\mathbf{g}|/|\mathbf{a}|$. A balloon will always float "upwards" in the local gravitational field. △

Example 6.2. Mass on an Oscillating Plate

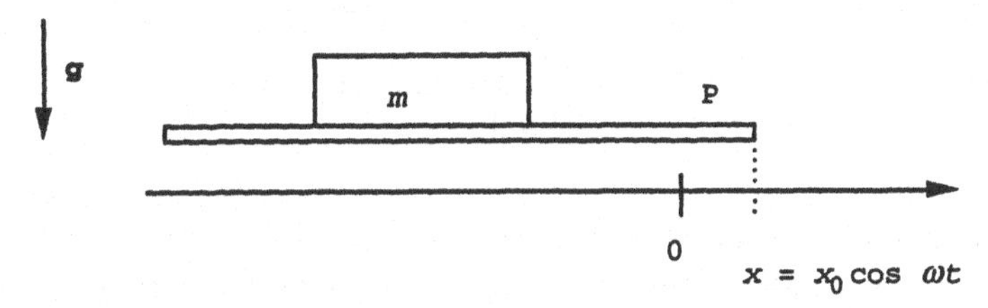

A body of mass m rests on a horizontal plate P. The plate oscillates horizontally, i.e., perpendicular to the gravitational field of the Earth. The amplitude in the oscillations is $x_0 = 15$ cm, and the period T is 1 second. The friction between the body and the plate is so that the body just stays on the plate without slipping. Find the coefficient of friction μ between the plate and the body. (Note: the maximal friction force between the plate and body is μN where N is the magnitude of the normal force).

Solution. In the reference frame where the plate is at rest we must include a fictitious gravitational field. This field is always opposite to the acceleration of the plate in the lab frame (an acceleration to the the right gives a gravity field to the left). The maximal value of the acceleration is $\omega^2 x_0$. The condition for the body just not to slip is then

$$\mu m g = m\omega^2 x_0 \ ,$$

$$\mu = \frac{x_0 \omega^2}{g} = 4\frac{\pi^2 x_0}{T^2 g} \ .$$

With the numerical values in the problem, $\mu = 0.6$. Try the experiment with a coin on a book. △

Example 6.3. Pendulum in an Elevator. A mathematical pendulum has a length such that the period of small oscillations in the surface gravitational field of the Earth is 1 second. Such a pendulum can be used as a watch. The pendulum is hung in an elevator which has a constant upward acceleration of $a = 1\,\mathrm{m\,s^{-2}}$ in the gravitational field of the Earth. How many "Earth" seconds will the clock gain per minute? Disregard changes in the external gravitational field.

Solution. On Earth the period of small oscillations is

$$T = 2\pi\sqrt{\frac{L}{g}} = 1 \text{ s},$$

where L is the length of the string. From this we conclude that the length is

$$L = \frac{gT^2}{4\pi^2} = 0.25 \text{ m}.$$

Inside the elevator the period of oscillation is:

$$T' = 2\pi\sqrt{\frac{L}{g+a}} = 0.95 \text{ s}.$$

In one minute of "Earth" time the pendulum oscillates $60/0.95 = 63$ times. It therefore gains 3 seconds "elevator time" per minute of "Earth time". Note that we had actually no need to compute the string length since we have directly that

$$\frac{T'}{T} = \sqrt{\frac{g}{g+a}}\,.$$

△

6.4 The Centrifugal Force

We will now examine the consequence of Equation (6.36) in a situation where $\dot{\boldsymbol{\omega}} = 0$ and $\ddot{\mathbf{R}}_0 = 0$ and where the particle in question is at rest in the accelerated reference frame S. With $\dot{\boldsymbol{\omega}} = 0$, S clearly has a constant angular velocity relative to the inertial frame I (and consequently relative to all other inertial frames). The Coriolis force of $-2m(\boldsymbol{\omega} \times \dot{\mathbf{r}})$ vanishes since the particle is at rest in S ($\dot{\mathbf{r}} = 0$). Equation (6.36) then has the form:

$$m\frac{\mathrm{d}^2\mathbf{r}}{\mathrm{d}t^2} = \mathbf{F} + m\omega^2\boldsymbol{\rho},$$

where $\rho = |\boldsymbol{\rho}|$ is the distance of the particle from the axis of rotation, and $\omega = |\boldsymbol{\omega}|$ is the angular velocity relative to the inertial frames. The term $m\omega^2\boldsymbol{\rho}$ is known as the centrifugal (center fleeing) force. It is familiar to anyone who has been on a merry-go-round. A particle at rest in S at the distance ρ from the rotation axis has the speed $v = \rho\omega$ relative to the inertial frame I. We can therefore say that the centrifugal force has the magnitude mv^2/ρ.

Let us first note that there has been considerable confusion about the concept of centrifugal force. We emphasize here that the *centrifugal force is a fictitious force* that acts on a given particle, when we study the motion of this particle in a coordinate frame in rotation relative to the inertial frames. No centrifugal forces arise within inertial frames.

Example 6.4. Earth's Orbit Around the Sun. Let us as a simple application of the concept of centrifugal force study the orbit of the Earth around the Sun.

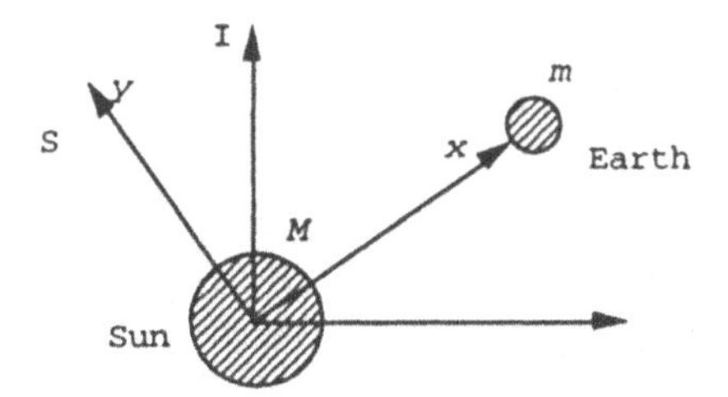

Fig. 6.7. Earth in its orbit around the Sun

The orbit of the Earth is approximately circular. Let us first describe the motion of the Earth in the heliocentric reference frame I (see Figure 6.7). The heliocentric reference frame is an inertial frame and therefore no fictitious forces act. The motion is completely described by the equation

$$m\frac{v^2}{r} = G\frac{Mm}{r^2}\,. \tag{6.44}$$

We are ignoring forces from the Moon and from other planets.

Let us next describe the motion from a reference frame S with its origin in the center of the Sun and an x-axis that always passes through the center of the Earth. S is rotating with an angular velocity equal to the orbital angular velocity of the Earth. Relative to S, the Earth is at rest. This means that the sum of the forces acting on the Earth seen from the rotating frame S is zero. The forces acting are: (1) the gravitational pull of the Sun, and (2) the fictitious forces. The fictitious force is the centrifugal force.

In this case, $\rho = r$, where r is the distance to the Sun. The gravitational force points in the negative x-direction, and we have:

$$-G\frac{Mm}{r^2} + m\omega^2 r = 0\,. \tag{6.45}$$

The gravitational force and the centrifugal force have the same magnitude and opposite directions. The two equations (6.44) and (6.45) are formally identical, but their interpretations are profoundly different. △

Example 6.5. Grass on a Rotating Disk. Let us build the following apparatus. On the rim of a horizontal disk of radius r we mount two walls and fill the space between them with soil (see Figure 6.8). We sow a rapidly growing variety of grass in the soil. The horizontal disk is now set in rotation with the constant angular velocity ω. After the grass has come up, we find that it has grown in the directions shown on the figure. This can be understood in terms of fictitious forces: the grass is growing in an accelerated reference frame. Within this frame, the grass "feels" a gravitational field composed of the gravitational field of the Earth, and the fictitious centrifugal field $\omega^2 r$. The grass grows in a direction opposite to the total gravitational field.

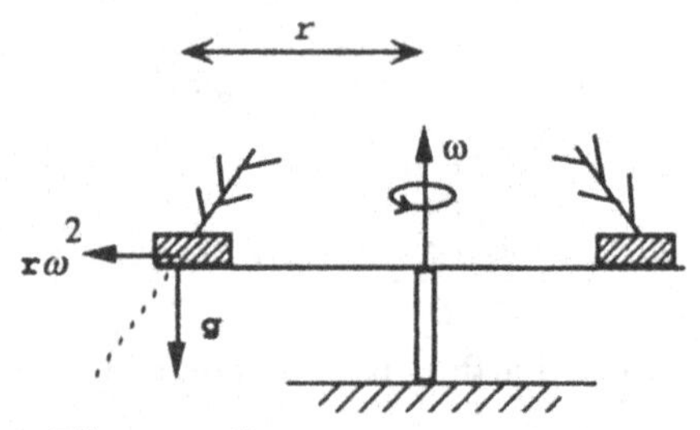

Fig. 6.8. Grass on a rotating disk

Note. The grass cannot distinguish how much of the gravitational field is due to the pull of the Earth and how much is because the disk is rotating.

As we shall see below, g itself (the local acceleration due to gravity) is composed of a contribution from mass attraction proper and a contribution from the centrifugal field that arises from the rotation of the Earth relative to the heliocentric frame.

Question. Let the radius of the above disk be 2 m. Determine the angular velocity of the disk such that the angle the grass forms with the vertical is 2°. [Answer: $\tan\theta = r\omega^2/g \Rightarrow \omega \approx 0.4\ \text{s}^{-1}$] △

Example 6.6. The Variation of g with Latitude. As yet another example we shall consider the variation of $\mathbf{g}$ with latitude. Consider a reference frame with its origin at the center of the Earth and axes fixed relative to the Earth. The reference frame thus rotates relative to the heliocentric reference frame. We consider a particle at rest within this frame. The particle has mass m and is hanging from a string as suggested in Figure 6.9. What forces act on the particle as seen from an observer standing next to it?

The Earth is not an inertial frame. We have to add fictitious forces to the "natural" forces acting on the particle. Let us ignore the acceleration of the Earth around the Sun, and consider only the rotation which the Earth performs in the heliocentric frame with a period of 23 hours and 56 minutes.

We shall – initially – assume that the Earth is a perfect sphere with radius R. The basic gravitational field of Earth will therefore act on the particle with a force $\mathbf{P}$ directed toward the center of the Earth. In addition to this, the

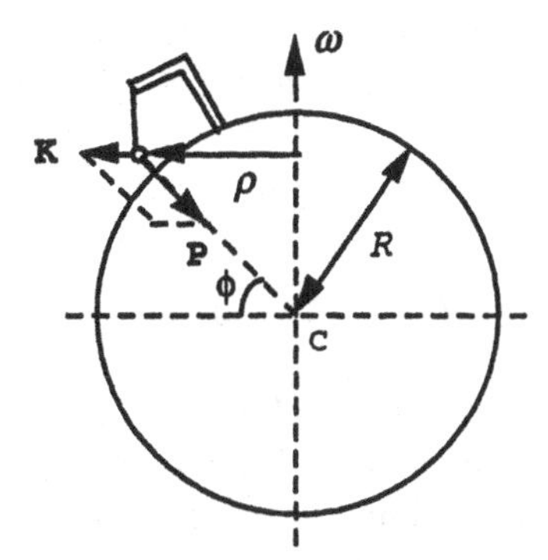

Fig. 6.9. The forces on a particle at rest relative to the Earth as seen in a frame fixed relative to the Earth

particle will be acted on by a centrifugal force:

$$\mathbf{K} = m\omega^2 \boldsymbol{\rho}, \tag{6.46}$$

where $\boldsymbol{\rho}$ is the vector shown on the figure. The particle is also acted upon by a string force $\mathbf{S}$. Since the particle is at rest, we must have:

$$\mathbf{P} + \mathbf{S} + \mathbf{K} = 0. \tag{6.47}$$

This tells us that the string will not be pointing exactly towards the center of the Earth. When we measure the acceleration due to gravity on the surface of the Earth, e.g., by a pendulum experiment ($T = 2\pi\sqrt{L/g}$), it is the acceleration produced by the sum of mass attraction and the centrifugal force that we measure.

Let us estimate the order of magnitude of these effects. We are assuming that the particle is at latitude ϕ. We then have:

$$\rho = |\boldsymbol{\rho}| = R\cos\phi\,. \tag{6.48}$$

Along the direction of the radius vector from the center of the Earth, the centrifugal force has a component of magnitude:

$$K_{||} = mR\omega^2\cos^2\phi\,. \tag{6.49}$$

This component will reduce the local value of the acceleration of a mass point towards the center of the Earth.

The component of the centrifugal force along the direction perpendicular to the radius vector is:

$$K_{\perp} = mR\omega^2\cos\phi\sin\phi\,. \tag{6.50}$$

This component will change the direction of the vertical, defined to be the direction of a string in which a heavy object is hanging at rest. Water at rest on the Earth will form a surface perpendicular to the vertical.

The component $K_{||}$ as given by (6.49) gives a contribution to the change with latitude of the acceleration due to "gravity". Let us estimate the order of magnitude of these effects:

$$\omega = 7.3 \times 10^{-5}\ \mathrm{s}^{-1}\,,$$

and

$$R = 6.37 \times 10^{3}\ \mathrm{km}.$$

We find:

$$R\omega^2 = 3.4\ \mathrm{cm\,s}^{-2}\,.$$

The calculation shows a difference of 3.4 $\mathrm{cm\,s}^{-2}$ between the acceleration of gravity at equator ($\phi = 0$) and at the poles ($\phi = \pi/2$). From experiments it is found that

$$\begin{aligned} g_{\mathrm{poles}} &= 983.2\ \mathrm{cm\,s}^{-2} \\ g_{\mathrm{equator}} &= 978.0\ \mathrm{cm\,s}^{-2} \end{aligned}$$

Note. Using the model of the Earth as a perfect sphere, we have calculated a difference in the acceleration of gravity between the equator and the poles having the magnitude 3.4 $\mathrm{cm\,s}^{-2}$. The observed difference is: $\Delta g = g_{\mathrm{p}} - g_{\mathrm{eq}} = 5.2\ \mathrm{cm\,s}^{-2}$. This deviation is due to the fact that the Earth is not a perfect sphere. The diameter from pole to pole is about 0.3 % less than the diameter through equator. This bulging at the equator is produced by the centrifugal force acting in the reference frame in which the Earth is at rest. The centrifugal force drives the mass of the Earth away from the axis of rotation. Such a concrete effect as the bulging of the Earth then, is produced by the "fictitious" centrifugal force. We shall later use the bulging of the Earth to confront two fundamentally opposed viewpoints in the theory of motion:

Isaac Newton: *The Earth bulges because it rotates in absolute space.*
Ernst Mach: *The Earth bulges because it rotates relative to the fixed stars.*
(fixed stars = remote matter in the universe)

These profoundly different viewpoints will be discussed in Chapter 7.

Note. If the Earth was liquid, the surface would everywhere be perpendicular to the plumb line. The Earth has (perhaps) once been partially liquified. It is a fact that the Earth is not spherical. We shall here not go into the interesting problem of the exact shape of the Earth. We shall just remark that a nonrigid spherical body set in rotation relative to the heliocentric system, will deviate from a spherical shape. Tensions will arise within the sphere. Its final shape will result from a balance between gravitation, tension and the centrifugal force, in the rotating frame.

Why is the heliocentric frame so fundamental that a planetary body rotating relative to this reference frame changes its shape? This is the most profound question in all theories of motion. We shall often return to it. △

6.5 Tidal Fields

In this section we shall consider a phenomenon which must be included in any discussion of the theory of motion: the tidal forces. As the name suggests, the well known phenomenon of the tides in the oceans has its origin in these force fields.

Tidal fields arise in reference frames (planets) that fall freely in an inhomogeneous gravitational field. In a reference frame that falls freely in a homogeneous gravitational field no tidal forces appear. The tidal forces are – as we shall see – what remains of gravitation when a reference frame falls freely in an inhomogeneous gravitational field.

The tidal forces are thus of crucial importance in the theory of gravitation, and they are with some right called "the true gravitational fields". Consider Figure 6.10. The motion of the Earth around the Sun can be perceived as composed of:

(1) the motion of the center of the Earth around the Sun in the course of a year, represented by ω_1;
(2) the rotation of the Earth around its axis once every 23 hours 56 minutes, represented by ω_2.

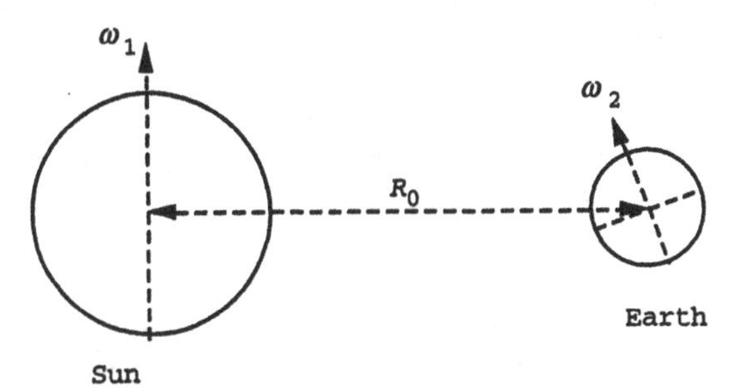

Fig. 6.10. Motion of the Earth around the Sun. The vectors 1 and 2 are not drawn to proportion

Since 1 year $\approx 3.16 \times 10^7$ s, we have:

$$\omega_1 = |\omega_1| = \frac{2\pi}{3.16 \times 10^7 \text{ s}} \approx 2.0 \times 10^{-7} \text{ s}^{-1} ,$$
$$\omega_2 = |\omega_2| = \frac{2\pi}{23 \text{ h } 56 \text{ min}} \approx 7.3 \times 10^{-5} \text{ s}^{-1} .$$

We shall now – by means of the principle of equivalence – consider the phenomenon caused by the yearly motion of the Earth around the Sun, described by ω_1. In other words we shall investigate the deviations from inertial motion relative to the Earth due to the fact that the Earth has a translational acceleration around the Sun.

Let us introduce the so-called *absolute geocentric reference frame*, i.e., a coordinate frame with its origin at the center of the Earth and axes that do not rotate relative to the heliocentric frame, or, to put it another way, the axes of the absolute geocentric reference frame always point towards the same three fixed stars (Note: most fixed stars are so far away that we cannot detect any changes in the directions of the axis over a year).

The Earth is in rotation relative to the absolute geocentric frame. The Earth is in free fall in the gravitational field of the Sun, with an acceleration **a**. This is the acceleration which the Sun would impose on a particle placed in the center of the Earth. All points in the absolute geocentric coordinate frame thus have an acceleration **a** towards the center of the Sun.

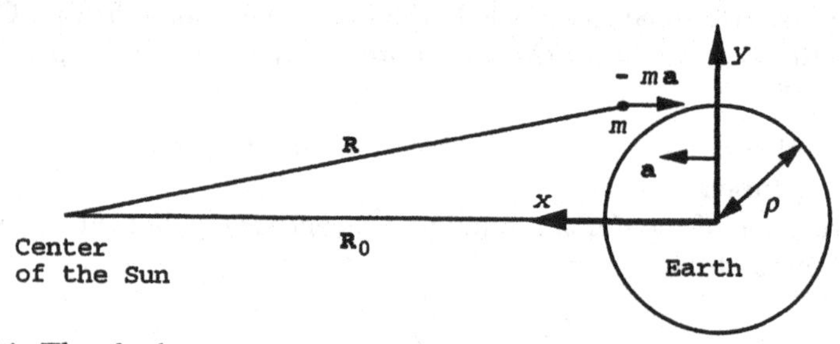

Fig. 6.11. The absolute geocentric coordinate system. The x-axis is chosen to be directed towards the Sun, at the time considered

Consider the motion of a particle within the absolute geocentric reference frame. Let the particle have mass m. Our reference frame has an acceleration **a**, so according to the equivalence principle we must add a fictitious force $m(-\mathbf{a})$ (see Figure 6.11). The particle is also acted on by gravitational forces from the Sun and from the Earth (we disregard the Moon in this discussion). We shall neglect the effects from the gravitation of the Earth, since we wish to focus our attention on the deviations from inertial motion caused by the free fall of the Earth in the gravitational field of the Sun.

If the particle in question was at the center of the Earth, the gravitational force from the Sun would be exactly canceled by the fictitious force (a particle at the center of the Earth is weightless in the absolute geocentric reference frame, see Example 3.1). If the particle is not in the center, the gravitational force from the Sun will be different in magnitude from the magnitude of the fictitious force $m(-\mathbf{a})$. In this case there is not complete cancellation of the two forces: gravitation from the Sun and the fictitious force. It is the difference between these two forces that causes the tidal field.

We shall now go through a few calculations to estimate the order of magnitude of these effects. The radius vector from the center of the Sun to the center of the Earth is called $\mathbf{R}_0$, and the radius vector from the center of the Sun to the particle is called $\mathbf{R}$ (see Figure 6.11). We have:

$$\mathbf{a} = -\frac{GM}{R_0^2}\frac{\mathbf{R}_0}{|\mathbf{R}_0|}, \tag{6.51}$$

where M is the Mass of the Sun.

The gravitational force from the Sun will cause an acceleration of the particle by the amount

$$\mathbf{g}_s = -\frac{GM}{R^2}\frac{\mathbf{R}}{R}.$$

The total force on the particle as seen from the geocentric system is the sum of the force of gravity and the fictitious force:

$$\mathbf{F} = -m\mathbf{a} + m\mathbf{g}_s,$$

or

$$\mathbf{F} = GmM\left(\frac{\mathbf{R}_0}{R_0^3} - \frac{\mathbf{R}}{R^3}\right). \tag{6.52}$$

The expression in (6.52) is called the tidal force field from the Sun; it is, along with a similar field from the Moon, the cause of the tides in the oceans. The corresponding acceleration is

$$\mathbf{b} = GM\left(\frac{\mathbf{R}_0}{R_0^3} - \frac{\mathbf{R}}{R^3}\right).$$

To simplify the computations we will consider only the points on the surface of the Earth that are on a straight line between the center of the Earth and the center of the Sun (see Figure 6.12).

The x-axis has its origin at the center of the Earth and a positive direction towards the center of the Sun. For points on this axis:

$$\frac{\mathbf{R}}{R} = \frac{\mathbf{R}_0}{R_0}. \tag{6.53}$$

The acceleration field is then:

$$\mathbf{b} = GM\frac{\mathbf{R}_0}{R_0}\left(\frac{1}{R_0^2} - \frac{1}{R^2}\right) = -\mathbf{a}\left(1 - \frac{R_0^2}{R^2}\right).$$

Note: $\mathbf{a}$ points towards the Sun, $-\mathbf{a}$ away from the Sun. For $R > R_0$ the field is directed away from the Sun; for $R < R_0$ it is directed towards the Sun.

Since $R = R_0 - x$ we can write:

$$\mathbf{b} = -\mathbf{a}\left(1 - \frac{R_0^2}{(R_0 - x)^2}\right) = -\mathbf{a}\left(1 - \frac{1}{(1 - x/R_0)^2}\right).$$

For points near the surface of the Earth, $x \ll R_0$. By expanding the expression in the parenthesis (in the small quantity x/R_0), we get:

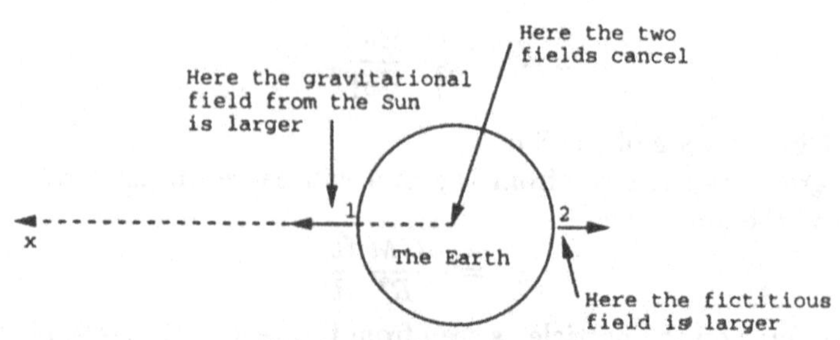

Fig. 6.12. The tidal field

$$\mathbf{b} \approx -\mathbf{a}\left[1-\left(1+2\frac{x}{R_0}\right)\right], \tag{6.54}$$

$$\mathbf{b} \approx \mathbf{a}\left(2\frac{x}{R_0}\right). \tag{6.55}$$

For $x > 0$, the vector **b** is directed towards the Sun, for $x < 0$ the vector **b** is directed away from the Sun (see Figure 6.12).

Let us denote the radius of the Earth by ρ. On the side of the Earth facing the Sun, the tidal field at the surface has the magnitude

$$b = \left(\frac{2a\rho}{R_0}\right), \tag{6.56}$$

and is directed towards the Sun. On the side facing away from the Sun, the tidal field has the same magnitude directed away from the Sun. The net effect is to introduce a distortion of the ocean levels as indicated (but grossly exaggerated) in Figure 6.13.

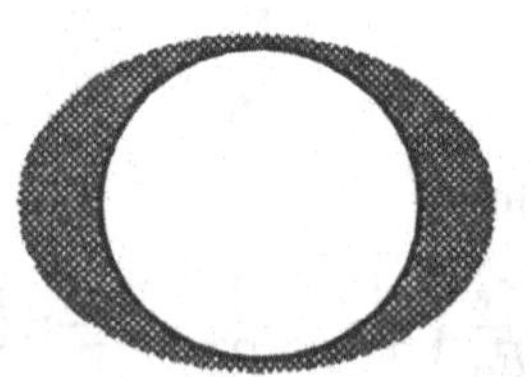

Fig. 6.13. The tidal distortion of the ocean surface levels

Since the Earth rotates with respect to the absolute geocentric reference frame, we have flood in the oceans about twice a day. It was a great triumph for Newtonian mechanics that it provided an explanation for the well known fact that flood and ebb appear twice each day.

At the center of the Earth the parameter b would be zero (compare (6.55)).

The magnitude of **a** is given by $a = R_0\omega_1^2$. Inserting values for R_0 and ω_1, we get

$$b \approx 5.1 \times 10^{-7}\ \mathrm{m\,s^{-2}}.$$

By using similar considerations on the Earth–Moon system, one finds that the magnitude of tidal acceleration from the Moon is about $2.5 \times b$. The tidal effect from the Moon is thus more than twice as large as that from the Sun. This is due to that fact that the effect varies with the distance R as $1/R^3$. Recall that

$$b = \frac{2a\rho}{R_0} = \frac{2GM\rho}{R_0^3} .$$

Since the distance of the Moon from the Earth is much less than the distance of the Sun from the Earth, the contribution from the Moon (which has an almost vanishing mass compared with the Sun) is nevertheless the greater of the two.

Note. The magnitude of the gravitational field from the Sun at the center of the Earth is

$$F_S = \frac{GM}{R_0^2} . \tag{6.57}$$

The magnitude of the derivative of F_S with respect to distance is

$$\left| \frac{dF_S}{dR_0} \right| = \frac{2GM}{R_0^3} .$$

The magnitude of the tidal field at the surface of the Earth is

$$b = \left| \frac{dF_S}{dR_0} \right| \rho = \frac{2GM\rho}{R_0^3} , \tag{6.58}$$

as we found above.

It is interesting that the gravitational field from the Sun near the Earth is 175 times as strong as the gravitational field from the Moon. The tidal force, however, is an expression of the fact that the tidal field – and the corresponding acceleration – is due to the variation with distance of the gravitational field from the Sun (and the Moon).

It is consequently the Moon that is the primary source of ebbs and floods in the oceans. If the Earth was not rotating, there would permanently be a small flood in the regions facing toward and away from the Moon, and a small ebb in the plane perpendicular to the direction to the Moon. Since the Earth is rotating, the tidal forces drive the ocean basins with a frequency of about $12\frac{1}{2}$ hour, so that this is the typical interval between floods. The extra half hour is due to the motion of the Moon around the Earth.

We shall not go into the geophysical aspects of tides here, but just notice that the amplitude of tides vary from less than one meter in the open ocean regions, to several meters in coastal regions, the most prominent of which is the Bay of Fundy in Canada. Here, the maximum difference of water level between ebb and flood is 15.4 m.

The theoretical significance of tidal forces is the following. In a laboratory on the surface of the Earth we make a very small error if we simultaneously disregard

(1) the gravitational fields of the Sun and the Moon; and
(2) the translational acceleration of the Earth.

If the Earth was not rotating relative to the heliocentric reference frame, a coordinate frame affixed to the Earth would nearly be an inertial frame. One can say that such a frame is inertial apart from the tidal fields. But we have also seen that the *tidal fields*, which ultimately are caused by the variations in the gravitational fields *can never be transformed away by letting the coordinate system fall freely*. The tidal forces are "the true gravitational forces". The "homogeneous" part of an arbitrary gravitational field can always be transformed away by going to a reference frame that falls freely in the local gravitational field.

We shall return to the relation between tidal forces and inertial frames in Section 6.7.

Example 6.7. The Roche Limit

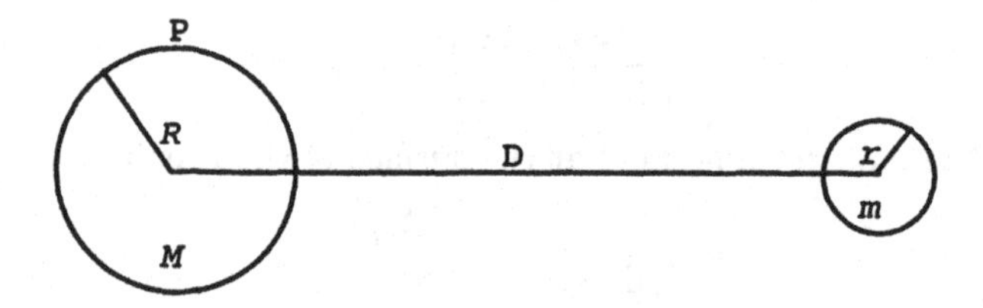

Consider the figure. P is a planet with mass M. It has a moon of mass m, and this moon is assumed to be held together by its own field of gravity. The moon is in free fall around the planet, e.g. in a circular orbit of radius D. Will the tidal force tear the moon apart?

The radius of the moon is r, and the radius of the planet is R. A crude estimate of the distance D_0 which the moon must have for the moon to be held together under its own gravity can be made as follows.

The gravitational force from the moon on a particle of mass μ (e.g., a rock on the surface) lying in the equatorial plane, is $Gm\mu/r^2$. The tidal force on the particle is $2GM\mu r/D_0^3$. The moon can just hold on to the rock when these forces balance:

$$2\frac{GM\mu}{D_0^3}r = \frac{Gm\mu}{r^2} . \tag{6.59}$$

If the orbital radius of the moon is less than D_0, the rock on the surface will be lifted away by the tidal forces. Let us assume that the average density of the planet is ρ_P and the average density of the moon is ρ_M. This gives:

$$M = \frac{4\pi}{3}R^3\rho_P , \tag{6.60}$$

$$m = \frac{4\pi}{3} r^3 \rho_M \, . \tag{6.61}$$

We then find, by substituting (6.60) and (6.61) into (6.59) that the critical distance can be expressed as:

$$D_0 = R \left(2 \frac{\rho_P}{\rho_M} \right)^{1/3} = 1.26 \cdot R \left(\frac{\rho_P}{\rho_M} \right)^{1/3} . \tag{6.62}$$

The orbital radius D_0 is known as the Roche limit. The calculation of the Roche limit D_0 is, when one takes into account the deformation of the participating bodies, rather complicated. One finds:

$$D_0 = 2.46 \; R \left(\frac{\rho_P}{\rho_M} \right)^{1/3}$$

which is (6.62) except for the numerical factor.

For the system of Earth–Moon, $D_0 = 2.9 R_E$, where R_E is the radius of the Earth. If the Moon came closer than $2.9 R_E$, rocks would be lifted away from the lunar surface. Now look at the picture below.

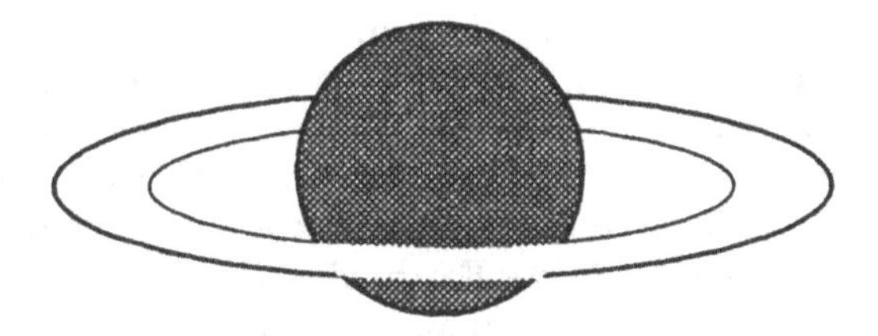

Most of Saturn's rings are inside the Roche limit for Saturn! Note how in this case the so-called fictitious forces are preventing the rocks in the rings of Saturn from assembling into a moon.

The rather dramatic statement that a body inside the Roche limit will be "torn apart" by the tidal forces is true only if the body is held together by its own field of gravity. Gravity does play an important role for example in the binding energy of the Moon, as can be seen directly from the fact that the Moon is spherical. A rock that you pick up from the ground is nowhere near spherical; its binding energy is determined by chemical bonds. Gravity plays no role in keeping cobblestones together.

Many of the satellites orbiting the Earth are inside the Roche limit of the Earth. These satellites are not torn apart by tidal forces, again because they are held together by chemical bonds, not gravity.

In fact, *you* yourself are inside the Roche limit of the Earth, but you are also held together by chemical bonds (essentially electrical interactions). When gravity is dominant in the binding energy of a given body, this body will have the shape of a sphere. △

6.6 The Coriolis Force

We shall consider the Coriolis force acting on a particle that moves along the surface of the Earth.

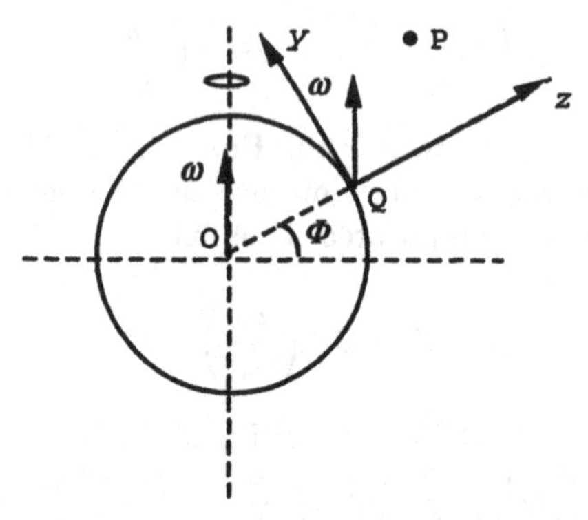

Fig. 6.14. A reference frame fixed on the surface of Earth

The natural reference frame here is a frame that is fixed relative to the surface of the Earth and which therefore is in a state of rotation relative to the heliocentric reference frame. We shall use a frame with origin on the surface of Earth (in stead of, say, the center of the Earth). Such a frame is commonly called the laboratory frame, and it is the frame from which you, standing still on the surface of the Earth, view the phenomena around you.

Look at Figure 6.14. The lab frame is oriented here in such a way that the y-axis points towards the north and the x axis towards the east; the z axis therefore points vertically upwards.

We briefly demonstrate that $\boldsymbol{\omega}$ describes the rotation vector relative to the heliocentric frame of any coordinate frame fixed relative to the Earth, no matter where the origin is located.

Consider the point P which is at rest relative to both a coordinate frame with its origin at the point O and the frame with its origin at the point Q. The "absolute" velocity of P, i.e., the velocity relative to the heliocentric frame, is:

$$\begin{aligned} \mathbf{v}_{\mathrm{P}} &= \mathbf{v}_{\mathrm{O}} + \boldsymbol{\omega} \times \dot{\mathbf{OP}} \\ &= \mathbf{v}_{\mathrm{O}} + \boldsymbol{\omega} \times (\mathbf{OQ} + \mathbf{QP}) \\ &= \mathbf{v}_{\mathrm{O}} + \boldsymbol{\omega} \times \mathbf{OQ} + \boldsymbol{\omega} \times \mathbf{QP} \\ &= \mathbf{v}_{\mathrm{Q}} + \boldsymbol{\omega} \times \mathbf{QP} . \end{aligned}$$

All coördinate frames at rest relative to a given frame share the same rotation vector.

Recall Equation (6.36). The Coriolis force on a particle with mass m, moving with velocity $\mathbf{v}$ relative to the surface of the Earth, is

$$\mathbf{F} = 2m(\mathbf{v} \times \boldsymbol{\omega}), \tag{6.63}$$

where $\boldsymbol{\omega}$ means the rotation vector relative to the heliocentric reference frame (the fixed stars).

Let us first consider a particle that has a horizontal velocity, and is located at latitude Φ. From Figure 6.14 we find that the rotation vector of the Earth has components

$$\boldsymbol{\omega} = (0,\ \omega \cos \Phi,\ \omega \sin \Phi),$$

in the lab frame.

The vector $\mathbf{F}$ has a horizontal component F_{H} and a vertical component F_{V}. The vertical component of $\boldsymbol{\omega}$ gives rise to the horizontal component of the force:

$$F_{\mathrm{H}} = 2mv\omega \sin \Phi,$$

and the horizontal component of $\boldsymbol{\omega}$ gives rise to the vertical component of the force:

$$F_{\mathrm{V}} = 2mv_x\omega \cos \Phi.$$

In geophysical considerations it is often the horizontal component of the Coriolis force that is of interest. This force influences wind and ocean currents. Since in the Northern hemisphere the vector $\boldsymbol{\omega}$ points out of the the Earth, we have the general rule that:

> In the Northern hemisphere where $\boldsymbol{\omega}$ points out of the ground, a particle moving with a horizontal velocity will be drawn to the right

This holds whether the particle in question is moving towards the north, south, east, or west. Similarly:

> In the Southern hemisphere, where $\boldsymbol{\omega}$ points into the ground, a particle moving with a horizontal velocity will be drawn to the left.

These rules are equivalent to counting Φ as positive on the northern hemisphere and negative on the southern hemisphere.

Let us now write the general expression for the force on a particle having an arbitrary velocity (relative to the lab frame):

$$\mathbf{v} \equiv \dot{\mathbf{r}} = (\dot{x}, \dot{y}, \dot{z}).$$

The components of $\mathbf{F}$ are then:

$$F_x = 2m\omega(-\dot{z}\cos\Phi + \dot{y}\sin\Phi), \tag{6.64}$$
$$F_y = -2m\omega\dot{x}\sin\Phi, \tag{6.65}$$
$$F_z = 2m\omega\dot{x}\cos\Phi. \tag{6.66}$$

Consider the following situation. A long vertical pipe is fastened to the ground in the northern hemisphere, and a particle can slide down the pipe with no friction. If we release the particle from the top of the pipe, it will slide down to the ground, its vertical motion being free fall.

The Coriolis force, however, will cause the particle to push against the side of the pipe. Since no friction forces act between the particle and the pipe, this will not slow the particle down, and the reaction force from the pipe will ensure that the velocity is vertical throughout the fall, i.e., $\dot{x} = \dot{y} = 0$. From (6.64), (6.65), and (6.66) we find that the Coriolis force then has only the x-component

$$F_x = -2m\omega\dot{z}\cos\Phi.$$

Note: $\dot{z}$ is negative. The particle is pushed by the Coriolis force in the positive x-direction, towards the east.

Let us estimate the magnitude of this force; suppose the pipe is 100 m high, and disregard air resistance.

We assume that the mass of the particle is 0.1 kg and that $\Phi = 45°$. The Coriolis force on the particle the instant it reaches the bottom of the pipe is then

$$F_x = 2m\omega\sqrt{2gh}\frac{1}{\sqrt{2}} = 4.6 \times 10^{-4}\ \mathrm{N}.$$

Compare this to the gravitational force on the particle from the Earth:

$$F_g = mg = 9.8 \times 10^{-1}\ \mathrm{N}.$$

It is now clear that if we remove the pipe, the particle will not fall along the vertical. To the lowest approximation it will receive a velocity component toward the east. In turn, this velocity component will give rise to a Coriolis force toward the south, but this effect is proportional to ω^2 and is typically much smaller than F_x.

Let us write the complete set of equations of motion for a particle in free fall near the surface of the Earth. The equations of motion are in cartesian coordinates relative to the lab frame. One finds:

$$m\ddot{x} = 2m\omega(-\dot{z}\cos\Phi + \dot{y}\sin\Phi), \tag{6.67}$$
$$m\ddot{y} = -2m\omega\dot{x}\sin\Phi, \tag{6.68}$$
$$m\ddot{z} = 2m\omega\dot{x}\cos\Phi - mg. \tag{6.69}$$

(The acceleration due to gravity is assumed to be constant throughout the fall.)

This system of coupled differential equations replaces the much simpler system which would hold if there was no rotation of the Earth relative to the heliocentric frame (i.e., $\omega = 0$):

$$\begin{aligned} m\ddot{x} &= 0\,, \\ m\ddot{y} &= 0\,, \\ m\ddot{z} &= -mg\,. \end{aligned}$$

We shall now integrate the coupled system (6.67)–(6.69) to find an expression for the motion of a particle near the surface of the Earth. Start by differentiating (6.67) with respect to time, then substitute (6.68) for $\ddot{y}$ and (6.69) for $\ddot{z}$. This gives us

$$\frac{\mathrm{d}^3 x}{\mathrm{d}t^3} + 4\omega^2 \frac{\mathrm{d}x}{\mathrm{d}t} = 2\omega g \cos\Phi\,. \tag{6.70}$$

This equation can be regarded as a second-order differential equation in $\dot{x}$:

$$\frac{\mathrm{d}^2 \dot{x}}{\mathrm{d}t^2} + 4\omega^2 \dot{x} = 2\omega g \cos\Phi\,. \tag{6.71}$$

The complete solution to (6.71) is, as we have seen,

$$\dot{x} = \frac{g\cos\Phi}{2\omega} + A\cos[2\omega(t - t_0)]\,, \tag{6.72}$$

where A and t_0 are constants of integration. Compare this with Example 2.3, where we solved a differential equation of the same form as (6.71). Show that (6.72) is a solution by substituting (6.72) into (6.71).

Now we are in a position to solve the coupled system completely. We integrate (6.72) to find $x(t)$. In turn, this solution can be used to find $y(t)$ and $z(t)$ from (6.68) and (6.69).

We shall not develop the most general solution; instead we shall consider some special cases of interest. Suppose the particle falls from height $z = h_0$. As initial conditions we then have:

$$\begin{aligned} (x, y, z) &= (0, 0, h_0)\,, \\ (\dot{x}, \dot{y}, \dot{z}) &= (0, 0, 0)\,. \end{aligned}$$

The constants of integration in (6.72) are then: $A = -g\cos\Phi/2\omega$ and $t_0 = 0$. Thus:

$$\dot{x} = \frac{g\cos\Phi}{2\omega}(1 - \cos 2\omega t)\,. \tag{6.73}$$

Now we use the fact that we typically are considering falls of a duration t such that $t \ll 1/\omega$. Since $\omega = 7.3 \times 10^{-5}\ \mathrm{s}^{-1}$, we have that $1/\omega = 1.3 \times 10^4$ s ≈ 3 hours. This leads us to use the small argument expansion of $\cos\Phi$:

$$\cos 2\omega t \approx 1 - \frac{1}{2}(2\omega t)^2 .$$

Then:

$$\dot{x} \approx \frac{g\cos\Phi}{2\omega}\left[1 - \left(1 - \frac{4\omega^2 t^2}{2}\right)\right],$$

or

$$\dot{x} \approx g\omega(\cos\Phi)t^2 . \tag{6.74}$$

We now integrate (6.74) again with the initial condition $x = 0$ at $t = 0$, resulting in

$$x \approx \frac{1}{3}g\omega(\cos\Phi)t^3 . \tag{6.75}$$

The result (6.75) shows how a particle falling mainly along the vertical will deviate, mainly toward the east (see Figure 6.14). That direction can of course be inferred directly from $\mathbf{F} = 2m(\mathbf{v} \times \boldsymbol{\omega})$.

From (6.68) and (6.74) we find:

$$\ddot{y} \approx -2g\omega^2(\cos\Phi\sin\Phi)t^2 .$$

By integrating twice and using the initial conditions $\dot{y} = 0$ and $y = 0$ for $t = 0$, we get

$$y \approx -\frac{1}{6}g\omega^2(\cos\Phi\sin\Phi)t^4 . \tag{6.76}$$

This shows that there is also a second-order (notice the factor ω^2!) effect resulting in a small deviation toward the south.

To illustrate the order of magnitude of these effects, let us take as an example a particle falling at 45° north. Suppose the fall lasts 10 s, and disregard air resistance. The total height through which the particle falls is $z \approx 1/2gt^2 \approx 500$ m.

From (6.75): $x \approx 0.17$ m. From (6.76): $y \approx -4.4 \times 10^{-5}$ m. Thus, the y-deflection is about 4 000 times smaller than the x-deflection! From (6.75) (and $h = (1/2)gt^2$) we thus conclude that in general, a particle in free fall through the height h will be deflected a distance

$$d \approx \frac{1}{3}\omega\cos\Phi\sqrt{\frac{8h^3}{g}}$$

toward the east, by the Coriolis force.

That the deflection of the particle towards the south is much smaller than the deflection towards the east is directly evident in the differential equations (6.67) and (6.68). The acceleration towards the south is proportional to $\dot{x}$ while the acceleration towards the east contains the term $\dot{z}$, which is by far the largest component of velocity.

The eastward deviation has been seen in experiments where balls have been dropped from towers and into mine shafts. Such experiments provide direct evidence for the rotation of the Earth relative to the heliocentric frame.

Example 6.8. Coriolis Force on a Train. A train with mass $M = 10^5$ kg is driving towards the north with a velocity of 108 km/hour. The track is straight and is located around $F = 60°$ northern latitude.

(1) Find the Coriolis force on the train.
(2) What would the Coriolis force be if the train was going towards the east?

Answer.

(1) $\mathbf{F} = (2Mv\omega \sin\Phi, 0, 0)$. $F_x = 379$ N (the "weight" of the train, i.e., the gravitational force on the train, is 9.8×10^5 N).
(2) $\mathbf{F} = (0, -2Mv\omega \sin\Phi, 2Mv\omega\cos\Phi)$, $F_y = -379$ N, $F_z = 219$ N.

△

Example 6.9. Particle on a Frictionless Disc

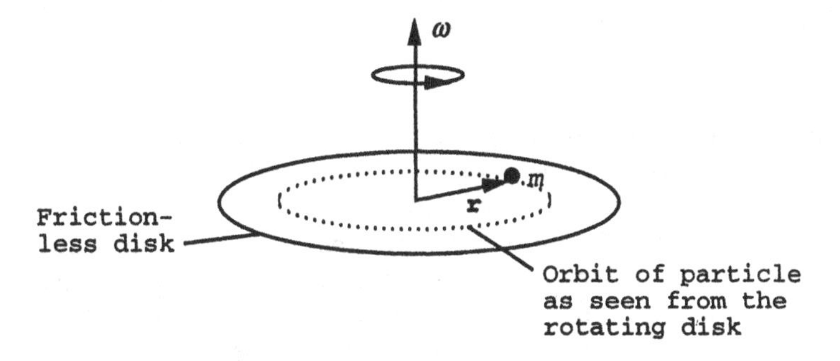

A horizontal disk is rotating about a vertical axis with the angular velocity ω. Let us study the fictitious forces assigned to a particle that is moving relative to the rotating disk. We put the particle carefully down at some distance r from the axis. If we assume that the disk is perfectly smooth such that no forces of friction act, the particle will remain at rest as seen in the lab frame. As viewed from a frame fixed on the disk however, the particle will perform a uniform circular motion. How is this motion explained in terms of the fictitious forces?

As seen from the disk, two forces must be taken into account: the Coriolis force and the centrifugal force. The centrifugal force is directed away from the center along a radius vector and has the magnitude

$$K_c = m\omega^2 r\,. \tag{6.77}$$

The Coriolis force has magnitude

$$|\mathbf{K}_C| = |2m(\mathbf{v}_{\text{rel}} \times \boldsymbol{\omega})| = 2m\omega^2 r\,. \tag{6.78}$$

Here, $\mathbf{v}_{\text{rel}}$ denotes the velocity of the particle relative to the disk. It is perpendicular to $\boldsymbol{\omega}$, and we have used $v = r\omega$.

The Coriolis force and the centrifugal force are oppositely directed (check the direction of the Coriolis force by using the right-hand rule), so the resulting force is directed toward the center of the disk and has the magnitude

$$K_{\text{total}} = m\omega^2 r = m\frac{v_{\text{rel}}^2}{r}\,. \tag{6.79}$$

This is exactly the force necessary to sustain uniform circular motion. One could say that the particle performs uniform circular motion relative to the disk "because" is has the relative velocity $v = \omega r$ as soon as it is placed on the frictionless rotating disk.

Question. A particle placed on a real disk with friction will not move in a circle but will spiral away from the axis. Why ?

Answer. The Coriolis force can no longer keep the particle in a circular orbit since, with friction dragging the particle, $v_{\text{rel}} < \omega r$. Since the centrifugal force is independent of velocity, it now "wins" and drives the particle outwards.

$\triangle$

Example 6.10. The Vertical Throw. At the time of Galileo an issue was much disputed: will the velocity of the Earth have any influence on the motion of a cannon ball shot vertically upwards? We now know that a uniform translational motion of the Earth relative to the heliocentric frame – no matter how large that velocity – would have no influence whatsoever on the trajectory of the cannon ball as seen from a reference frame fixed on the surface of the Earth. If the Earth had only a uniform velocity, the cannon ball would fall directly back to the place from which it was fired. But is this the case? No, for the Earth does *not* move in uniform translation, it is in rotation. A frame fixed on the surface of the Earth is not inertial. Let us work out the problem of the cannon ball.

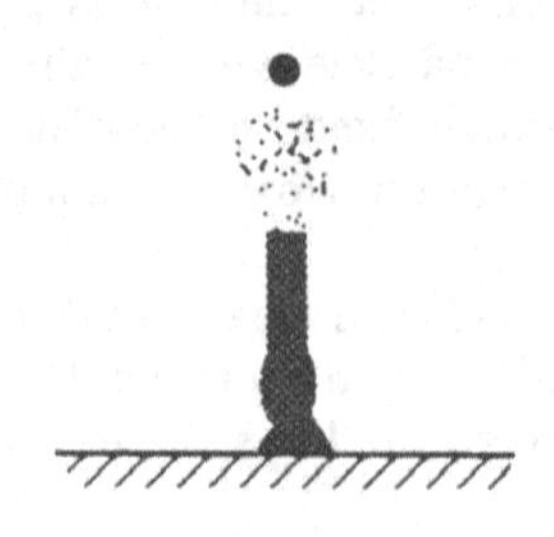

A particle is fired vertically upward from a point on the surface of the Earth, at $\Phi°$ northern latitude. Let us ignore resistance from the air, and consider only small vertical extensions, $h \ll \rho$ where ρ is the radius of the Earth.

We will show that the particle touches down at a point west of the launch point. We use the differential equations (6.67), (6.68), and (6.69), and ignore the quantities $\dot{x}$ and $\dot{y}$ in the equations. We then have, approximately,

$$\begin{aligned} \ddot{x} &\approx -2\omega\dot{z}\cos\Phi\,, \\ \ddot{z} &\approx -g\,. \end{aligned}$$

Let the time when the ball is launched be $t = 0$ and the initial velocity be v_0. Then

$$\dot{z} \approx v_0 - gt\,.$$

Now substitute this into the expression for $\ddot{x}$:

$$\ddot{x} \approx -2\omega(v_0 - gt)\cos\Phi\,.$$

Setting $\dot{x} = 0$ for $t = 0$ gives us

$$\dot{x} = -2\omega v_0(\cos\Phi)t + \omega g(\cos\Phi)t^2\,.$$

Finally, with $x = 0$ for $t = 0$, we find

$$x = \frac{1}{3}g\omega(\cos\Phi)t^3 - \omega(\cos\Phi)v_0t^2\,.$$

From the expression for $\dot{z}$ we find that the amount of time from when the ball leaves the cannon until it returns is $t = 2v_0/g = 2\sqrt{2h/g}$ where h is the maximal height the ball reaches (i.e., $1/2mv_0^2 = mgh$). Inserting this value for t in the above expression for $x(t)$ we get the x-deviation when the ball touches down:

$$d = -\frac{4}{3}\omega\cos\Phi\sqrt{\frac{8h^3}{g}}\,.$$

From the sign one can see that ball will touch down west of the launch point $x = 0$. △

6.7 Tidal Forces and Local Inertial Frames

In the Newtonian interpretation of classical mechanics (see Chapter 7) an inertial frame is either absolute space or a reference frame moving with constant velocity along a straight line relative to absolute space ("the astronomical space"). While Newton was of the opinion that "the astronomical inertial frame" forms the background on which mechanical phenomena take place,

Einstein introduced – for use in general relativity – the so-called *local* inertial frames.

Let us first describe an example of a local inertial frame. If the Earth was not rotating, the laboratory would be close to an inertial frame. The Earth falls freely in the local gravitational field, which is essentially a superposition of the gravitational field of the Sun and the Moon. Because of the free fall we feel only the tidal forces produced by the gravity of the Sun and the Moon.

> A reference frame, falling freely in the local gravitational field, and not rotating relative to the distant fixed stars is, apart from the tidal fields, an inertial frame

This definition actually rests upon the principle of equivalence. Since no gravitational field is completely homogeneous, one can say that the principle of equivalence is only "valid in a point". A freely falling, nonrotating coordinate frame is thus only an inertial frame in an infinitesimal neighborhood of the origin; i.e., a local inertial frame. One can say that the tidal fields reveal the presence of the real gravitational fields. The tidal fields cannot be transformed away by any choice of reference frame.

The reasons the heliocentric reference frame can be considered as a very good approximation to an inertial frame, are the following:

(1) it is in free fall toward the center of the galaxy (the Milky Way);
(2) it is not rotating relative to the fixed stars;
(3) the tidal fields within the heliocentric frame due to the gravitational field of the galaxy are so small that we cannot measure them.

The laboratory frame is not an inertial frame because the Earth is rotating relative to the heliocentric frame. For many practical applications in mechanics, the freely falling laboratory frame can be considered an inertial frame. This is true for motions over "short distances and short time intervals", i.e., for motions where we can disregard the Coriolis force. The decisive criterion for a specific coordinate frame to be considered inertial, is that the accuracy to be applied to the given mechanical problem makes the tidal forces undetectable.

The relativity principle of mechanics we formulate as follows. *There exist a threefold infinity of reference frames, all moving uniformly along straight lines relative to each other, within which the basic mechanical laws are valid in their simple, classical form. Such reference frames are called inertial frames.*

The basic mechanical laws are independent of the choice of reference frame, as long as the reference frame is chosen among the inertial frames.

In this formulation of the relativity principle of mechanics one very much needs to be given an example of just one inertial frame. If you have one, you have them all.

It is important to note the following. The relativity principle of mechanics, as it is discussed in Chapter 4, rests on Newton's laws. We shall see that while

Newtonian mechanics has not survived entirely, the relativity principle has – in an expanded form – become one of the pillars on which modern physics rests. The relativity principle is thus of a more fundamental character than the laws of Newtonian mechanics from which it originated.

Example 6.11. Global and Local Inertial Frames. Newton uses the absolute space (astronomical space) as an inertial frame. This is what might be called a global inertial frame, because the frame is inertial everywhere. Most presentations of Newtonian mechanics use the heliocentric frame as a global inertial frame. The heliocentric frame is nevertheless what Einstein calls a local inertial frame. The reason that we can use the heliocentric reference frame as an inertial frame even for the study of the motions of the planets is that the tidal forces that appear in the frame, because of its free fall in the gravitational field of the Milky Way, are so small that we cannot measure them.

It is interesting, that *in order to define an inertial frame it is necessary to refer to the fixed stars.* Apparently the global distribution of matter in the universe is decisive to any attempt to create a logically sound science of mechanics – or, is it ?

Go outside on a dark winter's night when the sky is clear, and look up at the stars. Carry with you a piece of string and tie a stone to the string. You can now make the following – and, to the understanding of mechanical physics, absolutely crucial – observation.

Take hold of one end of the string and release the stone. If the stars and the string seem at rest relative to one another, the string will hang along the vertical. Now set the string in rotation, i.e., give the stone an acceleration relative to the fixed stars. You will now discover that the string no longer is vertical. If you whirl the stone around sufficiently fast, the string can be almost horizontal. In the reference frame where the stone is at rest the fixed stars are rotating. In this reference frame there is a centrifugal force that, combined with the force of gravity and the string force, keeps the stone at rest and the string (almost) horizontal.

Can it be a coincidence that a reference frame, where the fixed stars seem at rest, is an inertial frame? $\triangle$

6.8 The Foucault Pendulum

In the year 1851 the French physicist Foucault carried out an experiment which demonstrated that a coordinate frame fixed on the surface of the Earth is not an inertial frame. Foucault studied the motion of a pendulum consisting of a spherical body of mass 28 kg hanging in a 69 m long metal wire. The experiment was located underneath the huge Pantheon dome in Paris. The suspension point of the pendulum was constructed in such a way that the pendulum could pivot freely in all directions; the period of oscillation was $(T = 2\pi\sqrt{L/g}) \approx 17$ s.

Foucault observed that the plane of oscillation for the pendulum slowly drifted around relative to the walls of the building (the "lab frame"). A full rotation took about 32 hours. Let us set up the equations of motion for the pendulum (see Figure 6.15).

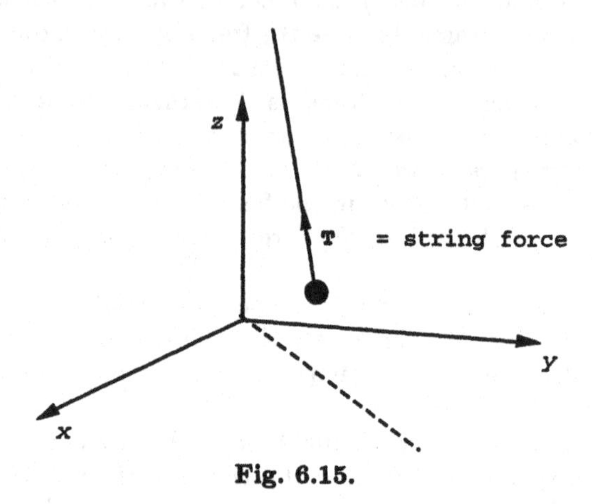

Fig. 6.15.

The pendulum is affected by the Coriolis force, and this is the cause of the precession – slow rotation – of the plane of oscillation. Let us introduce coordinates with the origin in the equilibrium point of the pendulum, the z-axis in the vertical direction, through the point of suspension. Since we are interested in the precession of the plane of oscillation, we project the motion into the xy-plane. Let us limit the motion to oscillations of small angular amplitude ($x/L \ll 1$ and $y/L \ll 1$ where L is the length of the pendulum string). For such a motion, $\dot{z}$ is small compared with $\dot{x}$ and $\dot{y}$, so we disregard $\dot{z}$.

Let us write the equations of motion in the chosen coordinates. Denote the magnitude of the string force by T. Then the components of the string force (recall that $x/L \ll 1$ and $y/L \ll 1$) are (see Figure 6.15):

$$\begin{aligned} T_x &\approx -T\frac{x}{L}, \\ T_y &\approx -T\frac{y}{L}, \\ T_z &\approx mg. \end{aligned}$$

The components of the Earth rotation vector $\boldsymbol{\omega}$ in the lab frame are (with Φ being the latitude of the experiment, see Figure 6.16):

$$\boldsymbol{\omega} = (-\omega \cos \Phi, 0, \omega \sin \Phi)$$

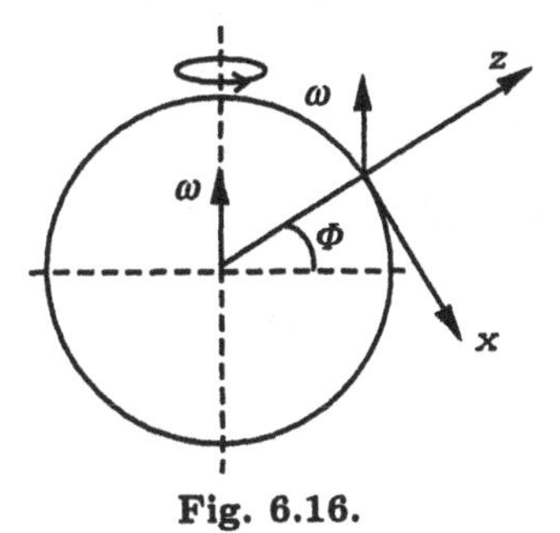

Fig. 6.16.

The Coriolis force is:

$$\mathbf{K} = -2m(\boldsymbol{\omega} \times \dot{\mathbf{r}}) = 2m(\mathbf{v} \times \boldsymbol{\omega})\,.$$

Since

$$\mathbf{v} = (\dot{x}, \dot{y}, \dot{z}) \approx (\dot{x}, \dot{y}, 0)\,,$$

we get

$$\mathbf{v} \times \boldsymbol{\omega} = (\dot{y}\omega \sin \Phi, -\dot{x}\omega \sin \Phi, \dot{y}\omega \cos \Phi)\,,$$

and the x and y equations of motions become:

$$\begin{aligned} m\ddot{x} &= -T\frac{x}{L} + 2m\dot{y}\omega \sin \Phi\,, \\ m\ddot{y} &= -T\frac{y}{L} - 2m\dot{x}\omega \sin \Phi\,. \end{aligned}$$

Assuming $T = mg$,

$$\begin{aligned} \ddot{x} + \frac{g}{L}x &= 2\omega \sin \Phi \dot{y}\,, \\ \ddot{y} + \frac{g}{L}y &= -2\omega \sin \Phi \dot{x}\,. \end{aligned} \tag{6.80}$$

These equations are not trivial to solve. We shall here use some intuitive and physically reasonable assumptions about the solution and then a posteriori verify that they are consistent.

Consider Figure 6.17. If the Coriolis force was not present, the pendulum would oscillate in a plane, with angular frequency (for small oscillations) given by $\alpha = \sqrt{g/L}$. For the angular frequency α we would find that $\alpha \gg \omega$, where ω is the angular frequency of the rotation of the Earth.

The presence of the Coriolis force causes the plane of oscillation to turn slowly, with an angular velocity β, which is of the same order of magnitude as ω (except very near the equator where β vanishes completely, see below). We have: $|\,\beta\,| \approx \omega \ll \alpha$.

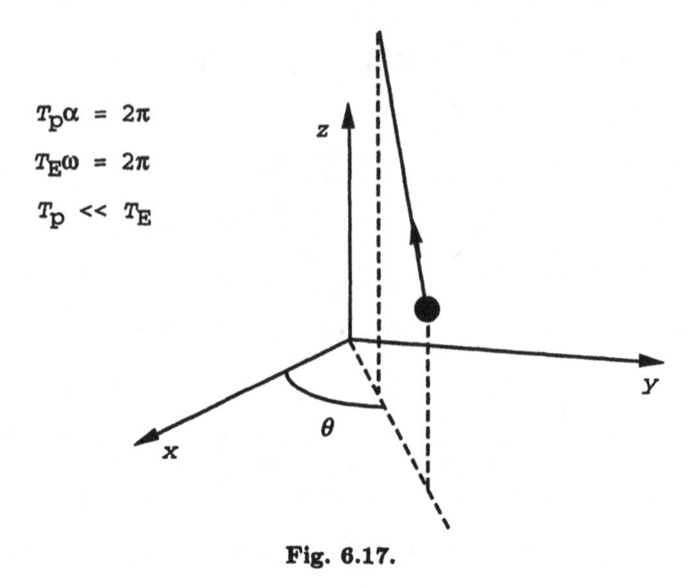

Fig. 6.17.

We consequently look for a motion of the pendulum of form

$$
\begin{aligned}
x &= \cos\alpha t \cos\beta t\,, \\
y &= \cos\alpha t \sin\beta t\,.
\end{aligned}
\tag{6.81}
$$

This set of equations describes a harmonic oscillator that oscillates with angular frequency α along a line whose direction angle θ turns with the angular frequency $\dot{\theta} = \beta$. Differentiating twice and rearranging, one gets, after some calculations:

$$
\begin{aligned}
\ddot{x} + (\alpha^2 - \beta^2)x &= -2\beta\dot{y}\,, \\
\ddot{y} + (\alpha^2 - \beta^2)y &= 2\beta\dot{x}\,.
\end{aligned}
$$

Comparing with the equation of motion (6.80), we find that the set (6.81) will be a solution if

$$\beta = -\omega \sin\Phi\,,$$

and

$$\alpha^2 - \beta^2 = \frac{g}{L}\,.$$

We may thus consider the motion of the pendulum to be a constant rotation of the plane of oscillation, superimposed on the harmonic oscillation of the pendulum with angular frequency $\alpha \approx \sqrt{g/L}$. The plane of oscillation turns with the angular velocity $\beta = -\omega\sin\Phi$ where Φ is the latitude of the geographical location. The period in the precessions is:

$$T_F = \frac{2\pi}{|\beta|} = \frac{2\pi}{\omega\sin\Phi} = \frac{23\ \text{h}\ 56\ \text{min}}{\sin\Phi}\,.$$

For Paris, $\sin\phi \approx 0.75$, and consequently $T_F \approx 32$ hours. Note: β is negative; the plane of the Foucault pendulum turns clockwise.

Let us remind ourselves of the basic reason for the precession of the plane of oscillation. Look at this sketch:

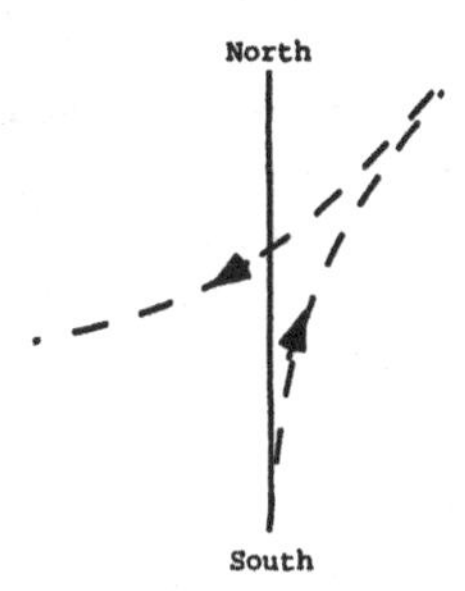

Fig. 6.18.

If the pendulum starts out to the south of the point of equilibrium, it travels towards the north, but is deflected to the right. As it returns it is again deflected towards the right of its velocity; the net result is that the plane of oscillation (here the average tangent to the curve) will drift slowly clockwise.

Consider the following illustration:

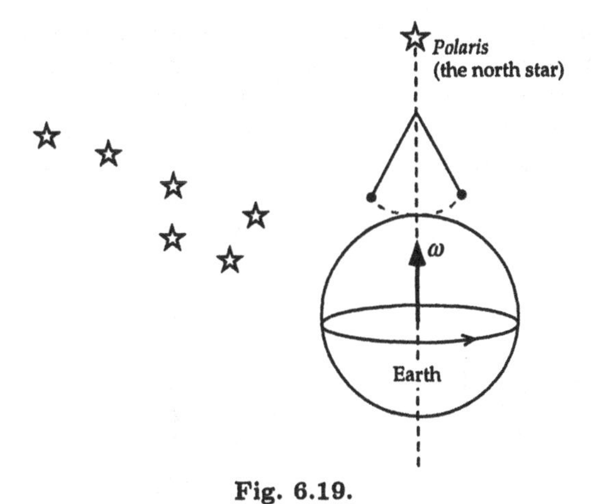

Fig. 6.19.

Assume that we perform the Foucault experiment on the North Pole of the Earth. In this case, the plane in which the pendulum swings turns clockwise over the ground, one full turn each 23 h and 56 minutes (one meridian day). Assume it is the winter period at the pole. Look up at the star Polaris, also known as the North Star, and the surrounding stars. You will then observe the stars majestically turning, keeping perfect step with the precession of the pendulum plane. The pendulum seems locked by the stars, while the Earth turns counterclockwise underneath.

We have placed this experiment on the North Pole for the following reason: in the northern hemisphere the suspension point of the pendulum remains fixed relative to the distant stars if and only if the point of suspension is directly above the Pole. Placed elsewhere the point of suspension would rotate with the Earth about the polar axis and this produces a counter rotation. It is this counter rotation that makes the rate of turning for the oscillation plane at other latitudes different from one meridian day (Paris: $T \approx 32$ h). At the equator a Foucault pendulum plane will not precess at all. Returning to the experiment at the North Pole:

(1) is it absolute space (astronomical space) that holds the plane of oscillation fixed while the Earth turns below; or
(2) is it the fixed stars – the totality of distant matter in the universe – that interacts with the mass of the pendulum in such a way that the pendulum swings in a plane that remains fixed relative to the stars?

We shall return to the problem in Chapter 7.

6.9 Newton's Bucket

It was particularly in the centrifugal force that Newton found his strongest arguments for the existence of an absolute "astronomical space" as a sort of stage upon which mechanical phenomena act. The unfortunate problem with absolute space is, as we have seen, that such a space can only be found to within a uniform translational motion. Does this observation make the whole idea of absolute space meaningless? Not quite, as we shall see. Newton gives a simple counter-argument based on a rotating bucket.

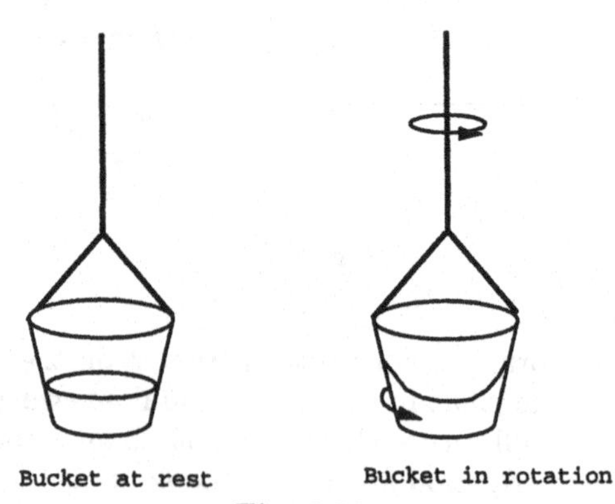

Fig. 6.20.

A bucket of water rotates with the angular velocity ω relative to the inertial frames. The water is forced away from the center by the centrifugal force (because it is constrained by friction to follow the bucket). Let us determine the exact shape of the surface. In the co-rotating frame (where the water is at rest), we consider a small surface element, of mass m.

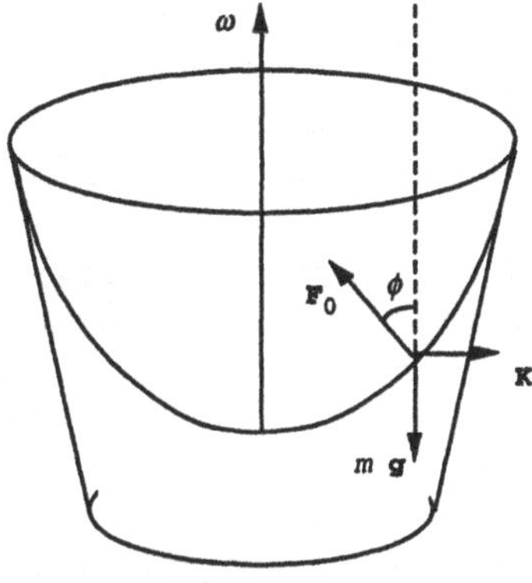

Fig. 6.21.

The surface element is at rest, so the force $\mathbf{F}_0$ on it from the surrounding fluid must be perpendicular to the surface (otherwise, the fluid element would move in response to a tangential tension). $\mathbf{F}_0$ must match the external (gravitational) forces on the element. Projecting $\mathbf{F}_0$ on to the vertical and horizontal directions we find (see Figure 6.21):

$$\begin{aligned} F_0 \cos \Phi &= mg \,, \\ F_0 \sin \Phi &= mr\omega^2 \,, \end{aligned}$$

or

$$\tan \Phi = \frac{r\omega^2}{g} \,.$$

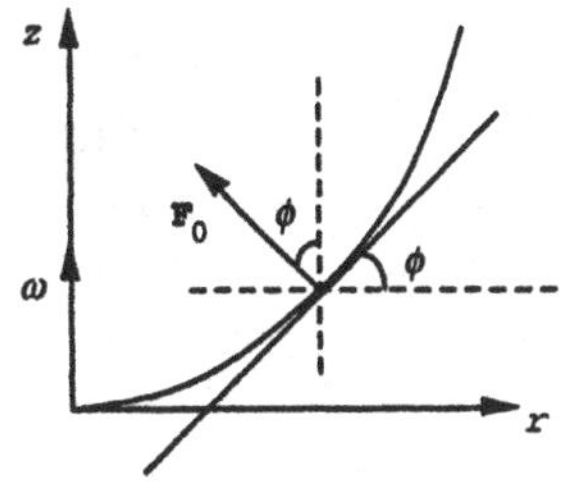

Fig. 6.22.

We seek the shape of the curve formed by the intersection between the surface and a vertical plane through the center of the bucket. We introduce a coordinate system as shown in Figure 6.22, and wish to determine the function $z = f(r)$. The slope of the tangent to the curve is at any point given by

$$\tan\Phi = \frac{\mathrm{d}z}{\mathrm{d}r} = \frac{\omega^2}{g}r\,.$$

Integrating this equation (with $z = 0$ for $r = 0$), we get

$$\int \mathrm{d}z = \frac{\omega^2}{g}\int r\mathrm{d}r \Rightarrow z = \frac{1}{2}\frac{\omega^2}{g}r^2\,.$$

The surface is consequently a parabola of revolution.

When the bucket rotates relative to the inertial frames (i.e., relative to the fixed stars), we observe the parabola of revolution. When the bucket is not rotating, the surface is planar. In both cases there is no *relative* motion between the water and the bucket.

Is it then the relative motion between water and the fixed stars which causes this effect? Or is it the rotation of the water through "absolute space", within which the stars seem to rest, that is decisive?

There can be no question that Newton was of the opinion that it is rotation relative to absolute space that changes the shape of the surface of the water. A change in the state of motion relative to absolute space requires a force.

If we make the bold thought experiment of keeping the bucket at rest and rotating all the fixed stars around it, Newton would predict no effect on the water, since the water would not be in rotation relative to absolute space.

Let us quote from *Principia*:

> The effects by which absolute and relative motions are distinguished from one another, are centrifugal forces, or those forces in circular motion which produce a tendency of recession from the axis. For in a circular motion which is purely relative no such forces exist; but in true and absolute circular motion they do exist, and are greater or less according to the quantity of the (absolute) motion.
> For instance. If a bucket, suspended by a long cord, is so often turned about that the cord is strongly twisted, then is filled with water, and held at rest together with the water; and afterwards by the action of a second force, it is suddenly set whirling about the contrary way, and continues, while the cord is untwisting itself, for some time in this motion, the surface of the water will at first be level, just as it was before the vessel began to move; but subsequently, the vessel, by gradually communicating its motion to the water, will make it begin sensibly to rotate, and the water will recede little by little from the middle and rise up at the sides of the vessel, its surface assuming a concave form. (This experiment I have made myself).

> At first, when the relative motion of the water in the vessel was the greatest, that motion produced no tendency whatever of recession from the axis; the water made no endeavor to move towards the circumference, by rising at the sides of the vessel, but remained level, and for that reason its true circular motion had not yet begun. But afterwards, when the relative motion of the water had decreased, the rising of the water at the sides of the vessel indicated an endeavor to recede from the axis; and this endeavor revealed the real circular motion of the water, continually increasing, till it had reached its greatest point, when relatively the water was at rest in the vessel... It is indeed a matter of great difficulty to discover and effectually to distinguish the true from the apparent motion of particular bodies; for the parts of that immovable space in which bodies actually move, do not come under observation of our senses.

Read the last four lines again and think about them for the rest of your life.

6.10 Review: Fictitious Forces

The equation of motion for a particle, seen from an inertial frame I is:

$$m\ddot{\mathbf{R}} = \mathbf{F} \text{ (Newton's second law).}$$

The equation of motion for the particle, referred to the accelerated frame S, is:

$$m\ddot{\mathbf{r}} = \mathbf{F} - m\ddot{\mathbf{R}}_0 + m\omega^2\rho - 2m(\boldsymbol{\omega} \times \dot{\mathbf{r}}) - m(\dot{\boldsymbol{\omega}} \times \mathbf{r}).$$

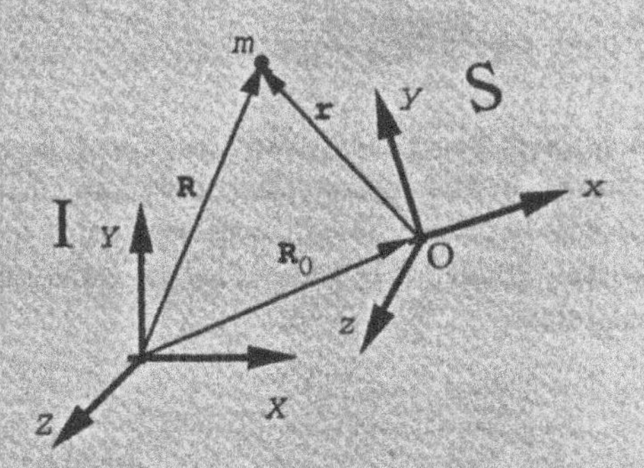

This equation shows that one may formally say that Newton's second law is true in an accelerated reference frame, if we interpret the last four terms on the right hand-side as forces acting on the particle besides the force $\mathbf{F}$. These forces are called fictitious forces. They may be distinguished from real forces by the fact that the fictitious forces are *independent* of the physical situation of the particle considered. They are dependent only on the motion of the reference frame and the position and velocity coordinates of the particle.

The term $m\omega^2\rho \equiv -m\,[\omega \times (\omega \times \mathbf{r})]$ is called the centrifugal force. The term $-2m(\omega \times \dot{\mathbf{r}})$ is called the Coriolis force. The other terms have no specific names. By the introduction of these fictitious forces the observer in S may say that the first and second law of Newton are correct. The third law, the law of action and reaction, does not hold for fictitious forces. There is no other body on which there is a reaction force.

The fictitious forces have one important property in common with gravitational forces: they are all proportional to the mass of the particle.

Forces like gravitational forces, electrical forces, elastic forces (string forces, tensions in springs) and so on, are sometimes – to distinguish them from fictitious forces – called natural forces.

6.11 Problems

Problem 6.1. A glass tube is filled with water and contains a piece of lead and a piece of cork (see the figure below). The tube is placed on a horizontal table and is accelerated toward the left by a blow from a hammer. How will the lead and the cork pieces move relative to the tube? Give a qualitative answer.

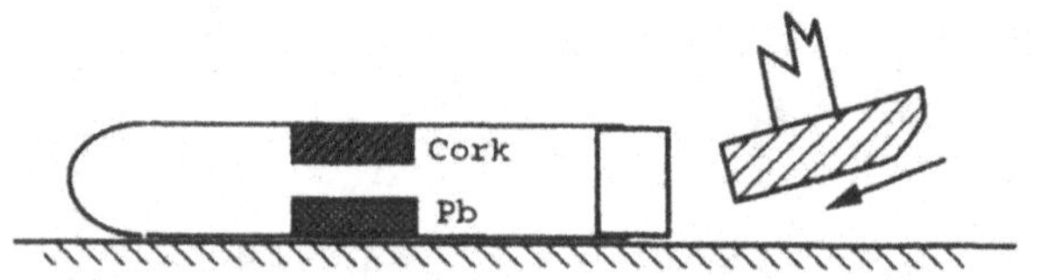

Problem 6.2. A wooden block B with the mass m is positioned on the front end of a car, as shown in the figure. The coefficient of friction between the car and the block is μ. The car has an acceleration of $\mathbf{a}$ (towards the right) relative to the inertial frames. Find the smallest value of $|\mathbf{a}|$ that will make the block stick to the car.

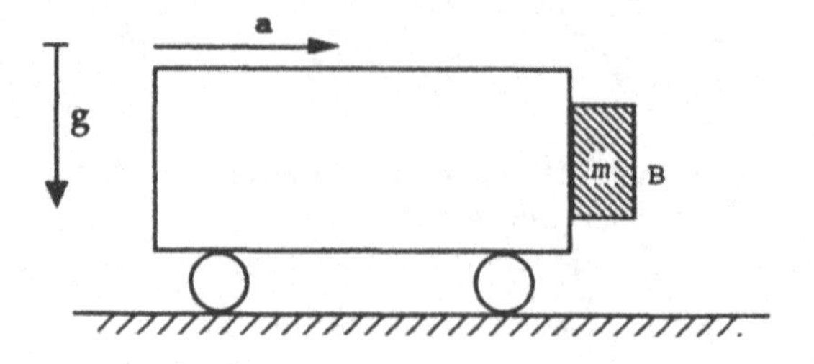

Problem 6.3. A satellite moves in an almost circular orbit around the Earth. A frictional force from the atmosphere acts on the satellite. This force acts against the motion of the satellite. Inside the satellite a small mass m is tied to a string, and the other end of the string is fastened to the satellite. You pick the point inside the satellite in which to fasten the string such that the mass can perform small pendulum oscillations.

(1) In which direction is the equilibrium position of the string? Make a small sketch showing the direction of the string and the point of attachment inside the satellite.
(2) The mass is now set into small oscillations. Which quantities determine the period T of oscillation? We are assuming that the satellite does not rotate relative to the Earth and that T is much smaller than the orbital period of the satellite.

Problem 6.4.

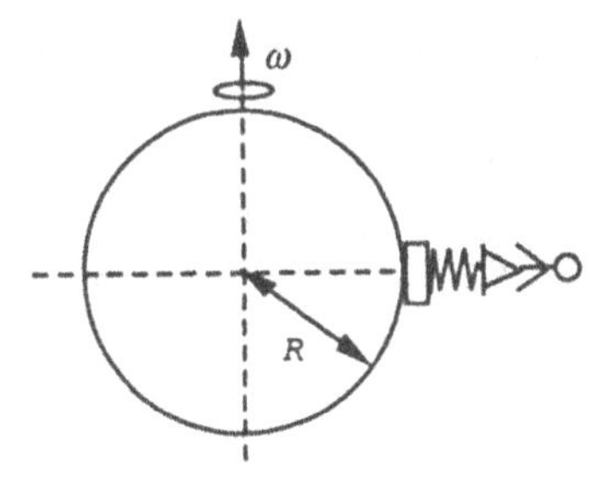

A man stands on a spring scale on the equator of the Earth. What should the length of a day be if the scale showed the man to be weightless?

Problem 6.5.

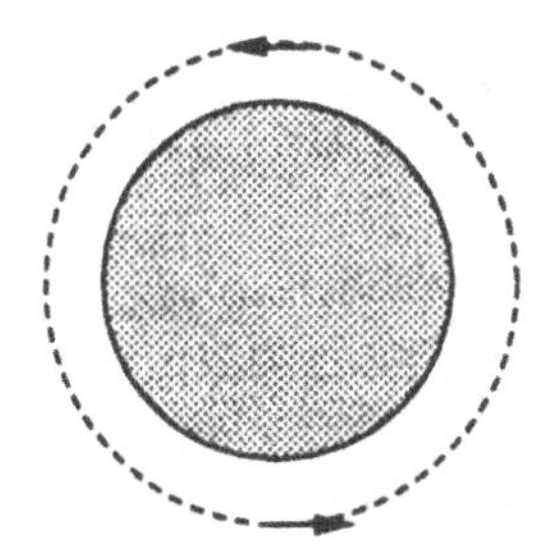

Consider a projectile in a circular orbit near the surface of the Earth. Disregard trivial effects such as air resistance, etc. (a) Find the period of revolution for the projectile. (b) For a reference frame in which the projectile is at rest, describe all forces acting on the projectile.

Problem 6.6. The so-called "Round Tower" in Copenhagen has a height of 36 m. A small, heavy sphere (a bullet) is dropped from the top of the Tower to the ground below. The sphere falls freely (neglect air resistance). Determine the distance and direction from the point directly beneath the drop point, to the point where the bullet actually hits the ground.

Problem 6.7. A rotor is a hollow cylindrical cabin able to turn around the vertical symmetry axis of the cylinder. A person goes into the rotor, closes the door and stands against the wall. The rotor is slowly set in rotation. At a certain angular velocity the floor is opened downwards. The passenger does not fall out but remains held against the interior wall.

The radius of the cylinder is R, the acceleration of gravity is g and the coefficient of friction between the passenger and the wall is μ .

(1) Find the least angular velocity ω such that the passenger does not fall downwards.

(2) The coefficient of friction between the man and the rotor has the value $\mu = 0.4$. The radius of the rotor is $R = 3\text{m}$. The angular velocity is as found in 1). Find the velocity of the passenger relative to the ground in this case. Consider the Earth to be an inertial frame for the motions in this problem.

Problem 6.8. A planar disk S made of wood is rotating relative to the lab frame, in a horizontal plane about the vertical axis through the center of the disk. The angular velocity has the magnitude ω. The direction of rotation is counter-clockwise. The figure shows the disk seen from above.

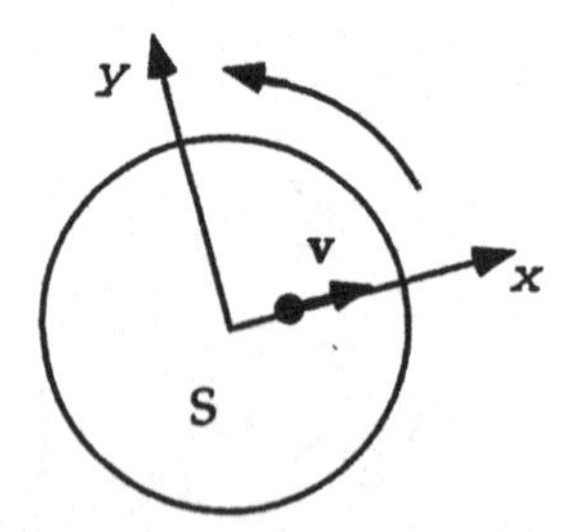

A small beetle of mass m is walking on the disk along a diameter. The constant velocity of the beetle relative to the disk is at all times $\mathbf{v}$. [Note: the angular velocity ω of the disk is maintained constant by an external source and is thus independent of the location of the beetle]. The acceleration of gravity is g and the lab frame is considered to be an inertial frame.

Let us consider a coordinate frame xyz, fixed relative to the disk (see the figure). This coordinate frame has origin in the center of the disk, and we arrange the frame in such a way that the beetle is moving out along the x-axis, and the z-axis is vertical. We call this frame 'the rotating frame'. Assume first that the friction between the disk and beetle is so large that the beetle does not slip on the disk.

(1) Find the magnitude and direction of the fictitious forces acting on the beetle as seen in the rotating coordinate frame.
(2) Show on a sketch the direction of the force F of friction with which the disk acts on the beetle.
(3) Now let the coefficient of friction between the disk and the beetle be μ. Determine the distance x_0 from the center where the beetle starts to slide.

Problem 6.9.

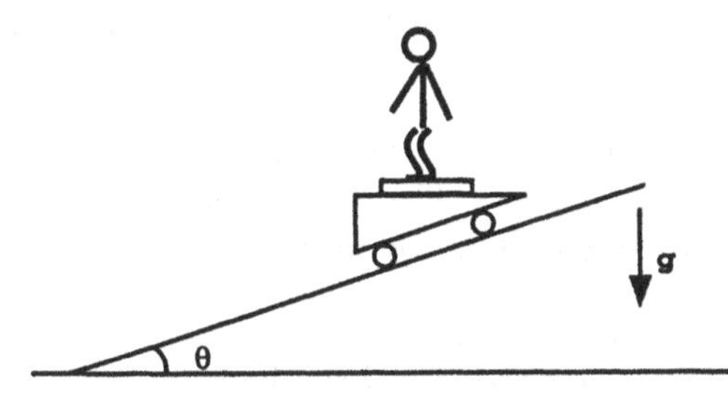

A man mounts a bathroom scale positioned on a skateboard such that it, as shown in the figure, can roll without friction down an inclined plane. He stands on the scale and reads off his weight as he is rolling down the inclined plane.

What is the slope θ of the inclined plane if the scale displays 45 kg during the descent and the actual mass of the man is 60 kg? Will the man stand as shown in the figure?

Problem 6.10.

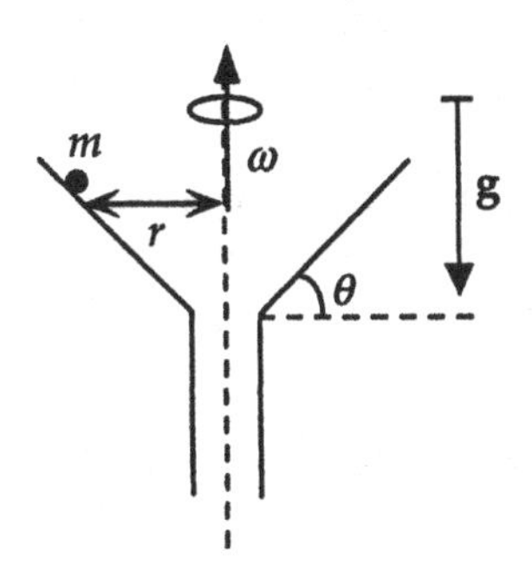

A very small bob (a particle!) of mass m is situated on the inside of a funnel. The funnel is rotating about a vertical axis with constant angular velocity corresponding to ν revolutions per second, i.e., $\omega = 2\pi\nu$. The walls of the funnel are at an angle θ to the horizontal.

The distance from the axis of rotation to the bob is r which is much bigger than the dimensions of the bob. The acceleration of gravity is $\mathbf{g}$.

(1) Assume first that there is no friction between the particle and the funnel. Determine the angular velocity ω_0 such that the particle does not move relative to the funnel.

Assume now that the coefficient of friction between the particle and the funnel is μ.

(2) Determine the smallest value of ν for which the bob starts to move upwards.
(3) Show that there is an angle $\theta = \theta_{\max}$ for which the bob cannot slide upwards if $\theta > \theta_{\max}$, no matter how large ω is. Determine $\theta_{\max}$.

Problem 6.11. A coordinate frame is affixed to the surface of the Earth some place on the southern hemisphere. The position has the latitude Φ (southern latitude). The coordinate axis have the following directions:

x-axis: positive towards the East
y-axis: positive towards the North
z-axis: vertical (i.e., along the local plum line).

The vector $\boldsymbol{\omega}$ describing the rotation of the Earth relative to the heliocentric system is considered constant in direction and magnitude.

A particle is at time $t = 0$ located at the origin, and the particle has the velocity $\mathbf{v} = (v, 0, 0)$. We consider motions only over distances small compared to the radius of the Earth. Air resistance is ignored.

We shall only regard the horizontal motion of the particle, i.e., we let $\dot{z} = 0$ (one can imagine the xy-plane to be a smooth horizontal plane which by its reaction force constrains motion to the plane).

(1) Show on a sketch the position of the vector $\boldsymbol{\omega}$ in the coordinate frame described, and work out the components of $\boldsymbol{\omega}$ in the frame.
(2) Write the equation of motion for the x-coordinate and for the y-coordinate of the particle.
(3) Find the distance of the particle from the x-axis for $x = 10$ km when $v = 500\ \mathrm{m\,s^{-1}}$ and $\Phi = 45°$ (southern) lattitude. Disregard the $\dot{y}$ term in the equation containing $\ddot{x}$.

Problem 6.12. The centrifugal force from the rotation of the Earth produces a change of the direction of a plumb line, from the direction of the gravitational force from the gravitating Earth. Let g_E be the acceleration that gravity "proper" would give a particle near the surface of the Earth. Assume that the Earth is spherical. Show that the deviation of the plum line from the direction of gravity "proper" at lowest order is given by an angle α, where

$$\alpha = \frac{\omega^2 \rho}{2 g_E} \sin 2\Phi \approx \frac{\omega^2 \rho}{2g} \sin 2\Phi ,$$

ρ = radius of the Earth,
Φ = lattitude,
ω = angular velocity of the Earth.
Determine the largest magnitude of α. [The deviation is towards the south on the northern hemisphere.]

Problem 6.13.

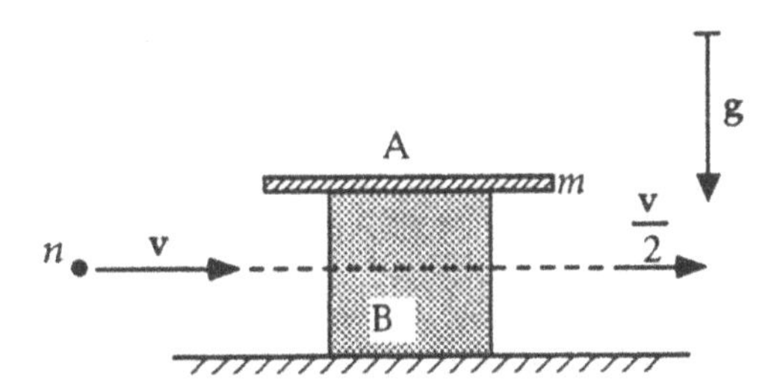

A cubic box B is made of wood and is filled with clay. The mass of the box plus clay is M. The box rests on a frictionless horizontal table. A thin plate A of mass m is placed on top of the box B as shown in the figure. The coefficient of friction between A and B is μ.

A small bullet with mass n is shot through B. Just before the bullet hits B the bullet has velocity $\mathbf{v}$. After the bullet has passed through B the bullet has velocity $\mathbf{v}/2$. Assume that the bullet moves along a horizontal line perpendicular to the end faces of B and passes through the centre of mass of B plus A.

It takes the bullet the time Δt to pass through B. We assume that the horizontal component of the force by which the bullet acts on the box is constant (independent of the velocity of the bullet). Find the smallest value of the time Δt such that the plate A does not move relative to B.

Problem 6.14. Consider a large horizontal frictionless area on the surface of the Earth, at $\phi°$ northern lattitude. Through an impulse a particle is set in motion with a velocity v and then left to move freely. Ignore all forms of friction. Find the period in the uniform circular motion of the particle. Compare with the period of precession for the Focault pendulum.

7. The Problem of Motion

In this chapter we will try to summarize some of the ideas and opinions on which Newtonian physics rests. These views have been substantially revised in the 20th century, largely due to the works of Einstein in his special and general theory of relativity.

It is outside the scope of this book to give a presentation of the works of Einstein. We shall merely outline the directions in which these theories point.

7.1 Kinematic and Dynamic Views of the Problem of Motion

It was an important event in the history of physics when Kepler published his famous three laws early in the 17th century. In these three laws the efforts of many centuries of study of the kinematical aspects of motion in the solar system culminated.

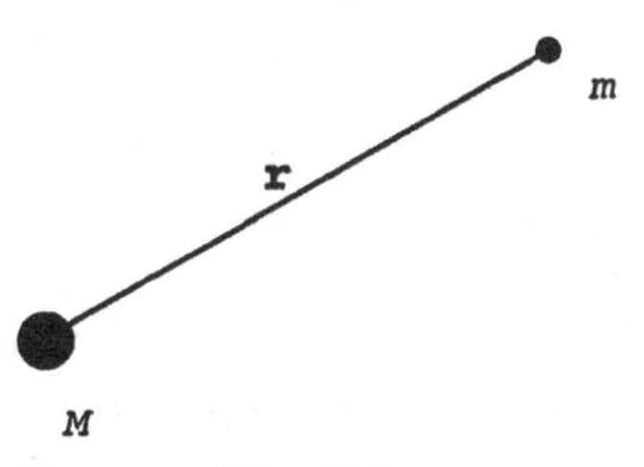

Fig. 7.1.

At the time of Leibniz and Newton the relativity of motion was a topic of discussion. Leibniz emphasized that it only made sense to discuss the motion of a body relative to other material bodies. According to Leibniz, motion is change of position of one body relative to another material body. He pointed out the inconsistency in declaring one particular body to be at rest since the "rest" of a material body can only be referenced to other bodies.

Let us label these views with the term *kinematic relativity.* It states that motion as a kinematic process, a change of spatial distances, is a relative concept. *The basic feature of this viewpoint is that the state of motion of a body can only be inferred by a comparison with other bodies.* No observation of a single body can reveal whether the body is in motion or not.

The cosmologies of Ptolemy and Kepler are kinematically equivalent. The epicycles traced out by the planets in the model of Ptolemy are kinematically equivalent to the ellipses in Kepler's model.

In the kinematical description of the motion of bodies we are concerned only with describing the motions as they are observed, not with any underlying cause.

Galileo began to look for causes of motion and thus founded *dynamics.* Newton used the ideas of Galileo to analyze the motion of celestial bodies and through an ingenious analysis of Kepler's laws (see Chapter 1) Newton arrived at the concept of *mechanical force as the cause* of motion for the planets.

This dynamical view of motions in astronomical space in turn necessitated precise assumptions of space and time to be introduced into mechanics. Such axioms appear first in the works of Newton, as explicit definitions (see Chapter 2).

Newton's discovery of the quantitative connection between the forces causing the motion and a kinematical quantity, the acceleration, enabled him to use force (dynamical quantity) as a measure of acceleration (kinematical quantity).

Newton was convinced that there was such a thing as *absolute motion* i.e., the motion of a body can be inferred without any comparison to other bodies. According to Newton it is not true that all observable phenomena are the same no matter which one of two bodies is considered to be at rest, since there are differences as soon as dynamical phenomena are included in the observations (centrifugal forces, Coriolis forces).

As mentioned above, Newton had to make very precise assumptions about time and space when formulating his laws. Without these assumptions even the simplest law of mechanics, the law of inertia, is without meaning.

Newton arrived at the conviction that an empirical reference frame, attached to material bodies – e.g. the Sun – could never be taken as a valid base for the law of inertia. This law with its close connection to the Euclidian concept of space where the basic element is a straight line, seems a natural starting point for *dynamics* in astronomical space. It is actually through the law of inertia that the element of Euclidian space, the line, manifest itself when we are far from reference bodies. Similar remarks can be made for time, the uniform progress of which is expressed in purely inertial motion. If one chose to take the rotational period of the Earth as the basis of time, the law of inertia would not hold, because there are small but finite variations in the rotation of the Earth.

Through these and similar considerations Newton came to the conclusion that an *absolute space* exists relative to which all kinematical quantities, in particular acceleration, had to be measured.

7.2 Einstein Speaks

It may seem peculiar that Newton, both in the axiom of absolute space and in the axiom of absolute time, states that these two phenomena exist "without relation to anything external". Newton himself often emphasized that he only wished to investigate phenomena that could be verified by observation. But that which exist "without relation to anything external" cannot be verified by observation. We shall return to this issue.

Newton has often been criticized for the phrase "without relation to anything external". In this connection it is instructive to quote from a speech made by Einstein in 1927, at the bi-centennial for Newton's death:

> Theoretical physics has grown from the Newtonian framework, which provided stability and intellectual guidance for natural science for 200 years.
> Newton's basic principles were so satisfactory from a logical point of view that motivation for questioning them could only arise from the strict demands of empiricism. Before I discuss this however, I wish to emphasize that Newton himself, far more than the generations of learned scientist following after him, was aware of the inherent weaknesses in his own intellectual construction. This feature has always aroused my deepest admiration, and I would like to dwell upon it for a few moments.
> I. Newton's efforts to manufacture his system as tightly bound to experience as possible are ubiquitous. In spite of this he nevertheless introduced the concepts of absolute space and absolute time. But actually Newton is here particularly consistent. He had realized that observable geometrical quantities (the distances between material points) and their way through time did not characterize motion in all its physical aspects. He demonstrated this in his famous experiment with the rotating bucket. *There must consequently besides masses and spatial variables be something else that determines motion.* This "something" is expressed in his notion of the absolute space. He realizes that space must possess a physical reality if the equation of motion are to make sense, a reality on equal footing with material points and their distances.
> The clear recognition of this reveals both the wisdom of Newton and the weak points of his theory. For the logical structure of the latter would undoubtedly be more satisfactory without this shadowy

concept; then only direct observables (mass-points, distances) would enter the laws.
II. Forces acting directly and immediately at a distance such as those introduced to describe gravity are not familiar from everyday processes. Newton meets this objection by stressing that his law of gravitational interaction is not meant to be the final explanation but is a rule derived from experience and observation.
III. Newton's theory contained no explanation of the remarkable fact that the inertia and the weight of a body are determined by the same quantity, the mass. Newton was quite aware of this puzzle.
None of these three points are severe logical stumbling blocks for the theory. In a sense they just express an unfulfilled desire in the struggle of the scientific mind to obtain a complete conceptual picture of natural phenomena.

In this quotation we have highlighted a few lines. As a comment to those lines we note the following ideas.

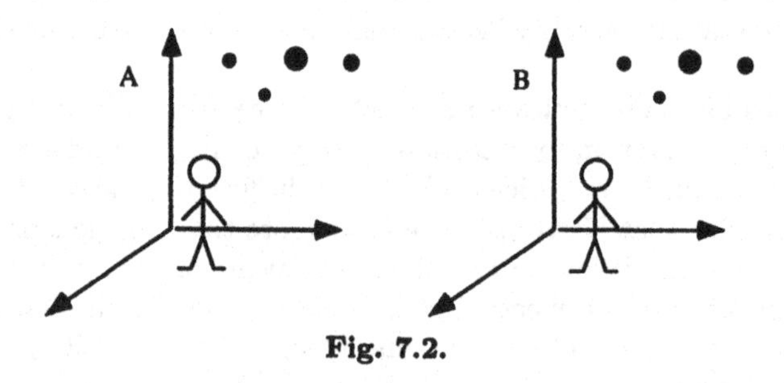

Fig. 7.2.

In Figure 7.2, the two physical systems – particles with given masses and mutual distances – are assumed to be kinematically equivalent at time $t = 0$, each within its reference frame. Since the two systems are identical, the particles have the same interactions. This does not, however, imply that the systems will evolve in the same manner within their respective reference frames. It will be the case if the two frames are in relative uniform linear motion. But if frame B is accelerated relative to frame A, the two systems will evolve differently within their frames. *There must consequently, in addition to masses and spatial variables, be something else that determines motion. This "something else" was to Newton the effect of absolute space.*

Before we go on to the description of the properties of the Newtonian equations of motion, we will briefly review what it means for a physical law to be symmetrical under a given transformation.

7.3 Symmetry

We will use Weyl's definition:

I. A body is symmetrical under some operation when it has the same appearance before and after the operation. Sometimes this property is called invariance under the operation in question. An often quoted example is a vase, symmetrical with respect to rotations around some axis A, as in Figure 7.3.

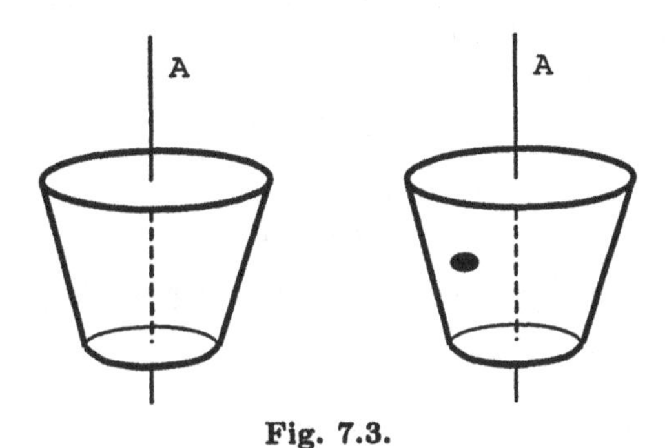

Fig. 7.3.

If the vase has a spot, it is asymmetrical under any rotation where the angle of rotation differs from an integer multiple of 360°.

II. A physical law is symmetrical under some transformation when the mathematical expression of the law looks the same before and after the transformation. This is often formulated as the physical law being invariant under the given transformation.

7.4 The Symmetry (Invariance) of Newton's 2nd Law

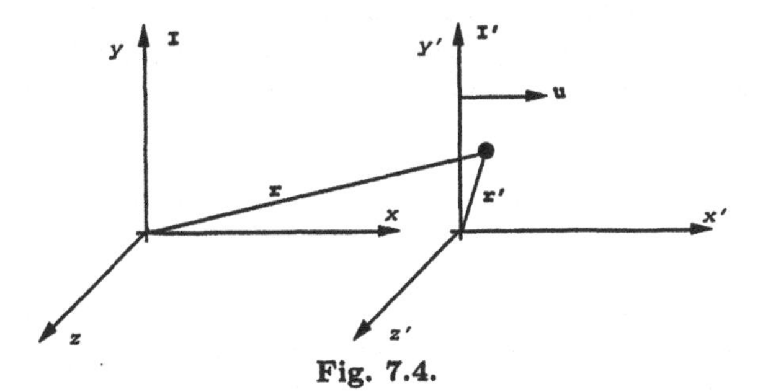

Fig. 7.4.

We demonstrated the invariance of Newton's second law under a Galilei transformation in Chapter 4, but it bears repeating here. A Galilei transformation is a relation between two inertial frames, I and I$'$, where the frame I$'$ moves relative to I with a constant velocity $\mathbf{u}$ (see Chapter 4). Time is common to the two frames. We have:

$$\begin{aligned} \mathbf{r}' &= \mathbf{r} - \mathbf{u}t\,, \\ t' &= t\,. \end{aligned}$$

To find the relation between the velocity as seen from frame I and the velocity relative to I$'$, we use the chain rule:

$$\frac{\mathrm{d}\mathbf{r}'}{\mathrm{d}t'} = \frac{\mathrm{d}\mathbf{r}}{\mathrm{d}t}\frac{\mathrm{d}t}{\mathrm{d}t'} - \mathbf{u}\frac{\mathrm{d}t}{\mathrm{d}t'}\,,$$

or

$$\mathbf{v}' = \mathbf{v} - \mathbf{u}\,,$$

as $t' = t$. We assume now that a force $\mathbf{F}$ is acting on the particle. Then, the equation of motion in I is

$$m\frac{\mathrm{d}^2\mathbf{r}}{\mathrm{d}t^2} = \mathbf{F}\,. \tag{7.1}$$

Differentiating the velocity relation further, we find:

$$\frac{\mathrm{d}^2\mathbf{r}'}{\mathrm{d}t'^2} = \frac{\mathrm{d}^2\mathbf{r}}{\mathrm{d}t^2}$$

Note: here is the essential feature of the Galilei transformation. When $\mathbf{u}$ is constant, the acceleration of a particle is the same when observed from the two frames, and acceleration becomes an absolute quantity.

Since furthermore $m' = m$ and $\mathbf{F}' = \mathbf{F}$, we conclude that if (7.1) is the equation of motion as seen in I, then the equation of motion in I$'$ is given by

$$m'\frac{\mathrm{d}^2\mathbf{r}'}{\mathrm{d}^2t'} = \mathbf{F}'\,. \tag{7.2}$$

Comparing (7.1) and (7.2) we see that Newton's second law is the same ("looks the same") in any two frames connected by a Galilei transformation. In other words, Newton's second law is invariant under a Galilei transformation.

7.5 Limited Absolute Space

The relativity principle of mechanics is the starting point for all the following considerations.

This principle is intimately connected with the concept of absolute space; in fact the relativity principle of mechanics sets limits on the physical reality of absolute space.

What do we mean by physical reality? *A concept has a physical reality if and only if the concept is associated with phenomena that can be verified by a physical measurement.*

We have here defined reality as the concept is commonly understood in physics. It follows that the concept of a point fixed in absolute space does not correspond to any physical reality.

If an observer claims that some particle is at rest relative to absolute space, another observer can with equal right claim that the particle is in uniform motion with respect to absolute space. No physical measurement, mechanical or otherwise, can decide the issue. Physically, it thus becomes a meaningless issue.

Because of the relativity principle of mechanics, Newton's absolute space loses some of its physical reality.

We frequently use the words "reference frame" in place of the word "space". We can thus formulate the relativity principle of mechanics as follows:

There is an infinite number of equivalent reference frames – the so-called inertial frames – that are mutually in uniform linear motion. Referred to these frames the mechanical laws are valid in their simple classical form.

Notice how closely connected space and mechanical laws are. Is it space that impresses the mechanical laws on the material bodies, or is it the mechanical laws that give rise to the entire concept of space?

7.6 The Asymmetry (Variance) of Newton's 2nd Law

We have seen that a single point in Newton's absolute space is a concept without physical reality. One may ask: what then remains of Newton's absolute space? The answer – from within the Newtonian world picture – is this: The resistance a body offers against acceleration, must be interpreted as an effect of absolute space. Any change in velocity relative to absolute space requires a force.

Reference frames that are accelerated relative to absolute space – and consequently relative to all inertial frames – are not equivalent to inertial frames or to each other. We can describe any motion relative to such frames, but Newton's second law is not invariant under a transformation to such frames.

From Chapter 6 we recall the following results on the transformation of Newton's second law when we go from an inertial frame I to a frame S accelerated relative to I.

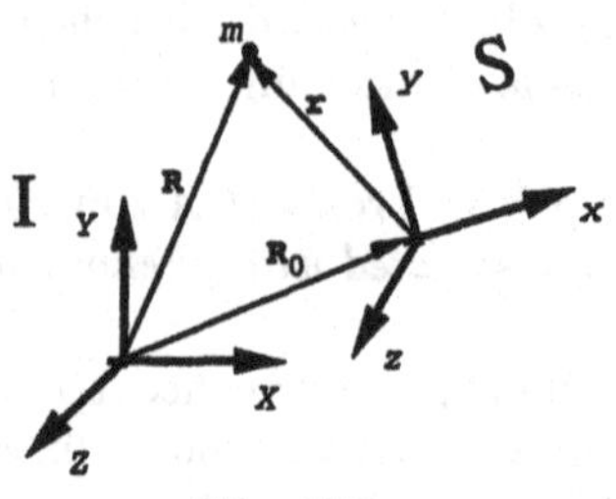

Fig. 7.5.

A particle of mass m is acted upon by a force $\mathbf{F}$ and moves through space. We can regard its motion either from the frame I or from the frame S. The motion described from I, obeys the simple law:

$$m\ddot{\mathbf{R}} = \mathbf{F}$$

The accelerated frame S is in a state of motion characterized by the translational acceleration $\ddot{\mathbf{R}}_0$ of its origin, and a rotation vector $\boldsymbol{\omega}$. The equation of motion of the particle, referred to S, takes the form:

$$m\ddot{\mathbf{r}} = \mathbf{F} - m\ddot{\mathbf{R}}_0 + m\omega^2\rho - 2m\boldsymbol{\omega} \times \dot{\mathbf{r}} - m\dot{\boldsymbol{\omega}} \times \mathbf{r}\,. \tag{7.3}$$

Here, $\mathbf{r}$ is the vector from the origin in S to the particle. $\mathbf{F}$, m, and the coordinate t are assumed to be invariant quantities. One can see that Newton's second law is variant (asymmetrical) with respect to a transformation to an accelerated frame. This asymmetry means that we can *by mechanical experiments alone decide if we are accelerated with respect to Newton's absolute space*, or with respect to the inertial frames.

Put differently: Newton argued from (7.3) that accelerated motion is absolute, and that this type of motion can be inferred without relation to anything external, in particular without refererence to other bodies. Let us assume that $\ddot{\mathbf{R}}_0 = 0$ and that $\dot{\boldsymbol{\omega}} = 0$. Then (7.3) is:

$$m\ddot{\mathbf{r}} = \mathbf{F} + m\omega^2\rho - 2m\boldsymbol{\omega} \times \dot{\mathbf{r}}\,. \tag{7.4}$$

It was particularly in the second term (the centrifugal force) that Newton found support for his doctrine of absolute space.

Newton devised the following procedure for the determination of absolute space. Consider two heavy bodies, each of mass M, connected with a thin elastic rod. This system is placed in empty space far from all galaxies. Now measure the tension in the rod. If there is no tension in the rod, the rod is

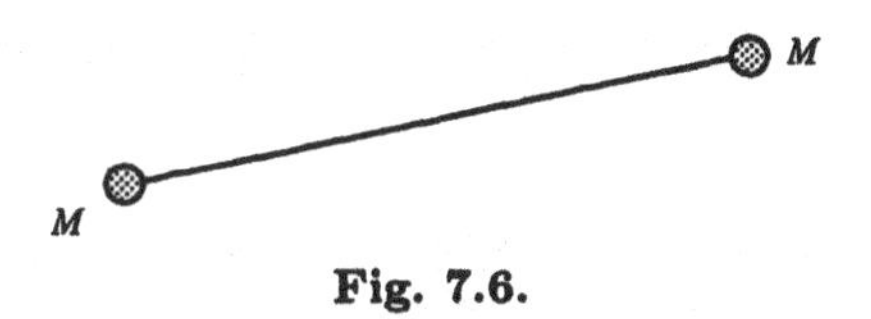

Fig. 7.6.

nonrotating relative to absolute space. If there is a tension T in the rod, we must have:

$$T = M\omega^2(L/2)$$

where L is the length of the rod. We have neglected the mass of the elastic rod. In principle therefore, we can measure $|\boldsymbol{\omega}|$, i.e., the angular velocity relative to absolute space by measuring the tension in the rod. Without relating the system to anything external, we have measured a rotational velocity relative to absolute space.

Nature has equipped us with a very basic example of a rotating system: Earth itself. The Earth is not an exact sphere. If the Earth had been at rest in absolute space it would, according to Newton, be spherical because only its own gravity acts on its masses. The fact that the Earth has the shape of an ellipse of revolution is proof, according to Newton, that the Earth rotates relative to absolute space.

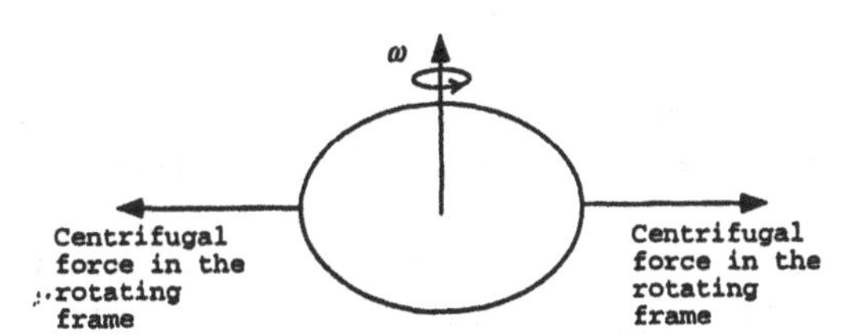

Fig. 7.7. The Earth as a body in rotation relative to the inertial frames

The centrifugal forces, acting in the rotating reference frame, are driving the masses away from the axis of rotation. Without referring to external bodies (e.g. stars) one can therefore deduce that the Earth is rotating relative to absolute space. The procedure is obvious: measure the distance from pole to equator. Then measure the circumference around equator and divide the result by four. The two results are not equal: the distance from the center of Earth to a pole is less than the distance from the center to the equator. This is an indication of rotation relative to absolute space.

It is important to be aware of the following, with respect to the Newtonian viewpoints. The above phenomena do not, according to Newton, arise because of rotation with respect to other masses (e.g. the fixed stars), but because of rotation with respect to absolute space. If the Earth was at rest, and the fixed stars revolved around it once every 23 hours 56 minutes, no centrifugal forces would act. The Earth would not be oblate, and the acceleration of gravity would not vary with latitude. Similarly, a Foucault pendulum would not show any drift.

The kinematical description is the *same* whether one regards the Earth as being stationary and the starry sky as rotating, or vice versa. According to Newton, *the dynamical laws of motion give rise to an observable difference between the two situations.*

Newton's idea of absolute space is based therefore on concrete facts. Fictitious forces are well known from everyday life here on Earth. The Earth is not spherical, but ellipsoidal. Observing the solar system, we find that planets trace out orbits where gravitational attraction is balanced by centrifugal forces. Remote binary star systems show that the law of centrifugal forces in rotating reference frames are valid thousands of light years away. Apparently the presence of centrifugal forces is a universal phenomenon. Newton concluded that fictitious forces result from a direct action of space itself.

7.7 Critique of the Newtonian View

One of the first to seriously criticize the above conclusions was Ernst Mach, late in the 19th century. In his work *The Science of Mechanics* (1883) Mach puts forward the point of view that mechanical experience can teach us nothing about absolute space. He insisted that only relative motions are verifiable and only these therefore have physical relevance. Newton's "proof" of absolute space based on dynamical evidence must therefore be an illusion.

The Newtonian point of view rests fundamentally on the assumption that Earth would not be ellipsoidal and the acceleration of gravity not less at equator than at the poles if it was the fixed stars that rotated in absolute space. Mach points out that this assumption exceeds the verifiable and criticizes Newton for having introduced nonverifiable elements into his mechanics.

Mach attempted to construct a mechanics without nonverifiable concepts. He was convinced that fictitious forces were directly related to interactions with the entire mass distribution in the universe. His attempt to construct a new mechanics failed. One reason for this is that he ignored an important property of fictitious forces: they are proportional to the mass of the body. Gravitational forces have this property as well. There thus seems to be a connection between the inertial mass (i.e., the property of a body that it resists acceleration) and the gravitational mass (i.e., the property of a body that causes it to attract other bodies).

Furthermore Mach seems to have ignored the new developments in the theories of electromagnetic phenomena in his efforts to construct a new mechanics. The relativistic theories of electromagnetic phenomena opened the possibility of eliminating the dogma of absolute time.

The new mechanics was constructed by Albert Einstein.

7.8 Concluding Remarks

Before we conclude our discussion of the Newtonian world picture we wish to restate the following observations.

Consider two railroad cars A and B. The cars are on a horizontal track; A is at rest and B has a constant velocity $\mathbf{v}$ relative to A.

The invariance of Newton's second law under a transformation between the frame where A is at rest and the frame where B is at rest means that all mechanical experiments will take the same course inside the two cars. A pendulum hanging from the ceiling will have a vertical equilibrium position and the period of its oscillation will be the same inside each car. In other words: by studying only mechanical phenomena inside the car, an observer in B cannot tell whether he is at rest or in uniform motion relative to A.

Note: the fact that the basic mechanical laws are the same in the two frames rests on firm experimental ground. One thing is to claim that all motion is relative and base this claim on purely kinematical considerations, e.g. that all motion is a change of distance relative to other bodies, and from this conclude that motion is relative. It is a completely different matter to verify by experiment that pendulum motion is the same in the two cars. The two cars are found to be not only kinematically but also dynamically equivalent. Is all motion then relative?

Suppose a car C has an acceleration $\mathbf{a}$ relative to car A. The two cars (A and C) are of course again kinematically equivalent, but Newton's second law is variant (asymmetrical) under transformation to a system accelerated relative to an inertial frame. The pendulum in C will have an equilibrium position not parallel to the end wall, and its period of oscillation changes by a factor $\sqrt{g/\sqrt{g^2+a^2}}$.

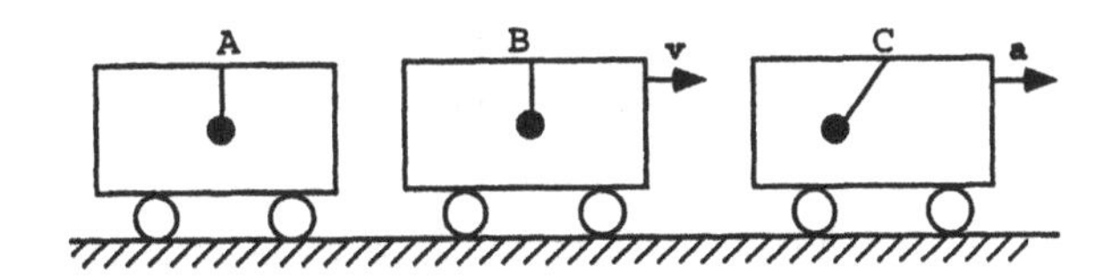

The equations of motion in the three reference frames are:

$$
\begin{aligned}
\text{In A:} \quad & m\ddot{\mathbf{R}} = \mathbf{F} \\
\text{In B:} \quad & m\ddot{\mathbf{R}} = \mathbf{F} \\
\text{In C:} \quad & m\ddot{\mathbf{R}} = \mathbf{F} - m\mathbf{a}
\end{aligned}
$$

"A spot has appeared on the vase"

Is acceleration absolute after all? One way out of the asymmetry is, as Mach did, to claim that C is accelerated relative to the average mass distribution of the universe (or the fixed stars as it is usually put) and that these distant bodies directly give rise to the fictitious forces. In other words the fixed stars are the direct cause of the fictitous forces that appear in accelerated reference frames.

It seems to have been Mach's belief that fictitious forces are acceleration-dependent actions of the distant stars. These actions should then be perceived on an equal footing with gravitation or electromagnetic forces. If the car C was at rest and the fixed stars were accelerating to the left, the equilibrium position of the pendulum would – according to Mach – be as shown, not parallel to the end wall. According to Newton it would be parallel to the end wall.

In the modern formulation: the fictitious forces are, according to Mach, acceleration dependent interactions with a field generated by the mass distribution of the universe. This should be compared with the velocity-dependent interactions between a moving charge and a magnetic field of distant currents. *A uniform translational velocity relative to the stars cannot be perceived. An acceleration relative to the stars can be detected. Why this difference between the first and the second derivative with respect to time?*

The view – that distant masses determine the inertial frames and thus the local inertial properties of matter – has been called Mach's principle.

Mach's principle has played a certain part in cosmology and it influenced Einstein in his construction of the general theory of relativity.

The about 10^{11} stars of our galaxy are in the shape of a flat disc (see Figure 7.8).

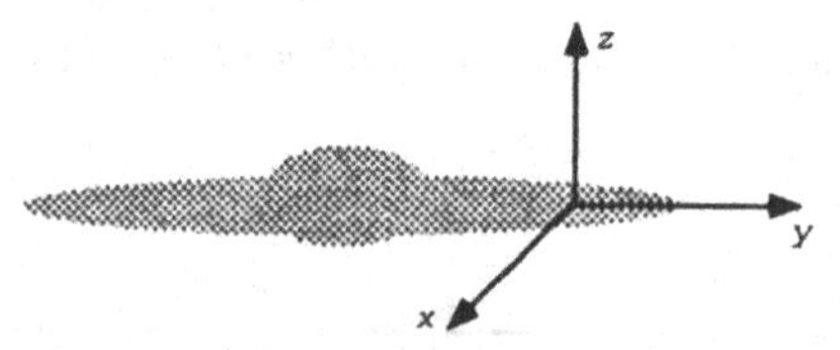

Fig. 7.8. The galaxy with the heliocentric reference frame

If there is some truth to Mach's principle, our galaxy participates in the determination of the inertia of a particle. The galaxy constitutes only a small fraction of the total mass in the universe, but it is very much closer than the rest. Since it is highly anisotropic in shape, one might imagine that the inertia of a particle would be anisotropic. For instance, for the coordinate frame shown, we might have:

$$\begin{aligned} m_x\ddot{x} &= F_x \, , \\ m_y\ddot{y} &= F_y \, , \\ m_z\ddot{z} &= F_z \, , \end{aligned}$$

where $m_x \neq m_y \neq m_z$.

Experiments using nuclear magnetic resonance have found that the inertial mass of a nucleus to a very high degree of accuracy is isotropic, i.e., it is independent of the direction of acceleration. This experimental observation has been interpreted by Mach principle proponents as simply stating that the inertial influence of our galaxy is vanishing compared to that of all other masses in the universe. We shall not dwell on the details of this discussion, and only point to the following considerations.

If one, from Mach's principle, believes that the overall mass distribution of the universe determines local inertial mass, it is logical to wonder if there are other connections between the global structure of the universe and basic physical laws here on Earth. Would, for instance, the charge, mass or spin of an electron change if the number of (or distribution of) protons and neutrons in the universe changed? Cosmology is a science in rapid evolution. Perhaps we shall some day know the answer.

We conclude this section with the following remark. *The objective importance of a discovery and the subjective interpretation it is given by the discoverer can be in contrast to one another. Newtonian dynamics became the cornerstone of theoretical physics. Newton's own interpretation has not survived. Another example is the Schrödinger equation. This equation became a cornerstone of quantum mechanics, but Schrödinger's own interpretation of his famous equation has not survived.*

8. Energy

We introduce the concepts of work, kinetic energy, and potential energy in a general setting. The fundamental importance of these concepts arise from the fact that they permit a formulation of a conservation law for mechanical energy.

8.1 Work and Kinetic Energy

Consider a particle moving under the influence of a force **F** and described from an inertial frame. The force **F** in question is arbitrary; it can depend on the position **r** of the particle, or on other variables in the problem. As the particle moves from the point A to the point B along some path (see Figure 8.1), work W_{AB} is done by the force. This work is defined through the scalar product and according to the curve integral:

$$W_{\mathrm{AB}} \equiv \int_{\mathrm{A}}^{\mathrm{B}} \mathbf{F} \cdot \mathrm{d}\mathbf{r} \tag{8.1}$$

Recall that $\mathbf{a} \cdot \mathbf{b} = |\,\mathbf{a}\,|\,|\,\mathbf{b}\,| \cos(\theta)$. The integral is understood to be taken along the orbit of the particle. If the displacement $\mathrm{d}\mathbf{r}$ is so small that the force can be considered constant throughout the displacement, the work performed along the displacement is the differential $\mathrm{d}W = \mathbf{F} \cdot \mathrm{d}\mathbf{r}$. Note that

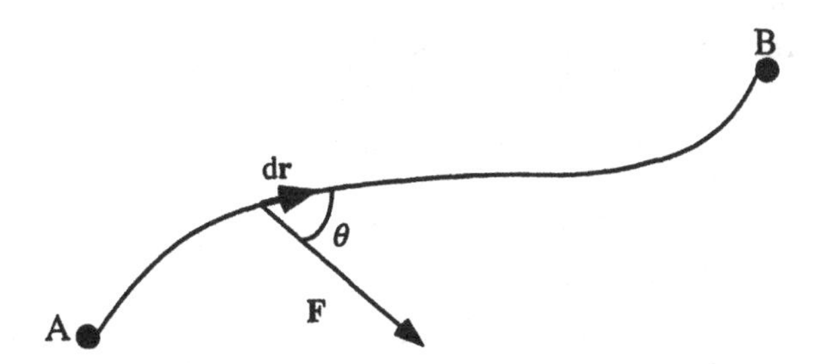

Fig. 8.1. Definition of work done by a force. A particle moves along the indicated curve from A to B and a force **F** acts on the particle

if the force vector is everywhere perpendicular to the tangent vector of the curve, no work is done by the force along the curve.

Assume that $\mathbf{F}$ denotes the vectorial sum of all forces acting on the particle. We may use Newton's second law to rewrite the defining equation (8.1), using m as the mass of the particle:

$$\begin{aligned} W_{\mathrm{AB}} &= \int_{\mathrm{A}}^{\mathrm{B}} \mathbf{F} \cdot \mathrm{d}\mathbf{r} = \int_{\mathrm{A}}^{\mathrm{B}} m\frac{\mathrm{d}\mathbf{v}}{\mathrm{d}t} \cdot \mathrm{d}\mathbf{r} = \int_{\mathrm{A}}^{\mathrm{B}} m\frac{\mathrm{d}\mathbf{v}}{\mathrm{d}t} \cdot \mathbf{v}\mathrm{d}t \\ &= \int_{\mathrm{A}}^{\mathrm{B}} m\mathbf{v} \cdot \mathrm{d}\mathbf{v} = \int_{\mathrm{A}}^{\mathrm{B}} \mathrm{d}\left(\frac{1}{2}mv^2\right) = \frac{1}{2}mv_{\mathrm{B}}^2 - \frac{1}{2}mv_{\mathrm{A}}^2 \, . \end{aligned}$$

The quantity $T = \frac{1}{2}mv^2$ is called the *kinetic energy* of the particle. Using this notation, we conclude that

$$W_{\mathrm{AB}} = \int_{\mathrm{A}}^{\mathrm{B}} \mathbf{F} \cdot \mathrm{d}\mathbf{r} = T_{\mathrm{B}} - T_{\mathrm{A}}. \tag{8.2}$$

The work done by all forces acting on a particle equals the increase in kinetic energy of the particle

Equation (8.2) can be expressed in differential form:

$$\mathrm{d}T = \mathbf{F} \cdot \mathrm{d}\mathbf{r} \, , \tag{8.3}$$

or

$$\frac{\mathrm{d}T}{\mathrm{d}t} = \mathbf{F} \cdot \frac{\mathrm{d}\mathbf{r}}{\mathrm{d}t} = \mathbf{F} \cdot \mathbf{v} \, . \tag{8.4}$$

Equations (8.2)–(8.4) are valid for all types of forces; the results are based only on Newton's second law. The scalar product $\mathbf{F} \cdot \mathbf{v}$ is called the *power* done by the force $\mathbf{F}$.

According to (8.4) the rate of change of the kinetic energy is the scalar product between force and velocity. If the particle happens to be acted upon by a force which is always perpendicular to the velocity vector (e.g., the Lorentz force on a charged particle in a magnetic field) there is no change in the kinetic energy.

It is interesting to compare the set of equations (8.2)–(8.3) for the change in kinetic energy to the set of equations describing the change in momentum:

$$\mathrm{d}\mathbf{P} = \mathbf{F}\mathrm{d}t \, ; \quad \Delta\mathbf{P} = \int_{t_{\mathrm{A}}}^{t_{\mathrm{B}}} \mathbf{F} \, \mathrm{d}t \, ; \quad \mathbf{P} = m\mathbf{v} \, .$$

Expressed in words the two sets of equations state that

- the work done by the force $\int \mathbf{F} \cdot \mathrm{d}\mathbf{r}$ gives the change in kinetic energy of the particle;
- the impulse of the force $\int \mathbf{F} \, \mathrm{d}t$ gives the change in momentum of the particle.

8.2 Conservative Force Fields

If we consider how to calculate the amount of work done directly from the definition (8.1) and Figure 8.1, a fundamental problem is immediately evident: in order to calculate the work done by the force it is necessary to know the orbit of the particle. Since the basic problem of dynamics is exactly to compute the orbit of the particle when the force is known, one must solve the entire problem first in order to compute the work done by the force. It may seem then, that the work integral $\int \mathbf{F} \cdot \mathbf{dr}$ is not a useful concept at all. In fact, as we shall see, the work integral is of immense value, in particular when the motion is caused by so-called conservative forces.

> A force field is called *conservative* if the force vector $\mathbf{F}$ of the field depends only on the position $\mathbf{r}$ of the particle and the work integral $\int_A^B \mathbf{F} \cdot \mathbf{dr}$ is independent of the path of integration, depending only on the initial point A and the final point B, of the path.

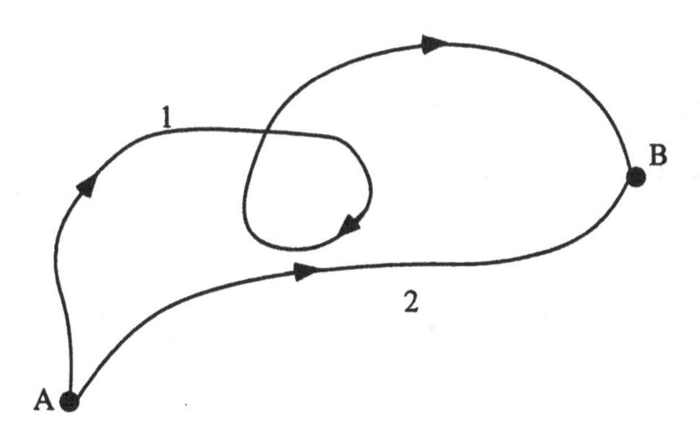

Fig. 8.2. Defining a conservative force field. The particle is moved from A to B along the path 1 or along the path 2

Figure 8.2 shows two points: A and B, and two different paths, 1 and 2, along which the particle can be moved (by you, for instance!) from A to B. The condition that the force field is conservative is:

$$\underset{\text{along 1}}{\int_A^B} \mathbf{F} \cdot \mathbf{dr} = \underset{\text{along 2}}{\int_A^B} \mathbf{F} \cdot \mathbf{dr} \,,$$

where path 1 and path 2 are arbitrary, and A and B can be any pair of points in the region of space where the force field acts.

An equivalent formulation of this condition is

$$\int_{\substack{A \\ \text{along } 1}}^{B} \mathbf{F} \cdot d\mathbf{r} - \int_{\substack{A \\ \text{along } 2}}^{B} \mathbf{F} \cdot d\mathbf{r} = 0,$$

or

$$\oint \mathbf{F} \cdot d\mathbf{r} = 0 .$$

This is the most common definition of a conservative force field: The integral of the force $\mathbf{F}$ around any closed curve in the field, is zero.

The force of friction is an example of a force that is *not* conservative. The frictional force always points against the direction of motion. If we drag a block around a closed curve on a table, the force of friction will do a finite amount of work on the block. In this case $\oint \mathbf{F} \cdot d\mathbf{r} \neq 0$ so friction is not conservative. Friction is not a fundamental interaction. Friction is ultimately a consequence of electromagnetic interactions between molecules in the block and molecules in the table. In dragging the block around the closed curve a certain amount of heat has been produced, i.e., the chaotic thermal motion of the atoms in the block and the table has – through frictional coupling – been increased. See Example 8.9.

8.3 Central Force Fields

A *central* field of force is a field where the force $\mathbf{F}(r)$ is everywhere directed towards (or away from) a center O, and where the magnitude of the force depends only on the distance r of the particle from O.

We proceed to show that all central fields of force are conservative. In Figure 8.3 we have chosen two points, A and B in the field, and drawn two arbitrary paths between them, path 1 and path 2. The field is conservative if the work integrals along such two arbitrary paths are equal.

As an aid to the computation we have drawn two dashed arcs centered in O, and with radii r and $r + dr$, respectively. These two arcs bound small segments $d\mathbf{r}_1$ on path 1 where the force is $\mathbf{F}_1$, and $d\mathbf{r}_2$ on path 2 where the force is $\mathbf{F}_2$. Since the two points have the same distance r from O, the magnitude of the forces is the same, $F = |\mathbf{F}_1| = |\mathbf{F}_2|$. The angles between $d\mathbf{r}_1$ and $\mathbf{F}_1$ and $d\mathbf{r}_2$ and $\mathbf{F}_2$ are denoted θ_1 and θ_2 as in Figure 8.3; consequently

$$|d\mathbf{r}_1| \cos(\theta_1) = |d\mathbf{r}_2| \cos(\theta_2) = dr ,$$

and thus

$$\mathbf{F}_1 \cdot d\mathbf{r}_1 = \mathbf{F}_2 \cdot d\mathbf{r}_2 = F dr .$$

The contribution to the work integral from the two segments is the same, and consequently

$$\int_{\substack{A \\ \text{along } 1}}^{B} \mathbf{F} \cdot d\mathbf{r} = \int_{\substack{A \\ \text{along } 2}}^{B} \mathbf{F} \cdot d\mathbf{r} .$$

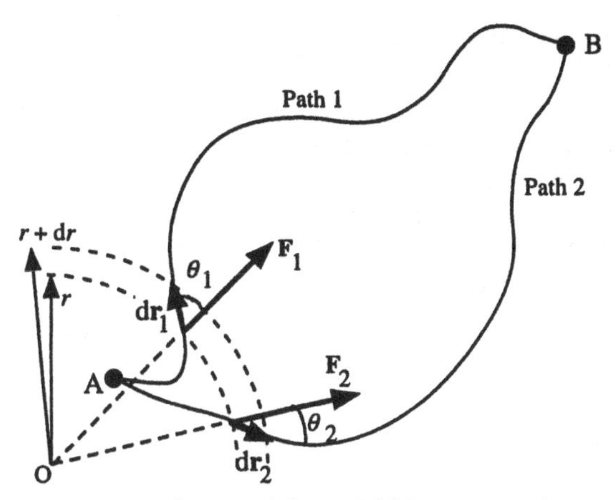

Fig. 8.3. A central force field is conservative

The work integral is independent of the path, and the field is thus conservative. In particular this means that the work done by the force along a closed loop is zero:

$$\underset{\text{along 1}}{\int_{\mathrm{A}}^{\mathrm{B}}} \mathbf{F}\cdot d\mathbf{r} + \underset{\text{along 2}}{\int_{\mathrm{B}}^{\mathrm{A}}} \mathbf{F}\cdot d\mathbf{r} = \oint \mathbf{F} d\mathbf{r} = 0\,.$$

The gravitational force on the Earth from the Sun is a central force. If we take the center of the Sun to be a fixed point in an inertial frame, the gravitational force field around the Sun is a central field. The electrostatic force on one proton from another proton is also a central force, a repulsive force.

8.4 Potential Energy and Conservation of Energy

For a particle in a conservative force field, we can introduce the concept of potential energy. The potential energy $U = U(\mathbf{r})$ is a scalar function of the position vector $\mathbf{r}$ of the particle.

Choose an arbitrary point P in the force field as the reference point for U. We define the value of U at this point to be zero, $U(\mathrm{P}) = 0$. The potential energy at any other point A is now defined through the equation:

$$U(\mathrm{A}) \equiv -\int_{\mathrm{P}}^{\mathrm{A}} \mathbf{F}\cdot d\mathbf{r}\,. \tag{8.5}$$

In words: the potential energy at the point A is equal to minus the work that the *force field* does on the particle when the particle is moved from the reference point P to the field point A.

Note that we do not have to specify along which path the particle was brought from P to A. The field is assumed to be conservative, so the work is independent of the path. This is precisely the reason why potential energy makes sense only for conservative force fields.

Furthermore, we can now establish the important theorem about conservation of mechanical energy in a conservative force field. Consider two arbitrary points A and B in the field (Figure 8.4). We have:

$$\int_{\mathrm{P}}^{\mathrm{A}} \mathbf{F} \cdot d\mathbf{r} = -U(\mathrm{A}) \quad \text{and} \quad \int_{\mathrm{P}}^{\mathrm{B}} \mathbf{F} \cdot d\mathbf{r} = -U(\mathrm{B}) \,.$$

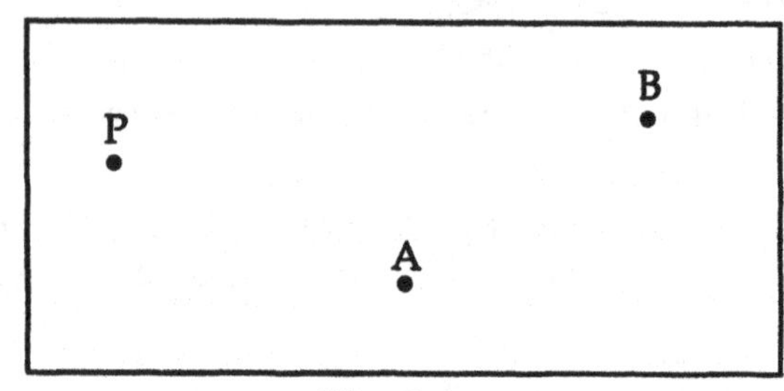

Fig. 8.4.

From this:

$$\int_{\mathrm{A}}^{\mathrm{B}} \mathbf{F} \cdot d\mathbf{r} = \int_{\mathrm{A}}^{\mathrm{P}} \mathbf{F} \cdot d\mathbf{r} + \int_{\mathrm{P}}^{\mathrm{B}} \mathbf{F} \cdot d\mathbf{r} = U(\mathrm{A}) - U(\mathrm{B}) \,.$$

Furthermore, from the theorem (8.2) on the work integral we have:

$$\int_{\mathrm{A}}^{\mathrm{B}} \mathbf{F} \cdot d\mathbf{r} = \frac{1}{2} m v_{\mathrm{B}}^2 - \frac{1}{2} m v_{\mathrm{A}}^2 \,.$$

The conclusion is that for two arbitrary points A and B:

$$\frac{1}{2} m v_{\mathrm{B}}^2 - \frac{1}{2} m v_{\mathrm{A}}^2 = U(\mathrm{A}) - U(\mathrm{B})$$

or, with the terms in a different order,

$$\frac{1}{2} m v_{\mathrm{B}}^2 + U(\mathrm{B}) = \frac{1}{2} m v_{\mathrm{A}}^2 + U(\mathrm{A}) \,.$$

We see, that the sum of the kinetic energy T and the potential energy U does not change when the force causes the particle to move from A to B. This sum $T + U$ is called the total mechanical energy of the particle, or just the mechanical energy. Since A and B were arbitrary, the mechanical energy is a constant of the motion. Or equivalently, the mechanical energy of a particle is *conserved* for motion in the conservative force field:

$$\boxed{T + U = E_0 = \text{constant}}$$

Under the influence of a conservative force field, the particle moves in such a way that the sum of the kinetic and potential energy has a constant value. The value of this constant depends only on the initial conditions.

Actually, all fundamental forces in nature are conservative. The notion of conservative forces has been introduced to exclude frictional forces. The frictional force, which is not a fundamental force, converts mechanical kinetic energy into heat. As is well known, heat is basically motion of the atoms constituting the warm body in question. If we could keep track of each and every atom, the law of conservation of energy would be generally true, and there would be no need for the introduction of friction forces, or nonconservative forces.

The theorem about conservation of mechanical energy is of immense importance in physics.

A Mathematical Note. In order to define the potential energy in (8.5) we considered the force **F** to be known as a function of position. In some cases it may happen that it is the potential energy function $U(\mathbf{r})$ which is known. The question is then, what force field **F** has given rise to $U(\mathbf{r})$.

According to (8.5), the potential energy results from an integration of the force. Since differentiation and integration are opposite operations, the force can be found by differentiation of the potential energy function. Let us first consider the one-dimensional case (two examples of this are in the next subsection); here, (8.5) has the form:

$$\int_{\mathrm{P}}^{x} F(x')\mathrm{d}x' = -U(x) .$$

From calculus we then know that

$$F(x) = -\frac{\mathrm{d}U(x)}{\mathrm{d}x} .$$

This result can be generalized to the three-dimensional case. Here, the force **F** is a vector with three components (say, in a cartesian coordinate system) F_x, F_y, and F_z, and the potential energy is (in general) a function of all the spatial coordinates, x, y, and z. The component F_x is then found by differentiating U with respect to x, considering y and z to be constants through the differentiation. This operation is known as *partial differentiation* and is denoted by the symbol $\partial U/\partial x$:

$$F_x = -\frac{\partial U(x,y,z)}{\partial x}$$

Similarly, one has for the other components:

$$F_y = -\frac{\partial U(x,y,z)}{\partial y}$$

and

$$F_z = -\frac{\partial U(x,y,z)}{\partial z}$$

The vector

$$\mathbf{grad}\, U = (\frac{\partial U}{\partial x}, \frac{\partial U}{\partial y}, \frac{\partial U}{\partial z})$$

is known as the gradient vector for U. It points in the direction where the change in U per unit of length is largest. In compact notation, $\mathbf{F} = -\mathbf{grad}\, U$.

For a one-dimensional motion of a particle the force is conservative if it is a function of position only. The question arises as to whether a corresponding statement is true of three-dimensional motion. Phrased differently: if the force acting on a particle is a function of the position coordinates only, is there then always a function $U(x,y,z)$ that can serve as the potential energy function?

The answer is no. The components of the force have to satisfy certain criteria if a potential function is to exist.

Assume that a potential function U does exist. Then we have:

$$\frac{\partial F_x}{\partial y} = -\frac{\partial^2 U}{\partial y \partial x}\,, \qquad \frac{\partial F_y}{\partial x} = -\frac{\partial^2 U}{\partial x \partial y}\,.$$

The order of differentiation can be interchanged, and the two expressions are equal. That means:

$$\frac{\partial F_x}{\partial y} = \frac{\partial F_y}{\partial x}\,, \qquad \frac{\partial F_y}{\partial z} = \frac{\partial F_z}{\partial y}\,, \qquad \frac{\partial F_z}{\partial x} = \frac{\partial F_x}{\partial z}\,. \tag{8.6}$$

These expressions are the necessary conditions on F_x, F_y, and F_z for a potential energy function to exist. These conditions express the condition that

$$\mathrm{d}W = F_x \mathrm{d}x + F_y \mathrm{d}y + F_z \mathrm{d}z$$

is a total (exact) differential.

In books on calculus it is shown that these conditions are also *sufficient*: if (8.6) holds at all points in space, the force components are derivable from some potential function $U(x,y,z)$.

8.5 Calculation of Potential Energy

When using the energy theorem, it is important to master the techniques of calculating potential energy. Below are a few examples of the calculation of U in simple situations.

Advice: use the definition (8.5): $U(\mathrm{A}) = -\int_{\mathrm{P}}^{\mathrm{A}} \mathbf{F}\cdot\mathrm{d}\mathbf{r}$, or $-U(\mathrm{A}) = \int_{\mathrm{P}}^{\mathrm{A}} \mathbf{F}\cdot\mathrm{d}\mathbf{r}$, i.e., use the work the field does on the particle and not the work "you" may happen to do "in moving the particle". We have not introduced that work at all.

Example 8.1. Constant Gravitational Field. We choose a z-axis with direction upwards and origin at the surface of the Earth. We choose $U = 0$ for $z = 0$. The force on a particle is parallel to the z-axis, and near the surface of the Earth it has the constant value: $F(z) = -mg$.

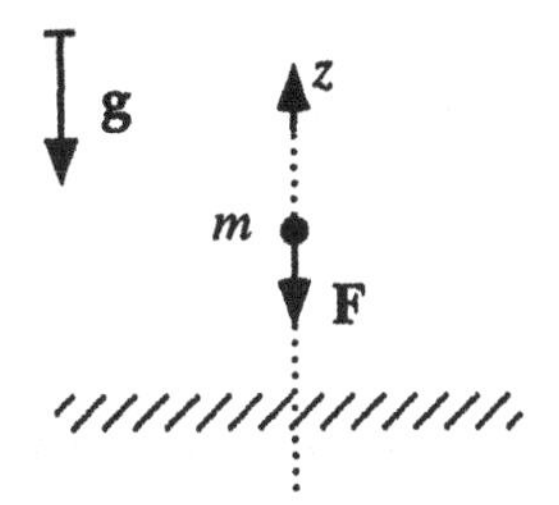

The potential energy at a height h above the surface is then found from:

$$\int_0^h (-mg)\mathrm{d}z' = -U(h)\,,$$

which gives

$$U(h) = mgh\,.$$

△

Example 8.2. Spring Force

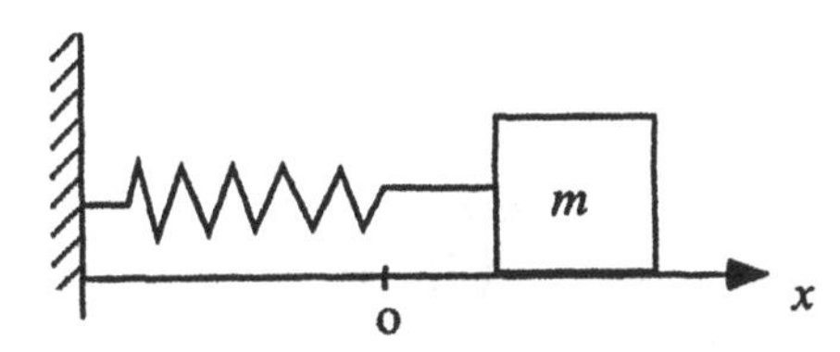

The horizontal x-axis has its origin at the equilibrium point of the spring. We choose $U = 0$ at $x = 0$. When the spring is stretched or compressed to a position x away from the equilibrium position it acts with a force $F = -kx$ on the mass, where k is the spring constant. We then have:

$$\int_0^x (-kx')\mathrm{d}x' = -U(x)\,,$$

or

$$U(x) = \frac{1}{2}kx^2\,.$$

△

Example 8.3. Gravity Outside a Homogeneous Sphere. We are assuming here that the gravitational field around a homogeneous sphere is known to be the same as if all the mass of the sphere were assembled in the center. This theorem will be proven in the next section.

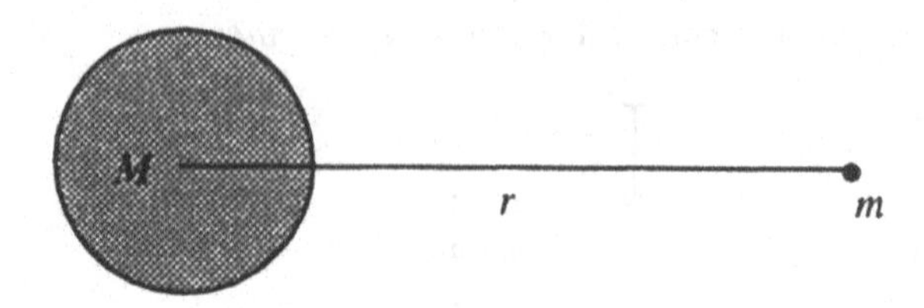

The force depends only on the distance r of the particle from the center. The value is

$$F(r) = -G\frac{Mm}{r^2} .$$

We set $U = 0$ at infinity. We then find:

$$-U(r) = -\int_{\infty}^{r} G\frac{Mm}{r'^2}\mathrm{d}r' .$$

From this we get

$$U(r) = \left[-G\frac{Mm}{r'}\right]_{\infty}^{r} = -G\frac{Mm}{r} .$$

△

8.6 The Gravitational Field Around a Homogeneous Sphere

We shall now demonstrate the important and often used fact that outside a homogeneous spherical distribution of mass the gravitational field is the same as if the entire mass was concentrated in the center of the sphere.

8.6.1 The Field Around a Spherical Shell

Let us begin by considering a thin spherical shell with its center in the point O, a radius R, and a total mass M_s (Figure 8.5). We are going to calculate the potential energy of a particle of mass m at the distance r from the center of the shell. That the shell is "thin" means that the thickness of the shell is negligible relative to the other length scales in the problem, R and r. Since we are dealing with a sphere only the distance of the mass from the point O matters.

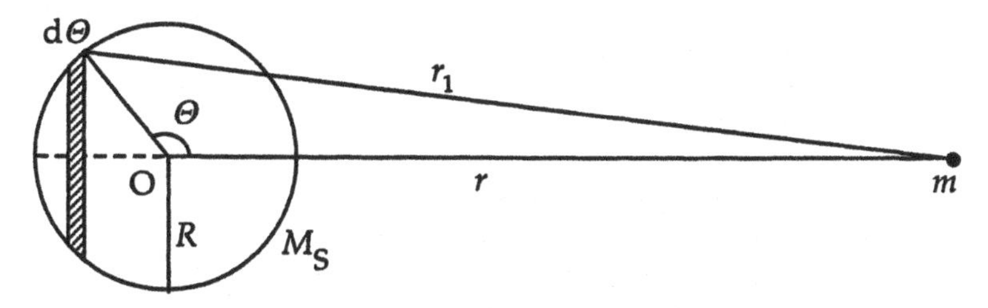

Fig. 8.5. Calculating the potential energy outside a thin spherical shell

A small surface element on the shell can be considered a point mass $\mathrm{d}M_s$. The distance to m is r_1, and we know that $\mathrm{d}M_s$ contributes to the potential energy of m with an amount

$$\mathrm{d}U(r) = -G\frac{\mathrm{d}M_s m}{r_1} \, .$$

This follows from Example 8.3 . The reference point for U is chosen as in Example 8.3, i.e., $U(\infty) = 0$.

Those points of the spherical shell that all have the distance r_1 from m form a circle perpendicular to the line through m and O, and with the center on that line. The angle between $\mathbf{r}$ and the vector to some point on the circle is denoted by Θ. The points on the shell with distance to m between r_1 and $r_1 + \mathrm{d}r_1$ constitute a ring, as indicated on Figure 8.5. Corresponding to the edges of the ring are the angles Θ and $\Theta + \mathrm{d}\Theta$. The width of the ring is then $R\mathrm{d}\Theta$ and the radius is $R\sin\Theta$. The total area of the ring is $2\pi R^2 \sin\Theta \mathrm{d}\Theta$.

Denote by σ the mass per unit of area of the shell:

$$\sigma = \frac{M_s}{4\pi R^2} \, .$$

The amount of mass in the ring is the area of the ring times σ. The contribution of the ring to the potential energy of m at r_1 is

$$\mathrm{d}U(r) = -G\frac{\sigma 2\pi R^2 \sin(\Theta)\mathrm{d}\Theta \cdot m}{r_1} \, . \tag{8.7}$$

We can now integrate the contributions from all the different rings that make up the shell. The variables r_1 and Θ are related by the cosine relation:

$$r_1^2 = R^2 + r^2 - 2rR\cos(\Theta) \, ,$$

which gives rise to the relation

$$r_1 \mathrm{d}r_1 = rR\sin(\Theta)\mathrm{d}\Theta.$$

This is inserted into (8.7) and one gets:

$$U(r) = -G\frac{\sigma 2\pi Rm}{r}\int_{r-R}^{r+R} dr_1 = -G4\pi R^2\sigma\frac{m}{r}\,.$$

Finally, using the definition of σ, we arrive at:

$$U(r) = -G\frac{M_s m}{r}, \; r > R\,.$$

The potential energy of the point mass m outside the spherical shell depends on r as if the entire mass of the shell was concentrated at the center!

It is also of interest to know the potential inside the shell, i.e., for points with $r < R$. The integration procedure is the same as before, only the lower limit is changed (see Figure 8.6):

$$U(r) = -G\frac{\sigma 2\pi Rm}{r}\int_{R-r}^{r+R} dr_1 = -G4\pi R\sigma m\,.$$

From this,

$$U(r) = -G\frac{M_s m}{R}, \quad r < R\,.$$

Inside the shell then, the potential energy is constant, maintaining the value which the outside potential assumes at the surface of the shell, $r = R$.

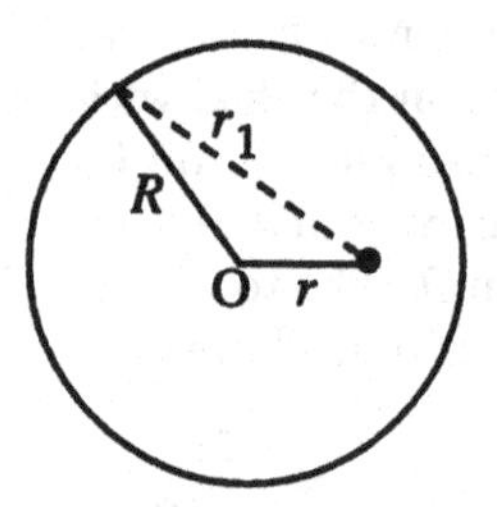

Fig. 8.6.

The force $\mathbf{F}$ on the particle is derived from the potential energy by differentiation with respect to the spatial coordinates. Expressed in terms of the unit radial vector $\mathbf{e}_r = \mathbf{r}/r$, we have:

$$\mathbf{F} = -\frac{dU}{dr}\mathbf{e}_r\,.$$

Using the two expressions for U inside and outside the shell, we get:

$$\mathbf{F} = \begin{cases} 0 & r < R; \\ -G(M_s m/r^2)\mathbf{e}_r & r > R\,. \end{cases}$$

Inside the shell, the particle does not feel any gravitational pull at all, and outside the shell the force is the same as it would be between two point masses, M_s and m respectively, separated by r. Figure 8.7 illustrates the dependence of U and $\mathbf{F}$ on r.

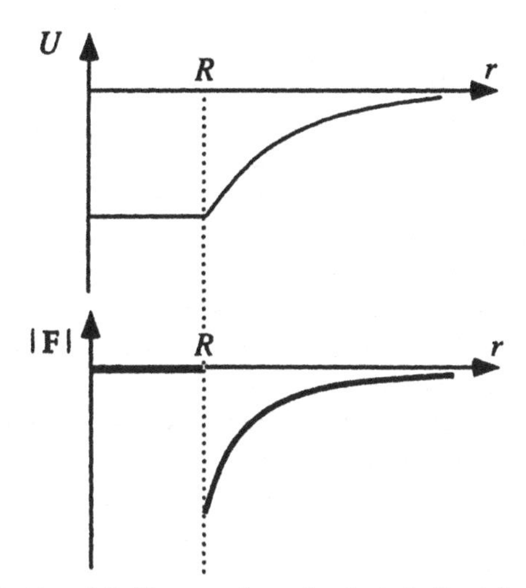

Fig. 8.7. The gravitational field around a spherical shell with radius R. Potential energy and magnitude of force as functions of distance from the center

8.6.2 A Solid Sphere

We now return to the original problem: the gravitational field around a homogeneous sphere. We divide (as the layers of an onion) the sphere into concentric shells all of which are so thin that the above calculation applies. Denote by m_i the mass of the ith shell and let there be N shells total. If the test particle is outside the sphere, $r > R$, the ith shell contributes to the potential energy of the particle by an amount

$$U_i(r) = -G\frac{m_i m}{r}, \qquad r > R,$$

and the total potential energy is

$$U(r) = \sum_{i=1}^{N} U_i(r) = -\sum_{i=1}^{N} G\frac{m_i m}{r};$$

and since $\sum_{i=1}^{N} m_i = M$, we get

$$U(r) = -G\frac{Mm}{r}, \qquad r > R. \tag{8.8}$$

Again, this is the potential energy the test particle would have if the entire mass of the sphere was concentrated in its center.

It is instructive to compute the potential energy inside the sphere as well. We can think of the particle moving through the surface of the sphere into the interior, perhaps moving in a narrow shaft. If we know the force $\mathbf{F}$ which acts on the particle at some distance r_1 from the center, we also know the potential energy as a function of distance:

$$\int_R^r \mathbf{F}(r_1)\,\mathrm{d}\mathbf{r}_1 = U(R) - U(r)\,.$$

In order to find the force at r_1, we divide the sphere into two regions. One is the shell with inner radius r_1 and outer radius R at the surface of the sphere. The rest is the small sphere with radius r_1. The particle is imagined to be located at the surface of the small sphere with radius r_1. The region outside the small sphere, i.e., the region with $r_1 < r < R$, may be thought of as composed of a number of thin shells, all of which gives rise to no force on a particle within them. The only force on the particle arises from the matter in the small sphere with radius r_1.

The matter is distributed homogeneously, so the inner sphere has the mass in proportion to its volume: $(r_1/R)^3 \cdot M$. The force $\mathbf{F}(r_1)$ is then

$$\mathbf{F}(r_1) = -G\left(\frac{r_1}{R}\right)^3 \frac{Mm}{r_1^2}\mathbf{e}_r = -G\frac{Mm}{R^3}\mathbf{r}_1, \qquad r < R\,.$$

The work integral can now be calculated:

$$\int_R^r \mathbf{F}(r_1)\cdot\mathrm{d}\mathbf{r}_1 = -G\frac{Mm}{R^3}\int_R^r r_1\mathrm{d}r_1 = -G\frac{Mm}{2R^3}\left(r^2 - R^2\right).$$

This expression equals $U(R) - U(r)$. We know from the previous calculation that $U(R) = -GMm/R$, so

$$-G\frac{Mm}{R} - U(r) = -G\frac{Mm}{2R}\left(\frac{r^2}{R^2} - 1\right),$$

i.e.,

$$U(r) = -G\frac{Mm}{2R}\left(3 - \frac{r^2}{R^2}\right). \tag{8.9}$$

Figure 8.8 shows the radial dependence of U and $|\mathbf{F}|$.

Note, that the quantities U and F computed above are the potential energy and force respectively on a test particle with mass m. If one is interested in characterizing the field around a spherical shell or a solid sphere, it is customary to introduce the *gravitational potential* Φ and the *field strength* $\mathbf{g}(r)$ by dividing out the mass m of the test particle:

$$\Phi(r) \equiv \frac{U(r)}{m},$$

and

$$\mathbf{g}(r) \equiv \frac{\mathbf{F}}{m}.$$

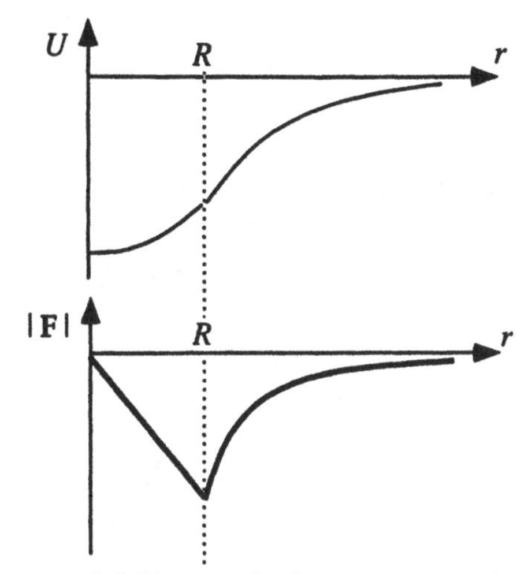

Fig. 8.8. The gravitational field around a homogeneous solid sphere with radius R. Potential energy and magnitude of force as functions of distance from center

8.7 Examples

Example 8.4. Particle on a Frictionless Curve in the Gravitational Field of the Earth

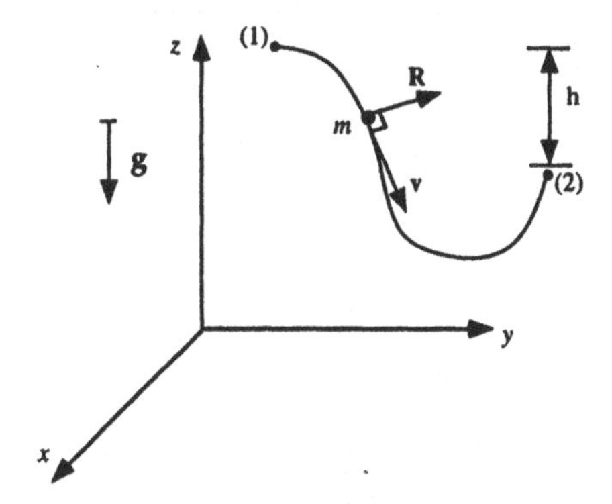

Remember: a frictionless curve means that the reaction force **R** (the force of constraint) is perpendicular to the curve at any point, which again means that $\mathbf{R} \cdot \mathbf{v} = 0$. The reaction force performs no work on the particle when the particle moves along the curve. The velocity of the particle when it reaches point (2) is independent of the shape of the curve and dependent only on the initial velocity and the height h through which the particle has fallen. If we start the particle in (1) with velocity zero, its velocity in (2) is determined by

$$\frac{1}{2}mv^2 = mgh, \quad v = \sqrt{2gh}\,.$$

△

Example 8.5. String Force in the Pendulum

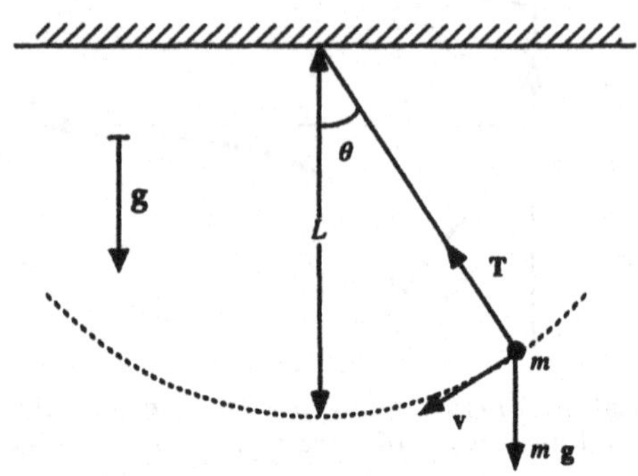

A pendulum has mass m and string length L. Find the string force at the lowest point of the orbit if the pendulum is started with velocity zero from a point where $\theta = 60°$. (The string is assumed to be massless and taut at all times). Note: the string force $\mathbf{T}$ performs no work, $\mathbf{T} \cdot \mathbf{v} = 0$ at every point.

Solution.

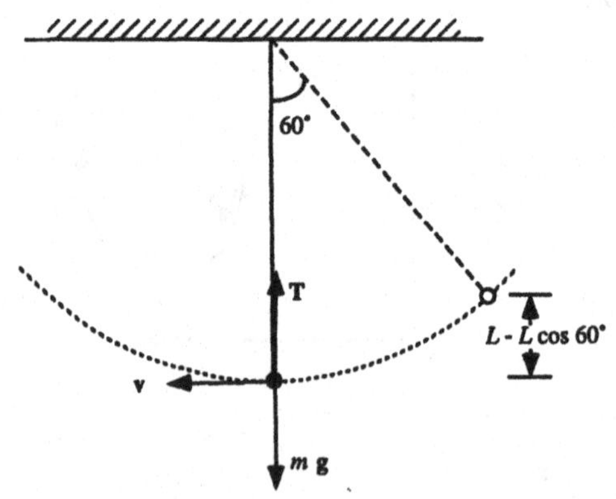

The motion of m is *not* uniform circular motion. The particle has both a centripetal and a tangential acceleration. The string force gives only a contribution to the centripetal acceleration.

At the lowest point of the orbit we have

$$m\frac{v^2}{L} = T - mg\,.$$

From conservation of energy,

$$\frac{1}{2}mv^2 = mg\left[L - L\cos(60°)\right] .$$

Combining the above two equations, we get

$$T = 2mg .$$

Questions. Are there any points on the orbit where the tangential acceleration is zero? Are there any points on the orbit where the centripetal acceleration is zero? △

Example 8.6. The Gravitational Potential Outside the Earth. The binding energy of a particle in the gravitational field of the Earth is the amount of energy that must be supplied to the particle in order that the particle is just able to escape the gravitational field of the Earth.

(1) What is the binding energy of a satelite with mass 100 kg at rest at the surface of the Earth. (Let us in this question assume that the Earth is a homogeneous nonrotating sphere.)
(2) Assume now that the satelite is at rest at the equator of the rotating Earth. Would the kinetic energy of the satelite (relative to the absolute geocentric frame) make any difference in its binding energy to the Earth?

Solution.

(1)

$$W = \int_R^\infty G\frac{mM}{r^2}dr = m\frac{GM}{R} = 6.3 \times 10^9 \text{ J}.$$

(2) The kinetic energy, relative to the absolute geocentric frame, of the satellite rotating along with the Earth (at the equator) is:

$$T = \frac{1}{2}m\left(\frac{2\pi R}{\tau}\right)^2 .$$

where $\tau = 23$ hours 56 min, and $R = 6.37 \times 10^6$ m. We find:

$$T = 1.1 \times 10^7 \text{ J} = 0.011 \times 10^9 \text{ J}.$$

As long as we want our binding energy with only one significant digit the kinetic energy of the satelite at rest relative to the Earth makes no contribution to the binding energy of the satelite.

△

Example 8.7. Potential Energy Due to Electric Forces. What is the ionization energy of the hydrogen atom in its ground state? An equivivalent question is: what is the binding energy of the electron in the ground state of the hydrogen atom?

Solution. In the ground state the average distance between the proton and the electron has the value $a = 0.53 \times 10^{-10}$ m. Classical mechanics may be used in the stationary states. In the classical picture of the ground state, the electron executes uniform circular motion around the proton. The binding energy is thus:

$$W = \frac{1}{4\pi\epsilon_0}\frac{q^2}{a} - \frac{1}{2}mv^2 ,$$

$$W = \frac{1}{4\pi\epsilon_0}\frac{q^2}{a} - \frac{1}{2}\left(\frac{1}{4\pi\epsilon_0}\right)\frac{q^2}{a} .$$

Inserting numerical values we get: $W = 13.6$ eV. (1 eV $= 1.6 \times 10^{-19}$ J; $q = 1.6 \times 10^{-19}$ C; $1/(4\pi\epsilon_0) \approx 9 \times 10^9\ \mathrm{N\,m^2\,C^{-2}}$.)

Note: the ratio of the electrical and the gravitational attraction between the proton and the electron in the hydrogen atom is:

$$\frac{F_{\mathrm{el}}}{F_{\mathrm{G}}} = \frac{q^2}{4\pi\epsilon_0}\frac{1}{Gm_pm_e} \approx 2.3 \times 10^{39} \approx 10^{40} .$$

This is one of the dimensionless numbers that a complete theory combining electrical and gravitational forces should be able to explain. We do not at present understand how such an enormous dimensionless number may appear. Does it have something to do with cosmology?

There are about $(10^{40})^2 = 10^{80}$ protons in the observable universe. The universe is believed to be about 4.7×10^{17} old. A second, however, is not a fundamental time scale. As a fundamental time scale, one could take the time it takes light to move the distance known as the "classical electron diameter": if an electron is considered a classical particle (as opposed to quantum mechanical), a diameter of 4.5×10^{-13} cm can be associated with the electron. The time for light to cross this distance is thus $4.5 \times 10^{-13}/3 \times 10^{10} = 1.5 \times 10^{-23}$ s. Measured in this unit of time, the age of the universe is approximately 10^{40} !

If you want to dig deeper you may consult *The Feynman Lectures* Vol I, Chapter 7, section 7-7. The question is certainly interesting and fundamental: Are there any relations between the properties of a single particle (proton, neutron, electron, and so on) and the rest of the Universe? The entire problem is also connected to Mach's principle, discussed in Chapter 7. △

Example 8.8. A Tunnel Through the Earth. Assume that a very narrow straight tunnel has been drilled through the Earth. The tunnel passes through the center of the Earth and has been drilled along the axis of rotation of the Earth. (No fictitious forces act on the particle in the tunnel.) The Earth is assumed to be a sphere with a homogeneous mass distribution and radius R.

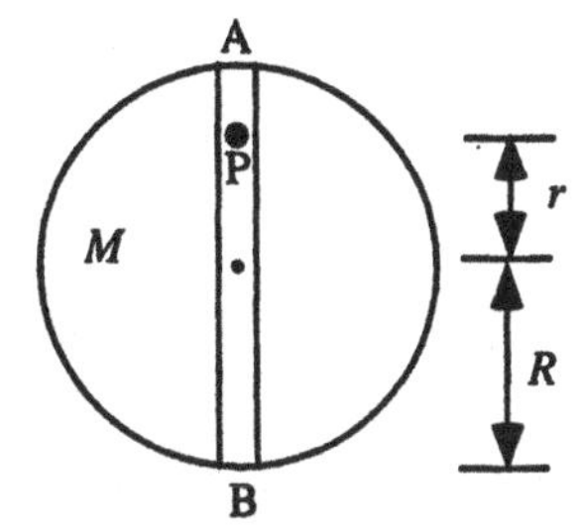

(1) Show that a particle P in the tunnel may perform harmonic oscillations. Neglect frictional effects.
(2) Let us assume that the particle is dropped with initial velocity zero at one end of the tunnel, e.g., at B. How long will it take until the particle for the first time appears at A ?

Solution.

(1) The tunnel is very narrow! The force on the particle is of magnitude $F = GMmr/R^3$ and always directed towards the center of the Earth. The particle thus has an equilibrium position at the center of the Earth and – as F is proportional to the displacement r – the particle will perform harmonic oscillations about the equilibrium position.
(2) The mass performs harmonic oscillations according to the equation of motion:

$$m\frac{\mathrm{d}^2r}{\mathrm{d}t^2} = -G\frac{mM}{R^3}r ,$$

or

$$m\frac{\mathrm{d}^2r}{\mathrm{d}t^2} = -kr , \quad k \equiv G\frac{mM}{R^3} .$$

The period of oscillation is

$$T = 2\pi\sqrt{\frac{m}{k}} .$$

The first time the particle emerges at A it has gone through one half oscillation. The time is

$$\begin{aligned} t &= \frac{T}{2} = \pi\sqrt{\frac{R^3}{GM}} = \pi\sqrt{\frac{R}{g}} \\ &= 42.2 \text{ minutes.} \end{aligned}$$

Note. If a particle "falls along" the surface of the Earth (we neglect frictional effects) we have:

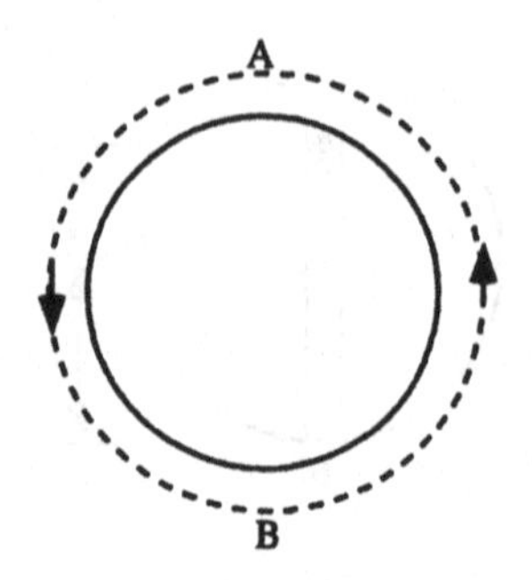

$$m\frac{v^2}{R} = G\frac{mM}{R^2} = mg\,,$$
$$v = \sqrt{Rg}\,.$$

The time it takes the particle to go from B to A is:

$$t = \frac{\pi R}{v} = \pi\sqrt{\frac{R}{g}} = 42.2 \text{ minutes.}$$

A harmonic motion is equivalent to the projection of a uniform circular motion onto a diameter. △

Example 8.9. The Asymmetry of Nature

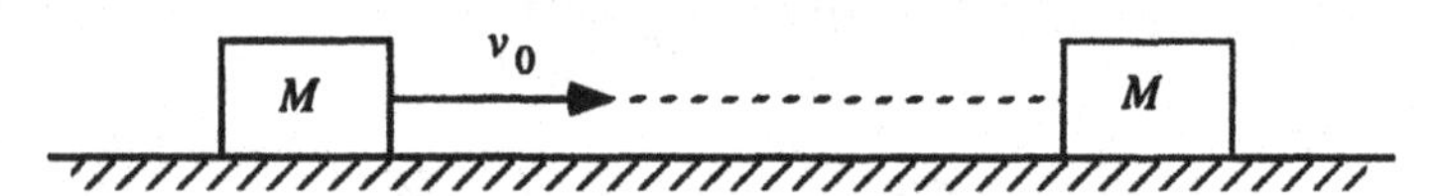

Consider a block of mass M on a horizontal table. The coefficient of friction between the block and the table is μ. The block is now given an initial velocity v_0. The block then slides across the table and stops after a distance D determined by:

$$\frac{1}{2}Mv_0^2 = \mu MgD\,.$$

The original macroscopic kinetic energy $T = \frac{1}{2}Mv_0^2$ has – through the work of the frictional forces – been converted into microscopic motion of the atoms in the block and the table.

The main outcome of the experiment is that the original macroscopic and "ordered" kinetic energy has been converted into "disordered" kinetic energy, also called heat.

What about the reverse process? Would it be possible that the block and the table cooled, and the energy thus liberated was converted into macroscopic ordered kinetic energy, i.e., would it be possible that the block at rest suddenly started to move back, while the temperature of the table dropped? Such a process would certainly not contradict the energy conservation theorem in its widest sense. If we kept track of each and every single atom in the block and the table, we would find that their kinetic energy had dropped by exactly the amount of energy the block had gained. Why do we then feel with certainty that the block will not start to move? This question touches one of the most fundamental theorems of theoretical physics: the second law of thermodynamics. We cannot of course formulate the law at this point. We may state that energy in the form of disordered (thermal) energy never spontaneusly (i.e., by itself) converts completely into ordered macroscopic kinetic energy. In other words: energy located on "many degrees of freedom" never collects spontaneusly on "few degrees of freedom". In due time you will learn – loosley speaking – that:

– ordered motion in its entirety can always be transferred into disordered (thermal) motion;
– disordered (thermal) motion will never spontaneously convert completely into ordered (macroscopic) motion.

In other words: work can be converted into heat, but conversion of heat into work needs deeper considerations. Some disordered (thermal) motion can be converted into ordered motion, i.e., macroscopic kinetic energy. For example: the combustion energy in a rocket engine is disordered (thermal) motion. Some of this thermal energy is converted into macroscopic kinetic energy in the majestic liftoff of the rocket! But you will find that not all of the energy liberated in the combustion is converted into macroscopic kinetic energy.

A large part of our civilization rests on the conversion of heat into work via combustion engines of various sorts. Even the engine inside our bodies – among other tasks – converts disordered kinetic energy (gained from the oxidation of reduced carbon compounds) into macroscopic kinetic energy when we move.

Actually, the entire Universe is expanding, and the energy in the Universe is continually being spread into more and more degrees of freedom. Life itself is a by-product of this process. Gravity and the second law of thermodynamics govern how the world works on the large scale – but it takes some time and effort to understand the law of gravity and its connection to the second law of thermodynamics. △

Remark. It is sometimes stated from government agencies that we should have as a goal to conserve energy. Taken litterally, the statement is without meaning: In any process, be it fire, wind, combustion, chemical reactions, or whatever, the energy is the same before, during, and after the process.

What is it then that we want to conserve? It is evidently not the energy. You will learn about this in detail, when you study thermodynamics. The essential point is that the constant energy of the Universe is distributed in a different manner before and after a natural process.

8.8 Review: Conservative Forces and Potential Energy

The kinetic energy T of a particle is given by $T = \frac{1}{2}mv^2$.

I: For all types of forces the following is always true for a particle:

$$\frac{dT}{dt} = \mathbf{F} \cdot \mathbf{v},$$

or, in integral form,

$$\Delta T = T_2 - T_1 = \int_1^2 \mathbf{F} \cdot \mathrm{d}\mathbf{r}.$$

II: A force field is called *conservative* if the work performed by the forces of the field when a particle is moved from position 1 to position 2 in the field, is independent of the path of the particle:

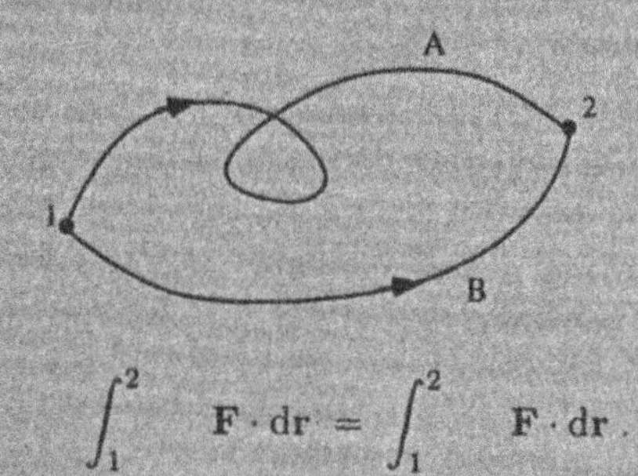

$$\underset{\text{along A}}{\int_1^2} \mathbf{F} \cdot \mathrm{d}\mathbf{r} = \underset{\text{along B}}{\int_1^2} \mathbf{F} \cdot \mathrm{d}\mathbf{r}.$$

An equivalent definition is: $\oint \mathbf{F} \cdot \mathrm{d}\mathbf{r} = 0$ along any closed path in the field.

III: For a conservative force field one can introduce the *potential energy* $U(x, y, z)$ of the particle, defined by:

$$\int_P^{(x,y,z)} \mathbf{F} \cdot \mathrm{d}\mathbf{r} = -U(x, y, z).$$

Here $-U(x, y, z) \equiv$ the work performed by the forces of the field when the particle is moved from an arbitrarily chosen zero position (P) to the point (x, y, z) in the field.

IV: In a conservative force field the conservation law for mechanical energy is valid:

$$T + U = \text{constant}.$$

In words: a particle moving under the influence of a conservative force moves in such a way that the sum of its kinetic energy and its potential energy has a constant value, determined by the initial conditions.

8.9 Problems

Problem 8.1. A water jet from a fire hose has a muzzle velocity $u = 25$ $\mathrm{m\,s^{-1}}$. How high would the water jet go if it were directed vertically upwards from the surface of the Earth?

Problem 8.2. Calculate the escape velocity for the planet Jupiter. In 1994 fragments of the comet Shoemaker-Levy hit the surface of Jupiter with a velocity of about 60 $\mathrm{km\,s^{-1}}$.

Problem 8.3.

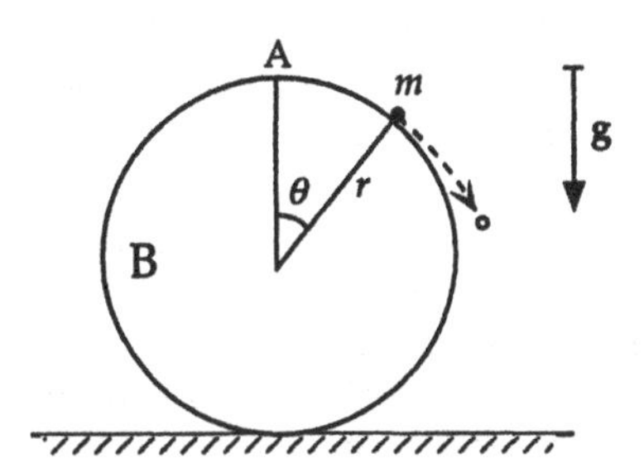

A small particle of mass m slides without friction on a large sphere B. The sphere B is fastened to a horizontal table in the gravitational field of the Earth, such that B cannot move. To begin with the mass is at rest at the highest point A of the sphere. We now give the mass a very small push and it starts to slide down the sphere B. Find the angle θ at which the mass leaves the sphere B (neglect the kinetic energy the mass has when it starts to move).

Problem 8.4. Consider Example 8.8.

(1) With what velocity v_0 should a projectile be started at the center of the Earth to reach the surface of the Earth?
(2) How much time t_0 would it take a particle to drop through a straight frictionless tunnel drilled through the Earth between Copenhagen and Los Angeles? (Show first that the particle may perform harmonic oscillations in the tunnel.)

Problem 8.5. A cube of glass (B) has a spherical cavity of radius r. A particle P of mass m is placed at the bottom point (O) of the hollow sphere. We assume that P can slide without friction on the glass surface.

(1) Let us first assume that B is at rest on a table in the gravitational field of the Earth. Assume that P is started from O with a horizontal velocity v_0. Find the reaction R from the glass cube on the particle as a function

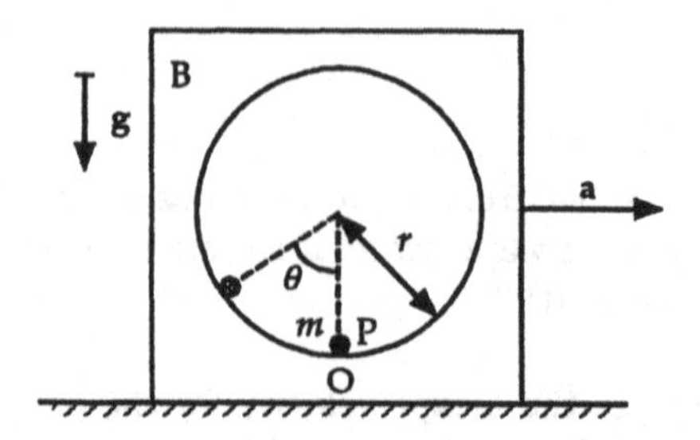

of θ, where θ is the angle the radius vector of the particle forms with the vertical through O. (Note: the mass of the glass cube is assumed to be much greater than m, so that the motion of P does not cause B to move.) Also, determine v_0 such that the particle just gets around the sphere.

(2) We now give the cube an acceleration **a** to the right along the table. Assuming $a = g\sqrt{3}$, find the new equilibrium position of P.

(3) Let us start the particle at the point O with velocity zero relative to the glass cube (the glass cube remains accelerated to the right with acceleration $a = g\sqrt{3}$). Find the reaction force F from the glass cube on the particle P, when P passes through its new equilibrium point D.

(4) Let us assume that the acceleration a is equal to g. Where is the equilibrium position now?

Problem 8.6. A particle P with mass m can slide down an inclined plane and into a circular loop with radius R.

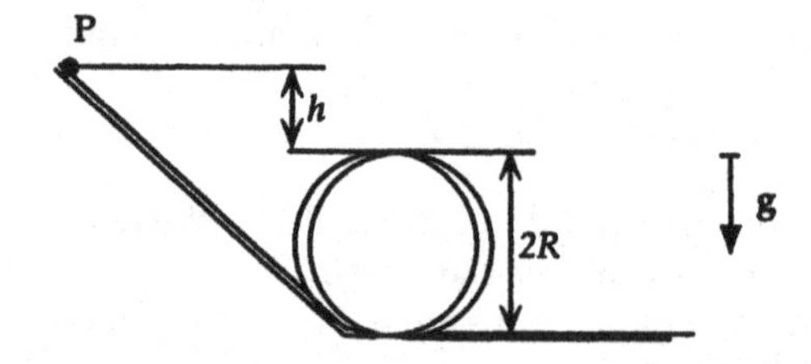

(1) The particle starts at a height h above the top of the loop. The value of h is such that the particle can just move around the circular loop. Find h.

(2) Assume that the particle instead starts at a height $2h$ above the top of the loop. Find in this case the force **F** with which the loop acts on the particle P when the particle is at the height R above the lowest point of the loop.

Problem 8.7.

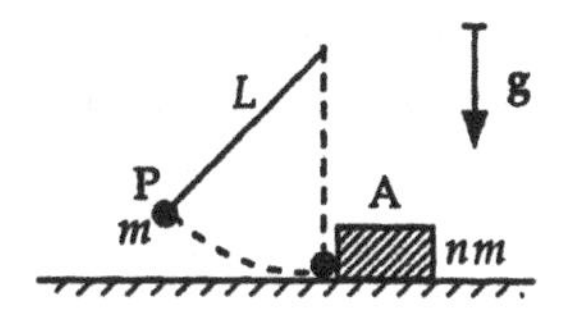

A particle P of mass m is fastened to the end of a string with length L. The particle starts from rest, in a position where the string is taut, and the direction of the string forms an angle of 60° with the vertical. In the lowest point of the circular arc, the particle P hits the body A.

Before the collision the body A is at rest on a frictionless horizontal table. The body A has mass nm where n is some positive number.

The central collision between P and A is completely elastic. The acceleration of gravity is g, and we neglect air resistance. The known quantitites are thus n, m, L, and g.

(1) Just before the collision the particle P has kinetic energy T_0. Find T_0.
(2) Determine the velocity of the body A and the particle P just after the collision. Furthermore, determine that fraction q of the kinetic energy T_0 of the particle that is transferred to the body A.
(3) Examine the limiting value of q as $n \to \infty$.

After the collision the particle P swings back.

(4) Determine $\cos \Theta_0$, where Θ_0 is the angle between the string and the vertical, when P is at the highest point of its orbit after the collision.
(5) Investigate the expression for $\cos \Theta_0$ for $n \to \infty$. Comment on the result.

Problem 8.8.

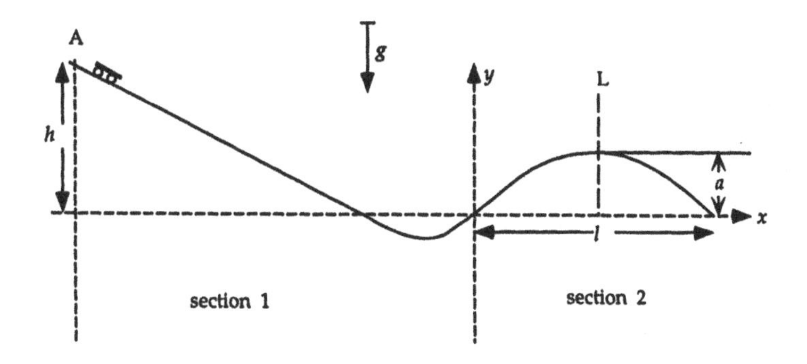

A roller coaster is contained in a vertical plane and has the shape shown on the figure. Section 2 of the roller coaster is symmetrical about the line L, and has the shape of a sine curve, $y = a\sin(x\pi/l)$. The dimensions of the car are so small that the car can be considered a point particle. The track is assumed to be frictionless, and the car starts at the point marked A (see the figure).

Determine the maximum value that h can have if we require that the car does not leave the track at the top point in section 2.

Problem 8.9.

(1) Let us assume that the Sun is a homogeneous sphere with mass M and radius R. Show that the gravitational self energy of the Sun is

$$U = -\frac{3}{5}\frac{GM^2}{R}.$$

(U may also be called the gravitational binding energy of the Sun, i.e., the energy that is needed to move all the particles of the Sun away to infinity.)

(2) The Sun radiates the power $P = 4\times10^{26}\ \mathrm{J\,s^{-1}}$. Assume that the radiated energy per second stems from the liberation of gravitational energy. From the so-called virial theorem it is known that half the liberated gravitational energy would go into an increase of the temperature of the Sun; the other half would be radiated into space.
Estimate, on the above assumption, the order of magnitude of the age of the Sun (this is known to be about 4.6×10^9 years).

9. The Center-of-Mass Theorem

In this chapter we investigate the motion of the center of mass for a system of particles, acted upon by external forces. As an important special case we shall consider the motion of the center of mass of the system of particles when no external forces act on the system.

9.1 The Center of Mass

Consider a system consisting of N point masses (Figure 9.1). The system is described relative to some inertial frame I.

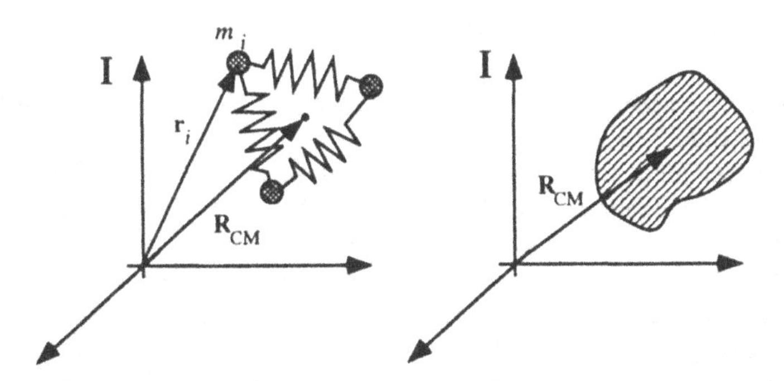

Fig. 9.1. Definition of the center of mass of a system of particles. The system could be a rigid body. The center of mass point is denoted CM

The N point masses may be mutually interacting, and they may even form a so-called rigid body. If the forces of interaction are so strong that the relative positions of the masses do not change during the motion, the system is called a *rigid body*. The following derivations hold for rigid as well as nonrigid bodies.

The position vector $\mathbf{R}_{\mathrm{CM}}$ for the center of mass is defined through the equation

$$M\mathbf{R}_{\mathrm{CM}} = \sum_{i=1}^{N} m_i \mathbf{r}_i \,, \tag{9.1}$$

where m_i and $\mathbf{r}_i$ are the mass and the position vector of the ith particle, and $M = \sum_i m_i$ is the total mass of the system. We use here and in the following the symbol $\sum_i$ to denote a sum over all the particles in the system.

Differentiating (9.1) with respect to time, we find

$$M\dot{\mathbf{R}}_{\mathrm{CM}} = \sum_i m_i \dot{\mathbf{r}}_i = \sum_i \mathbf{p}_i \equiv \mathbf{P} . \tag{9.2}$$

The total momentum of the system, $\sum_i \mathbf{p}_i$, is denoted $\mathbf{P}$. Equation (9.2) is then the statement that:

> The total momentum $\mathbf{P}$ of a system of particles is the same as that of a particle with mass M moving with the velocity of the center of mass.

The theorem is often put in the form:

$$\mathbf{P} = M\mathbf{v}_{\mathrm{CM}} . \tag{9.3}$$

where $\mathbf{v}_{\mathrm{CM}}$ is the velocity of the center of mass (CM).

We shall now study the motion of the center of mass when external forces are acting on the system of particles. Differentiating (9.2) once more with respect to time we get

$$M\ddot{\mathbf{R}}_{\mathrm{CM}} = \sum_i m_i \ddot{\mathbf{r}}_i . \tag{9.4}$$

According to Newton's second law, $m_i\ddot{\mathbf{r}}_i$ is equal to the total force acting on the ith particle.

The total force acting on the ith point mass may be divided into two parts: the *external* forces and the *internal* forces. The internal forces originate in all the other particles of the system; the external forces represent forces originating outside the system being considered, e.g. gravitational forces.

Let us denote the force by which particle i acts on particle j by $\mathbf{F}_{ij}$. Accordingly, the force with which particle j acts on particle i must be labeled $\mathbf{F}_{ji}$. From Newton's third law these two forces are equal in magnitude but opposite:

$$\mathbf{F}_{ij} = -\mathbf{F}_{ji} .$$

From this equation we conclude that the sum over all internal forces vanish; they cancel each other pairwise. What remains in (9.4) is the sum over external forces, i.e.:

$$M\ddot{\mathbf{R}}_{\mathrm{CM}} = \sum_i \mathbf{F}_{\mathrm{ext}} .$$

With $\sum_i \mathbf{F}_{\mathrm{ext}} \equiv \mathbf{F}^{\mathrm{ext}}$, the equation can thus be written:

$$M\ddot{\mathbf{R}}_{\mathrm{CM}} = \mathbf{F}^{\mathrm{ext}} . \tag{9.5}$$

The content of (9.5) is often expressed in the following way:

The center of mass of a system of particles – rigid or non-rigid – moves *as if* the entire mass were concentrated in that point, and all external forces act there.

This is the center-of-mass theorem. Sometimes called the center-of-gravity theorem because the center of mass is also a center of gravity when the system is in a homogeneous gravitational field (see Section 10.5).

We can write (9.5) in several equivalent ways, i.e.:

$$M \frac{\mathrm{d}\mathbf{v}_{\mathrm{CM}}}{\mathrm{d}t} = \mathbf{F}^{\mathrm{ext}} , \tag{9.6}$$

or

$$\frac{\mathrm{d}\mathbf{P}}{\mathrm{d}t} = \mathbf{F}^{\mathrm{ext}} . \tag{9.7}$$

Newton's second law guides the motion of the center of mass just as it guides the motion of a single mass point. This is the justification for the often used approximation of an extended body as a mass point.

From (9.7) we obtain the important theorem of conservation of momentum $\mathbf{P} = \sum_i \mathbf{p}_i$.

If no external forces act on a system of particles the total momentum $\mathbf{P}$ of the system is a constant vector.

A system of particles on which no external forces act is often called a *closed system.*

The two theorems, the theorem of conservation of momentum for a closed system, and the center-of-mass theorem have the same physical content.

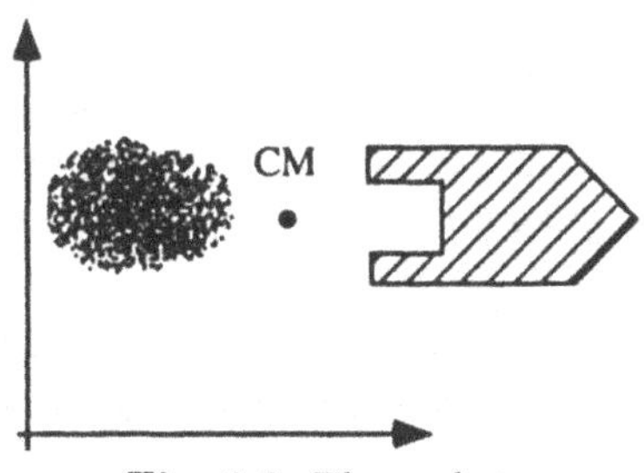

Fig. 9.2. The rocket

As an application, consider a rocket initially at rest in an inertial frame, and not acted upon by external forces. To begin with the rocket – and with it the center of mass – is at rest. Now the rocket engine starts. The forces that act during the combustion of the rocket fuel are all internal forces, and internal forces cannot provide acceleration of the center of mass. The rocket head moves forward, expelling gasses behind it; the center of mass of the system remains at rest (see Figure 9.2).

You cannot lift yourself by your hair! Explain why you are able to jump.

From the moment a gymnast leaps off the floor, she can no longer guide her center of mass. No matter how the body twists and turns, the center of mass follows a parabola in the homogeneous gravitational field.

Fig. 9.3. A shell explodes in mid-flight. Fragments fly off in different directions. The center of mass follows a parabola before, during and after the explosion. The forces that cause the shell to explode are internal forces of the system

See Figure 9.3; if we ignore air resistance, the center of mass of a shell fired from the ground will follow a parabola. Even if the shell explodes in flight, the center of mass – often called CM – will continue to follow the same parabola. The forces acting during the explosion are all internal forces for the system and as such they cannot contribute to the acceleration of the center of mass.

A man standing still on a perfectly frictionless horizontal surface cannot by walking move himself along the surface. The only forces acting on the man are gravity and the reaction from the frictionless surface. Both forces are vertical. No matter what motions the man performs he can not move his center of mass in the horizontal direction. The CM can only move along a vertical.

Question. Assume the man is on the frictionless icy surface of a frozen lake. If the man carries an object in his pocket, can he get to the shore? How?

The reason you are able to walk on a floor is friction. Your walking makes the floor exert a friction force on your foot, with a horizontal component. This component accelerates your CM in the horizontal direction.

9.2 The Center-of-Mass Frame

Let us assume that our system of particles is not subject to any external forces. The center-of-mass point CM will then – in some inertial frame I – move with the constant velocity

$$\mathbf{v}_{\text{CM}} = \frac{\mathbf{P}}{M} \,.$$

The reference frame where the total momentum of the particle system is zero is called the *center-of-mass frame*. The center-of-mass frame can also be characterized as the frame where the center of mass of the system in question is at rest.

The origin of the center-of-mass coordinate frame is usually chosen to coincide with the center of mass, and it moves with the velocity $\mathbf{v}_{\text{CM}}$ relative to our chosen inertial frame.

The center of mass frame will in the following be abbreviated to the "CM frame". This expression is widely used in elementary particle physics.

We shall demonstrate the following important theorem:

The kinetic energy of a particle system, as seen in an inertial frame, can be written as a sum: the kinetic energy of the particles relative to the CM frame, plus the kinetic energy of a particle with mass equal to the total mass of the system and moving with the CM velocity.

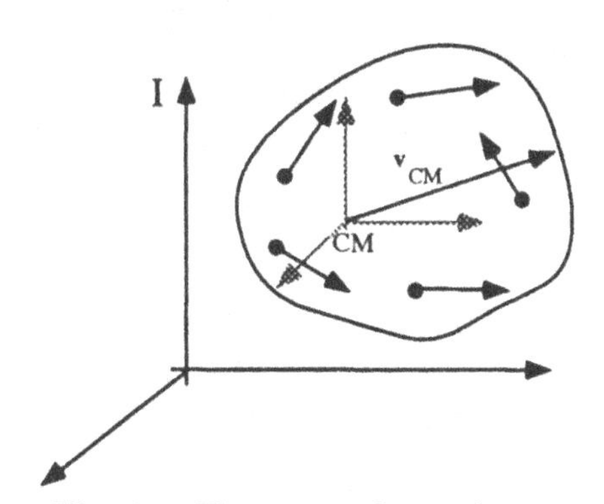

Fig. 9.4. The center-of-mass frame

Measured in the inertial frame I, the total kinetic energy of the particle system is:

$$T = \sum_i \frac{1}{2} m_i v_i^2 \,.$$

Here, v_i is the speed of the ith particle relative to I.

Consider now the motion of the ith particle relative to the CM frame. Let the velocity of the ith particle relative to CM be $\mathbf{v}_{ri}$ (r for relative). From the Gallilei transformation follows:

$$\mathbf{v}_i = \mathbf{v}_{ri} + \mathbf{v}_{\mathrm{CM}} .$$

Consequently the kinetic energy relative to I can be written

$$T = \sum_i \frac{1}{2} m_i (\mathbf{v}_{ri} + \mathbf{v}_{\mathrm{CM}})^2 ,$$

or

$$T = \sum_i \frac{1}{2} m_i (v_{ri}^2 + v_{\mathrm{CM}}^2 + 2\mathbf{v}_{ri} \cdot \mathbf{v}_{\mathrm{CM}}) ,$$

from which

$$T = \frac{1}{2}\left(\sum_i m_i\right)(\mathbf{v}_{\mathrm{CM}}^2) + \sum_i \frac{1}{2} m_i v_{ri}^2 + \left(\sum_i m_i \mathbf{v}_{ri}\right) \cdot \mathbf{v}_{\mathrm{CM}} .$$

Now, $\sum_i m_i \mathbf{v}_{ri} = 0$. (Why? Because this is how the CM frame was defined!) The expression is reduced to

$$T = \frac{1}{2} M v_{\mathrm{CM}}^2 + \sum_i \frac{1}{2} m_i (v_{ri}^2) \tag{9.8}$$

or

$$T = \frac{1}{2} M v_{\mathrm{CM}}^2 + T_r ,$$

where $T_r \equiv \sum_i \frac{1}{2} m_i (v_{ri}^2)$ means the kinetic energy of the particles relative to the CM frame. This quantity is known as the relative kinetic energy. The first term, $\frac{1}{2} M v_{\mathrm{CM}}^2$, is called the translational energy. Equation (9.8) is the mathematical content of the above theorem. The result is sometimes called *König's theorem.*

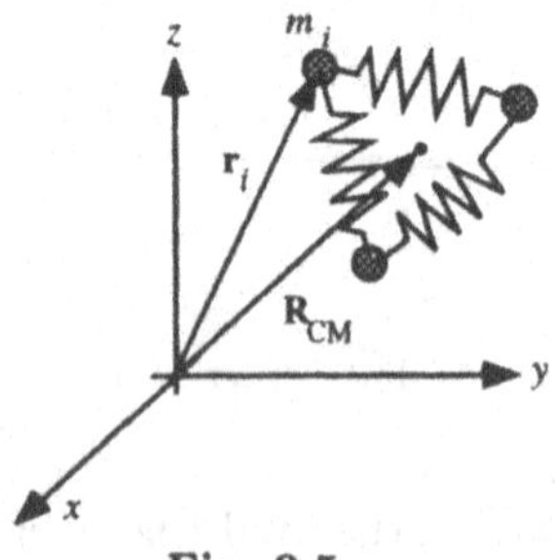

Fig. 9.5.

As an illustration of (9.8) we consider a system consisting of three masses interconneted with springs (Figure 9.5). At some initial time the springs are compressed and the entire system is given a velocity and released in the gravitational field of the Earth. No matter how the individual particles move relative to one another, the CM of the system will fall along a parabola. The total kinetic energy of the system can be written as

$$T = \frac{1}{2}Mv_{\mathrm{CM}}^2 + T_{\mathrm{vibr}} \,,$$

where T_{vibr} means the kinetic energy associated with the vibrations of the particles relative to one another, or more precisely relative to the CM of the system. If, moreover, the system is in rotation around the CM, the relative energy will contain an additional term T_{rot}, so that $T_{\mathrm{r}} = T_{\mathrm{vibr}} + T_{\mathrm{rot}}$. We shall return to the study of rotations in Chapter 11.

9.3 Examples

Example 9.1. Two Masses Connected with a Spring

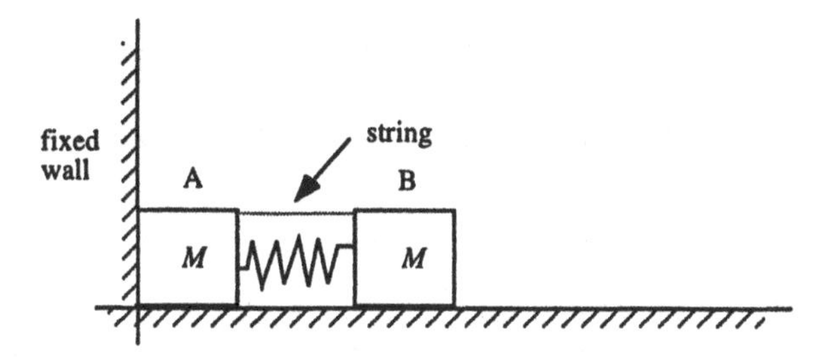

Two blocks A and B each have mass M. They are connected with a spring and rest on a frictionless surface. The spring is compressed an amount Δl by means of a string connected to A and B. The spring constant is k . Initially, block A is touching a wall. At $t = 0$ the string breaks.

(1) Calculate the accelerations $a_{\mathrm{A}}, a_{\mathrm{B}}$, and a_{CM} of block A, block B and the center of mass, respectively, just after the string breaks, i.e., before block B starts to move. Calculate the reaction force from the wall just after the string breaks.
(2) Calculate the time $t = t_1$ when block A begins to move (assuming this happens as soon as the spring reaches its equilibrium length for the first time).
(3) Calculate the magnitude P of the total momentum for the system and the magnitude of the velocity of the center of mass v_{CM} for $t > t_1$.
(4) Calculate the translational kinetic energy of the system for $t > t_1$.

(5) Find the maximum compression of the spring in the harmonic oscillations the system performs around the center of mass after A has left the wall, and determine the period of these oscillations.

Solution.

(1)

$$a_{\mathrm{A}} = 0; \quad a_{\mathrm{B}} = \frac{k\Delta l}{M}; \quad a_{\mathrm{CM}} = \frac{k\Delta l}{2M} .$$

The external force which gives the CM its acceleration is the reaction force from the wall. If the wall was not there CM would remain stationary while the blocks oscillated on either side of it. $R = k\Delta l$.

(2)

$$t_1 = \frac{\tau_1}{4} = \frac{\pi}{2}\sqrt{\frac{M}{k}} .$$

(3) For $t \geq t_1$ no external horizontal forces act on the system. Consequently $P = |\,\mathbf{P}\,|$ is constant, and P equals the momentum of the system for $t = t_1$. For $t = t_1$, block A is still motionless and B has the momentum Mv_{B}. The velocity v_{B} may for instance be found from

$$\frac{1}{2}Mv_{\mathrm{B}}^2 = \frac{1}{2}k(\Delta l)^2 \Rightarrow v_{\mathrm{B}} = \Delta l\sqrt{\frac{k}{M}} .$$

From this: $P = M\Delta l\sqrt{k/M}$, and $v_{\mathrm{CM}} = P/2M = (\Delta l/2)\sqrt{k/M}$.

(4)

$$T_{\mathrm{trans}} = \frac{1}{2}(2M)v_{\mathrm{CM}}^2 = \frac{1}{4}k(\Delta l)^2 .$$

(5) The energy in the oscillations is $E_{\mathrm{osc}} = \frac{1}{2}k(\Delta l)^2 - T_{\mathrm{trans}} = \frac{1}{4}k(\Delta l)^2$. The maximum compression d of the spring occurs when the entire oscillatory energy is potential:

$$\frac{1}{2}kd^2 = \frac{1}{4}k(\Delta l)^2 \Rightarrow d = \frac{\Delta l}{\sqrt{2}} .$$

In the CM frame, both masses perform harmonic oscillations. Suppose one mass at some time is a distance x away from the equilibrium position and the other mass a distance y from its equilibrium position. Since in the CM frame the center of mass is stationary, we must have $Mx = My$ (had the masses been different this equation would have been $M_{\mathrm{A}}x = M_{\mathrm{B}}y$; see below). Here, we have $x = y$; the total stretching of the spring is therefore always given by $2x$. In the CM frame, each block moves according to the equation of motion

$$M\ddot{x} = -k(2x) ,$$

from which

$$\tau_2 = 2\pi\sqrt{\frac{M}{2k}} .$$

Note carefully the following concerning the translational kinetic energy $T_{\text{trans}} = (1/2)(2M)v_{\text{CM}}^2$ of the system. From the center-of-mass theorem it follows that in order for the velocity of the CM to change, an external force must act. This implies that only an external force can change the translational kinetic energy. However, one can*not* conclude that the change in T_{trans} is the work done by the external force. This would be true if the external force de facto acted in the center of mass.

The external force which gives the CM its acceleration is the reaction force from the wall. The work done by the external force on the system is zero, since the point in which the force acts does not move. While the force acts on the system, T_{trans} changes from 0 to $(1/4)k(\Delta l)^2$.

Internal forces can change the total kinetic energy, but they can never change the total momentum of the system.

In the combustion of rocket fuel some of the chemical (potential) energy of the fuel is converted into macroscopic kinetic energy of the rocket head. Internal forces have also here produced mechanical kinetic energy.

Questions.

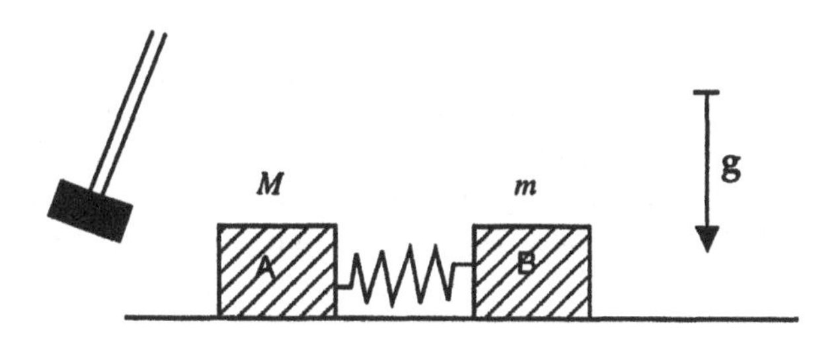

Two cubic blocks A and B of the same size but different masses (M and m, respectively) are connected with a spring of spring constant k (see the figure). The mass of the spring is negligible.

Initially, the blocks are at rest on a horizontal frictionless table. A is now hit by a blow from a hammer. If the blow acts with the force $\mathbf{F}(t)$ during the short time interval Δt, the system receives the total momentum

$$\int_0^{\Delta t} \mathbf{F}(t)\mathrm{d}t = \mathbf{P}\,.$$

Here $\mathbf{F}(t)$ is perpendicular to the end face of block A and a line through $\mathbf{F}(t)$ goes through the center of mass of block A, along the spring and through the center of mass of block B. The interval Δt is assumed to be so short that the block A does not move appreciably during the blow. After the blow the entire system begins to move along the table.

(1) Calculate the maximum compression Δl of the spring. *Hint:* find first the initial kinetic energy of A; next, the translational kinetic energy of the system; finally the energy in the oscillations.

(2) Calculate the period τ in the oscillations of the system after the blow. *Hint:* in the CM:

$$mx = My; \quad \Delta l = x + y; \quad m\ddot{x} = -k\Delta l; \quad M\ddot{y} = -k\Delta l;$$

$$m\ddot{x} = -k\left(1 + \frac{m}{M}\right)x; \quad M\ddot{y} = -k\left(1 + \frac{M}{m}\right)y.$$

Answer.

(1)

$$\Delta l = P\sqrt{\frac{m}{(m+M)Mk}}.$$

(2)

$$\omega = \sqrt{\frac{k}{\mu}}; \quad \tau = 2\pi\sqrt{\frac{\mu}{k}}; \quad \mu = \frac{mM}{m+M},$$

where μ is known as the reduced mass of the system.

△

Example 9.2. Inelastic Collisions. We consider a completely inelastic collision : A bullet (mass m, velocity v) is shot horizontally into a block of wood with mass M. The block is at rest on a frictionless horizontal table. It is assumed that the bullet is stopped completely by the block and that the system of the bullet plus block moves as a single body after the collision.

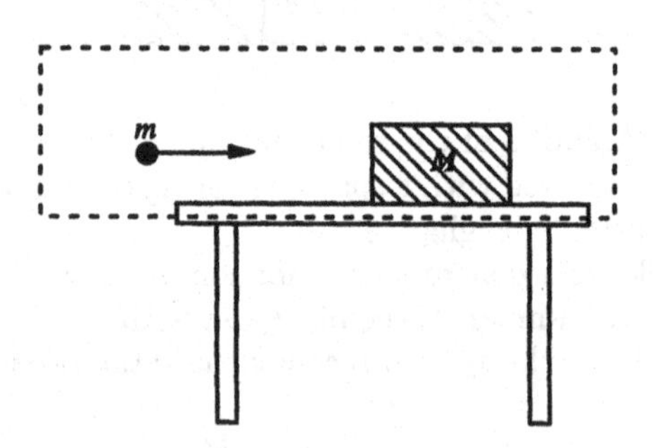

(1) Find the velocity u of the system after the collision.
(2) Find the mechanical kinetic energy T_a after the collision.
(3) Calculate the amount Q of initial kinetic energy which has been converted into heat.
(4) Find the translational kinetic energy in the lab frame before and after the collision.
(5) Find the relative kinetic energy T_r before the collision. T_r is the kinetic energy in the CM frame.

Solution.

(1) No external forces act in the horizontal direction, so the initial momentum is conserved:

$$mv = (M + m)u\,,$$

or

$$u = \frac{mv}{m + M}\,.$$

(2)

$$T_a = \frac{1}{2}(m + M)u^2 = \frac{m^2v^2}{2(m + M)}\,.$$

(3)

$$Q = \frac{1}{2}mv^2 - T_a = \frac{Mmv^2}{2(m + M)}\,.$$

(4) No external horizontal forces act on the system. The velocity of the CM is therefore the same before, during, and after the collision:

$$v_{CM} = \frac{P}{m + M} = \frac{mv}{m + M}\,.$$

From this we get:

$$T_{trans} = \frac{1}{2}(m + M)v_{CM}^2 = \frac{P^2}{2(M + m)} = \frac{m^2v^2}{2(M + m)}\,.$$

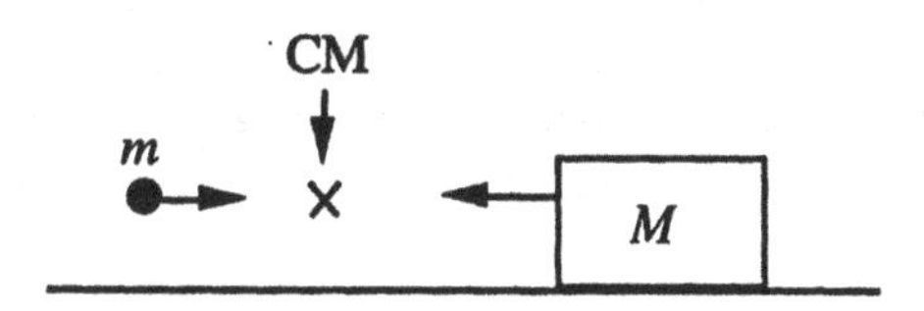

(5)

$$T_r = \frac{1}{2}mv^2 - T_{trans} = \frac{mMv^2}{2(m + M)}\,.$$

Note. The heat energy Q generated in the collision is equal to T_r ,i.e., the kinetic energy in the motion relative to the CM. The overall translational energy in the system cannot be converted into heat. This fact is closely related to conservation of linear momentum, since

$$T_{trans} = \frac{1}{2}(m + M)v_{CM}^2 = \frac{P^2}{2(M + m)}\,,$$

where P is the total momentum in the lab frame. Since P remains constant, we see that the translational part of the kinetic energy is also constant. Of the total amount of kinetic energy T available in the lab frame $T = T_{trans} + T_r$, only T_r can be converted into other forms of energy (in this instance, heat) whereas T_{trans} remains as translational mechanical energy before, during, and after the collision.

If we look at the collision from the CM frame, the CM is stationary, and in particular there is no mechanical energy after the collision. This is the *definition* of a perfectly inelastic collision! The kinetic energy T_r has been completely converted into heat.

This simple consideration – and its relativistic version – is of enormous importance in accelerator physics, where (in particle collisions) only the kinetic energy in the CM frame can be converted into new particles! △

Example 9.3. The Collision Approximation

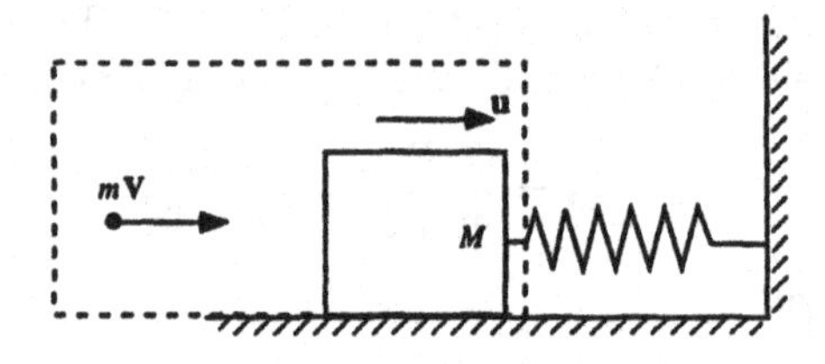

A box filled with clay (total mass M) rests on a frictionless horizontal table in the lab frame. A spring is attached to one end of the box, and to a solid vertical wall. A bullet of mass m is now shot horizontally into the box and is stopped after a time Δt. The time interval Δt is assumed so short that the box does not move appreciably during the time Δt. The velocity of the bullet just before it hits the box is horizontal and has the magnitude v. The spring constant is k.

(1) Calculate the maximum compression d of the spring.
(2) Numerical example: $M = 10$ kg, $m = 5$ g, $k = 40\ \text{N m}^{-1}$, $v = 400\ \text{m s}^{-1}$.
(3) What condition must Δt satisfy if we can really ignore the motion of the box while the bullet is being stopped?

Solution.

(1) While the bullet stops its motion relative to the box, nonconservative forces act (heat is produced). Thus, during this process the theorem of conservation of mechanical energy does not apply. We have, however, assumed that the distance the box moves during the collision can be ignored. This means that during the collision no horizontal external forces act on the box (the spring has not yet begun to compress). Momentum is therefore conserved in the collision. Denote the velocity of box plus bullet just after the collision by u (see the figure) then:

$$mv = (M+m)u\,.$$

When the bullet has stopped its motion relative to the box, the nonconservative forces no longer act. From this point on, the energy theorem may be used to determine the maximum compression of the spring:

$$\frac{1}{2}(M+m)u^2 = \frac{1}{2}kd^2\,,$$

or

$$d = \frac{mv}{\sqrt{k(M+m)}}\,.$$

The result may be found in a more cumbersome (but instructive) way by computing the total translational kinetic energy T_{trans} just before the collision. It is this kinetic energy which is converted into compression of the spring. The kinetic energy T_r relative to CM is converted into heat. The calculations proceed like this:

$$v_{\text{CM}} = \frac{mv}{M+m} \;\Rightarrow\; v - v_{\text{CM}} = v\frac{M}{M+m}\,.$$

$$T_r = \frac{m}{2}\left(\frac{Mv}{M+m}\right)^2 + \frac{M}{2}\left(\frac{mv}{M+m}\right)^2 = \frac{mM}{2(M+m)}v^2\,.$$

$$T_{\text{trans}} = T - T_r = \frac{1}{2}mv^2 - T_r = \frac{m^2v^2}{2(M+m)} = \frac{1}{2}(M+m)v_{\text{CM}}^2\,.$$

$$T_{\text{trans}} = \frac{1}{2}kd^2\,; \quad d = \frac{mv}{\sqrt{k(M+m)}}\,.$$

(2) $d = 10$ cm.

(3) The main point is that the momentum delivered to the box during the collision by the bullet ($m\mathbf{v}$) should be large compared to the momentum given to the box by the spring during the time Δt. The momentum delivered by the spring force to the box during the collision is given by

$$\mathbf{P} = \int_0^{\Delta t} \mathbf{F}(t)\mathrm{d}t\,,$$

where $\mathbf{F}(t)$ is the spring force as a function of time during the collision. As we do not know the details of forces acting between the bullet and the box we do not know the spring force as function of time either. Thus we cannot compute $\mathbf{P}$. But we can estimate the *order of magnitude* of $\mathbf{P}$. The distance the box moves towards the wall during the collision will be of the order of magnitude

$$\Delta x \approx u\Delta t\,.$$

The spring force will thus be of the order of magnitude

$$F \approx k\Delta x \approx ku\Delta t\,.$$

The order of magnitude of $\mathbf{P}$ is then

$$P \approx ku(\Delta t)^2 .$$

This should be compared to the momentum delivered by the bullet

$$mv = (M+m)u .$$

The condition that must be satisfied (if we apply conservation of momentum during the collision) is thus:

$$mv = (M+m)u \gg ku(\Delta t)^2 ,$$

that is

$$\Delta t \ll \sqrt{\frac{M+m}{k}} .$$

Note. The period T in the harmonic oscillations the box performs under the action of the spring force is

$$T = 2\pi\sqrt{\frac{M+m}{k}} .$$

In the quoted numerical example we find:

$$\sqrt{\frac{M+m}{k}} = \frac{1}{2}\ \mathrm{s}.$$

The time Δt should thus be much less than 1/2 s. Is that reasonable? The approximation in which we neglect $\Delta x \approx u\Delta t$ is called the *collision approximation.*

Question. A *ballistic pendulum* can be used to determine the velocity of bul-

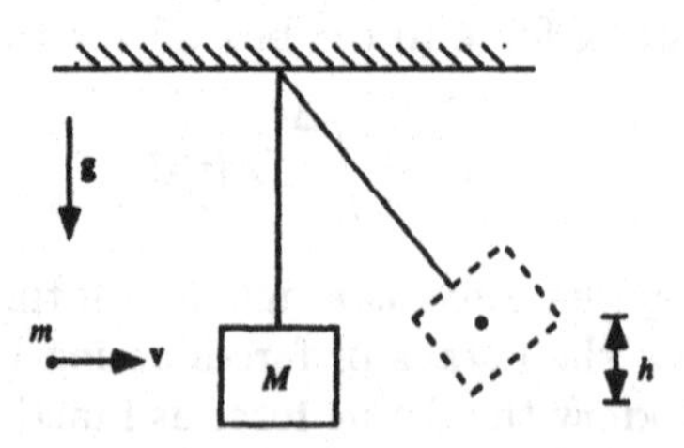

lets. The pendulum consists of a wooden box filled with sand. The mass of box plus sand is M. A bullet of mass m with a horizontal velocity $\mathbf{v}$ strikes the pendulum and remains embedded in it. The time required for the bullet to come to rest with respect to the box is Δt.

(1) $M = 2$ kg, $m = 5 \times 10^{-3}$ kg and the maximum height h to which the pendulum swings is given by $h = 10$ cm. Find the velocity v of the bullet just before impact.
(2) How many percent of the original mechanical energy of the bullet is converted into heat?
(3) State the condition Δt must satisfy in order that the collision approximation may be used.

Answer.

(1)

$$v = \frac{M+m}{m}\sqrt{2gh} = 560 \text{ m s}^{-1} .$$

(2)

$$f = \frac{M}{M+m} .$$

That is, more than 99.8% of the original initial kinetic energy is converted into heat. If M were not free to move, all the original kinetic energy would be converted into heat. What does it mean that M is not free to move? It means that M is fastened to the Earth, e.g. by leaning against a wall. The body that takes up the momentum of the bullet is now M plus the whole Earth. Such a massive body may absorb the bullet without gaining any kinetic energy $\left(T = p^2/2M, \quad M = \infty\right)$

△

Example 9.4. Freely Falling Spring

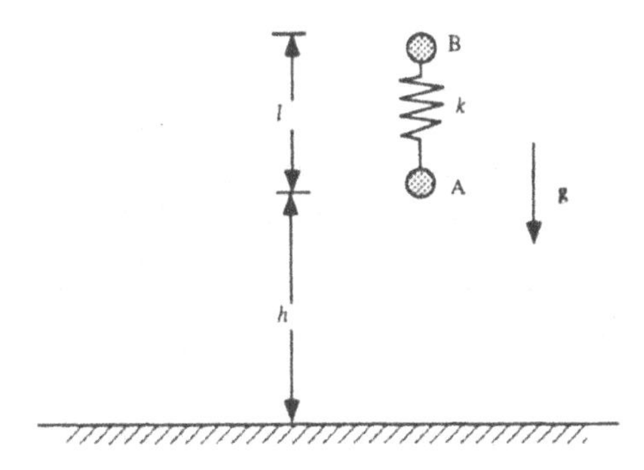

A mechanical system consists of two equal nonelastic particles A and B each of mass m. A and B are connected by a spring with force constant k and natural length l. The mass of the spring is negligible compared to m.

The system is allowed to fall freely through a height h in the gravitational field of the Earth. It is assumed that $h \gg l$. During the free fall the direction of the spring is vertical and the spring maintains its natural length.

The lower particle A now meets a fixed horizontal plane. The collision is completely inelastic and A thus stays at rest during the collision. The particle B does not stop until it has compressed the spring a certain distance $d < l$. The spring will now expand and the system will jump upwards.

During the collision, i.e., during the time interval where A is in contact with the horizontal plane, we neglect the gravitational force.

(1) Find the kinetic energy T of the system just before the collision.
(2) Find d.

We now consider the system at the time $t = t_0$, when the spring again has its natural length.

(3) Find the total kinetic energy T_0, the total momentum P_0 and the translational kinetic energy K_0 of the system at the time $t = t_0$.
(4) From the time $t = t_0$ the gravitational force acts again. The center of mass of the system will rise to a height h_1 above the horizontal surface before it stops. Find h_1.

Solution.

(1) $T = 2mgh$.
(2) The kinetic energy of A is lost in the inelastic collision. The kinetic energy of B is used to compress the spring:

$$\frac{1}{2}kd^2 = mgh,$$

$$d = \sqrt{\frac{2mgh}{k}}.$$

(3)

$$\begin{aligned} T_0 &= mgh, \\ P_0 &= mv_A + mv_B = m \cdot 0 + m \cdot \sqrt{2gh} = m\sqrt{2gh}, \\ K_0 &= \frac{1}{2}(2m)v_{CM}^2 = \frac{P_0^2}{2(2m)} = \frac{1}{2}mgh. \end{aligned}$$

(4) The translational energy K_0 is converted into potential energy in the gravitational field of the Earth.

$$K_0 = 2mgh_1 \Rightarrow h_1 = \frac{1}{4}h.$$

Questions.

(1) Obtain the last result $h_1 = \frac{1}{4}h$ without using energy conservation, i.e., by direct use of the CM theorem.
(2) Show that the condition for neglecting gravity during the collision is: $mg \ll kd$.

△

Example 9.5. The Wedge

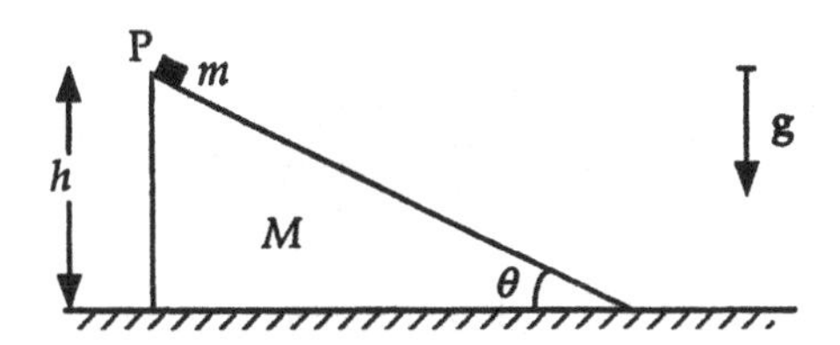

A wedge of mass M is at rest on a horizontal frictionless table. A particle P of mass m is placed on the wedge. The figure shows a vertical section containing P and the CM of the wedge. There is no friction between the particle and the wedge. The height of the wedge is h.

The particle starts to slide down the wedge at $t = 0$, and the wedge accelerates to the left (P slides along the straight line contained in the vertical plane through the CM of the wedge and P).

(1) Find, by applications of the conservation laws, the velocity u of the wedge relative to the table when P reaches the table.
(2) Find, by direct application of the equation of motion, the time interval t it takes P to reach the table. Then find u by direct application of the equation of motion.

Solution.

(1) We apply energy and momentum conservation in a reference frame fixed relative to the table. The velocity of P *relative to the wedge* at the moment

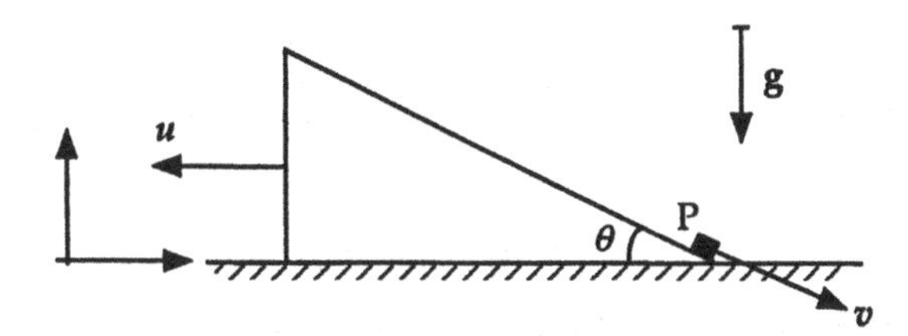

P touches the table is $\mathbf{v}$. The velocity of the wedge relative to the table at the same instant is $\mathbf{u}$.
Conservation of momentum in the horizontal direction gives

$$Mu = m(v\cos\theta - u)\,. \tag{9.9}$$

(There are no external forces in the horizontal direction acting on the system). Conservation of energy (no heat is produced!) gives:

$$mgh = \frac{1}{2}Mu^2 + \frac{1}{2}m\left[(v\cos\theta - u)^2 + (v\sin\theta)^2\right] . \tag{9.10}$$

From (9.9) and (9.10) we find:

$$u^2 = \frac{2mgh}{(M+m)\left(\frac{M+m}{m\cos^2\theta} - 1\right)} . \tag{9.11}$$

Check the dimensions!
Let us examine (9.11) for various limiting cases:

$$\begin{aligned} M &\to \infty \Rightarrow u^2 \to 0 ; \\ \theta &\to 0, \Rightarrow u^2 \to 0, \qquad \text{because } h \to 0 ; \\ \theta &\to \frac{\pi}{2} \Rightarrow u^2 \to 0, \qquad \text{because } \cos\theta \to 0 . \end{aligned}$$

(2) While P is moving down the wedge, the wedge is accelerating to the left: Seen from a reference frame fixed relative to the wedge, a fictitious force

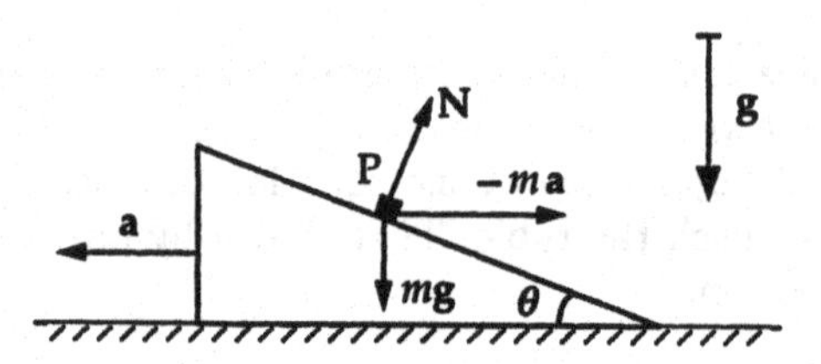

acts on P. Let the acceleration of the wedge relative to the table be a, and the acceleration of P relative to the wedge be α. Then (see the figure) the equation of motion for P in its motion relative to the wedge is:

$$m\alpha = mg\sin\theta + ma\cos\theta . \tag{9.12}$$

The equation of motion for P perpendicular to the surface of the wedge is:

$$0 = N + ma\sin\theta - mg\cos\theta . \tag{9.13}$$

The equation of motion of the wedge along the horizontal surface is:

$$Ma = N\sin\theta . \tag{9.14}$$

This provides us with three equations in three unknowns (a, α, and N). From (9.13) and (9.14) we find a by eliminating N:

$$a = \frac{mg\cos\theta\sin\theta}{M + m\sin^2\theta} . \tag{9.15}$$

(Check: for $M \to \infty$, $a \to 0$, and for $\theta \to \pi/2$, $a \to 0$.) From (9.12) we then find α:

$$\alpha = g\sin\theta + \frac{mg\cos^2\theta\sin\theta}{M + m\sin^2\theta}\,. \tag{9.16}$$

(Check: for $\theta \to \pi/2$, $\alpha \to g$.)
The distance S the particle has moved relative to the wedge is given by

$$S = \frac{h}{\sin\theta}\,. \tag{9.17}$$

We thus have, for T,

$$\frac{1}{2}\alpha T^2 = \frac{h}{\sin\theta}\,. \tag{9.18}$$

From (9.18) we find:

$$T = \sqrt{\frac{2h}{\alpha\sin\theta}}\,. \tag{9.19}$$

To find the velocity u, we use, since the accleration a is constant,

$$u = aT\,. \tag{9.20}$$

Writing (9.20) as $u^2 = a^2T^2$, and using (9.15) and (9.16), we find:

$$u^2 = \frac{2mgh}{(M+m)\left(\frac{M+m}{m\cos^2\theta} - 1\right)}\,. \tag{9.21}$$

Question. Find the reaction force $\mathbf{R}$ from the table on the wedge during the motion.

Answer.

$$R = \frac{Mg(M+m)}{M + m\sin^2\theta}\,. \tag{9.22}$$

△

9.4 Review: Center of Mass and Center-of-Mass Theorems

Definitions

Center of mass, CM :	$M\mathbf{R}_{\mathrm{CM}} = \sum_i m_i \mathbf{r}_i\,, \quad M = \sum_i m_i$
Total momentum:	$\mathbf{P} = \sum_i \mathbf{p}_i = \sum_i m_i \mathbf{v} i$

Results

Momentum:	$\mathbf{P} = M\mathbf{v}_{\mathrm{CM}}$
Center-of-mass theorem:	$M\mathrm{d}^2\mathbf{R}/\mathrm{d}t^2 = \sum \mathbf{F}^{\mathrm{ext}}$ $M\mathrm{d}\mathbf{v}_{\mathrm{CM}}/\mathrm{d}t = \sum \mathbf{F}^{\mathrm{ext}}$ $\mathrm{d}\mathbf{P}/\mathrm{d}t = \sum \mathbf{F}^{\mathrm{ext}}$
König's theorem:	$T = \frac{1}{2}Mv_{\mathrm{CM}}^2 + T_{\mathrm{r}}$
Conservation theorems:	For all closed systems the total momentum $\mathbf{P}$ is conserved. For all closed systems with *conservative forces*, the total mechanical energy $T + U$ is conserved.

9.5 Comments on the Conservation Theorems

These theorems are not explicitly stated in Newton's *Principia.* One may claim that the theorems of energy and momentum conservation indirectly are contained in Newton's third law.

The rate of change of the momentum of a particle equals the force acting on the particle. This force has its origin in some other particle on which the reaction force acts. As an amount of momentum "flows" into a given particle, an equal amount must – according to Newton's third law – "flow" out of (at least) one other particle; here momentum is thought of as a vector quantity. The overall amount of momentum in a Newtonian universe can never change.

If we similarly take into account every (elementary) particle in the universe, we can say that energy can neither be created nor destroyed, only transferred from one place to another. The law of action and reaction ensures that if energy flows into one particle, the same amount is given up elsewhere.

In mechanical physics we have restricted ourselves to considerations of mechanical energy and we have found it to be conserved only in systems with conservative forces. The reason is that we cannot keep track of the detailed motion of all particles (atoms, electrons) in the universe or for that matter in the smaller mechanical systems under consideration.

9.6 Problems

Problem 9.1.

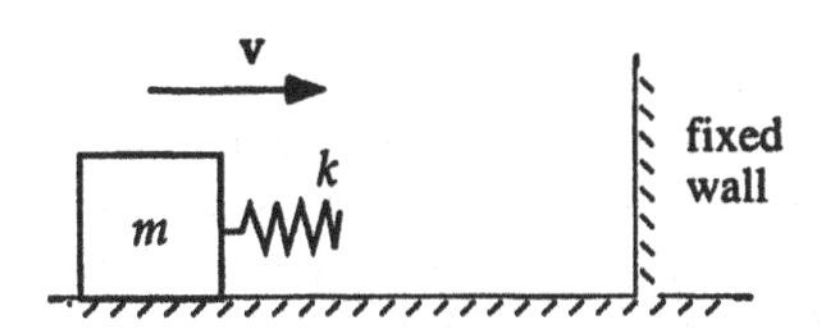

A block of mass m slides along a frictionless horizontal plane. The block has velocity $\mathbf{v}$. A massless spring with spring constant k is attached to the block as shown on the figure above. The block collides with the fixed wall, and it is assumed that the spring does not bend during the collision.

(1) Find the maximum compression of the spring.

Let us now assume that the block instead of hitting a fixed wall hits another block of mass M initially at rest on the table. See the figure below.

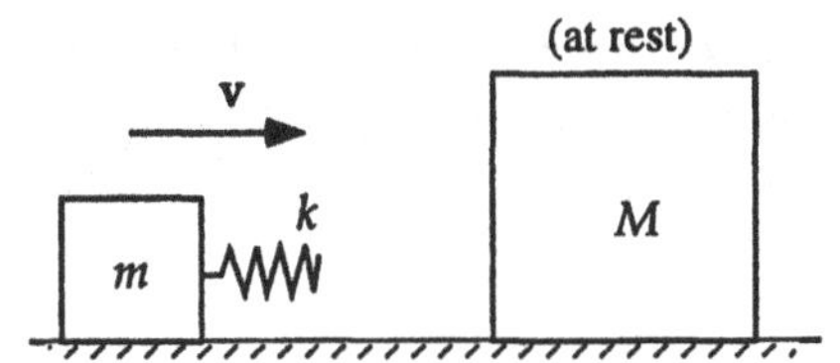

(2) What is now the maximum compression of the spring?

Problem 9.2.

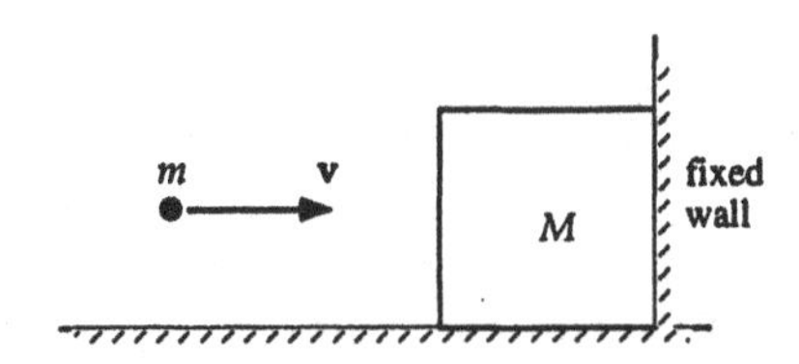

A wood block of mass M rests on a horizontal frictionless table in the lab frame. One end of the block rests against a solid vertical wall. A bullet of mass m is shot horizontally into one end of the block. The bullet moves a distance D into the block. Assume that the force acting on the bullet while it is being stopped is independent of its velocity.

How far would the bullet move into the block if the experiment is repeated, now without the wall, i.e., if the block is free to slide along the table?

Problem 9.3.

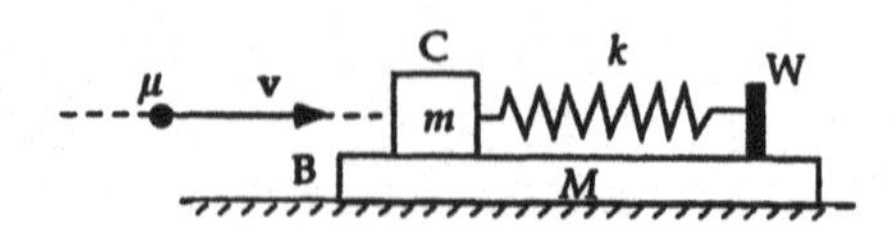

A physical system consists of a wood block B of mass M. B is at rest on a frictionless, horizontal table. A small vertical wall W is placed near one of the ends of B. The wall is fastened to B. A spring with spring constant k connects the wall to a smaller block C of mass m (see the figure).

There is no friction between C and B. Initially both C and B are at rest relative to the lab frame. A pistol bullet of mass μ is now shot into (and absorbed by) C. Just prior to hitting C the bullet has velocity $\mathbf{v}$ parallel with the spring.

We assume that the bullet is stopped in such a short time Δt that we may neglect the distance that C has moved relative to B in the time Δt.

The mass of the wall and of the spring may be neglected.

(1) Find the maximum compression d of the spring, in terms of v, k, M, m, and μ.
(2) Find the time τ it takes from the time the bullet is stopped till the spring reaches for the first time its maximal compression.
(3) Find the maximum compression of the spring if the pistol bullet instead of being shot into C is shot into B.

Problem 9.4.

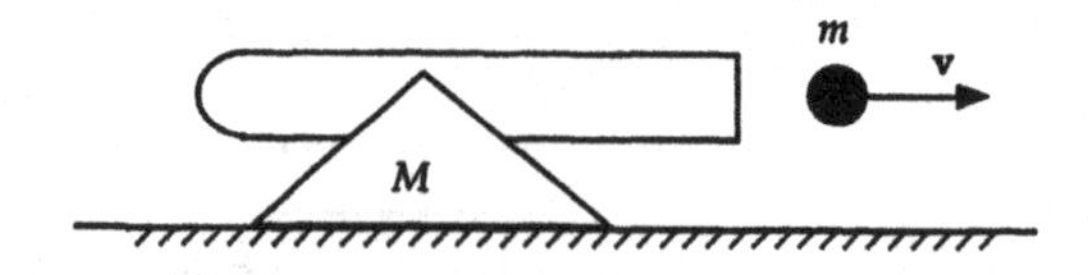

An old cannon is standing at rest on a horizontal plane in the gravitational field of the Earth. The cannon can slide (recoil) on the ground, and the coefficient of friction is μ. The mass of the cannon is M and it is loaded with a bullet of mass m. The cannon is fired, and the bullet leaves the cannon with a velocity $\mathbf{v}$ *relative to the cannon.*

(1) Find the distance S that the cannon recoils. It is assumed that the time Δt it takes the bullet to leave the muzzle is so short that the cannon has moved a negligible distance in that time.

(2) Suppose that $v = 600\ \mathrm{m\,s^{-1}}$, $m = 1.2$ kg, $M = 300$ kg, $\mu = 0.1$, and that the length of the muzzle is $L = 2$ m. Show that the order of magnitude of the momentum Δp given to the cannon by the frictional force during the time it takes the bullet to pass through the muzzle is small compared to the recoil momentum imparted to the cannon from the firing.

Problem 9.5. A big sledge is standing at rest on a horizontal icy surface. A rail is placed in such a way that the sledge can move only along a straight line. It is assumed that the sledge moves without friction. At one end of the sledge n persons stand. They all have the same mass.

These n persons can now leave the sledge in two different ways:

(a) All n persons run to the front of the sledge and jump off simultaneously, all with velocity $\mathbf{v}$ relative to the sledge. This velocity is along the direction in which the sledge is constrained to move.
(b) One at a time, a person runs to the end of the sledge and jumps off with velocity $\mathbf{v}$ relative to the sledge while the remaining persons stand still. This process continues until they have all left the sledge.

Demonstrate which of the two methods imparts to the sledge the highest velocity relative to the ice.

Problem 9.6.

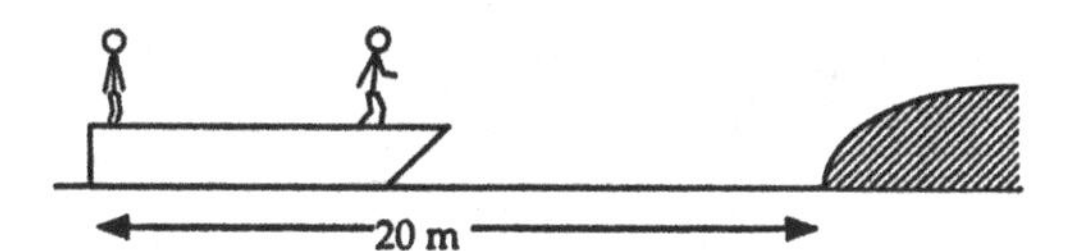

A sledge of mass 120 kg stands at rest on the horizontal surface of an icy lake. A man of mass 70 kg stands at one end of the sledge so that initially the distance from the man to the shore is 20 m. The man now walks 3 m relative to the sledge, towards the shore. Then he stops. Assuming that the sledge moves without friction on the ice, how far from the shore is the man when he stops?

Problem 9.7.

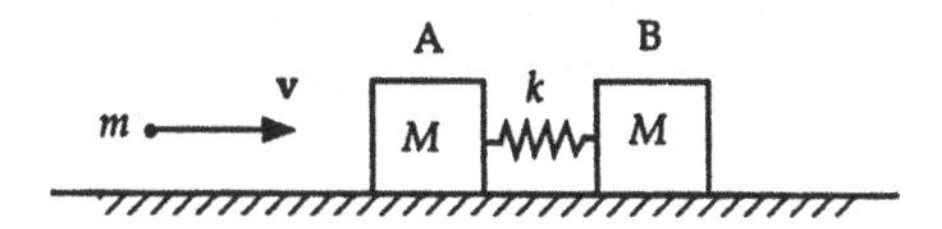

Two identical wood blocks, A and B, both of mass M, are fastened to the ends of a spring. The spring constant is k.

Both blocks are placed on a frictionless horizontal table. Initially, both blocks are at rest and the spring has its unstretched length.

A bullet of mass m ($m \ll M$) is now fired into A with velocity $\mathbf{v}$, and is absorbed there. This causes A to begin moving in the direction of the spring, towards B (see the figure). It is assumed that the time it takes the bullet to come to rest is so short that A has not moved appreciably before the bullet has come to rest relative to A.

After the bullet is stopped, we ignore all frictional effects. Furthermore, we neglect the mass of the spring. Find the following quantities:

(1) the velocity $\mathbf{u}$ of A just after the bullet is stopped relative to A;
(2) the total momentum $\mathbf{P}$ of the system following the impact;
(3) the velocity $\mathbf{v}_{\text{CM}}$ of the center of mass of the system after the impact;
(4) the translational kinetic energy K of the system, after the impact;
(5) The total mechanical energy E of the system, after the impact (the potential energy due to gravity is constant and may be set equal to zero);
(6) The maximum compression Δl of the spring;
(7) The period τ of the oscillations of the spring.

Problem 9.8.

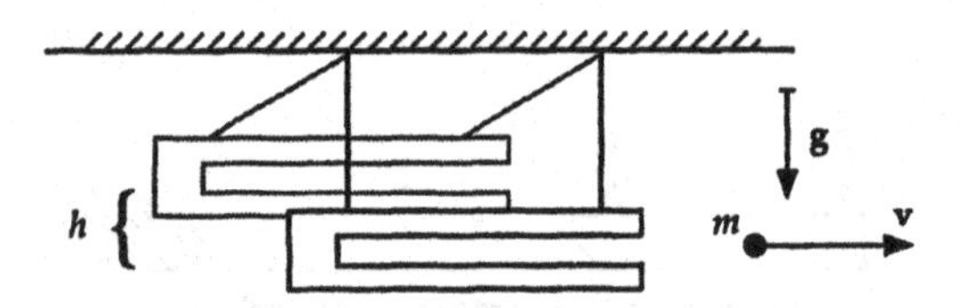

The American physicist Benjamin Thomson (Count of Rumford, 1753–1814) measured the velocity of cannon balls.

The cannon hangs horizontally in he gravitational field of the Earth (field strength g, see the figure). After firing, the cannon swings like a pendulum. The center of mass (CM) of the cannon reaches a height h. The value of h is measured. The unloaded mass of the cannon is M and the cannon ball has mass m. We neglect the mass of the gunpowder. At the moment the cannon ball leaves the barrel the velocity of the ball is horizontal, and the magnitude of the velocity relative to the cannon is v.

We neglect the distance that the cannon has moved during the firing, i.e., during the time in which the cannon ball passes through the barrel. We also neglect air resistance. Determine v as a function of M, m, h, and g.

Problem 9.9. Two blocks, both of mass m, are fastened to a spring as shown in the figure. The length of the unforced spring is L, the spring constant is k. We neglect the mass of the spring. The system is attached to the ceiling

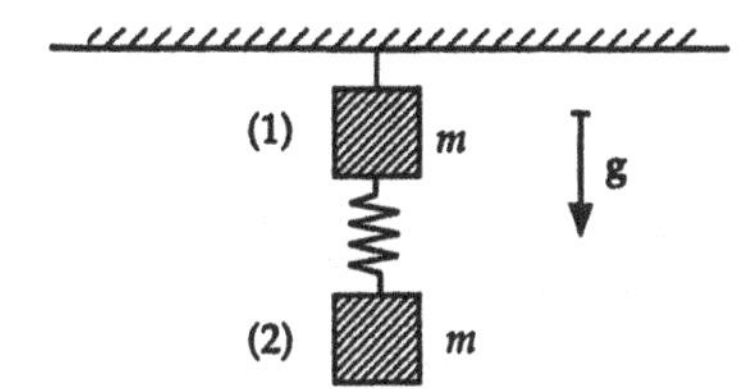

of the laboratory with a string. The acceleration of gravity is g. Initially, the system hangs at rest. At time $t = 0$ the string is cut, and the system falls freely in the gravitational field. We neglect air resistance.

(1) Determine the position of the center of mass (CM) at times $t > 0$. The position should be given as the distance y_{CM} from the CM position at $t = 0$.
(2) Determine the acceleration $\mathbf{a}_1$ of the top mass (1) and the acceleration $\mathbf{a}_2$ of the bottom mass (2) just after the string has been cut.

We now refer the motion for $t > 0$ to a reference frame that has its origin in the CM. Denote in this frame the position of the lower mass (2) by z.

(3) Determine z as a function of t (consider only values of t such that (2) has not hit the floor).

10. The Angular Momentum Theorem

The previous chapters have dealt with problems of translational motion. We now begin an investigation of motions where rotation is an important ingredient. We start by associating with one particle the so-called angular momentum, defined with respect to some reference point often taken to be at rest in an inertial frame. These considerations will then be extended to a system of particles. Forming the angular momentum for a system of particles with respect to the center of mass (CM), we shall see that the equation of motion for this angular momentum has a simple form, even if CM is accelerating.

10.1 The Angular Momentum Theorem for a Particle

Consider a particle of mass m, moving with velocity vector $\mathbf{v}$ relative to some inertial frame I (Figure 10.1). The origin of the coordinate frame is denoted O. The particle has the momentum $\mathbf{p} = m\mathbf{v}$ relative to the inertial frame.

We define the angular momentum $\mathbf{L}$ relative to O as:

$$\mathbf{L} \equiv \mathbf{r} \times \mathbf{p} = \mathbf{r} \times m\mathbf{v}\,. \tag{10.1}$$

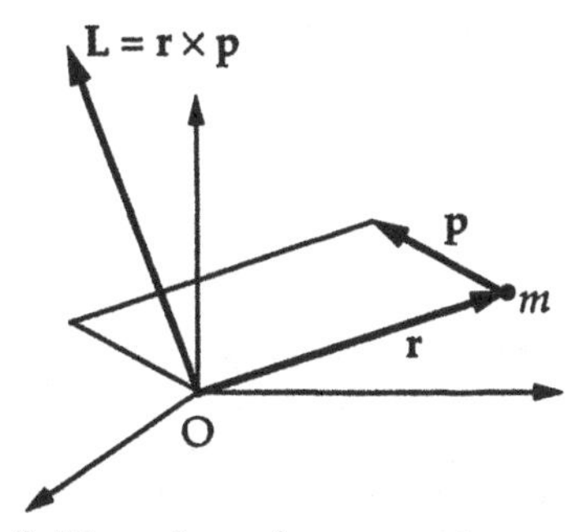

Fig. 10.1. The definition of angular momentum around the point O

The symbol "×" between two vectors describes the vector cross product. Since $\mathbf{L}$ depends on the reference point O one talks about "the angular momentum with respect to O", or "the angular momentum around O".

Differentiate **L** with respect to time:

$$\frac{d\mathbf{L}}{dt} = \frac{d}{dt}(\mathbf{r} \times \mathbf{p}) = \frac{d\mathbf{r}}{dt} \times \mathbf{p} + \mathbf{r} \times \frac{d\mathbf{p}}{dt} .$$

On the right-hand side, the first term equals zero since $d\mathbf{r}/dt = \mathbf{v}$, and the two vectors are parallel. Thus:

$$\frac{d\mathbf{L}}{dt} = \mathbf{r} \times \frac{d\mathbf{p}}{dt} .$$

Assume that the particle is accelerated by a force **F**. From Newton's second law, $d\mathbf{p}/dt = \mathbf{F}$, so the equation of motion for **L** becomes

$$\frac{d\mathbf{L}}{dt} = \mathbf{r} \times \mathbf{F} .$$

The product $\mathbf{N} = \mathbf{r} \times \mathbf{F}$ (Figure 10.2) is called the torque on the particle with respect to (or around) the point O.

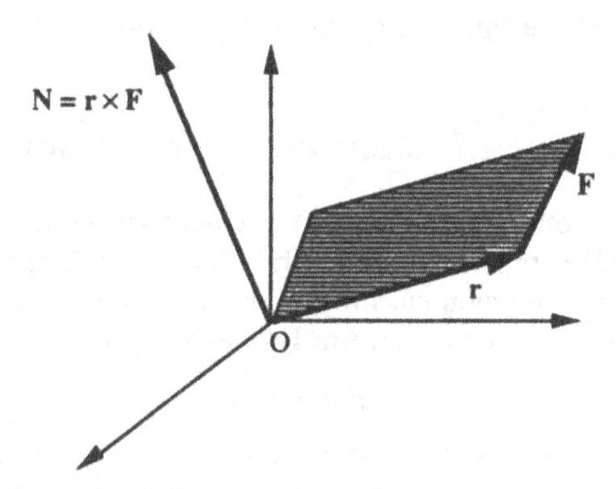

Fig. 10.2. The definition of torque **N** around the point O

With this notation:

$$\frac{d\mathbf{L}}{dt} = \mathbf{N} . \qquad (10.2)$$

This is the content of the *angular momentum theorem*:

The rate of change of the angular momentum of a particle around some point O equals the torque on the particle, with respect to O.

10.2 Conservation of Angular Momentum

Equation (10.2) implies the theorem of conservation of angular momentum:

> If no torque acts on a particle, the angular momentum of that particle with respect to O is constant in time.

A classic example is the motion of a particle in a central force field.

> A central force field is a field where the force on the particle is always directed towards (or away from) a point fixed in an inertial frame. This point is called the force center.

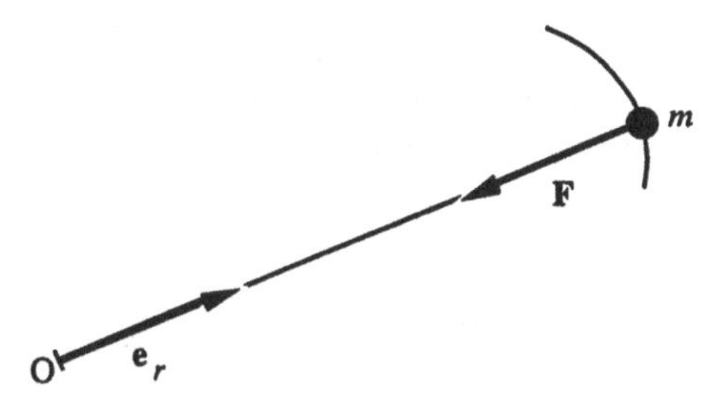

Fig. 10.3. A central force has no torque around the force center

Consider a particle being accelerated by a force of type $\mathbf{F} = f(r)\mathbf{e}_r$, where $\mathbf{e}_r$ is a unit vector along the position vector $\mathbf{r}$ relative to the coordinate center O, and $f(r)$ is some function depending only on the magnitude of $\mathbf{F}$. Such a force has no torque around O (Figure 10.3):

$$\mathbf{N} \equiv \mathbf{r} \times \mathbf{F} = \mathbf{r} \times f(r)\mathbf{e}_r = \mathbf{0}\,.$$

From the angular momentum theorem it then follows that the angular momentum of the particle around O is conserved. The angular momentum associated with a particle moving in a central force field is a constant vector. Angular momentum is a key concept in many physical problems because several important models of physical systems involve central forces.

Consider a planet orbiting the Sun. Let us assume that the Sun is at rest in an inertial frame and use the center of the Sun, O, as origin of our coordinate system. We have just seen that the angular momentum of the planet relative to O is a constant vector. From this it follows that the position vector and the velocity vector always remain in the same plane (Figure 10.4). The orbits of the planets around the Sun are planar.

We can now demonstrate that Kepler's second law follows from the fact that angular momentum of a planet is conserved (see Figure 10.4). The infinitesimal increment in area marked by an increment dr along the orbit is:

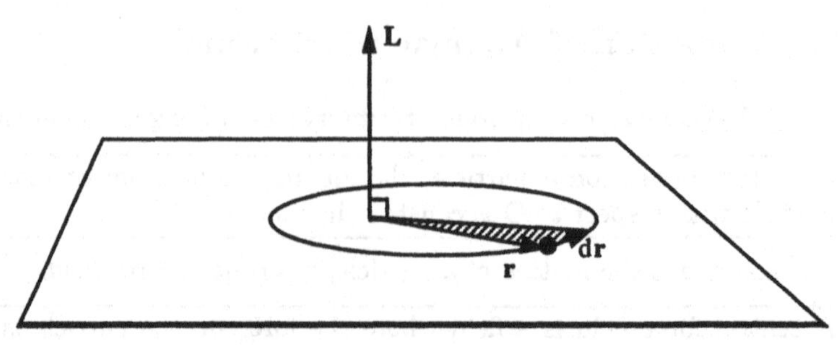

Fig. 10.4. Planet orbiting the Sun. The orbit remains in a plane perpendicular to the angular momentum vector **L**

$$\mathbf{dA} = \frac{1}{2}\mathbf{r} \times \mathbf{dr}\,.$$

We have represented the area swept out by **r** as a vector **A**. The rate of change of **A** is obtained by dividing by dt:

$$\frac{\mathbf{dA}}{dt} = \frac{1}{2}\mathbf{r} \times \frac{\mathbf{dr}}{dt} = \frac{1}{2}\mathbf{r} \times \mathbf{v}\,.$$

Let m denote the mass of the planet. We then have:

$$\frac{\mathbf{dA}}{dt} = \frac{\mathbf{r} \times m\mathbf{v}}{2m} = \frac{\mathbf{L}}{2m}\,.$$

Since the (mass and the) angular momentum vector **L** is constant, the area swept out by the radius vector of a planet changes with a constant rate. This is Kepler's second law. Note how this follows simply from the nature of the central force field. The result does not contain any information about how the force varies with distance.

The planets in our solar system do not exactly follow Kepler's 2nd law. The most important reason for this is that each planet is not only attracted towards the Sun, but also perturbed by the gravitational field from the other planets.

Additionally, even for a single planet and the Sun, the center of the Sun is not fixed but orbits the common center of mass of the planet and the Sun. In other words, the Sun is also accelerated towards the planet, and a reference frame with its origin at the center of the Sun is not an inertial frame. A true "heliocentric" frame should have its origin in the common center of mass for the entire solar system.

The Sun has an enormous mass, 330 000 times the mass of the Earth. The center of mass of the solar system is quite close to the surface of the Sun. The issue will be considered again in Chapter 14.

It is important to note that the concept of angular momentum is not solely associated with closed orbits. Figure 10.5 illustrates motion with constant angular momentum along a nonclosed orbit; a free particle of mass m moves along a straight line with velocity $\mathbf{v}$, as seen from an inertial frame. We consider the angular momentum relative to some point O not on the line, and use the notation of the figure.

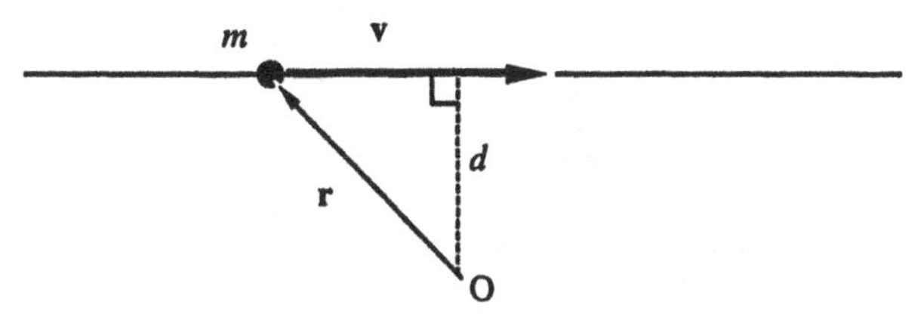

Fig. 10.5. Angular momentum for a nonclosed orbit

There is no force acting on the particle, so the torque around any point (in particular O) is zero. The constant angular momentum around O is given by

$$\mathbf{L}_O = \mathbf{r} \times m\mathbf{v},$$

$\mathbf{L}_O$ is perpendicular to the plane of the paper and points into the plane of the paper. The magnitude of $\mathbf{L}_O$ is

$$L_O = |\mathbf{L}_O| = mvd,$$

where d is the perpendicular distance from the line to O. The quantity d is sometimes called the arm of the momentum.

10.3 Torque and Angular Momentum Around an Axis

We have defined torque and angular momentum relative to a point. One can also consider torque and angular momentum with respect to an axis (a line). This is simply the projection of the (torque or angular momentum) vector on that line.

The torque $\mathbf{N} \equiv \mathbf{r} \times \mathbf{F}$, in cartesian coordinates, is

$$\begin{aligned} N_x &= yF_z - zF_y , \\ N_y &= zF_x - xF_z , \\ N_z &= xF_y - yF_x . \end{aligned}$$

The projection of the torque $\mathbf{N}$ on, say the z-axis, is the component N_z. This is called the torque around the z-axis. Figure 10.6 shows intuitively how N_z emerges: The component F_y acts over the arm x and "creates a rotation in the positive direction around the z-axis". The component F_x acts over the arm y and "turns in the negative direction around the z-axis".

For angular momentum we talk in a similar way about the *angular momentum around some axis*; this is the projection of $\mathbf{L}$ onto the axis.

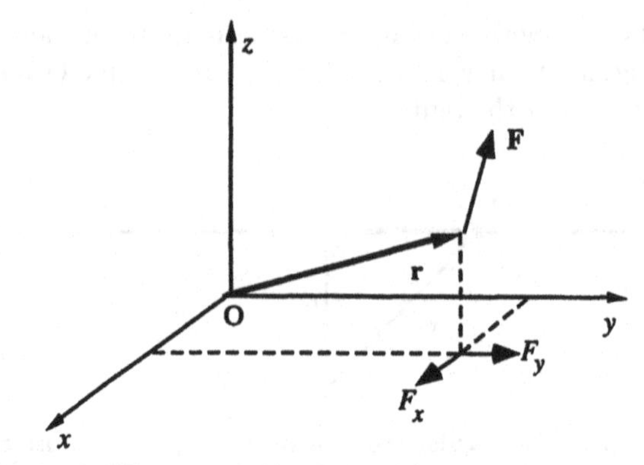

Fig. 10.6. The torque N_z of the force **F** around the z-axis

10.4 The Angular Momentum Theorem for a System of Particles

We shall now develop the angular momentum theorem for a system of particles. Consider first the total torque (with respect to O) of the internal forces for the system (see Figure 10.7). The force on particle i from particle j is denoted $\mathbf{F}_{ij}$ and the force on particle j from particle i is denoted $\mathbf{F}_{ji}$. We assume that the force is directed along a line connecting the two particles. From Newton's third law we then have that $\mathbf{F}_{ij} = -\mathbf{F}_{ji}$. Clearly, the contributions to the torque around O from such a pair vanishes:

$$\begin{aligned} \mathbf{N}_{ij} &= \mathbf{r}_i \times \mathbf{F}_{ij} + \mathbf{r}_j \times \mathbf{F}_{ji} \\ &= (\mathbf{r}_i - \mathbf{r}_j) \times \mathbf{F}_{ij} \\ &= \mathbf{0}, \end{aligned}$$

since the vectors $(\mathbf{r}_i - \mathbf{r}_j)$ and $\mathbf{F}_{ij}$ are parallel. This is also obvious from Figure 10.7 since the two forces have the same arm but the torques are in opposite directions.

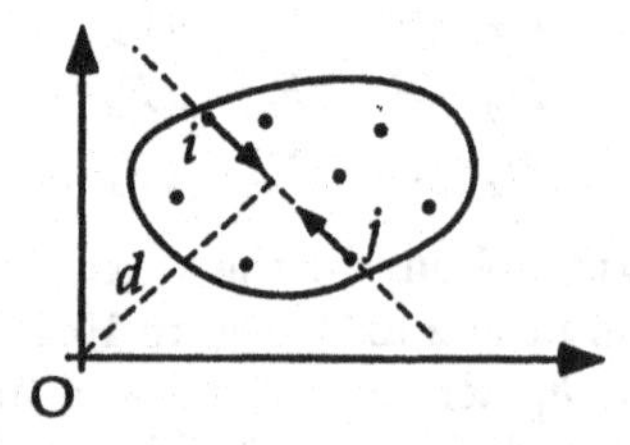

Fig. 10.7. The angular momentum theorem for a system of particles. There is no net torque from internal forces

Similarly the contributions from the forces between any other pair of particles vanish. The torque from internal forces is zero.

Now let us assume that in addition to the internal forces *external* forces act on the system. We can write the angular momentum theorem for each particle:

$$\frac{d\mathbf{L}_i}{dt} = \mathbf{N}_i \,.$$

Here, $\mathbf{N}_i$ is the total torque around O from all forces acting on particle i, external as well as internal. Summing over all particles:

$$\sum_i \frac{d\mathbf{L}_i}{dt} = \sum_i \mathbf{N}_i \,,$$

or

$$\frac{d}{dt}\sum_i \mathbf{L}_i = \sum_i \mathbf{N}_i \,.$$

The contribution to the total torque from the *internal* forces is, as we have seen, zero. Thus the only contribution to the total torque is from the *external* forces. Letting $\mathbf{L}^{\text{tot}} \equiv \sum_i \mathbf{L}_i$, we then have the following result:

$$\frac{d}{dt}\mathbf{L}^{\text{tot}} = \sum_i \mathbf{N}_i^{\text{ext}} \,.$$

This is the *angular momentum theorem* for particle systems:

> The time derivative of the total angular momentum for a particle system relative to a point O fixed in an inertial frame, equals the sum of the torques of the *external* forces, around O.

An important special case of the angular momentum theorem is obtained when there are no external torques on the system: $\mathbf{L}^{\text{tot}}$ = constant vector.

> When the sum of external torques around a point O (fixed in an inertial frame) vanishes, the total angular momentum of the particle system relative to the point O is a constant of the motion.

When no external forces act on a system of particles, we talk about a *closed* system. Thus:

> For a closed system the total angular momentum is conserved.

When we developed the center of mass theorem we saw that internal forces can never change the total momentum of a system. We now have the analogous result that internal forces can never change the total angular momentum of a system.

As an application of the theorem of conservation of angular momentum we consider the system in Figure 10.8. A wheel may turn without friction around an axis A which passes through the center of the wheel. The axis is fixed in the lab. A rope is passed over the wheel as shown. Near one end of the rope is a monkey of mass M. In the other end a bob is attached, also of mass M. Initially the system is at rest. Disregard the mass of the rope and the mass of the wheel.

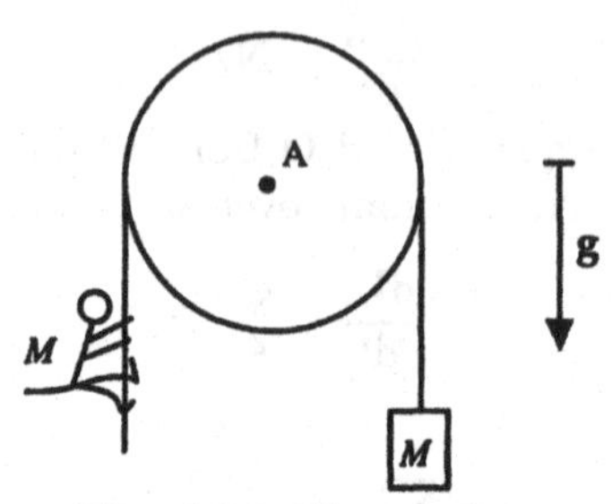

Fig. 10.8. The monkey

Question. The monkey now begins to climb up the rope with velocity v. How will the bob move?

Answer. We consider the wheel, rope, bob, and monkey to be one system. Initially the system is at rest, and there is no angular momentum around A. The external forces on the system are: The reaction force on the pivot A and the gravitational forces on the monkey and the bob. The torques around A of the latter two add to zero. The reaction force acts in A, i.e., it has no arm and therefore no torque. There is thus no external torque around the axis A, and so the angular momentum around A will remain zero. The bob will therefore also move upward with velocity v, cancelling the contribution to the angular momentum around A from the monkey.

The forces between monkey and rope – complicated as they are – are internal forces for the system, and internal forces can never generate any amount of total angular momentum. The center of mass of the system moves upwards. Where is the external force that produces the acceleration of the CM? [Answer: the reaction force in A.]

As another example, consider Figure 10.9. A man stands on a potter's wheel which can turn without friction around a vertical axis. He holds a stick at the end of which a disk has been mounted so that it can turn without friction around the stick. Initially the potter's wheel is at rest in the lab, the stick is horizontal and the disk is spinning as indicated in the figure.

Now the man turns the stick into a vertical position. Since the potter's wheel turns without friction, no external torque around a vertical axis can arise on the system "man + potter's wheel + stick". The vertical component of the total angular momentum is conserved, and since it was zero initially it will remain so. When the stick is turned to a vertical position the pot-

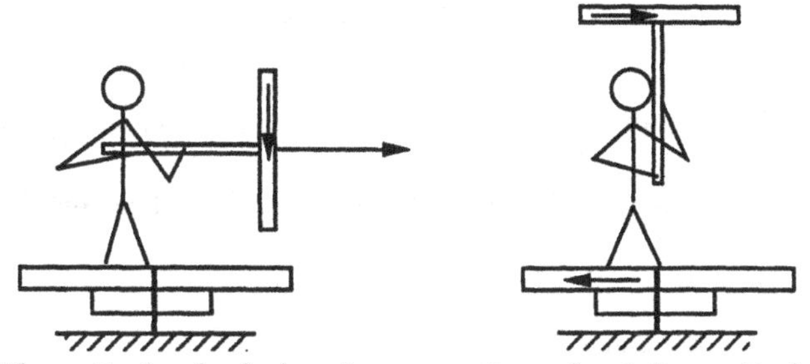

Fig. 10.9. The potter's wheel. Angular momentum about the vertical axis is conserved

ter's wheel will start spinning in the opposite direction with such an angular velocity that the vertical component of the total angular momentum still vanishes.

The total angular momentum has, however, been changed since the initial horizontal component has vanished. When the man turns the stick, he feels a strong reaction force from the potter's wheel, preventing him from starting a rotation around a horizontal axis. If he was floating in free space (!!) he would indeed start spinning, both around a horizontal axis and around a vertical axis as he attempts to turn the stick.

Since he is standing on the potter's wheel, strong reaction forces (from the bearings that keep the wheel moving horizontally) prevent him from rotating. These forces are external to the system (man + stick). They only influence the horizontal components of the angular momentum vector.

10.5 Center of Gravity

Consider a body in the gravitational field of the Earth (Figure 10.10). We assume that the extent of the body is such that the acceleration due to gravity, $\mathbf{g}$, can be considered as constant. The total torque of gravity around some fixed point O is then

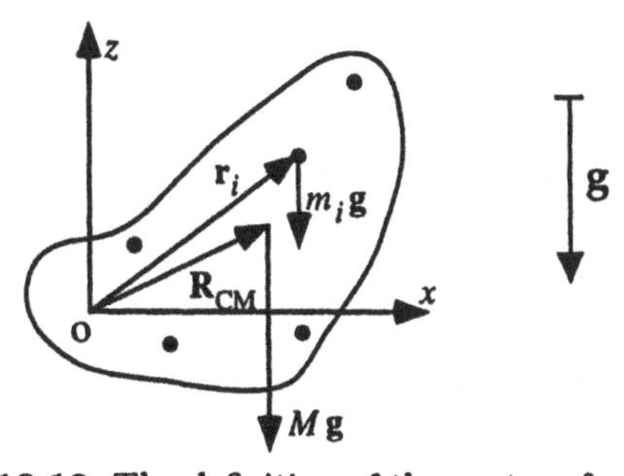

Fig. 10.10. The definition of the center of gravity

$$\mathbf{N}_O = \sum_i m_i \mathbf{r}_i \times \mathbf{g},$$

where m_i is the mass of the volume element (mass point) with position vector $\mathbf{r}_i$ (from O).

From the definition of the position vector $\mathbf{R}_{CM}$ of the center of mass we find:

$$\sum m_i \mathbf{r}_i = M\mathbf{R}_{CM},$$

where M is the total mass of the body. Thus,

$$\mathbf{N}_O = \mathbf{R}_{CM} \times M\mathbf{g}.$$

The total torque of gravity on a body can be calculated by assuming the entire mass to be concentrated at the center of mass. This is the reason the center of mass is often called the center of gravity.

In particular, the torque of gravity about the CM is zero, and thus a body in a homogeneous gravitational field and supported at the CM is in equilibrium for all positions of the body.

10.6 Angular Momentum Around the Center of Mass

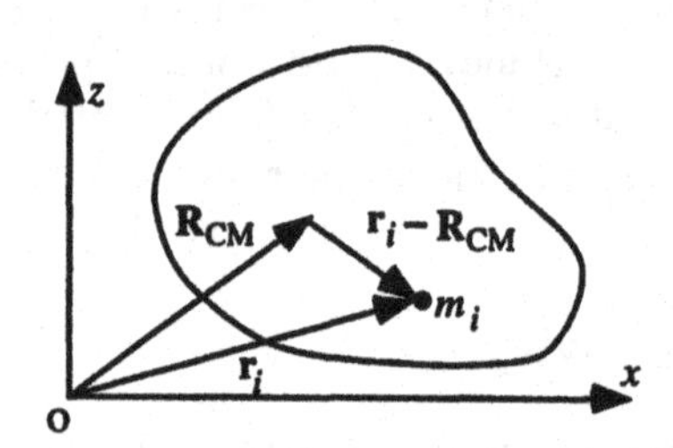

Fig. 10.11. Defining the angular momentum around the CM

For a system of particles the total angular momentum with respect to some fixed point O (Figure 10.11) is defined as:

$$\mathbf{L}_0 = \sum \mathbf{r}_i \times m_i \mathbf{v}_i.$$

The value of $\mathbf{L}_0$ depends on which point O is chosen as reference point.

Let $\mathbf{R}_{CM}$ be the position vector of the center of mass. We can now rewrite $\mathbf{L}_0$ in the following way: we add and subtract the vector

$$\sum \mathbf{R}_{CM} \times m_i \mathbf{v}_i,$$

in the defining equation for $\mathbf{L}_0$:

$$\mathbf{L}_0 = \sum (\mathbf{r}_i - \mathbf{R}_{CM}) \times m_i \mathbf{v}_i + \sum \mathbf{R}_{CM} \times m_i \mathbf{v}_i.$$

The first term in the above expression is the total angular momentum in the motion around the CM, i.e., the angular momentum computed with CM as the reference point. The second term can be written $\mathbf{R}_{\mathrm{CM}} \times \mathbf{P}$ where $\mathbf{P} = M\mathbf{v}_{\mathrm{CM}}$ is the total momentum. Thus,

$$\mathbf{L}_0 = \mathbf{L}_{\mathrm{CM}} + \mathbf{R}_{\mathrm{CM}} \times \mathbf{P}.$$

In words: the total angular momentum around some point O is the angular momentum around CM, $\mathbf{L}_{\mathrm{CM}}$, plus "the angular momentum of CM" around the point O, $\mathbf{R}_{CM} \times \mathbf{P}$. The "angular momentum of CM" is an abbreviation of "the angular momentum of a particle with the mass of the entire system, moving with CM".

Note that the angular momentum of CM (often called the orbital angular momentum) depends on the choice of origin, O. $\mathbf{L}_{\mathrm{CM}}$ (often called the intrinsic angular momentum or spin) is independent of the choice of origin, O.

The angular momentum theorem with respect to the origin of the inertial frame is:

$$\frac{d\mathbf{L}_0}{dt} = \mathbf{N}_0^{\mathrm{ext}} .$$

It is often useful to take the angular momentum relative to the center of mass:

$$\frac{d\mathbf{L}_{\mathrm{CM}}}{dt} = \mathbf{N}_{\mathrm{CM}}^{\mathrm{ext}} .$$

There is an important difference between the above two equations. The first is valid when the angular momentum and the torque are calculated relative to a point O at rest in an inertial frame. The second is valid no matter how CM moves. If the center of mass is accelerated we should include a homogeneous fictitious gravitational field in the system where CM is at rest. But a homogeneous gravitational field has no torque around CM. Therefore the angular momentum theorem is valid in its simple form even if CM is accelerated.

When the external torque about CM vanishes the spin–angular momentum of the system of particles (the body) is a constant of the motion.

If we assume that the solar system is isolated, i.e., if we disregard the forces from the other masses in the galaxy, the total angular momentum of all the planets in the system remains constant.

Consider our own planet, Earth. The Earth stays in rotation around an axis through its center of mass, with an angular momentum which is almost constant. This is due to the fact that most external forces acting on the Earth (from the Sun, the Moon, and the other planets) have no (or a very small) torque around the CM.

Later we shall see that due to external forces the Earth's angular momentum vector, which currently points toward the North Star, slowly changes direction. With a period of about 25 800 years the rotational axis of Earth sweeps out a cone with an interior angle of about 23.5°. In some 14 000 years the axis of the Earth will point toward the bright star Vega in the constellation The Lyre. The phenomenon, which is further described in Chapter 14 is called the precession of the Earth's axis.

10.7 Review: Equations of Motion for a System of Particles

We have seen that the fundamental equations governing the motion of a particle system are:

$$\frac{d\mathbf{P}}{dt} = \sum \mathbf{F}^{\text{ext}}, \quad \mathbf{P} = M\mathbf{v}_{\text{CM}}, \text{ the center of mass theorem;}$$

and

$$\frac{d\mathbf{L}_{\text{CM}}}{dt} = \sum \mathbf{N}_{\text{CM}}^{\text{ext}}, \text{ the angular momentum theorem.}$$

One may claim that these two equations, together with the definition of an inertial frame, contain the science we call classical mechanics. The first equation describes the translational motion of a particle system (which may be a galaxy, a rocket or an atom). The other equation governs the rotation of the system around its center of mass.
The angular momentum theorem is deceptively simple in appearance. In the following chapters we shall see that only in special cases can the angular momentum of a system be expressed in simple terms. Even for a rigid body, the angular momentum theorem is quite complicated. Before we go into the details of rigid body motion we end the chapter with a few applications of the theorem of conservation of angular momentum.

10.8 Examples of Conservation of Angular Momentum

Example 10.1. Particle in Circular Motion. A particle with mass m is tied to a string which goes through a hole in a smooth horizontal table. The particle moves in uniform circular motion with a velocity v_0. The radius of the circle is R_0. By pulling the string one can slowly diminish the radius of the circular motion to a new value, R_1.

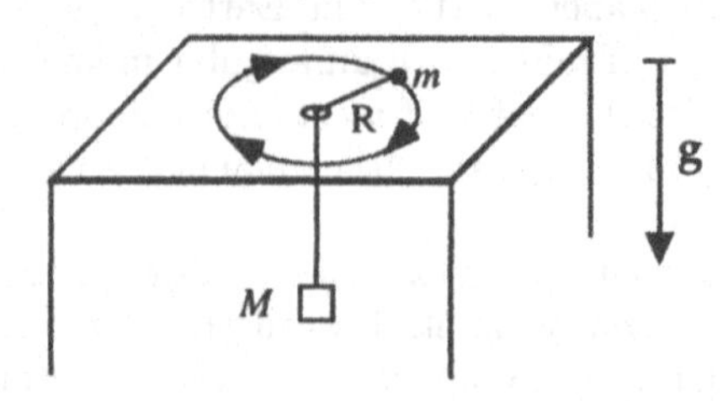

(1) Calculate the new velocity, v_1.
(2) Calculate the increase ΔT in the kinetic energy of the particle.
(3) Show that the work done by the string force equals the increase in kinetic energy of the particle.

Solution.

(1) The string force has no torque around the center of the circle. Thus, the angular momentum about the center is constant:

$$mv_0R_0 = mv_1R_1 .$$

(2)

$$\Delta T = \frac{1}{2}m\left(v_1^2 - v_0^2\right) = \frac{1}{2}mv_0^2\left(\frac{R_0^2}{R_1^2} - 1\right) .$$

(3)

$$W = \int_{R_0}^{R_1} \mathbf{F}\cdot \mathrm{d}\mathbf{r} = -\int_{R_0}^{R_1} m\frac{v^2}{r}\mathrm{d}r = \frac{1}{2}mv_0^2\left[\frac{R_0^2}{R_1^2} - 1\right] .$$

△

Example 10.2. Rotation of Galaxies, Solar Systems, etc

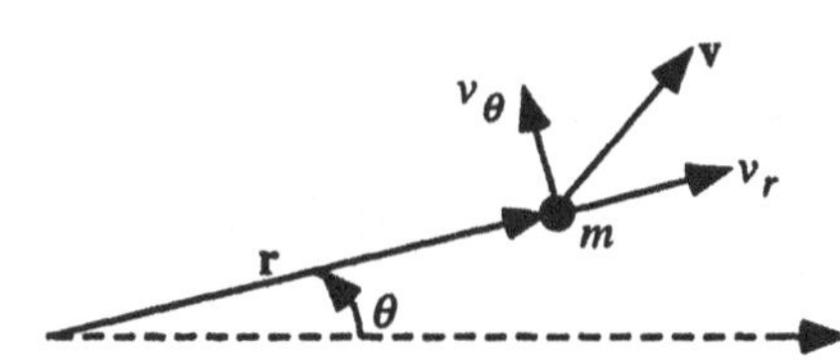

A particle moving in a central field has a constant angular momentum around the force center. An important example of this type of motion is a planet orbiting the Sun. In this example we shall discuss some of the implications of conservation of angular momentum in gravitating systems.

The orbit of a planet lies in a plane perpendicular to the angular momentum vector. The motion is two dimensional and it is useful to describe it in polar coordinates with the force center as the origin. We resolve the velocity vector in two components: a radial component v_r and a transversal component v_θ (see the figure). The potential energy, $U(r)$, of the planet in the force field depends only on the distance r to the force center. The total energy E is constant, and can be written in the following way:

$$E = U(r) + \frac{1}{2}mv_r^2 + \frac{1}{2}mv_\theta^2$$

This expression is valid for any particle moving in a central force field.

For a planet of mass m in the gravitational field of the Sun, we have $U(r) = -GMm/r$ where M is the mass of the Sun and G is the gravitational constant. The magnitude of the angular momentum about the force center is $L = mrv_\theta$; the radial component v_r of the velocity does not contribute to the angular momentum about O. The "angular" part of the kinetic energy is thus:

$$\frac{1}{2}mv_\theta^2 = \frac{L^2}{2mr^2} ,$$

and the total energy can be written as

$$E = U(r) + \frac{L^2}{2mr^2} + \frac{1}{2}mv_r^2 .$$

Since L is a constant of the motion, we can collect the first two terms in a so-called effective potential energy U_{eff}, depending only on r:

$$U_{\text{eff}}(r) = U(r) + \frac{L^2}{2mr^2} .$$

The second term is sometimes denoted the centrifugal potential energy.

The figure below shows U_{eff} as a function of r for the gravitational potential. $U_{\text{eff}}(r)$ is the sum of the two terms. The constraint of keeping L constant appear as a repelling potential which adds a positive term to the *effective potential energy.*

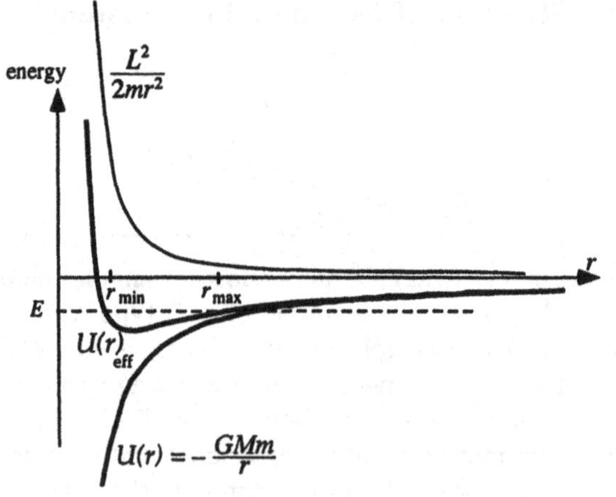

For a bound particle (a planet!), the total energy E is negative. One sees from the figure how for a given amount of total energy, the particle can never come closer that r_{min} (perihelion) and never be further than r_{max} (aphelion) from the force center. The value of r will in fact oscillate between these two values.

As a further illustration we shall briefly describe the formation of a galaxy. Consider a gas cloud of astronomical dimensions. The cloud has mass M and total angular momentum $\mathbf{L}$. No external torques act on the cloud, so $\mathbf{L}$ is a constant vector. The gas cloud begins to contract under its own gravity. When the volume decreases, the angular velocities increase, keeping $\mathbf{L}$ constant, and the cloud now contains more kinetic energy. The increase in kinetic energy comes from the work done by the gravitational forces as the particles fall towards the center of the cloud.

The formation of galaxies is a complicated process, not understood in details. The following qualitative arguments are to be considered a very crude sketch.

Consider a particle with mass m, in the outer regions of the cloud, near its "equator". The particle will have gravitational potential energy of order $U(r) = -GMm/r$, where r here is a measure of the distance to the center of the cloud. The particle will experience an effective potential energy of form:

$$U_{\text{eff}}(r) = \frac{l^2}{2mr^2} - \frac{GMm}{r} \, ,$$

where l is the magnitude of the orbital angular momentum about the center of the cloud. This expression for the potential energy is far from exact; since the mass of the cloud is not uniformly distributed, the distance dependence will be more complicated than the above. We shall, however, base our crude arguments on the above form of U_{eff}.

If the particle is nearer to the center of the cloud it must interact with other particles in the neighbourhood. Its orbit around the center of mass (which would otherwise have been an ellipse) is now perturbed. Let us assume that the particle ends up in an approximately circular orbit with a constant distance to the rotational axis of the cloud.

If we further assume that the radius r of the orbit corresponds to a minimum of the effective potential energy, we can find r from the equation

$$0 = \frac{\mathrm{d}U_{\text{eff}}(r)}{\mathrm{d}r} = -\frac{l^2}{mr^3} + \frac{GMm}{r^2} \, ,$$

i.e.,

$$r = \frac{l^2}{GMm^2} \, .$$

This value of the radius for the orbit of the particle is where the gravitational attraction from the cloud just provides the centripital force for the circular motion:

$$m\frac{v^2}{r} = \frac{GMm}{r^2} \, .$$

In such orbits then, the radius depends on the initial value of the angular momentum of the individual particle.

This is a crude picture of the evolution of the gas cloud in the direction perpendicular to the rotational axis. The contraction parallel to the axis is a different story. The cloud can collapse parallel to $\mathbf{L}$ without any minimum parallel distance to the center of mass. The net result is that the cloud flattens and forms a disk shaped galaxy.

When interstellar clouds collapse, as in our own Milky Way, the same process happens although on a smaller scale. The contraction perpendicular to the rotational axis is halted by centrifugal forces, but contraction continues in the direction parallel to **L**. The result is a star surrounded by gas and dust in an equatorial plane. In such a cloud of gas and dust surrounding the newborn Sun, the planets Mercury, Venus, Earth, Mars, etc. were formed. These processes took place some 4.6×10^9 years ago. The solar system is shaped like a disk, perpendicular to the original angular momentum vector. All the planets move in the same direction around the Sun, the original direction of rotation. The solar system "remembers the process of birth", i.e., the initial conditions.

The formation of planets around newborn stars is much studied, theoretically (including computer experiments), and observationally. Figure 10.12 shows the cloud of gas and dust around the star β-pictoris. Planets are probably forming in this cloud, and perhaps life will emerge on one of these planets. One of the most essential tasks of natural science in the 21st century will be to determine whether the Earth is unique in the Milky Way. Related to this is the problem of the origin of life and its cosmic abundance.

Note the decisive role conservation of angular momentum plays in the formation of planets and thus of life itself. If the initial gas cloud has no angular momentum the result will be a spherical star with no planets, and no life.

From the point of view of mechanics we might ask; what is the original gas cloud rotating relative to? Relative to astronomical space, or relative to all the other masses in the universe? In modern physics, absolute space has been abandoned, and rotation is thought of as rotation relative to the other masses in the universe. It is worth reflecting upon the fact that uniform rotation relative to the stars is felt, but uniform linear motion cannot be detected by any physical experiment at all! $\triangle$

11. Rotation of a Rigid Body

The general treatment of the motion of a rigid body is rather involved and we shall in this book consider only certain special cases. The insights to be gained can be applied in virtually all of theoretical physics, and have important technological ramifications. Applications of the theorems derived stretch from studies of the spin of the electron and rotating atomic nuclei, to investigations of the motion of planets and galaxies.

11.1 Equations of Motion

A body is said to be *rigid* if the distances between the mass elements (the particles of the body) remain constant throughout the motion. The most general laws of motion for particle systems are of course true for rigid bodies in particular:

In *translational motion*, the center-of-mass theorem holds:

$$M\frac{d\mathbf{v}_{\mathrm{CM}}}{dt} = \sum \mathbf{F}^{\mathrm{ext}} . \tag{11.1}$$

In *rotational motion* around the center of mass, the angular momentum theorem holds:

$$\frac{d\mathbf{L}_{\mathrm{CM}}}{dt} = \sum \mathbf{N}^{\mathrm{ext}}_{\mathrm{CM}} . \tag{11.2}$$

The simplest case of rotation of a rigid body is found when the body rotates around an axis fixed in inertial space and fixed relative to the body. This case will be given the most attention but we shall also briefly discuss the case where the body rotates around an axis which changes its orientation in the inertial frame in question and in relation to the rigid body.

In several important cases – e.g., the gyroscope – the body under consideration will move with one point O fixed in the inertial frame. The equation of motion for the rigid body in this case is:

$$\frac{d\mathbf{L}_{\mathrm{O}}}{dt} = \sum \mathbf{N}^{\mathrm{ext}}_{\mathrm{O}} \tag{11.3}$$

In (11.3) we have used O as the point relative to which the angular momentum of the body and the torque of the external forces will be calculated.

Before we move on to apply (11.1)–(11.3) we shall demonstrate an important property of the vector that describes the rotation of a rigid body.

11.2 The Rotation Vector

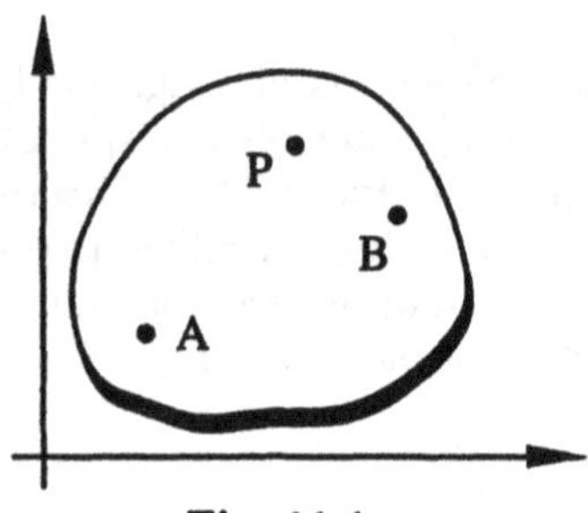

Fig. 11.1.

From geometry it is known that the most general motion of a rigid body can always be separated into a translational motion of a point A in the body, and a rotation about an axis through A. We thus need six coordinates to specify the position of a body relative to a given coordinate system: three coordinates to pin down the point A, two coordinates to specify the direction of the axis of rotation, and 1 coordinate to specify the angle through which the body is rotated. We say that a rigid body has *six degrees of freedom.*

Let us for simplicity assume that the body shown in Figure 11.1 is a plane disk. The velocity of an arbitrary point P in the body relative to the chosen coordinate frame may be written as

$$\dot{\mathbf{r}}_{\mathrm{P}} = \dot{\mathbf{r}}_{\mathrm{A}} + \boldsymbol{\omega}_{\mathrm{A}} \times \mathbf{r}_{\mathrm{AP}} , \tag{11.4}$$

where $\boldsymbol{\omega}_{\mathrm{A}}$ is the vector describing the rotation about the axis through A. The vector $\dot{\mathbf{r}}_{\mathrm{A}}$ is the translational velocity of A, and $\mathbf{r}_{\mathrm{AP}}$ is the vector from A to P.

Let the rotation vector $\boldsymbol{\omega}_{\mathrm{A}}$ be perpendicular to the plane of the paper. If we choose another point B in the body instead of A, we may describe the motion of the body as a translation of the point B plus a rotation around an axis through B. The rotation vector in this case is denoted $\boldsymbol{\omega}_{\mathrm{B}}$, and the velocity of the point P is:

$$\dot{\mathbf{r}}_{\mathrm{P}} = \dot{\mathbf{r}}_{\mathrm{B}} + \boldsymbol{\omega}_{\mathrm{B}} \times \mathbf{r}_{\mathrm{BP}} , \tag{11.5}$$

where

$$\dot{\mathbf{r}}_B = \dot{\mathbf{r}}_A + \boldsymbol{\omega}_A \times \mathbf{r}_{AB} \, .$$

But then, inserting $\dot{\mathbf{r}}_B$ in (11.5):

$$\dot{\mathbf{r}}_P = \dot{\mathbf{r}}_A + \boldsymbol{\omega}_A \times \mathbf{r}_{AB} + \boldsymbol{\omega}_B \times \mathbf{r}_{BP} \, . \tag{11.6}$$

Since the right-hand sides of (11.4) and (11.6) must be equal, we get:

$$\boldsymbol{\omega}_B \times \mathbf{r}_{BP} + \boldsymbol{\omega}_A \times \mathbf{r}_{AB} = \boldsymbol{\omega}_A \times \mathbf{r}_{AP} \, ,$$

or,

$$\boldsymbol{\omega}_B \times \mathbf{r}_{BP} = \boldsymbol{\omega}_A \times (\mathbf{r}_{AP} - \mathbf{r}_{AB}) = \boldsymbol{\omega}_A \times \mathbf{r}_{BP} \, .$$

We conclude that $\boldsymbol{\omega}_A = \boldsymbol{\omega}_B \equiv \boldsymbol{\omega}$. The point B was arbitrary. Therefore the motion of the rigid body relative to a chosen frame is given by specifying the translational velocity of an arbitrary point A in the body, and the rotation vector $\boldsymbol{\omega}$. *Both the magnitude and the direction of* $\boldsymbol{\omega}$ *are independent of the choice of A.*

11.3 Kinetic Energy

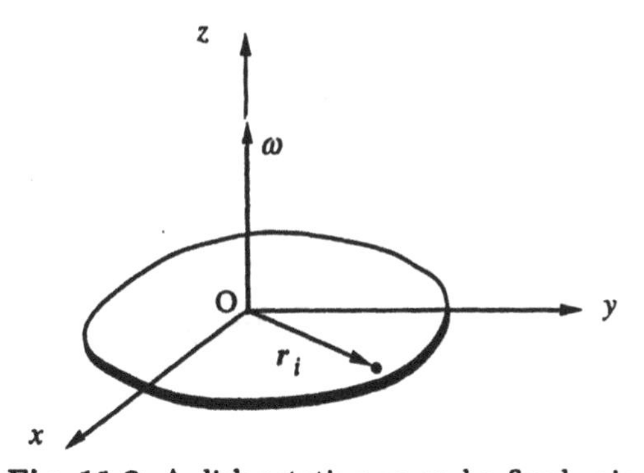

Fig. 11.2. A disk rotating around a fixed axis

Let us start by considering a rigid body in the form of a thin disk confined to the xy-plane. The disk is rotating around the z-axis, with rotation vector $\boldsymbol{\omega} = (0, 0, \omega)$.

A mass element m_i in the disk at the position $\mathbf{r}_i$ has the velocity

$$\mathbf{v}_i = \boldsymbol{\omega} \times \mathbf{r}_i \, .$$

The magnitude of $\mathbf{v}_i$ is:

$$v_i = \omega r_i \, .$$

The total kinetic energy T of the rotating disk is found by summing over all the mass elements:

$$T = \sum_i \frac{1}{2} m_i v_i^2 = \sum_i \frac{1}{2} m_i r_i^2 \omega^2 = \frac{1}{2} \left(\sum_i m_i r_i^2 \right) \omega^2 .$$

The quantity in the paranthesis is denoted by I_z, and called the *moment of inertia* around the z-axis. The kinetic energy for the rotating disk is thus

$$T = \frac{1}{2} I_z \omega^2 , \tag{11.7}$$

where

$$I_z = \sum_i m_i r_i^2 .$$

Before we proceed with the study of rigid bodies in rotation we need two important results concerning the calculation of moments of inertia.

11.3.1 The Parallel Axis Theorem

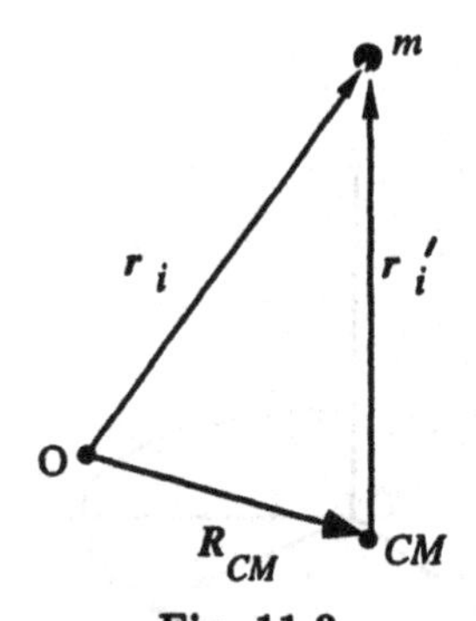

Fig. 11.3.

The vectors in Figure 11.3 all lie within a disk in the xy-plane, and we seek to calculate I_z. The z-axis is perpendicular to the paper and passes through the point O, and CM denotes the center of mass of the disk. The moment of inertia about the z-axis may be written as

$$\begin{aligned} I_z &= \sum_i m_i \mathbf{r}_i^2 = \sum_i m_i \left(\mathbf{R}_{\mathrm{CM}} + \mathbf{r}_i' \right)^2 \\ &= \sum_i m_i \mathbf{R}_{\mathrm{CM}}^2 + \sum_i m_i \mathbf{r}_i'^2 + 2\mathbf{R}_{\mathrm{CM}} \cdot \sum_i m_i \mathbf{r}_i' . \end{aligned}$$

The cross term $2\mathbf{R}_{\mathrm{CM}} \cdot \sum_i m_i \mathbf{r}_i'$ vanishes because of the definition of the center of mass. The second term is the moment of inertia I_{CM} about an axis through CM parallel to the z-axis. Thus, the moment of inertia about the z-axis may be written as

$$I_z = I_{CM} + MR_{CM}^2 .$$

where $M = \sum_i m_i$ is the total mass of the disk. This is the parallel axis theorem.

> The moment of inertia about an arbitrary axis z equals the moment of inertia about an axis parallel to z through the center of mass, plus the total mass M of the body times the square of the distance d between the two axes:
>
> $$I_z = I_{CM} + Md^2 .$$

The parallel axis theorem is used to determine moments of inertia about axes that do not pass through CM or possibly not through the body at all. Note that an immediate consequence of the parallel axis theorem is that the moment of inertia about an axis through CM is smaller than about any other axis parallel to this. We shall see later that the parallel axis theorem is valid in general, not just for bodies with a planar mass distribution (a disk).

The *kinetic energy* for a disk rotating about the z-axis can – using the parallel axis theorem – be written as

$$T = \frac{1}{2}I_z\omega^2 = \frac{1}{2}I_{CM}\omega^2 + \frac{1}{2}MR_{CM}^2\omega^2 ,$$

or, since $v_{CM} = R_{CM}\omega$,

$$T = \frac{1}{2}I_{CM}\omega^2 + \frac{1}{2}Mv_{CM}^2 = T_{rot} + T_{trans} . \tag{11.8}$$

Equation (11.8) is a special case of König's theorem (see Section 9.2).

The *angular momentum* of the rotating disk around the origin O is:

$$\mathbf{L}_O = \sum_i \mathbf{r}_i \times m_i\mathbf{v}_i ; \quad \mathbf{v}_i = \boldsymbol{\omega} \times \mathbf{r}_i .$$

Since $\mathbf{r}_i$ and $\mathbf{v}_i$ are mutually perpendicular we find that $\mathbf{L}_O$ is along the z-axis (the rotational axis) and has the magnitude

$$L_O = L_z = \sum_i m_i r_i^2\omega = I_z\omega .$$

The parallel axis theorem shows that the magnitude of $\mathbf{L}_O$ can be written as:

$$\begin{aligned} L_O = L_z &= I_{CM}\omega + MR_{CM}^2\omega \\ &= I_{CM}\omega + Mv_{CM}R_{CM} . \end{aligned}$$

For a planar mass distribution we thus have: the angular momentum about some point O equals the angular momentum about CM plus "the angular momentum of a particle with the mass M, located in and moving with the CM". See Section 10.6.

11.3.2 The Perpendicular Axis Theorem

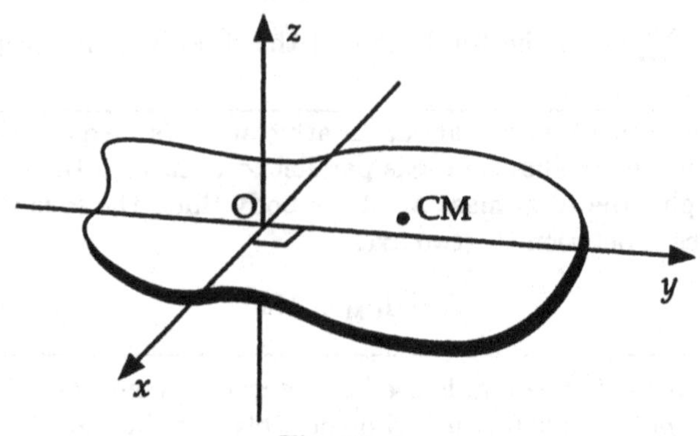

Fig. 11.4.

This theorem concerns moments of inertia about axes that are perpendicular to each other. We still consider a thin disk (see Figure 11.4). In contrast to the parallel axis theorem, this theorem is valid *only* for a planar mass distribution. Consider the moment of inertia about the x-axis:

$$I_x = \sum m_i y_i^2 ,$$

and similarly about the y-axis:

$$I_y = \sum m_i x_i^2 .$$

Furthermore:

$$I_z = \sum m_i r_i^2 = \sum m_i \left(x_i^2 + y_i^2\right) = \sum m_i x_i^2 + \sum m_i y_i^2 ,$$

or

$$I_z = I_x + I_y . \tag{11.9}$$

This is the perpendicular axes theorem:

> For a planar mass distribution (a disk), the moment of inertia about an axis perpendicular to the disk equals the sum of the moments of inertia about two mutually perpendicular axes in the plane of the disk and intersecting where the perpendicular axis passes through the disk.

11.4 An Arbitrary Rigid Body in Rotation Around a Fixed Axis

Consider a rigid body rotating about the z-axis with rotation vector $\boldsymbol{\omega} = (0, 0, \omega)$, assumed to be constant in time and fixed relative to the body. The angular momentum about the origin is

$$\mathbf{L}_O = \sum_i \mathbf{r}_i \times m_i \mathbf{v}_i \; .$$

We decompose $\mathbf{r}_i$ into two vectors $\mathbf{a}_i$ and $\mathbf{b}_i$ (see Figure 11.5). The vector $\mathbf{a}_i$ is perpendicular to the rotation axis while $\mathbf{b}_i$ is parallel to it (Figure 11.5).

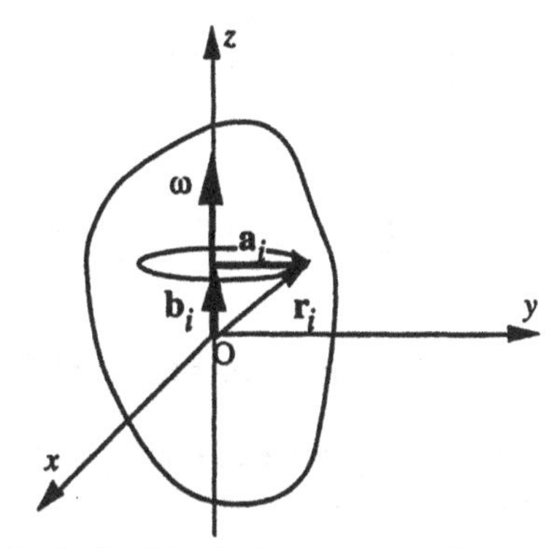

Fig. 11.5. A rigid body in rotation about a fixed axis

Since

$$\mathbf{r}_i = \mathbf{a}_i + \mathbf{b}_i \; ,$$

we have that

$$\mathbf{v}_i = \boldsymbol{\omega} \times \mathbf{r}_i = \boldsymbol{\omega} \times \mathbf{a}_i \; ,$$

and

$$\mathbf{L}_O = \sum_i m_i \left[\mathbf{r}_i \times (\boldsymbol{\omega} \times \mathbf{a}_i)\right] \; .$$

Since for three vectors $\mathbf{A}$, $\mathbf{B}$, and $\mathbf{C}$ in general,

$$\mathbf{A} \times (\mathbf{B} \times \mathbf{C}) = \mathbf{B}\,(\mathbf{A} \cdot \mathbf{C}) - \mathbf{C}\,(\mathbf{A} \cdot \mathbf{B}) \; ,$$

we find that $\mathbf{L}_O$ may be written as

$$\begin{aligned}
\mathbf{L}_O &= \sum_i m_i \boldsymbol{\omega}(\mathbf{r}_i \cdot \mathbf{a}_i) - \sum_i m_i \mathbf{a}_i (\mathbf{r}_i \cdot \boldsymbol{\omega}) \\
&= \boldsymbol{\omega} \sum_i m_i \left[(\mathbf{a}_i + \mathbf{b}_i) \cdot \mathbf{a}_i\right] - \sum_i m_i \mathbf{a}_i (\mathbf{b}_i \cdot \boldsymbol{\omega}) \\
&= \boldsymbol{\omega} \sum_i m_i a_i^2 - \sum_i (\omega b_i m_i) \mathbf{a}_i \; .
\end{aligned}$$

$\mathbf{L}_0$ thus has a component *parallel* to ω (i.e. parallel to the z-axis) and a component *perpendicular* to ω. The component parallel to the z-axis is $I_z\omega$, where I_z is the moment of inertia about the z-axis.

Note that the moment of inertia is defined by means of the perpendicular distance of the mass point from the axis of rotation, not the distance from O. As we shall see in due time, the fact that $\mathbf{L}_0$ is not necessarily parallel to ω makes the general problem of rotation complicated. For the translational motion of a body, the momentum is always parallel to $\mathbf{v}$, $\mathbf{P} = M\mathbf{v}_{\mathrm{CM}}$.

The kinetic energy is

$$T = \frac{1}{2}\sum_i m_i \dot{r}_i^2 .$$

We find that

$$|\dot{\mathbf{r}}_i| = |\omega \times \mathbf{r}_i| = \omega a_i ,$$

and consequently

$$T = \frac{1}{2}\left(\sum_i m_i a_i^2\right)\omega^2 = \frac{1}{2}I_z\omega^2 .$$

11.4.1 The Parallel Axis Theorem in General Form

Assume next that we know the moment of inertia of a body about an axis A parallel to the z-axis and through CM (see Figure 11.6).

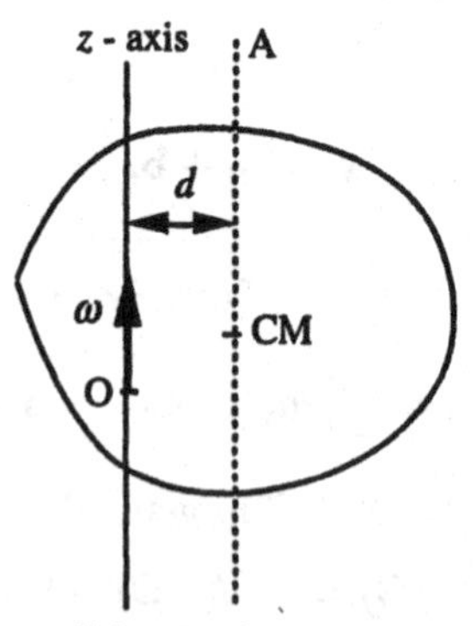

Fig. 11.6. The parallel axis theorem for an arbitrary body

The kinetic energy of the body in its rotation about the axis A is then

$$T_{\mathrm{A}} = \frac{1}{2}I_{\mathrm{CM}}\omega^2 .$$

Note that ω is the same for rotation about the z-axis as for rotation about A. The motion of the body can be considered either as a pure rotation about the z-axis or as a translation of the CM plus a rotation about A. In both views the rotation vector is $\boldsymbol{\omega}$.

Using König's theorem, we may write the total kinetic energy of the body in rotation about the z-axis as

$$T = \frac{1}{2} M \mathbf{v}_{\mathrm{CM}}^2 + \frac{1}{2} I_{\mathrm{CM}} \omega^2 .$$

The velocity of CM is $v_{\mathrm{CM}} = \omega d$ (where d is the perpendicular distance from the rotation axis through O to CM). Thus,

$$\begin{aligned} T &= \frac{1}{2} M d^2 \omega^2 + \frac{1}{2} I_{\mathrm{CM}} \omega^2 \\ &= \frac{1}{2} \left(M d^2 + I_{\mathrm{CM}} \right) \omega^2 . \end{aligned}$$

If the motion is described as a pure rotation about the z-axis we have:

$$T = \frac{1}{2} I_z \omega^2 .$$

Comparing these results we find:

$$I_z = I_{\mathrm{CM}} + M d^2 .$$

Using König's theorem we have demonstrated the parallel axis theorem for an arbitrary rigid body in rotation about a fixed axis.

11.5 Calculation of the Moment of Inertia for Simple Bodies

11.5.1 Homogenous Thin Rod

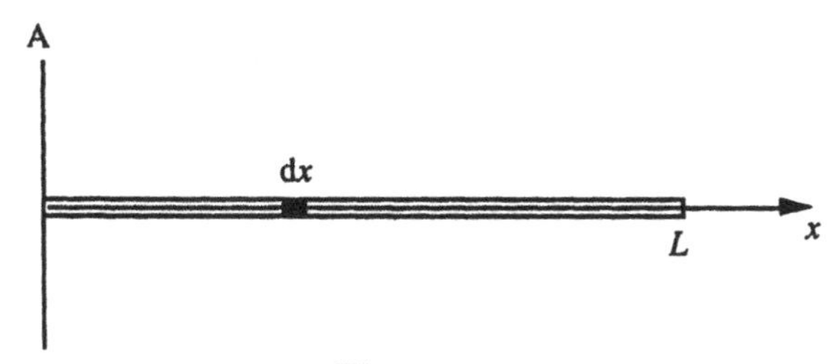

Fig. 11.7.

A homogeneous thin rod has length L and total mass M. A small segment of length $\mathrm{d}x$ has the mass $(M/L)\mathrm{d}x$. The segment, at distance x from the rotational axis A, will contribute to the moment of inertia with an amount

$$dI = x^2(M/L)dx$$

The total moment of inertia for the homogeneous rod is now found by integrating over the range of x-values:

$$I_A = \frac{M}{L}\int_0^L x^2 dx = \frac{1}{3}ML^2 .$$

By applying the parallel axis theorem we find that for an axis through the center (of mass) for the rod:

$$I_{\text{CM}} = \frac{1}{3}ML^2 - M\left(\frac{L}{2}\right)^2 = \frac{1}{12}ML^2 .$$

The moment of inertia about the axis through the center could also have been calculated as the sum of two moments of inertia for rods of mass $M/2$ and length $L/2$:

$$I_{\text{CM}} = \frac{1}{3}\frac{M}{2}\left(\frac{L}{2}\right)^2 + \frac{1}{3}\frac{M}{2}\left(\frac{L}{2}\right)^2 = \frac{1}{12}ML^2 .$$

For a body of complicated shape the moment of inertia can be found by piecing the body together from "simple" parts with known moments of inertia.

11.5.2 Circular Disk

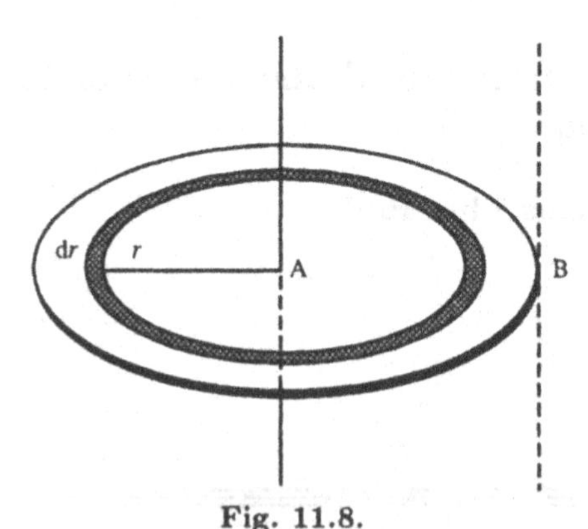

Fig. 11.8.

A homogeneous circular disk has mass M and radius R. The axis A of rotation is perpendicular to the plane of the disk and passing through the center. A ring of thickness dr at distance r from the center has the mass

$$dM = \frac{2\pi r dr}{\pi R^2}M = M\frac{2r dr}{R^2} .$$

The contribution of the ring to the moment of inertia about A for the disk is

$$\mathrm{d}I_{\mathrm{A}} = (\mathrm{d}M)r^2 = M\frac{2r\mathrm{d}r}{R^2}r^2 = \frac{2M}{R^2}r^3\mathrm{d}r\,.$$

Consequently the total moment of inertia for the homogeneous disk is

$$I = \frac{2M}{R^2}\int_0^R r^3\mathrm{d}r = \frac{1}{2}MR^2\,.$$

From the parallel axis theorem we find that the moment of inertia about an axis perpendicular to the plane of the disk but at the edge of the disk (the point B on the figure) is

$$I_{\mathrm{B}} = \frac{1}{2}MR^2 + MR^2 = \frac{3}{2}MR^2\,.$$

From the perpendicular axis theorem we can find the moment of inertia about a diameter of the disk:

$$I_D + I_D = \frac{1}{2}MR^2\,, \quad \text{or} \quad I_D = \frac{1}{4}MR^2\,.$$

A homogeneous cylinder can be considered as a stack of disks sharing the same axis and adding their mass. Thus for a cylinder around its axis of symmetry:

$$I = \frac{1}{2}MR^2\,.$$

If the cylinder has height h one finds for the moment of inertia about an axis perpendicular to the cylinder axis and through CM:

$$I = \frac{1}{4}M\left(R^2 + \frac{h^2}{3}\right)\,.$$

A cylinder is a rod made up from disks!

11.5.3 Thin Spherical Shell

A homogeneous thin spherical shell has mass M and radius R. From the symmetry of the shell follows that $I_x = I_y = I_z$. We have:

$$\begin{aligned}
I_x &= \int (y^2 + z^2)\,\mathrm{d}M\,,\\
I_y &= \int (z^2 + x^2)\,\mathrm{d}M\,,\\
I_z &= \int (x^2 + y^2)\,\mathrm{d}M\,.
\end{aligned}$$

The moment of inertia around an arbitrary diameter can thus be written as

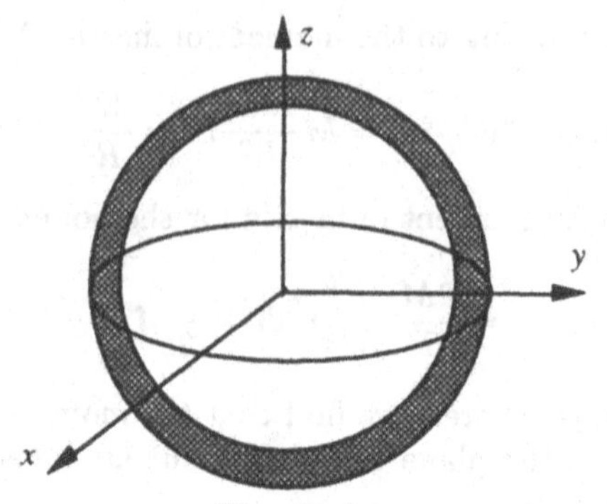

Fig. 11.9.

$$I = \frac{1}{3}(I_x + I_y + I_z) ,$$

or

$$\begin{aligned} I &= \frac{1}{3}2 \int (x^2 + y^2 + z^2)\,\mathrm{d}M \\ &= \frac{2}{3}R^2 \int \mathrm{d}M = \frac{2}{3}MR^2 . \end{aligned}$$

11.5.4 Homogenous (Solid) Sphere, Mass M and Radius R

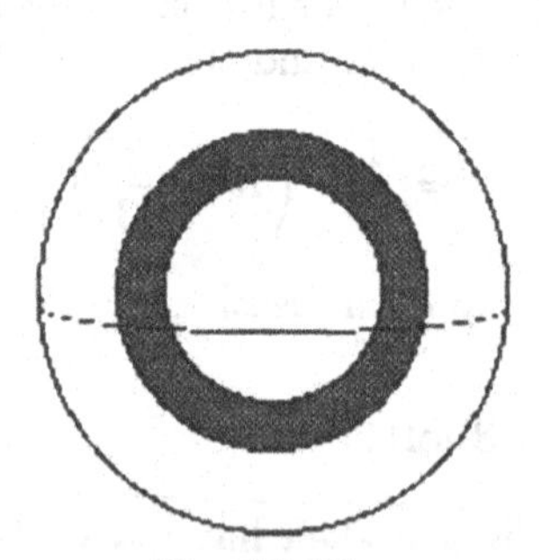

Fig. 11.10.

Inside the solid sphere consider first a spherical shell with radius r and thickness $\mathrm{d}r$. This shell has mass:

$$\mathrm{d}M = \frac{4\pi r^2 \mathrm{d}r}{\frac{4}{3}\pi R^3} M = \frac{3r^2 \mathrm{d}r}{R^3} M .$$

The spherical shell contributes to the moment of inertia around a diameter by the amount:

$$dI = \frac{2}{3}\left(\frac{3r^2 dr}{R^3} M\right) r^2 .$$

The total moment of inertia around a diameter for the entire homogeneous sphere is then

$$I = \frac{2M}{R^3}\int_0^R r^4 dr = \frac{2}{5} M R^2 .$$

The moment of inertia for a homogeneous sphere around an axis which is tangential to the surface of the sphere is

$$I_{\tan} = \frac{2}{5} M R^2 + M R^2 = \frac{7}{5} M R^2 .$$

11.5.5 Rectangular Plate

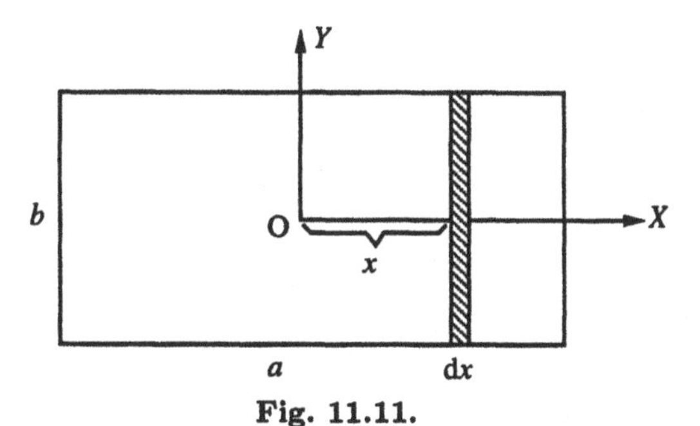

Fig. 11.11.

A homogeneous thin rectangular plate has sides a and b as shown on the sketch. The entire mass of the plate is M. We calculate the moment of inertia around the Z-axis which goes through the CM of the plate and is perpendicular to the plate.

Consider the plate to be made up from rods of thickness dx (see the sketch).

$$dM = \frac{M}{ab} dx \cdot b = \frac{M}{a} dx .$$

The total moment of inertia of the plate with respect to the Z-axis is

$$I_z = \frac{M}{a}\int_{-a/2}^{+a/2}\left(\frac{b^2}{12} + x^2\right) dx = \frac{M}{12}\left(a^2 + b^2\right) .$$

A different way of obtaining this result is the following. Since the plate can be seen as a stack of rods each of length b, $I_x = \frac{1}{12} M b^2$ and similarly $I_y = \frac{1}{12} M a^2$. From this, $I_z = I_x + I_y = \frac{1}{12} M \left(a^2 + b^2\right)$.

We conclude this section with a table of some moments of inertia (Table 11.1).

Table 11.1. Moments of inertia for selected bodies

Body of total mass M		Moment of Inertia
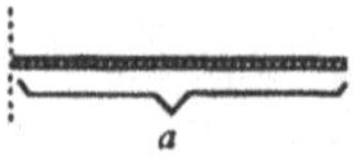 thin rod (axis at end, length a)	thin rod	$\frac{1}{3}Ma^2$
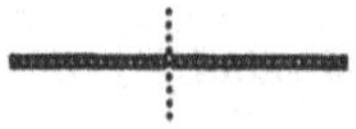 thin rod (axis through center)		$\frac{1}{12}Ma^2$
 circular disk (axis tangent, radius R)	circular disk	$\frac{3}{2}MR^2$
 circular disk (axis through center, perpendicular)		$\frac{1}{2}MR^2$
 circular disk (axis along diameter)		$\frac{1}{4}MR^2$
 R	thin spherical shell	$\frac{2}{3}MR^2$
 R	homogenous sphere	$\frac{2}{5}MR^2$
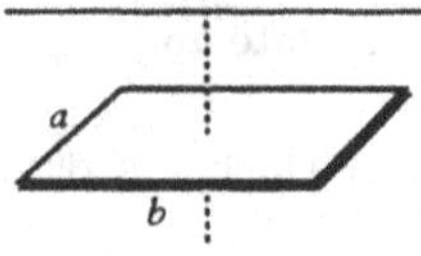 a, b	Rectangular plate	$\frac{1}{12}M\left(a^2+b^2\right)$

11.6 Equation of Motion for a Rigid Body Rotating Around a Fixed Axis

We are now in a position to apply the angular momentum theorem to the study of the rotation of a rigid body around a fixed axis which we choose as the z-axis (see Figure 11.12). The equation of motion is (regarding the lab frame as an inertial frame)

$$\frac{d\mathbf{L}_O}{dt} = \sum \mathbf{N}_O^{ext} ,$$

where the angular momentum $\mathbf{L}_O$ and the torque $\mathbf{N}_O^{ext}$ are both taken about the point O which is on the axis of rotation. That the axis of rotation is fixed means that it is stationary in the inertial frame we use (the lab frame) and is fixed relative to the rigid body.

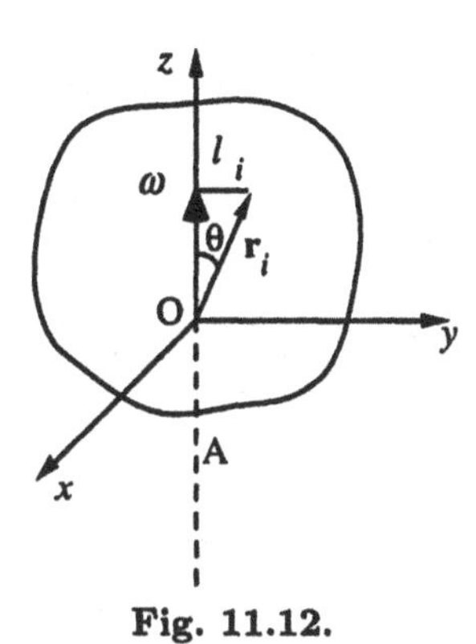

Fig. 11.12.

When considering rotation about a fixed axis, we can re-write the equations of motion in a simpler form.

Since the axis of rotation, A, is stationary in an inertial frame and relative to the body, the inertial properties of the rigid body with respect to A are constant. We shall later see that it is just this fact which makes rotation about a fixed axis a particularly simple problem.

In order to describe the motion we need to consider only the component along the z-axis of the angular momentum vector, and the corresponding component of the torque of the external forces along the z-axis. In many instances those two vectors will have components only along the z-axis. In particular this will be the case for bodies with a high degree of symmetry (e.g. a cylinder or a disk) when the rotational axis coincides with the axis of symmetry.

We now demonstrate that the equation of motion for a rigid body in *rotation about a fixed axis is a scalar equation.* The vector $\boldsymbol{\omega}/|\boldsymbol{\omega}|$ is a unit vector along the axis of rotation. The component of $\mathbf{L}_O$ along the axis of rotation is:

$$\begin{aligned} L_A &= \mathbf{L} \cdot \frac{\boldsymbol{\omega}}{|\boldsymbol{\omega}|} = \frac{1}{\omega} \left[\sum_i \mathbf{r}_i \times m_i \mathbf{v}_i \right] \cdot \boldsymbol{\omega} \\ &= \frac{1}{\omega} \left[\sum_i m_i \mathbf{r}_i \times (\boldsymbol{\omega} \times \mathbf{r}_i) \right] \cdot \boldsymbol{\omega} . \end{aligned}$$

Using the identity $\mathbf{A} \times (\mathbf{B} \times \mathbf{C}) = \mathbf{B}\,(\mathbf{A} \cdot \mathbf{C}) - \mathbf{C}\,(\mathbf{A} \cdot \mathbf{B})$, we get:

$$\begin{aligned} L_{\mathrm{A}} &= \frac{1}{\omega}\left[\sum_i m_i \left[r_i^2 \boldsymbol{\omega} - (\mathbf{r}_i \cdot \boldsymbol{\omega})\,\mathbf{r}_i\right]\right] \cdot \boldsymbol{\omega} \\ &= \frac{1}{\omega}\left[\sum_i m_i \left[r_i^2 \omega^2 - r_i^2 \omega^2 \cos^2(\theta)\right]\right] \\ &= \frac{1}{\omega}\left[\sum_i m_i r_i^2 \omega^2 \sin^2(\theta)\right] . \end{aligned}$$

The distance of the particle m_i from the rotation axis is $l_i = r_i \sin(\theta)$. We find

$$L_{\mathrm{A}} = \frac{1}{\omega}\left[\sum_i m_i l_i^2\right] \omega^2 ,$$

or

$$L_{\mathrm{A}} = I_{\mathrm{A}} \omega ,$$

where I_{A} is the moment of inertia of the body with respect to the axis of rotation. L_{A} is called the angular momentum of the body with respect to the rotation axis. The component along the rotation axis of the torque of the external forces, denoted N_{A}, is given by

$$N_{\mathrm{A}} = \mathbf{N} \cdot \frac{\boldsymbol{\omega}}{\omega} .$$

The *equation of motion* describing the rotation of a rigid body around a fixed axis A is then

$$\frac{\mathrm{d}L_{\mathrm{A}}}{\mathrm{d}t} = N_{\mathrm{A}} ,$$

or, since the moment of inertia about A is constant,

$$\boxed{I_{\mathrm{A}} \frac{\mathrm{d}\omega}{\mathrm{d}t} = N_{\mathrm{A}} .}$$

An important special case of the angular momentum theorem for rotation around a fixed axis results when the torque of the external forces vanishes:

When the external torque around a given fixed axis A is zero, the angular momentum about the axis A is a constant of the motion.

11.6.1 Conservation of Angular Momentum

Consider Figure 11.13. A man stands on a turntable which may turn, without friction, about a vertical axis. The turntable (and the man along with it) is now given an angular velocity ω_1 about a vertical axis while the man holds two heavy weights strecthed away from his body as in (1) on the sketch. While in rotation the man pulls the weights towards his body, as in (2). We consider the man plus the turntable as our system. As mentioned, the turntable rotates without friction, i.e., the vertical axis is well oiled. No external torques act about the vertical axis of the system. Therefore the angular momentum about this axis is conserved.

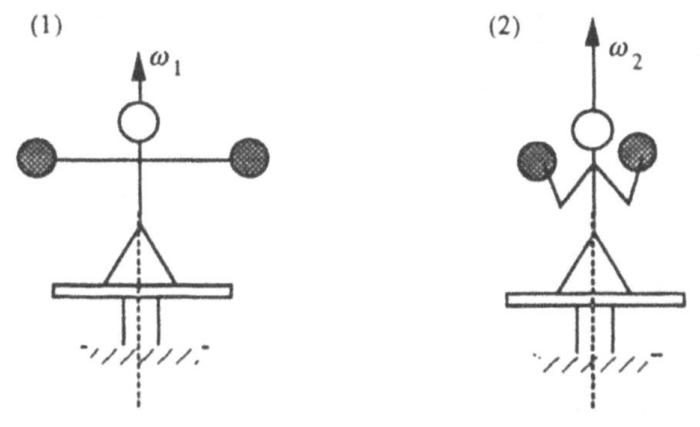

Fig. 11.13. Man on a turntable. Illustration of the conservation of angular momentum

Suppose the moment of inertia about the vertical axis initially is I_1. When the man rotates with his arms stretched out, the angular momentum is

$$L = L_1 = I_1\omega_1 .$$

When the man pulls the weights towards his body he changes (decreases) the moment of inertia to a new value, I_2 , where $I_2 < I_1$. The angular velocity will increase in such a way that the angular momentum is kept constant:

$$I_1\omega_1 = I_2\omega_2 ,$$

so $\omega_2 > \omega_1$.

Consider the change in the mechanical kinetic energy. The kinetic energy of a system rotating about an axis is

$$T = \frac{1}{2}I\omega^2 = \frac{1}{2}L\omega .$$

Consequently

$$T_1 = \frac{1}{2}L\omega_1 ,$$
$$T_2 = \frac{1}{2}L\omega_2 .$$

The angular momentum L is unchanged, so we have

$$T_2 > T_1 \quad \text{since} \quad \omega_2 > \omega_1 .$$

The mechanical kinetic energy of the system has increased. The man has done work against the centrifugal force in the co-rotating frame. Compare with Example 10.1.

Internal forces can produce a change in macroscopic mechanical kinetic energy (compare with the rocket). However, internal forces *cannot* produce a change in the total angular momentum.

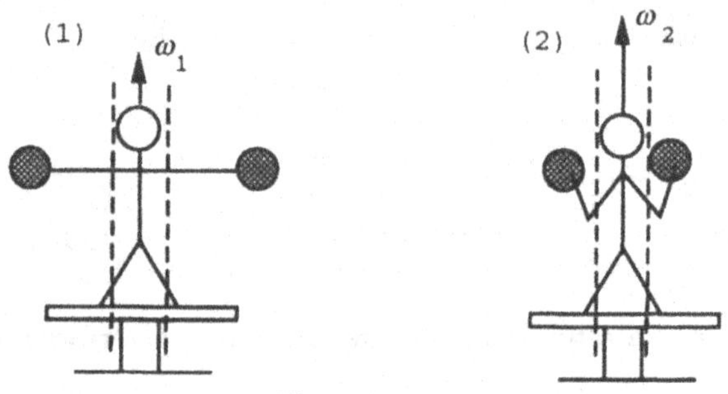

Fig. 11.14.

Let us finally consider the body of the man without the arms as our system (see Figure 11.14). The man's body rotates faster after he has pulled the weights towards his body. The moment of inertia of his body is unchanged, so the angular momentum of the body has increased. This increase is caused by a torque from the fictitious forces in the co-rotating frame. The man is assumed to be fixed to the turntable. The fictitious forces are: the Coriolis force $[-2m(\boldsymbol{\omega} \times \mathbf{v})]$ and the force due to $\dot{\boldsymbol{\omega}}$, $[-m(\dot{\boldsymbol{\omega}} \times \mathbf{r})]$.

11.7 Work and Power in the Rotation of a Rigid Body Around a Fixed Axis

A rigid body may rotate around the z-axis. An external force $\mathbf{F}$, is acting at the point A which has position vector $\mathbf{r}$ (see Figure 11.15). Consider now an infinitesimal rotation of the body. In polar coordinates

$$x = r\cos(\theta) , \quad y = r\sin(\theta) .$$

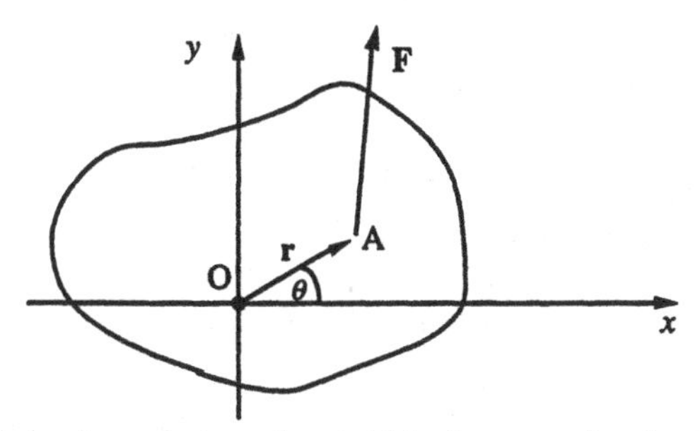

Fig. 11.15. Work in the rotation of a rigid body around a fixed axis (the z-axis, out of the plane of the paper)

We find that

$$\begin{aligned} dx &= -r\sin(\theta)d\theta = -y d\theta\,, \\ dy &= r\cos(\theta)d\theta = x d\theta\,. \end{aligned}$$

The work done by the force **F** when the body turns through the angle $d\theta$ is thus

$$dW = F_x \cdot dx + F_y \cdot dy = (xF_y - yF_x)d\theta$$

or

$$dW = N_z d\theta,$$

where N_z is the torque about the z-axis. The work equation for a rigid body rotating around a fixed axis is then

$$\frac{dW}{dt} = N_z \frac{d\theta}{dt} = N_z \omega\,.$$

11.7.1 Torsion Pendulum

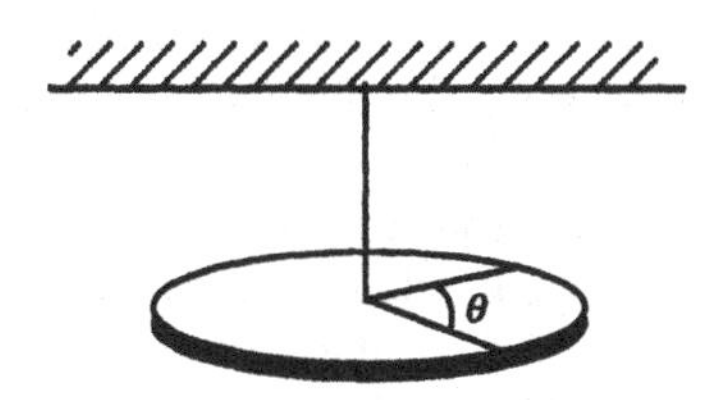

Fig. 11.16. The torsion pendulum

A torsion pendulum consists of a circular disk, suspended by a steel wire, as shown in the sketch. When the disk turns, the steel wire will be twisted about its axis. A torque is now acting to turn the disk back towards the equlibrium position. When the angle θ is small, the torque is, to a good approximation, proportional to θ, i.e.,

$$N = -C\theta ,$$

where C is called the torsion constant.

If the disk is turned away from its equlibrium position and released, it will carry out a simple harmonic oscillation, governed by the equation

$$I\frac{d^2\theta}{dt^2} = -C\theta .$$

Here, $I = \frac{1}{2}MR^2$ is the moment of inertia of the disk with respect to the vertical axis through its center.

The angular frequency in the oscillation becomes

$$\omega = \sqrt{\frac{C}{I}} ,$$

and consequently the period of the oscillation is

$$T = 2\pi\sqrt{\frac{I}{C}} .$$

These relations permit a measurement of the torsion constant C for a given steel wire.

As for the linear harmonic oscillator one can introduce potential energy:

$$U(\theta) = -\int_0^\theta N_z d\theta = -\int_0^\theta -C\theta d\theta = \frac{1}{2}C\theta^2 .$$

11.8 The Angular Momentum Theorem Referred to Various Points

We have demonstrated how the rotational motion of a body may be described by means of the angular momentum theorem. Before we end this section, we shall briefly discuss how to refer the angular momentum theorem to various points inside or outside the body.

A rigid body moves under the influence of forces relative to some given inertial frame I. An arbitrary particle P in the body is acted upon by external and internal forces, $\mathbf{F}_P^{ext}$ and $\mathbf{F}_P^{int}$ respectively.

We now consider a point Q, which moves with velocity $\mathbf{v}_Q$ relative to the inertial frame I. We wish to use Q as center for calculating the torque and the angular momentum. The point Q may be inside or outside the body.

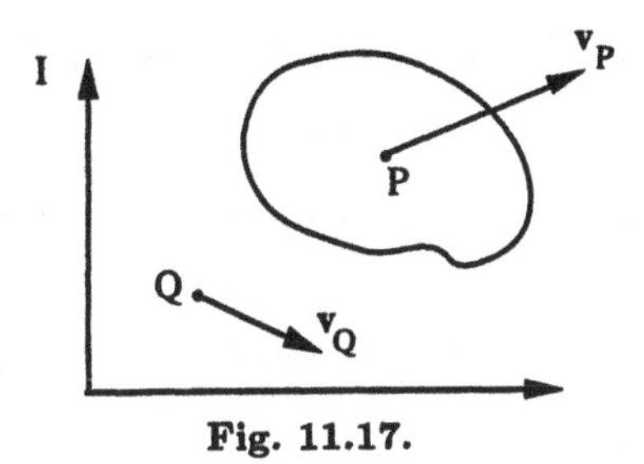

Fig. 11.17.

The total torque of the forces about Q is

$$\mathbf{N}_Q = \sum_P \left(\mathbf{QP} \times \mathbf{F}_P^{\text{ext}}\right) ,$$

and the angular momentum about Q is

$$\mathbf{L}_Q = \sum_P \left(\mathbf{QP} \times m_P \mathbf{v}_P\right) .$$

By differentiating $\mathbf{L}_Q$ with respect to time we obtain

$$\begin{aligned} \dot{\mathbf{L}}_Q &= \sum_P \left(\mathbf{QP} \times m_P \dot{\mathbf{v}}_P\right) \\ &+ \sum_P \left(\dot{\mathbf{QP}} \times m_P \mathbf{v}_P\right) \end{aligned}$$

Now,

$$\begin{aligned} \sum_P \left(\mathbf{QP} \times m_P \dot{\mathbf{v}}_P\right) &= \sum_P \left(\mathbf{QP} \times \mathbf{F}_P^{\text{ext}}\right) + \sum_P \left(\mathbf{QP} \times \mathbf{F}_P^{\text{int}}\right) \\ &= \sum_P \left(\mathbf{QP} \times \mathbf{F}_P^{\text{ext}}\right) , \end{aligned}$$

since the total torque of the internal forces vanish.

Furthermore,

$$\dot{\mathbf{QP}} = \mathbf{v}_P - \mathbf{v}_Q ,$$

i.e.,

$$\dot{\mathbf{L}}_Q = \sum_P \left(\mathbf{QP} \times \mathbf{F}_P^{\text{ext}}\right) + \sum_P \left(\mathbf{v}_P - \mathbf{v}_Q\right) \times m_P \mathbf{v}_P ,$$

or, since $\sum_P m_P \mathbf{v}_P = M\mathbf{v}_{\text{CM}}$,

$$\boxed{\dot{\mathbf{L}}_Q = \mathbf{N}_Q - \mathbf{v}_Q \times M\mathbf{v}_{\text{CM}} .}$$

The angular momentum theorem is valid in the simple form

$$\dot{\mathbf{L}}_Q = \mathbf{N}_Q \, ,$$

whenever $\mathbf{v}_Q \times \mathbf{v}_{CM} = \mathbf{0}$, that is, only if one of the following conditions are satisfied:

(1) $\mathbf{v}_Q = \mathbf{0}$ (Q is a point at rest in the inertial frame);
(2) Q = CM (Q coincides with the center of mass);
(3) $\mathbf{v}_Q$ is parallel to $\mathbf{v}_{CM}$.

11.9 Examples

Example 11.1. Rotating Cylinder. A circular disk with radius R may turn without friction about a horizontal axis A. The mass of the disk is M

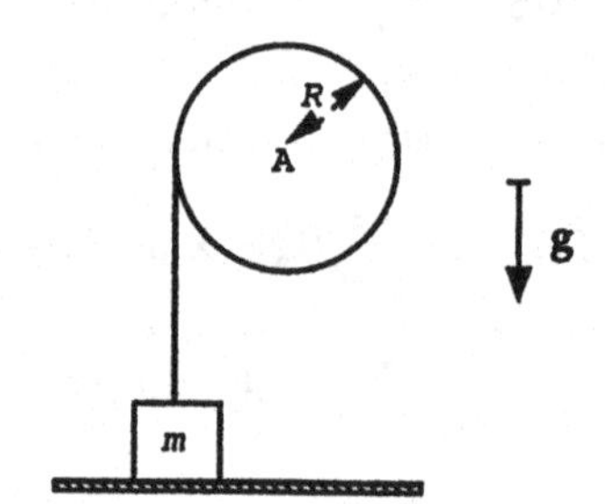

and its moment of inertia about the axis A is thus $I = \frac{1}{2}MR^2$. A light string is wound around the side of the disk. A mass m is tied to one end of the string and supported. The string is taut. At the time $t = 0$, the support of the mass m is released and the mass m begins to fall.

Find the angle θ that the disk has turned as a function of the time t.

Solution.

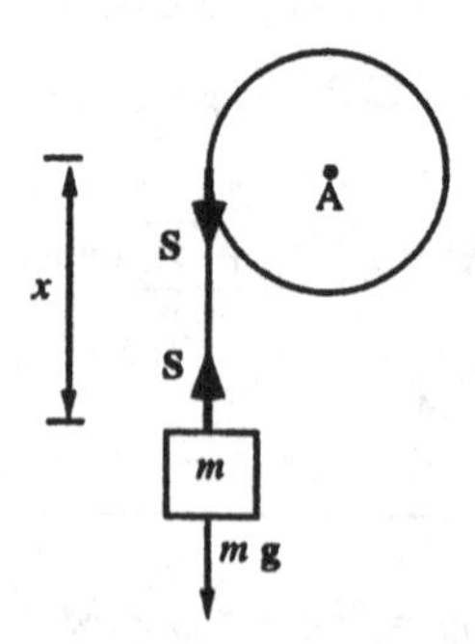

The equation of motion for m is

$$m\frac{\mathrm{d}^2x}{\mathrm{d}t^2} = mg - S\,. \tag{11.10}$$

The equation of motion for the disk is

$$I\frac{\mathrm{d}^2\theta}{\mathrm{d}t^2} = SR\,. \tag{11.11}$$

Since we are assuming that the string does not slip on the disk we have the following *geometrical constraint* on the motion:

$$x = R\theta + \text{constant} \;\; \Rightarrow \dot{x} = R\dot{\theta} \;\; \Rightarrow \ddot{x} = R\ddot{\theta} = R\frac{\mathrm{d}\omega}{\mathrm{d}t}\,. \tag{11.12}$$

Combining (11.10) and (11.11) with the geometrical constraint we find:

$$\left(I + mR^2\right)\frac{\mathrm{d}^2\theta}{\mathrm{d}t^2} = mgR\,.$$

By integration (and the use of the initial conditions $x = 0$ and $\theta = 0$ for $t = 0$) we find

$$\theta = \frac{1}{2}\frac{mR}{(I + mR^2)}gt^2\,, \quad I = \frac{1}{2}MR^2\,.$$

Questions.

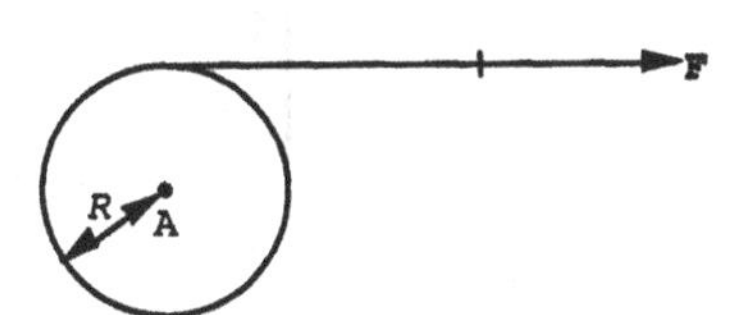

Assume that we pull with a constant horizontal force **F** in the string and that there are frictional forces resisting the rotation of the disk about the fixed horizontal axis A. We assume that the frictional forces can be represented by a torque N about the axis A, where N at all times is proportional to the instantanous angular velocity of the disk. We denote the constant of proportion γ, i.e., $N = \gamma\omega$.

(1) The disk will approach a constant angular velocity ω_f. Determine ω_f.
(2) Determine the angular velocity ω of the disk as a function of time. Initial conditions: $\omega = 0$ at $t = 0$.
(3) Determine the power P yielded by the force F when ω_f is reached.

Answers.

(1)

$$\omega_f = \frac{FR}{\gamma}\,.$$

(2)

$$\omega = \omega_f \left[1 - \exp\left(-\frac{\gamma}{I}t\right)\right],$$

where I is the moment of inertia about A.

(3)

$$P_F = F\frac{dx}{dt} = FR\omega_f = \frac{F^2R^2}{\gamma},$$
$$P_N = N\frac{d\theta}{dt} = \gamma\omega_f^2 = \frac{F^2R^2}{\gamma}.$$

When ω_f is reached, the power yielded by F will be converted into heat by the frictional forces. Check the dimensions (Js^{-1}). △

Example 11.2. Falling Cylinder. A light string is wound around a homogeneous cylinder, with mass M and radius r. The other end of the string is attached to the ceiling of the laboratory. Initially the system is held at rest with the string vertical (see the figure). At the time $t = 0$, the cylinder is allowed to fall under the influence of gravity. The initial velocity of the cylinder is 0. Disregard the mass of the string.

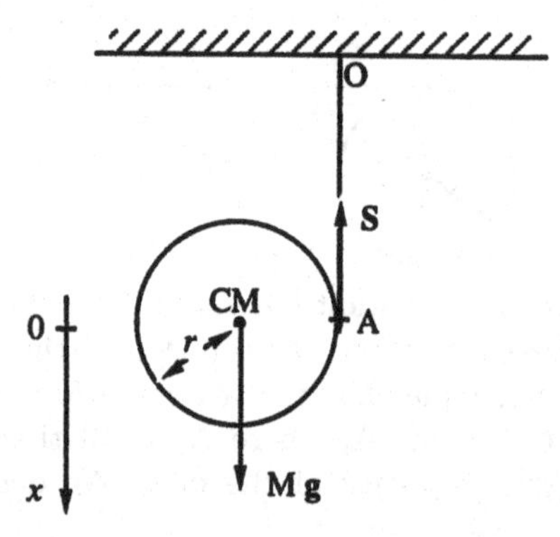

The string force acts on the cylinder in the point A, where the free end of the string is a vertical tangent to the cylinder.

(1) Find the translational acceleration of the center of mass of the cylinder and find the string force S.
(2) Determine the acceleration of the center of mass by using the work equation in the lab frame.

Answers.

(1) The center-of-mass theorem: $M d\mathbf{v}_{CM}/dt = \sum \mathbf{F}^{ext}$, shows that CM will move along a vertical line, because the external forces have no component in the horizontal direction. We find that

$$M\ddot{x} = Mg - S\,. \tag{11.13}$$

Note: S is *not* acting in the CM. This does not matter. The CM moves *as if* all the external forces were acting in the CM. We apply the angular momentum theorem about CM:

$$\frac{Mr^2}{2}\ddot{\theta} = Sr\,. \tag{11.14}$$

Since the string is taut throughout the motion there is the following geometrical constraint:

$$x = r\theta + \text{constant},$$

or

$$\ddot{x} = r\ddot{\theta} = r\frac{\mathrm{d}\omega}{\mathrm{d}t} \tag{11.15}$$

From (11.13)–(11.15) we get:

$$\ddot{x} = \frac{2}{3}g\,, \quad S = \frac{1}{3}Mg\,.$$

(2) Note: the string force acts on a particle which is instantaneously at rest in the lab frame. Loosely speaking, the particle participates in two motions, a downward translation with velocity $\dot{x} = v_{\text{CM}}$ and a rotation which directs the particle upwards. The velocity in the rotation is $v_{\text{rot}} = r\dot{\theta} = r\dot{x}/r = \dot{x}$. The total instantaneous velocity of the particle at A is thus zero, seen from the lab frame (the acceleration of the particle at A is not zero, or the particle would not move at all). The motion of the cylinder can thus be seen as a pure rotation about the axis A. Such an axis is called the instantanous axis of rotation. Since the string force acts on a particle instantanously at rest, the string force performs no work. The only force that does perform work is the force of gravity. We find

$$Mg\dot{x} = \frac{\mathrm{d}}{\mathrm{d}t}\left\{\frac{1}{2}M\dot{x}^2 + \frac{1}{2}\cdot\frac{1}{2}Mr^2\dot{\theta}^2\right\},$$

or, since $\dot{x} = r\dot{\theta}$, $Mg\dot{x} = \frac{3}{2}M\dot{x}\ddot{x}$. Thus, $\ddot{x} = \frac{2}{3}g$

Remarks. The angular momentum theorem can be applied about two more axes (points):

(1) about A (in instantaneous rest):

$$\left(Mr^2 + \frac{1}{2}Mr^2\right)\ddot{\theta} = Mgr\,,$$
$$\left(Mr^2 + \frac{1}{2}Mr^2\right)\frac{\mathrm{d}\omega}{\mathrm{d}t} = Mgr\,.$$

[Note: the rotation vector $\boldsymbol{\omega}$ is the same.]

(2) about the point of suspension, O (at rest in the inertial frame).

$$\begin{aligned} L_\mathrm{O} &= Mv_\mathrm{CM}r + \frac{1}{2}Mr^2\omega \\ &= M\dot{x}r + \frac{1}{2}Mr^2\omega, \end{aligned}$$

$$\frac{\mathrm{d}}{\mathrm{d}t}\left(M\dot{x}r + \frac{1}{2}Mr^2\omega\right) = Mgr\,.$$

The two equations expressing the angular momentum theorem with reference to O and A are the same [$\ddot{x} = r\ddot{\theta} = r(d\omega/dt)$]. When $\ddot{x}$ has been determined from one of these equations, the string force S may be determined from the center-of-mass theorem.

Questions. Let us assume that the point O is not fixed. A person is holding the string (a yo-yo!). The string is now brought vertically upwards in such a way that the center of mass (CM) for the cylinder is not falling while the string is unwinding.

Determine the string force and determine the work that has been performed on the cylinder when it has reached the angular velocity ω.

Answer. $S = Mg$; $W = Mr^2\omega^2/4$. △

Example 11.3. The Atwood Machine

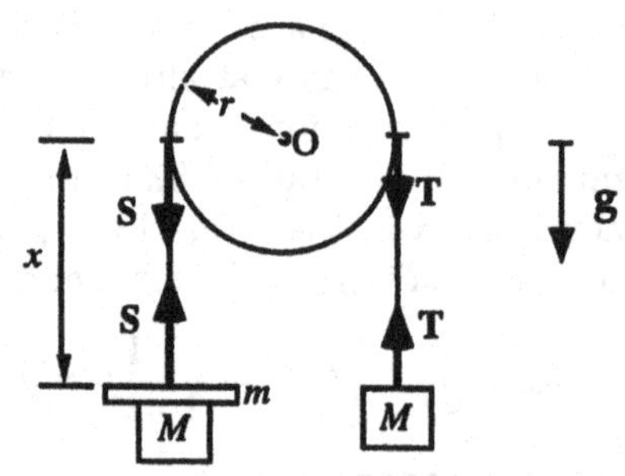

We shall again consider the Atwood machine (see Example 2.6), including now the moment of inertia I of the disk. The string force is not the same on either side of the disk. These string forces are internal forces for the system as a whole, but external forces for the disk. The string forces provide the torque necessary to give the disk an angular acceleration.

We have three bodies, and we write the equation of motion for each one:

$$(M+m)\,\ddot{x} = (M+m)\,g - S\,, \tag{11.16}$$

$$M\ddot{x} = T - Mg\,, \tag{11.17}$$

$$I\alpha = (S-T)\,r\,, \tag{11.18}$$

where α is the angular acceleration of the disk. In addition we have the geometric constraint that

$$\ddot{x} = r\ddot{\theta} = r\alpha\,. \tag{11.19}$$

Note: among the forces acting on the disk only the string forces have a torque about the axis O. The weight of the disk and and the reaction force from O have no torque about O.

From (11.16)–(11.19) we find

$$\begin{aligned} \ddot{x} &= \frac{m}{2M + m + I/r^2} g\,, \\ S &= (M+m)\,g\left(1 - \frac{m}{2M + m + I/r^2}\right), \\ T &= Mg\left(1 + \frac{m}{2M + m + I/r^2}\right). \end{aligned}$$

For $I = 0$ we recover $S = T$.

A different way of solving the same problem is to use the angular momentum theorem about the axis O. The only external forces that have a torque about O are the forces of gravity on the masses M and $m + M$.

The magnitude of the angular momentum about the axis O is

$$L = (M+m)\,\dot{x}r + M\dot{x}r + I\frac{\dot{x}}{r}$$

The angular momentum theorem now gives that

$$\frac{\mathrm{d}}{\mathrm{d}t}\left((M+m)\,\dot{x}r + M\dot{x}r + I\frac{\dot{x}}{r}\right) = (M+m)\,gr - Mgr\,.$$

From this equation $\ddot{x}$ may be determined.

There is a third way to determine $\ddot{x}$: The rate of change of kinetic energy for the system equals the power from the external forces (the strings remain taut throughout the motion, so no energy is stored in the strings!). Therefore

$$\frac{\mathrm{d}}{\mathrm{d}t}\left(\frac{1}{2}(M+m)\dot{x}^2 + \frac{1}{2}M\dot{x}^2 + \frac{1}{2}I\left(\frac{\dot{x}}{r}\right)^2\right) = (M+m)\,g\dot{x} - Mg\dot{x}\,.$$

From this equation $\ddot{x}$ may be determined.

Question. Determine the reaction force R from the axis O when the system is in motion. The mass of the disk is (of course) $2I/r^2$. [Answer: $R = S + T + 2Ig/r^2$.] △

Example 11.4. The Physical Pendulum. A rigid body, which is able to oscillate about a horizontal axis O that does not pass through the center of mass, is called a physical pendulum (see the figure). Let CM be the center of

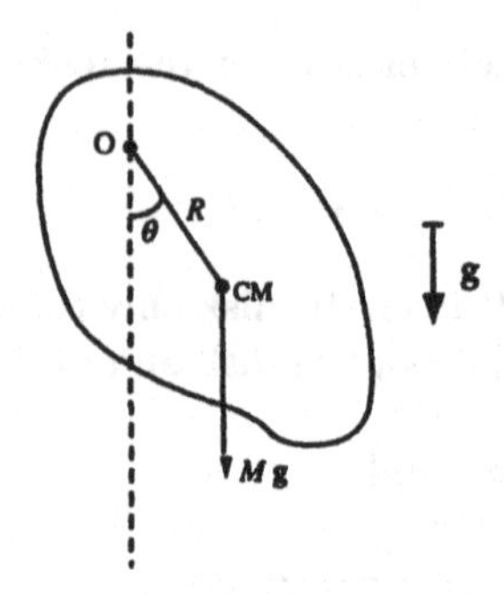

mass. When the line segment OCM forms an angle θ with the vertical, the force of gravity has a torque about the axis O. The moment of inertia about O is denoted I. The distance from O to CM is R.

Show that for small oscillations ($\sin(\theta) \approx \theta$) the physical pendulum will exercise harmonic oscillations. Determine the period of oscillation.

Solution.

$$I\frac{\mathrm{d}^2\theta}{\mathrm{d}t^2} = -MgR\sin(\theta) \approx -MgR\theta .$$

The solutions to this equation are harmonic oscillations with a period of

$$T = 2\pi\sqrt{\frac{I}{MgR}} .$$

For $R \to 0$, $T \to \infty$. How can this be interpreted ?

Often the so-called arm of inertia (k) is introduced. For a body of mass M, able to rotate about a given axis O, k is defined by $I \equiv Mk^2$. If this is inserted into the expression for the period we get

$$T = 2\pi\sqrt{\frac{k^2}{gR}} .$$

Note. The period is independent of the mass. This is due to the fact that the gravitational and inertial mass are equal (proportional).

Question. Consider larger amplitudes of the pendulum. Show that the angular velocity $\dot{\theta} \equiv \omega$ is given by

$$\dot{\theta}^2 = \frac{2MgR}{I}\left[\cos(\theta) - \cos(\theta_0)\right] ,$$

where θ_0 is the maximum angle of deflection. △

Example 11.5. The Rod. A rod of mass M and length l can turn in a horizontal plane and without friction around a vertical axis O through the

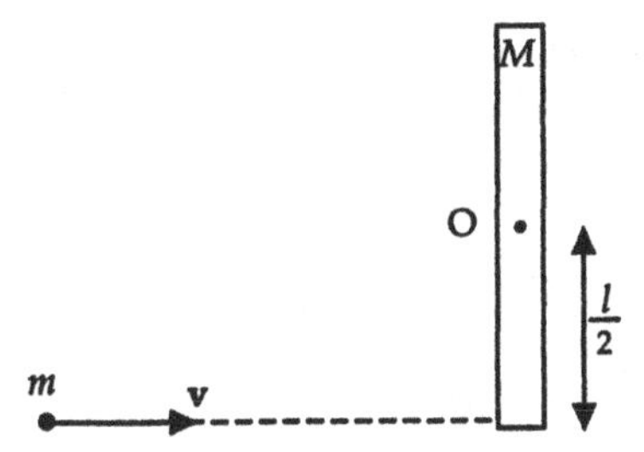

center of mass. Initially the rod is at rest, as shown in the figure. A bullet of mass m is now shot into one end of the rod and remains embedded in the rod. The velocity of the bullet just before it hits the rod is $\mathbf{v}$ (see the figure). Find the angular velocity ω of the rod after the bullet is stopped.

Solution. The only external forces acting on the system are gravity and the reaction force from the axle O. None of these external forces have any torque around O: $\sum N_{\text{ext}} = 0$. Thus, the angular momentum of the system around O is conserved. The moment of inertia of the rod around O is denoted I.

L before the collision $= mv(l/2)$ (L is the angular momentum around O).

L after the collision $= I\omega + m(l/2)^2\omega$.

Conservation of angular momentum around O means that

$$mv(l/2) = I\omega + m(l/2)^2\omega\,. \tag{11.20}$$

Furthermore

$$I = \frac{1}{12}Ml^2\,. \tag{11.21}$$

(11.20) and (11.21) now give

$$\omega = \frac{6mv}{Ml + 3ml}\,.$$

Questions. A homogeneous rod of length $2l$ and mass M can turn without friction in a horizontal plane and about a fixed vertical axis O through the center of mass of the rod. Initially, the rod is at rest. A particle of mass m collides in a short, completely elastic collision with one end of the rod. It is assumed that the collision time Δt is so short that the rod does not turn during the collision. The velocity of the particle (both before and after the collision) is horizontal, and perpendicular to the original direction of the rod. The magnitude of the velocity of m before the collision is u.

(1) Find the velocity of m and the angular velocity ω of the rod after the collision.
(2) It is assumed that the force F by which the particle acts on the rod is constant during the collision. Find F. Numerical example: $l = 0.30$ m, $M = 0.3$ kg, $m = 0.04$ kg, $u = 20\ \text{m}\,\text{s}^{-1}$, $\Delta t = 10^{-2}$ s.

Answers. Use conservation of angular momentum about O and – because the collision is elastic – conservation of kinetic energy.

(1)

$$v = \frac{3m - M}{3m + M} u \,,$$
$$\omega = \frac{6mu}{(M + 3m)l} \,.$$

Note: if $M > 3m$, the rod is so massive that it reflects m. For $M = 3m$ the mass m is stopped by the rod.

(2)

$$F = \frac{2mMu}{\Delta t(M + 3m)} \,.$$

$$|\mathbf{v}| = 8.6 \,\mathrm{m\,s^{-1}} \,,$$
$$\omega = 38 \,\mathrm{s^{-1}} \,,$$
$$F = 1.14 \times 10^2 \,\mathrm{N}.$$

△

11.10 Review: Comparison Between Linear Motion and Rotation About a Fixed Axis

The mathematical description of the rotation of a rigid body about a fixed axis is equivalent to the mathematical description of the motion of a particle along a straight line.
By a fixed axis we shall mean an axis that is kept fixed relative to the inertial frame used as reference, and fixed relative to the rotating body. All methods and results can, as shown in the table below, be taken over directly from linear motion of a particle to rotation of a rigid body about a fixed axis.

The angular momentum theorem may be applied to the following types of axes:

(a) any fixed axis;
(b) any axis through the center of mass (CM) even if the axis is accelerated (homogeneous gravitational fields, fictitious or not, have no torque about CM);
(c) any axis that has a velocity $\mathbf{v}_Q$ parallel to the velocity of the center of mass, $\mathbf{v}_{CM}$.

Linear motion of a particle	Rotation of a rigid body about a fixed axis
position: x	angular position: θ
velocity: $v = \dot{x}$	angular velocity: $\omega = \dot{\theta}$
acceleration: $a = \dot{v} = \ddot{x}$	angular acceleration: $\alpha = \dot{\omega} = \ddot{\theta}$
force: F	torque: N
mass: m	moment of inertia: I
momentum: $p = m\dot{x} = mv$	angular momentum: $L = I\dot{\theta} = I\omega$
Equation of motion:	**Equation of motion:**
$\frac{\mathrm{d}p}{\mathrm{d}t} = F$	$\frac{\mathrm{d}L}{\mathrm{d}t} = N$
$m\frac{\mathrm{d}v}{\mathrm{d}t} = F$	$I\frac{\mathrm{d}\omega}{\mathrm{d}t} = N$
$m\frac{\mathrm{d}^2x}{\mathrm{d}t^2} = F$	$I\frac{\mathrm{d}^2\theta}{\mathrm{d}t^2} = N$
Kinetic energy:	**Kinetic energy:**
$T = \frac{1}{2}mv^2 = \frac{p^2}{2m}$	$T = \frac{1}{2}I\omega^2 = \frac{L^2}{2I}$
Potential energy:	**Potential energy:**
$U(x) = -\int_{x_o}^{x} F(x)\mathrm{d}x$	$U(\theta) = -\int_{\theta_o}^{\theta} N(\theta)\mathrm{d}\theta$
Work:	**Work:**
$W = \int_1^2 \mathbf{F}\mathrm{d}x$	$W = \int_1^2 N(\theta)\mathrm{d}\theta$
Power:	**Power:**
$\frac{\mathrm{d}W}{\mathrm{d}t} = \mathbf{F}\cdot\mathbf{v}$	$\frac{\mathrm{d}W}{\mathrm{d}t} = N\omega$
Impulse:	**Impulse:**
$\int \mathbf{F}\mathrm{d}t = \Delta\mathbf{p}$	$\int N\mathrm{d}t = \Delta L$

We furthermore want to call attention to the following facts.

(1) For a system – rigid or nonrigid – on which no external torques are acting, angular momentum is conserved.

(2) Parallel axis theorem: the moment of inertia of a body of mass M about an axis through an arbitrary point O is equal to the sum of the moment of inertia about the parallel axis through the CM and Md^2, where d is the distance between CM and the axis through O.

11.11 Problems

Problem 11.1. A homogeneous disk of radius $r = 0.20$ m can oscillate as a physical pendulum around a horizontal axis O located 0.10 m ($r/2$) from the

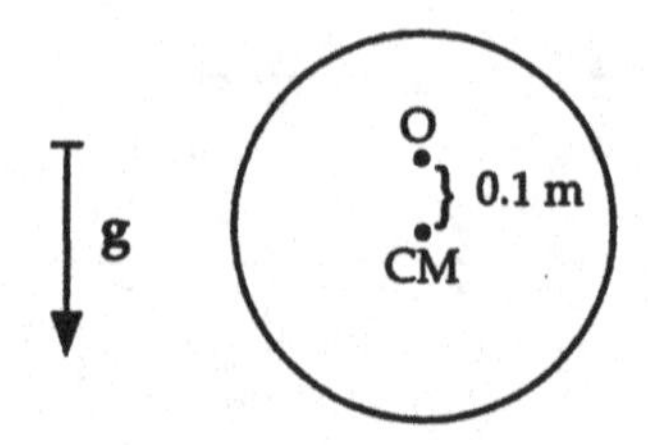

center of mass of the disk. The disk is perpendicular to O. Find the period for small oscillations of the disk.

Problem 11.2.

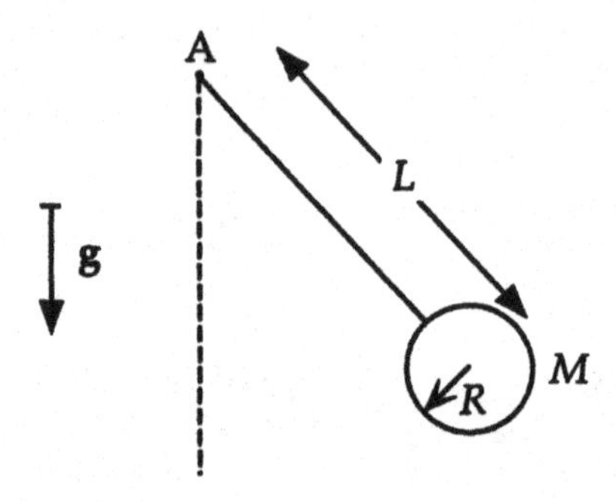

A heavy circular disk with radius R and mass M is fastened to a light stiff rod. The mass of the rod is negligible compared to the mass of the disk. The system can oscillate as a physical pendulum about a fixed horizontal axis A. The distance from A to the center of the disk is L (see the figure).

(1) Determine the period for small oscillations when the disk is fastened to the rod as shown, i.e., when the disk swings in the plane of the paper.
(2) Assume now that the plane of the disk is perpendicular to the plane of the paper. Determine the period for small oscillations.

Problem 11.3.

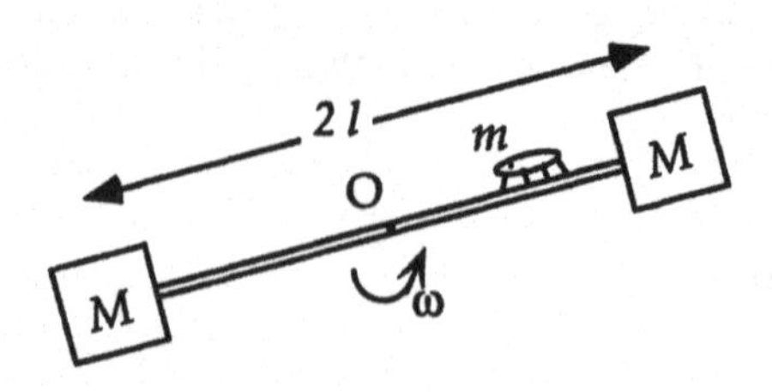

Two weights, each of mass M are fastened to the ends of a rod of length $2l$ and negligible mass. The dimensions of the weights are much smaller than

l. The rod can rotate without friction in the horizontal plane about a fixed vertical axis O through the center of the rod.

A beetle of mass m sits at rest on one of the weights. The rod is set in rotation about O with angular velocity ω_1. The system is then left to itself.

(1) Find the kinetic energy $T_1$of the system.

The beetle now crawls along the rod towards O. When it reaches O, it stops.

(2) Find the angular velocity ω_2 and the kinetic energy T_2 of the rod, when the beetle has reached O.

The beetle then slides out along the rod until it hits one of the endpoints of the rod, where it then remains without motion relative to the rod.

(3) Find the angular velocity ω_3 and the kinetic energy T_3 of the system after the beetle has come to rest.
(4) Let us assume that we can neglect the friction between the beetle and the rod. Find the speed v of the beetle relative to the rod as the beetle reached the endpoint of the rod.

Problem 11.4.

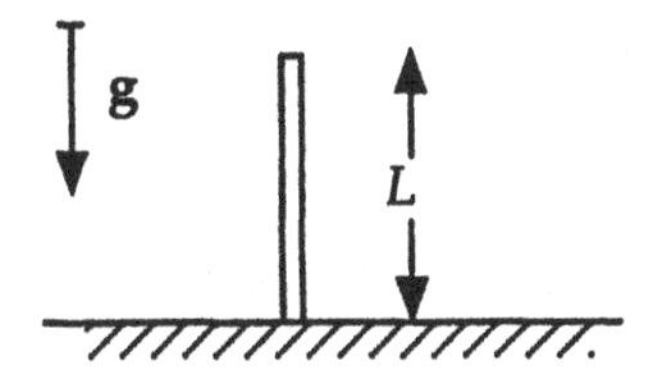

A thin homogeneous rod has length L and mass M. Initially the rod is vertical and at rest in the laboratory. One end of the rod is in contact with the floor, which is assumed to be frictionless. The acceleration of gravity is $\mathbf{g}$.

The rod is now given a slight push and begins to fall in the gravitational field. The initial value of the velocity of the CM is zero.

(1) Describe qualitatively the curve along which the CM moves.

Consider the rod at the moment when the CM of the rod is a distance y below the initial position. The acute angle which the rod forms with the vertical is denoted by θ.

(2) Write an expression for the kinetic energy T and the potential energy U of the rod, as functions of M, L, g, θ, and $\dot{\theta}$.
(3) Determine $\dot{y}$ as a function of L, g, and θ.

Problem 11.5.

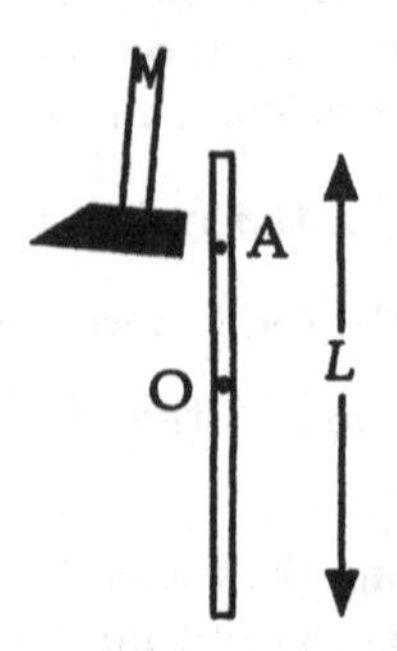

A homogeneous thin rod has length L and mass M. Initially the rod is at rest on a horizontal frictionless table in the laboratory.

The rod is then hit with a hammer, at a point A which is at the distance d from the center point O of the rod (see the sketch). The blow is in the direction perpendicular to the rod. The blow transfers a momentum $\mathbf{P}$ to the rod. It is assumed that the duration of the blow is so short that we may neglect the motion of the rod during the blow. The known quantities are thus L, M, d, and $\mathbf{P}$.

(1) Determine the velocity of the center of mass (CM) of the rod after the blow.
(2) Determine the kinetic energy of the rod after the blow.

We assume that there is a point C on the rod which – just after the blow – is at rest relative to the table.

(3) Determine the distance OC. Under what conditions is C a point on the rod ?

Problem 11.6.

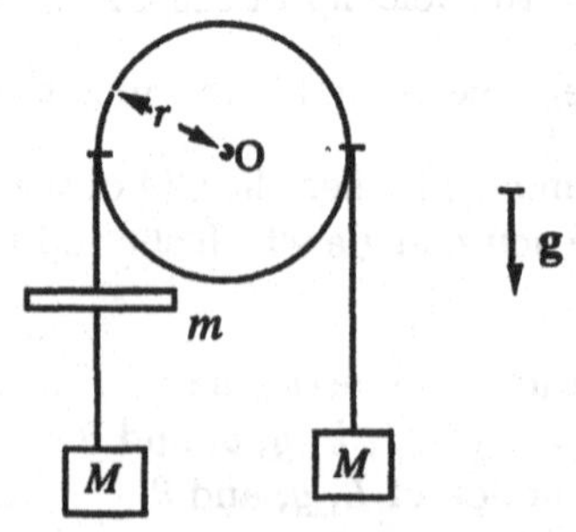

An Atwood machine consists of a cylinder that may turn around a fixed, horizontal symmetry axis, O. The radius of the cylinder is r, and the moment of inertia of the cylinder around O is I. A string passes over the cylinder and the string holds a mass M at each end.

Initially the two masses are at rest. A small disk, with a hole in the center, has mass m. This disk may slide without friction along the string. The small mass m now falls freely in the gravitational field, through a height h. The initial velocity of the disk is zero. The disk then hits one of the masses M in a completely inelastic collision. The duration of the collision can be neglected.

The string is taut and the string cannot slide on the cylinder. The acceleration of gravity is g.

(1) Determine the velocity u of the total mass $M+m$ just after the collision. The velocity u should be given as a function of m, M, g, h, I, and r.
(2) Determine the amount Q of the mechanical kinetic energy that is converted into heat during the collision. Q should be given as a function of m, M, g, h, I, and r.

Problem 11.7.

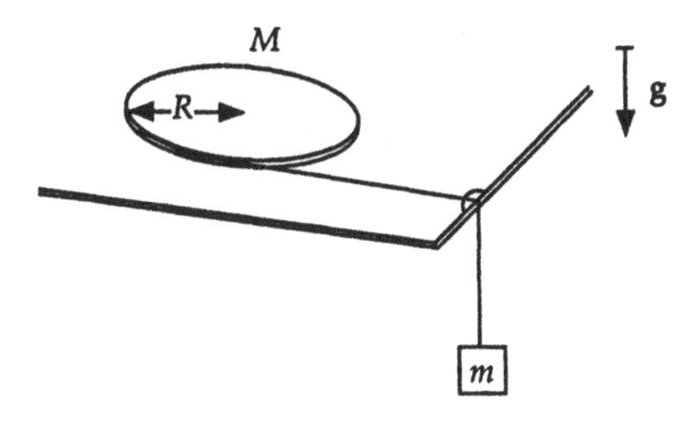

A plane, homogeneous disk may slide on a frictionless horizontal table. A light cord is fastened to the disk and wrapped many times around the rim of the disk. The cord goes without friction through a small eyelet at the edge of the table and is connected to a mass m (see the figure). The mass of the disk is M and its radius is R. The system is started from rest and the cord is assumed to be taut throughout the motion. Ignore the mass of the cord.

(1) Find the velocity v of the center of mass of the disk as a function of time (before the disk hits the eyelet or the mass m hits the floor).
(2) Find the distance S the disk has moved, and the distance d the mass m has been lowered when a length l of the cord has been unwound.

12. The Laws of Motion

This chapter contains nothing new. Through a series of examples we wish to illustrate the application of the basic theorems of mechanics. We start by reviewing classical mechanics in the form of ten "laws" or theorems.

12.1 Review: Classical Mechanics

Motion of Particles

(1) $m\mathrm{d}^2\mathbf{r}/\mathrm{d}t^2 = \mathbf{F}_{\text{nat}}$ inertial frames

(2) $m\mathrm{d}^2\mathbf{r}/\mathrm{d}t^2 = \mathbf{F}_{\text{nat}} - m\ddot{\mathbf{R}}_0 + m\omega^2\rho - 2m\,(\omega \times \dot{\mathbf{r}}) - m\,(\dot{\omega} \times \mathbf{r})$ noninertial frames

Motion of Particle Systems

(3) $M\mathrm{d}^2\mathbf{R}_{\text{CM}}/\mathrm{d}t^2 = \sum \mathbf{F}^{\text{ext}}$ CM theorem

(4) $\mathrm{d}\mathbf{L}_{\text{O}}/\mathrm{d}t = \sum \mathbf{N}_{\text{O}}^{\text{ext}}$; $\mathrm{d}\mathbf{L}_{\text{CM}}/\mathrm{d}t = \sum \mathbf{N}_{\text{CM}}^{\text{ext}}$ angular momentum theorem

Conservation Theorems

(5) $T + U =$ constant; for a closed *conservative* system

(6) $\sum_i \mathbf{p}_i =$ constant vector; for a closed system

(7) $\sum_i \mathbf{L}_i =$ constant vector; for a closed system

Auxiliary Theorems

(8) $T = \frac{1}{2}Mv_{\text{CM}}^2 + T_{\text{CM}}$; $\mathbf{P} = M\mathbf{v}_{\text{CM}}$

(9) $\mathbf{L}_O = \mathbf{L}_{\text{CM}} + \mathbf{R}_{\text{CM}} \times M\mathbf{v}_{\text{CM}}$

(10) rotation about a fixed axis:
$\mathrm{d}L_{\text{A}}/\mathrm{d}t = N_{\text{A}}$; $I_{\text{A}}\mathrm{d}\omega/\mathrm{d}t = N_{\text{A}}$, since $L_{\text{A}} = I_{\text{A}}\omega$

12.2 Remarks on the Three Conservation Theorems

The formulation of the three conservation theorems (energy, momentum, and angular momentum) is one of the main results of our development of classical mechanics. We have seen how the conservation theorems follow from the basic laws of Newtonian mechanics, in particular from a combination of Newton's second law and Newton's third law (the law of action and reaction).

In the 20th century the world witnessed a revolution in the sciences, a revolution which is comparable to that brought about by Brahe, Kepler, Galilei, and Newton. The revolution has resulted in the recognition that Newtonian mechanics has a limited range of applicability. The Newtonian laws are now seen as an approximation, valid for macroscopic bodies, and for velocities much less than the speed of light. Deeper and more precise laws are formulated in the theories of quantum mechanics and relativity.

The inner dynamics of atoms, i.e., the motion of electrons around the nucleus, and the motion of nucleons within the nucleus are governed by the laws of quantum mechanics and special relativity.

A perhaps surprising point is the following: the three fundamental conservation theorems, i.e., conservation of energy, conservation of momentum, and conservation of angular momentum are valid also in the world of quantum mechanics and relativity.

In these, the most fundamental theories of the structure of matter known to us, the three conservation laws still hold.

The conservation theorems reveal some of the most essential aspects of how nature works.

12.3 Examples

Example 12.1. Conservation of Angular Momentum. The Collision Approximation

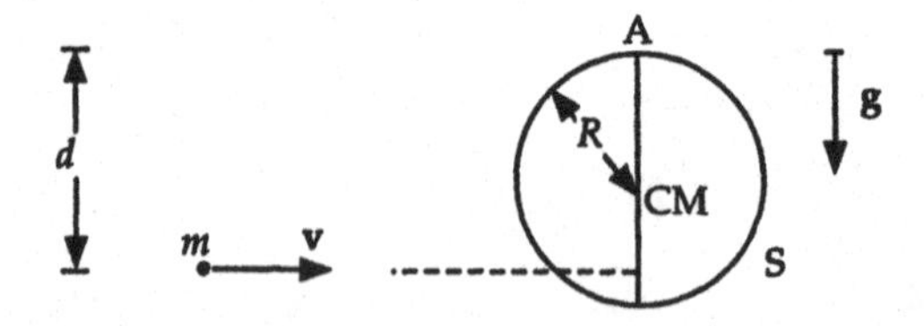

The physical content of this example is closely related to the physical content of Example 9.3.

A homogeneous sphere S, with mass M and radius R hangs at rest in the gravitational field of the Earth (see the figure). The sphere S may turn without friction about the horizontal axis A which is a tangent to the top of the sphere and perpendicular to the plane of the paper. A small pistol bullet with mass m is now shot into the sphere. The velocity of the bullet is horizontal and perpendicular to the vertical plane that contains the axis A and the center of mass (CM) of the sphere S. The pistol bullet comes to rest inside the sphere S at a point which is directly underneath the point A, a distance d from A.

We assume that $m \ll M$ and that the time Δt in which the bullet comes to rest relative to S is so small that we can disregard the angle through which S turns in the time interval Δt (the collision approximation).

(1) Determine the angular velocity ω_0 with which the sphere S begins to turn about the axis A, i.e., determine ω_0 just after the collision is over.
(2) Determine the maximum value θ_0 of the angle θ that the sphere S turns after the collision.
(3) Determine the condition Δt must satisfy in order that we may disregard the angle through which the sphere S turns during the collision.
(4) Determine the conditions for which the linear momentum $\mathbf{P}$ is conserved in the collision.

Solution. We start with some general considerations.

If the sphere S had not been fastened to the axis A we should have been able to compute the velocity of CM after the collision. We could have used the law of conservation of the momentum for a closed system. In the present case, however, the sphere S is connected to the fixed axis A, and reaction forces from the axis will in general act during the collision. Thus: in general the momentum will not be conserved in the collision. We do not know the reaction forces, but we do know the geometrical constraints on S, and that the reaction forces determine these constraints.

By applying the theorem of conservation of angular momentum about the axis A we circumvent the problem of the unknown reaction forces acting at A. *The reaction forces have no torques about the axis* A.

If the collision time Δt was not sufficiently short, the sphere S would turn a finite angle before the collision was over. When the CM is no longer vertically underneath A, gravity will produce a torque about A, and the angular momentum of the system is no longer conserved. Do *not* use conservation of mechanical energy when heat is produced.

(1) The angular velocity ω_0 is determined from conservation of angular momentum during the collision:

$$\begin{aligned} mvd &= (I_\mathrm{A} + md^2)\omega_0\,, \\ I_\mathrm{A} &= I_\mathrm{CM} + MR^2 = \frac{2}{5}MR^2 + MR^2 = \frac{7}{5}MR^2\,, \\ \omega_0 &= \frac{mvd}{(7/5)MR^2 + md^2}\,. \end{aligned} \tag{12.1}$$

(2) When the collision is over there are no more nonconservative forces in the system (we disregard friction in the pivot at A). We can then use the theorem of conservation of mechanical energy. The kinetic energy with which the sphere S begins to turn is:

$$T = \frac{1}{2}\left(\frac{7}{5}MR^2 + md^2\right)\omega_0^2 .$$

We now assume that S, as a result of the collision, turns an angle θ_0 away from the equilibrium position. The center of mass of S is then raised an amount

$$h_1 = R[1 - \cos(\theta_0)] .$$

Assuming the bullet to be placed as indicated on the figure, the bullet is lifted an amount

$$h_2 = d[1 - \cos(\theta_0)] .$$

We therefore have the following equation to determine $\cos(\theta_0)$:

$$\frac{1}{2}\left(\frac{7}{5}MR^2 + md^2\right)\omega_0^2 = MgR\,[1 - \cos(\theta_0)] + mgd\,[1 - \cos(\theta_0)] \; . \quad (12.2)$$

From (12.1) and (12.2) we find:

$$\cos(\theta_0) = 1 - \frac{(mvd)^2}{2g\,(MR + md)\left(\frac{7}{5}MR^2 + md^2\right)} .$$

Check the dimensions!

(3) The condition for Δt to be ignorable is that the amount of angular momentum imparted to the sphere S by the bullet in the collision (i.e., mvd) is large compared to the angular momentum the sphere received from the torque of gravity during Δt. In the estimates below we shall use that $m \ll M$.

The magnitude of the angular momentum about A which the torque of gravity supplies to the sphere S during the collision is:

$$L = \int_0^{\Delta t} N \mathrm{d}t = \int_0^{\Delta t} MgR\sin[\theta(t)]\mathrm{d}t .$$

We do not know in detail the forces acting between the bullet and the sphere S during the collision. We are thus unable to calculate L exactly. But we can estimate the *order of magnitude* of L. The angle through which S turns during the collision has the order of magnitude,

$$\theta \approx \omega_0 \Delta t ,$$

because the angular velocity is of order ω_0 during the collision. Since θ is a small angle, we let $\sin(\theta) \approx \theta$. The torque of gravity is therefore of magnitude

$$N \approx MgR\omega_0 \Delta t\,.$$

This torque acts in the time interval Δt. The order of magnitude of L is therefore

$$L = MgR\omega_0 (\Delta t)^2\,.$$

The magnitude of L is to be compared to the angular momentum imparted to the sphere S by the bullet:

$$mvd \approx I_A \omega_0\,.$$

The condition for conservation of the angular momentum about A during the collision is then:

$$\omega_0 I_A \gg MgR\omega_0 (\Delta t)^2\,,$$

or

$$\Delta t \ll \sqrt{\frac{I_A}{MgR}} = \sqrt{\frac{7}{5}\frac{R}{g}}\,.$$

The period T of oscillation for the small oscillations which the sphere S can perform under the influence of gravity is given by

$$T = 2\pi\sqrt{\frac{I_A}{MgR}} = 2\pi\sqrt{\frac{7}{5}\frac{R}{g}}\,.$$

The condition may thus be expressed simply as $\Delta t \ll T$.

(4) We calculate the change ΔP in the horizontal linear momentum during the collision, still assuming that $m \ll M$. The velocity of the center of mass just after the collision is known:

$$v_{CM} = R\omega_0\,.$$

The horizontal component of the linear momentum just after the collision is then

$$P_{\text{after}} = MR\omega_0 \approx \frac{MRmvd}{I_A}\,.$$

The linear momentum before the collision is

$$P_{\text{before}} = mv\,.$$

Therefore the change in linear momentum in the collision is

$$\Delta P = P_{\text{after}} - P_{\text{before}} = mv\left[\frac{MRd}{I_A} - 1\right]\,.$$

We may conclude that in general the linear momentum is not conserved during the collision, even if $\Delta t \to 0$. The finite change in the linear momentum is produced by the external force on the system: the reaction force from the axis A. The average value of the horizontal reaction force from A during the collision is

$$\langle F_r \rangle = \frac{\Delta P}{\Delta t} .$$

From the above expression for ΔP we see that $\Delta P = 0$ if $MRd/I_{\mathrm{A}} = 1$ or,

$$d = \frac{I_{\mathrm{A}}}{MR} = \frac{7}{5}R,$$

that is, linear momentum is conserved when $d = (7/5)R$. When the pistol bullet hits the sphere S at a distance $d = (7/5)R$ from the axis A, the axis is affected with a minimal amount.

△

Example 12.2. Rotating Rod

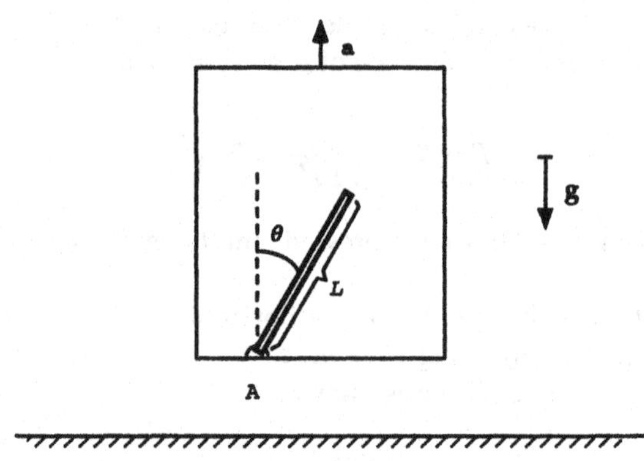

An elevator is moving vertically upwards with the constant acceleration **a** in the gravitational field **g** of the Earth. A horizontal axis A is fastened to the bottom of the elevator. A homogeneous rod with mass M and length L can turn without friction about A. Initially the rod is at rest and vertical. At a certain time the rod begins to turn in the vertical plane perpendicular to the axis A. The rod begins its motion with (near) zero angular velocity.

(1) Determine the angular velocity ω and angular acceleration α as functions of the angle θ formed with the vertical.
(2) Determine the force **F**, with which the axis A acts on the rod when the rod forms the angle θ with the vertical. The force **F** may be decomposed into suitable components.

Solution.

(1) The magnitude of the gravitational field in the elevator is $g + a$. From

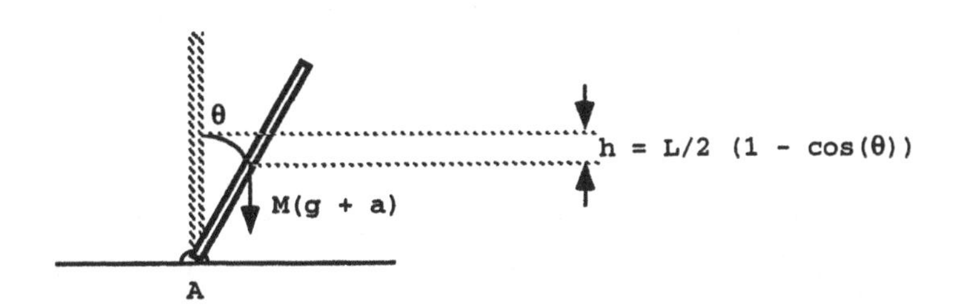

the conservation of energy theorem, we find ω as a function of θ:

$$\frac{1}{2}\left(\frac{1}{3}ML^2\right)\omega^2 = M(g+a)\frac{L}{2}\left[1-\cos(\theta)\right]$$

$$\Rightarrow \quad \omega = \sqrt{\frac{3(g+a)[1-\cos(\theta)]}{L}} \, .$$

The angular acceleration can be determined from the angular momentum theorem:

$$\begin{aligned} \left(\frac{1}{3}ML^2\right)\alpha &= M(g+a)\frac{L}{2}\sin(\theta) \, , \\ \alpha &= \frac{3}{2}\frac{g+a}{L}\sin(\theta) \, . \end{aligned}$$

A different way of determining α is by differentiation:

$$\alpha = \frac{d\omega}{dt} \, .$$

The angular velocity ω is known. We find, by differentiating the expression for ω^2:

$$2\omega\frac{d\omega}{dt} = \frac{3(g+a)}{L}\sin(\theta)\frac{d\theta}{dt} \, ,$$

or, since $d\theta/dt \equiv \omega$,

$$\frac{d\omega}{dt} = \alpha = \frac{3(g+a)\sin(\theta)}{2L} \, .$$

(2) The reaction force from the axis A gives, together with the effective gravity $M(g+a)$, the acceleration of the center of mass of the rod. The center of mass does not perform a uniform circular motion; the acceleration has two components. A radial acceleration directed along the rod towards A, and a tangential acceleration perpendicular to the rod.

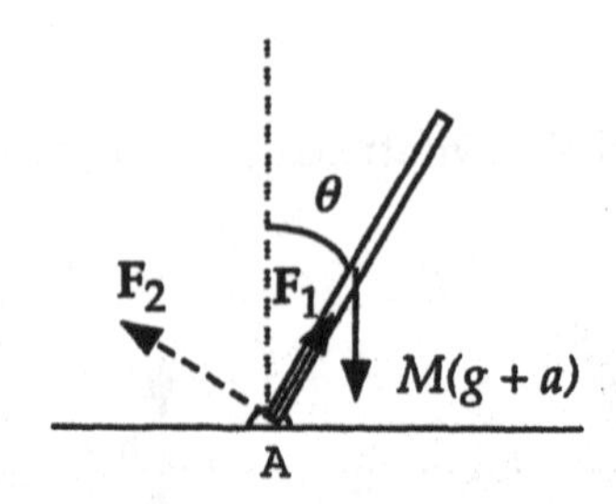

We decompose **F** into two components: $\mathbf{F} = \mathbf{F}_1 + \mathbf{F}_2$ where $\mathbf{F}_1$ is along the rod and $\mathbf{F}_2$ is perpendicular to the rod. The component $\mathbf{F}_1$ is calculated as follows:

$$M\frac{L}{2}\omega^2 = M(g+a)\cos(\theta) - F_1 ,$$
$$F_1 = M(g+a)\frac{5\cos(\theta) - 3}{2}$$

(check for $\theta = 0$!)
The orthogonal component $\mathbf{F}_2$ is calculated as follows:

$$M\frac{L}{2}\alpha = M(g+a)\sin(\theta) - F_2 ,$$
$$F_2 = \frac{1}{4}M(g+a)\sin(\theta)$$

(check the magnitude of F_2 in the limit $\theta = 0$!).

△

Example 12.3. Man on Disk

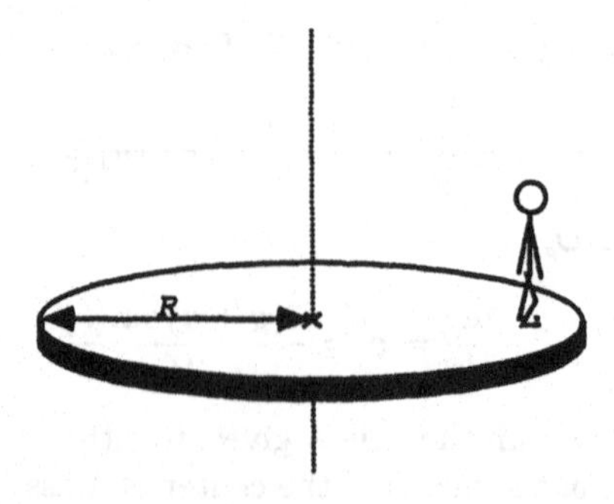

A large horizontal circular and homogeneous disk with radius R and mass M can rotate without friction about a vertical axis O through the center of mass of the disk. The disk is at rest and a man with mass m stands near the edge of the disk.

The man now walks around the periphery of the disk until he returns to his starting point on the disk. There he stops. Find the angle θ_0 which the disk has turned relative to the ground, when the man stops.

Solution. The angular momentum about O is conserved since there are no

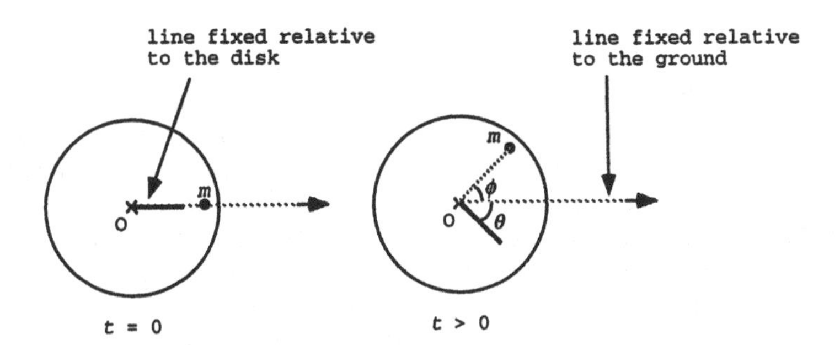

external torques about O. Internal forces cannot produce a total angular momentum. As the man walks along the circumference of the disk he exerts a force on the disk equal to, but oppositely directed to the force the disk exerts on the man. The total angular momentum about O therefore remains zero:

$$L = mR^2 \frac{\mathrm{d}\phi}{\mathrm{d}t} - I \frac{\mathrm{d}\theta}{\mathrm{d}t} = 0\,,$$

where $I = (1/2)MR^2$ is the moment of inertia of the disk about the axis O.

Integration, using the initial values $\theta = \phi = 0$ at $t = 0$, gives:

$$2m\phi = M\theta\,.$$

When the man has finished his walk along the periphery of the disk

$$\phi_0 + \theta_0 = 2\pi\,.$$

From this we find

$$\theta_0 = \frac{2\pi}{1 + M/2m}\,.$$

△

Example 12.4. The Sprinkler. A garden sprinkler consists mainly of a straight tube that can turn about a fixed vertical axis O. Through an opening at O in the middle of the tube, water is fed into the tube, exiting at the ends in a direction perpendicular to the tube (and perpendicular to the axis through O).

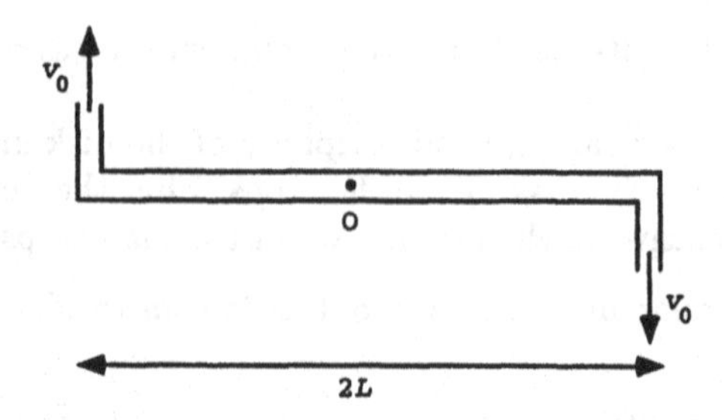

We assume that the mass per unit time, μ, which passes through a cross section of the tube, is independent of the motion of the tube. μ is thus constant in time. We furthermore assume that the water, at the instant it leaves the tube, has a constant velocity v_0 relative to the tube.

In the motion about the axis O the tube is subject to frictional forces. We assume that these forces can be described by a torque N which is proportional to the angular velocity of the tube. The constant of proportionality is denoted γ. The total length of the tube is $2L$ and the moment of inertia of the tube with respect to the axis O, when the tube is filled with water, is I. We assume the tube starts its motion at $t = 0$.

Determine the angular velocity $\omega(t)$ of the tube in its motion about the axis O.

Solution. When the angular velocity of the tube is ω, the magnitude of the velocity of the water relative to the earth is $v_0 - L\omega$. The differential equation governing the rotation of the tube about the axis O is:

$$I\frac{d\omega}{dt} = 2\mu\,(v_0 - L\omega)\,L - \gamma\omega\,, \tag{12.3}$$

$$\frac{d\omega}{dt} = -\frac{2\mu L^2 + \gamma}{I}\left(\omega - \frac{2\mu L v_0}{2\mu L^2 + \gamma}\right). \tag{12.4}$$

Note that the angular acceleration vanishes when the tube has reached the angular velocity ω_f, given by

$$\omega_f = \frac{2\mu L v_0}{2\mu L^2 + \gamma}\,.$$

By integration of $d\omega/dt$ we find:

$$\omega = \omega(t) = \omega_f\left[1 - \exp\left(-\frac{2\mu L^2 + \gamma}{I}t\right)\right].$$

Questions.

(1) Determine the final angular velocity ω_f for the case $\gamma = 0$. What is the physical interpretation of this result?

(2) Assume that the water is suddenly turned off at a time after ω_f has been reached. How long would it take for the tube to come to rest, for finite values of γ ? The mass of the empty tube is set to m, and we assume that the tube is empty a very short time after the water has been turned off.

Answers. (1) $\omega = v_0/L$, (2) $\omega = \omega_f \exp\left(-3\gamma t/mL^2\right)$. △

Example 12.5. Rolling. When a body of circular cross section is *rolling*, it is rotating about an axis while at the same time in translational motion. Think about a ball or a cylinder. Both in rolling and in sliding the body is in contact with the supporting surface.

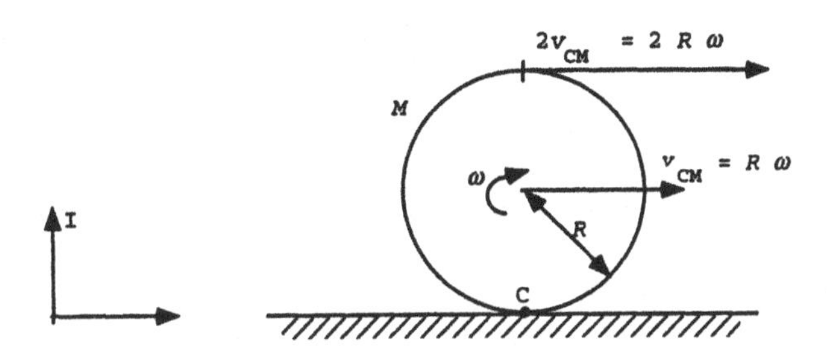

Pure rolling, i.e., rolling without sliding, is characterized by the fact that *the part of the body in contact with the supporting surface is instantaneously at rest relative to the surface.* In pure rolling we have the following relation between the translational velocity of the center of mass v_{CM}, the radius R of the circular cross section, and the angular velocity ω in the rotation of the body:

$$v_{\mathrm{CM}} = R\omega\,. \tag{12.5}$$

Loosely speaking, the point or line of contact participates in two motions, one forwards and one backwards, both with speed v_{CM}. Let us use an inertial frame I at rest relative to the supporting surface. The parts of the rolling body in contact with the surface thus has velocity zero relative to I (pure rotation).

The motion of the body can be considered in two different ways:

(1) a translation of the center of mass, CM, plus a rotation about an axis through CM; or
(2) a pure rotation about the instantaneous axis of rotation, which passes through the point C of contact between the body and the surface.

Note. In a rotation about a fixed axis C, the axis C is characterized by the fact that the velocity of each point of C is zero (relative to the chosen inertial frame). The angular velocity ω is independent of the axis chosen to describe the motion.

The kinetic energy of a rolling body of revolution can be written in different forms. If we choose to describe the motion as in point (1) above, we have:

$$T = \frac{1}{2}Mv_{\mathrm{CM}}^2 + \frac{1}{2}I_{\mathrm{CM}}\omega^2 . \tag{12.6}$$

If we choose to describe the rolling as in (2), i.e. as pure rotation about the "contact axis" C, we find

$$T = \frac{1}{2}I_{\mathrm{C}}\omega^2 , \tag{12.7}$$

where ω has the same value in (12.6) and (12.7). From the parallel axis theorem we have

$$I_C = MR^2 + I_{\mathrm{CM}} .$$

From this it can be seen that the right hand sides of (12.6) and (12.7) have the same value.

Let us sketch a concrete case.

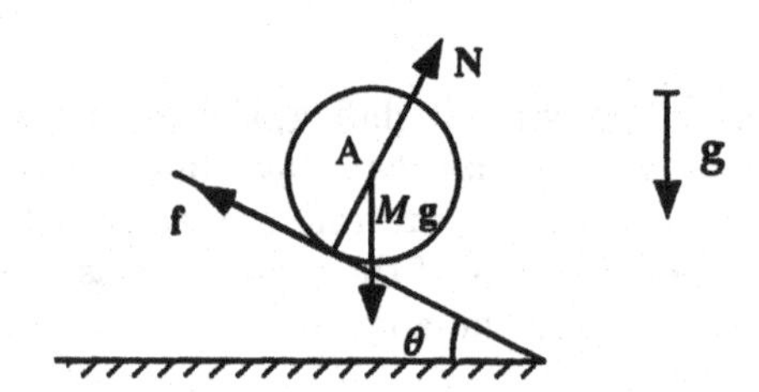

The figure shows a homogeneous symmetrical body with a circular cross section (cylinder, sphere, cylindrical shell, spherical shell, etc.). The moment of inertia about the axis of symmetry and perpendicular to the plane of the paper is I. Let us denote this axis by A. The mass of the body is M and the radius of the circular cross section is R.

If the body is placed at rest on an inclined plane, as shown in the figure, and if there is *no friction* between the body and the inclined plane, the body will slide down the inclined plane. For, in case of a frictionless inclined plane there is no torque about the axis A through CM and consequently the body will not begin a rotation.

Let us now assume that there is friction between the supporting surface (the inclined plane) and the body. The force of friction is **f** (see the figure). Other forces acting on the body are: the normal reaction **N** from the surface, and the force of gravity, $M\mathbf{g}$.

We write the equations of motion for the body.

The center-of-mass theorem. The acceleration of CM along the inclined plane is $\dot{v}_{\mathrm{CM}} = a$. Thus

$$Ma = Mg\sin(\theta) - f\,. \tag{12.8}$$

Note: the CM accelerates as if all the external forces were acting in CM.

The angular momentum theorem. The angular acceleration about the axis A is $\dot{\omega} = \alpha$. We have

$$I\alpha = fR\,. \tag{12.9}$$

The axis A is accelerated, but the angular momentum theorem can be applied about an accelerated axis without adding fictitious forces, when the axis passes through the center of mass.

We stated that if $f = 0$ the body will (if started from rest) slide down the inclined plane, without rolling. This is a consequence of (12.9) which, for $f = 0$, gives $\alpha = 0$.

Loosely speaking, the frictional forces acting from a surface on the body at the point of contact (or in general line of contact, the axis C) are in such a direction that they "oppose sliding". Just as we speak of the ideal case of a perfectly smooth (frictionless) surface, we speak in theoretical mechanics of the ideal case of a "perfectly rough" surface. A perfectly rough surface is a surface over which a body cannot slide, only roll.

Let us first assume that the inclined plane is perfectly rough so that the body performs pure rolling down the inclined plane. For pure rolling we have (differentiate (12.5) with respect to time) the following relation between a and α:

$$a = R\alpha\,. \tag{12.10}$$

From (12.8)–(12.10)

$$a = g\sin(\theta)\frac{1}{1 + I/MR^2}\,. \tag{12.11}$$

We introduce the so-called *arm of inertia*, k, which is defined by:

$$I = Mk^2\,, \tag{12.12}$$

$$a = g\sin(\theta)\frac{1}{1 + k^2/R^2}\,. \tag{12.13}$$

If a body *slides* down a frictionless inclined plane, the acceleration of the body will be $g\sin(\theta)$.

Equation (12.13) shows that the acceleration of the center of mass (CM) of a body that rolls down an inclined plane is less than the acceleration of the same body when it slides down a frictionless inclined plane with the same slope.

Furthermore, the mass of the body does not appear in (12.13). The acceleration is determined completely by the geometry of the particular body, through the arm of inertia, k. All bodies fall, regardless of their overall mass, with the same acceleration.

Examples.

(1) Hollow cylinder with thin walls:

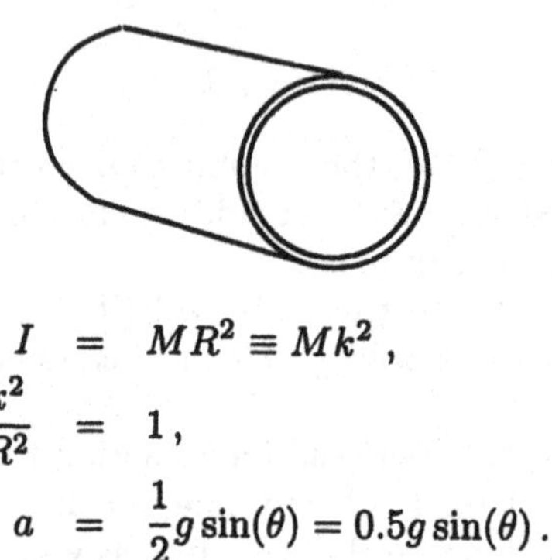

$$\begin{aligned} I &= MR^2 \equiv Mk^2 , \\ \frac{k^2}{R^2} &= 1 , \\ a &= \frac{1}{2} g \sin(\theta) = 0.5 g \sin(\theta) . \end{aligned}$$

(2) Homogeneous cylinder:

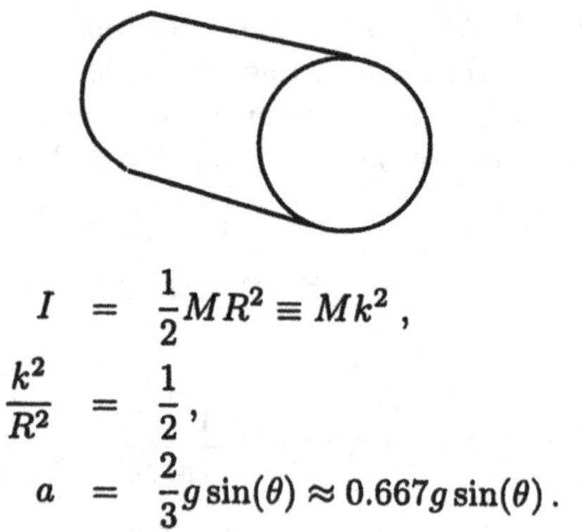

$$\begin{aligned} I &= \frac{1}{2} MR^2 \equiv Mk^2 , \\ \frac{k^2}{R^2} &= \frac{1}{2} , \\ a &= \frac{2}{3} g \sin(\theta) \approx 0.667 g \sin(\theta) . \end{aligned}$$

(3) Homogeneous sphere:

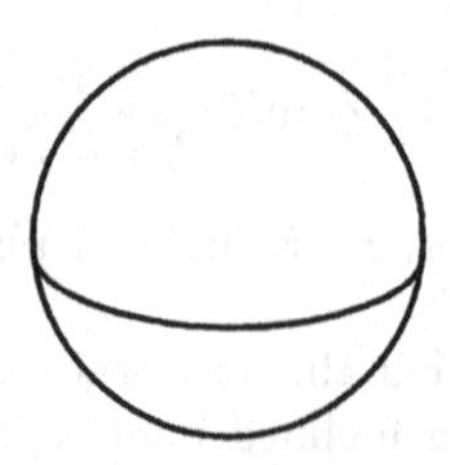

$$
\begin{aligned}
I &= \frac{2}{5}MR^2 \equiv Mk^2 , \\
\frac{k^2}{R^2} &= \frac{2}{5} , \\
a &= \frac{5}{7} g \sin(\theta) \approx 0.714\, g \sin(\theta) .
\end{aligned}
$$

Note. The radius R does not appear in these expressions. Consequently, all hollow cylinders, regardless of their radius, roll with the same acceleration down a rough inclined plane. Any sphere will roll faster than any homogeneous cylinder, which in turn will roll faster than any hollow cylinder.

Consider the case where the surface is not completely rough, and ask the question: what condition must the coefficient of friction μ satisfy in order that the body rolls without sliding down the inclined plane?

From (12.8)–(12.10), which describe pure rolling, we find

$$
f = Mg \sin(\theta) \frac{k^2}{k^2 + R^2} . \tag{12.14}
$$

This result means that *if* a body is to roll without sliding, f must have the value given by (12.14). The expression (12.14) shows that **f** is always directed up along the inclined plane, independent of whether the body rolls upwards or downwards. The friction force is a reaction force which tries to hold the contact point at rest, i.e., tries to prevent the body from sliding.

The question now is: can a given surface provide the friction necessary to keep the body rolling without sliding?

The answer is: the maximal friction force that a surface, with coefficient of friction μ can provide is

$$
f_{\max} = \mu N = \mu Mg \cos(\theta) . \tag{12.15}
$$

The condition for a body to roll without sliding is thus

$$
Mg \sin(\theta) \frac{k^2}{k^2 + R^2} \leq \mu Mg \cos(\theta) , \tag{12.16}
$$

$$
\mu \geq \frac{k^2}{k^2 + R^2} \tan(\theta) . \tag{12.17}
$$

Equation (12.17) expresses the fact that the surface must be so rough that it can provide the friction force necessary for rolling without sliding. When $\theta \to 90°$ a very large coefficient of friction is necessary for pure rolling to be maintained.

If the condition given by (12.17) is not satisfied, the motion down the inclined plane will be both rolling and sliding.

Consider again pure rolling. When a body rolls without sliding down an inclined plane, the friction force does no work. The friction force acts on a body particle which is instantaneously at rest. The power done by the friction force therefore is zero.

Let us use the energy theorem to describe the motion of a body that rolls without sliding down an inclined plane. The power, i.e., the work per time done by the normal force, is zero. (Why?) The gravitational field is conservative, so

$$T_{\text{rot}} + T_{\text{trans}} + U = \text{constant}.$$

We wish to determine the translational velocity of the body when it has rolled a distance d down the inclined plane. We assume that the body starts from a state of rest.

The initial position is denoted (1) and the final position (2). We set the potential energy equal to 0 in the final position. Using the above equation, we have

$$T_{\text{rot}}(1) + T_{\text{trans}}(1) + U(1) = T_{\text{rot}}(2) + T_{\text{trans}}(2) + U(2),$$

$$0 + 0 + Mgd\sin(\theta) = \frac{1}{2}I\omega^2 + \frac{1}{2}Mv_{\text{CM}}^2 + 0. \tag{12.18}$$

Since $v_{\text{CM}} = R\omega$, we find that

$$v_{\text{CM}}^2 = 2gd\sin(\theta)\frac{1}{1 + I/MR^2}. \tag{12.19}$$

Alternatively one can determine the velocity of the center of mass when the body has rolled part way down the inclined plane by applying (12.13).

Rolling friction. Apart from what could be called sliding friction, a body

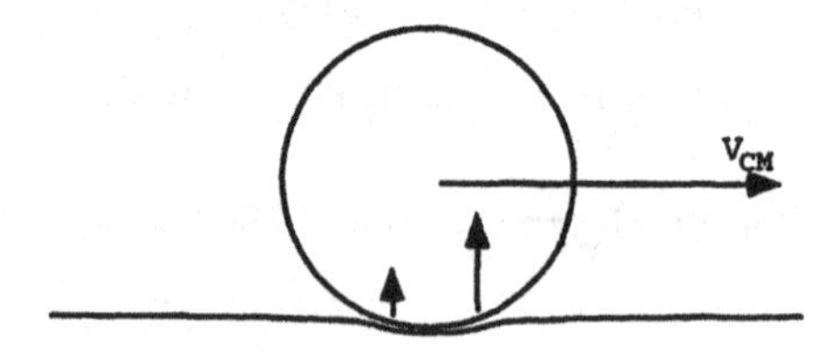

rolling on a level surface will experience rolling friction. We shall not deal with the phenomenon in any detail, but merely suggest the underlying causes.

Consider a cylinder rolling without sliding on a horizontal surface. Because of the elastic deformation of the body and the supporting surface, there is a finite size surface of contact between the two bodies. When the cylinder rolls, there will be an asymmetry in the contact forces (due to the elastic lag forces). This asymmetry in the distribution of the forces from the surface will produce a slight torque in a direction to counteract the rotation. A detailed knowledge of rolling friction – and material properties – is essential in the manufacture of ball bearings.

Question. A car is at rest on a horizontal road. The car now begins to accelerate along the road. The forces released by the combustion of the gasoline are internal forces for the system (the car). What gives rise to the external force that produces the acceleration of the center of mass of the car? △

Example 12.6. Yo-Yo on the Floor. A disk shaped body of revolution (Yo-Yo) is provided with a deep groove in the median plane perpendicular to the axis of symmetry A which passes through the center of mass of the body (see the sketch). A string is wound around the cylindrical shaft, which has radius r. The moment of inertia with respect to A is I and the total mass of the body (the Yo-Yo) is M. The free end of the string is held by a hand which pulls lightly on the string with force $\mathbf{F}$.

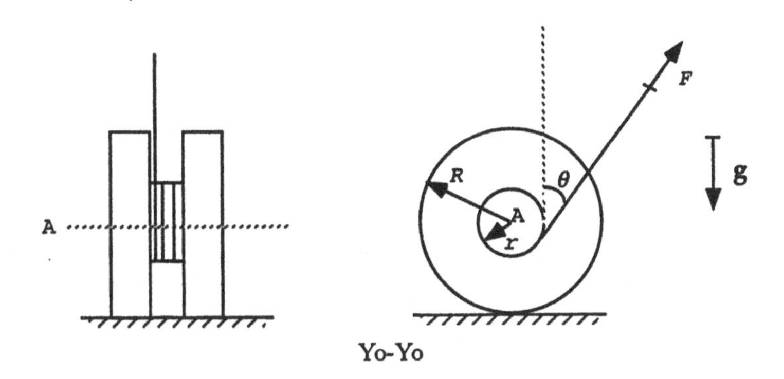

Yo-Yo

The coefficient of friction between the body and the horizontal surface (a table) is μ. Let us assume that the body is always in contact with the table and that the body initially is at rest on the table.

(1) Determine the initial angular acceleration α of the Yo-Yo as a function of θ, assuming that the Yo-Yo does not slide.
(2) Determine the value $\theta = \theta_0$ for which the body cannot perform pure rolling. To which side does the body roll for $\theta > \theta_0$ and $\theta < \theta_0$, assuming no sliding?
(3) What condition must the coefficient of friction μ satisfy in order that the motion, for given values of $F = |\mathbf{F}|$ and θ, can be pure rolling?

Solution.

(1) Orientation: V_{CM} is positive towards the right. The positive direction of rotation is chosen in such a way that a positive value of ω gives a positive value of V_{CM}.

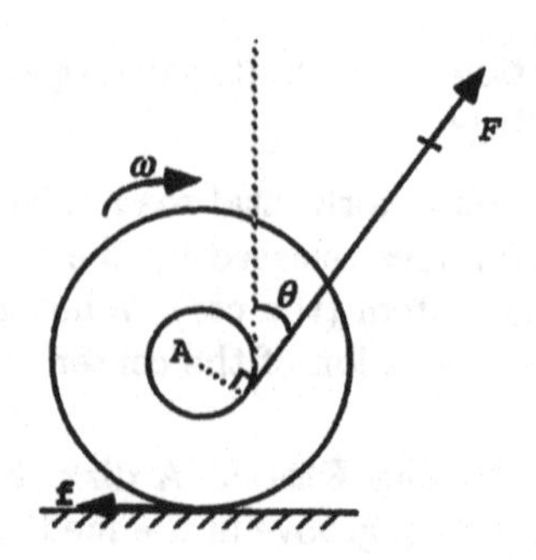

$$M\frac{dV_{CM}}{dt} = F\sin(\theta) - f, \tag{12.20}$$

$$I\alpha = -Fr + fR. \tag{12.21}$$

The geometric condition for pure rolling is

$$R\alpha = \frac{dV_{CM}}{dt} \equiv a. \tag{12.22}$$

From (12.20)–(12.22):

$$\alpha = \frac{F}{I + MR^2}[R\sin(\theta) - r]. \tag{12.23}$$

This value of the angular acceleration can be found directly by using the angular momentum theorem with respect to the axis C (the contact axis – see the sketch).

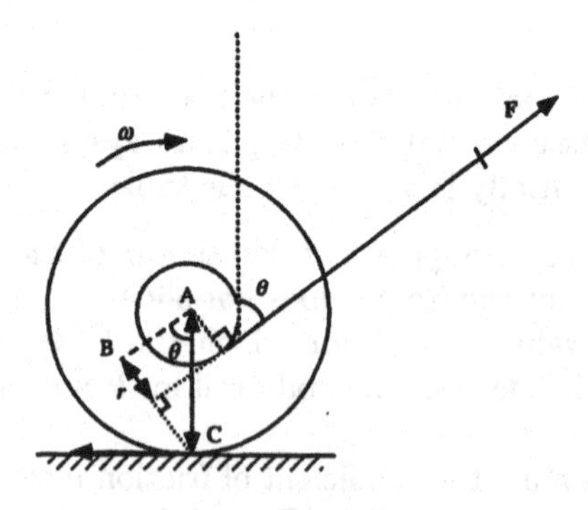

The segment CB equals $R\sin(\theta)$. Using the angular momentum theorem with respect to C we get

$$(MR^2 + I)\alpha = F[R\sin(\theta) - r]. \tag{12.24}$$

This leads directly to (12.23).

(2) θ_0 is determined by the condition $\alpha = 0$:

$$\begin{aligned} R\sin(\theta_0) - r &= 0, \\ \sin(\theta_0) &= r/R, \\ \theta > \theta_0 &\Rightarrow \text{towards the right,} \\ \theta < \theta_0 &\Rightarrow \text{towards the left.} \end{aligned}$$

The last two results can also be found by inspection of the figure below.

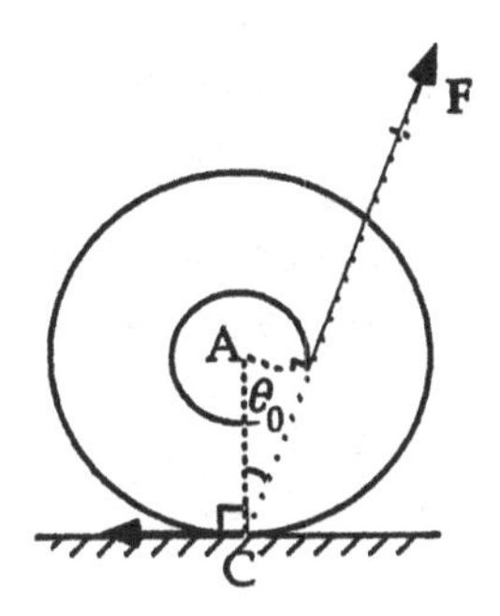

When the extension of **F** passes through C, the Yo-Yo cannot roll (rotate about C). When the extension of **F** falls to the right of C, i.e., $\theta < \theta_0$, the body rolls towards the left.

(3) The largest reaction force on the Yo-Yo that the surface can provide is $f_{max} = \mu N = \mu[Mg - F\cos(\theta)]$. For pure rolling, we have from (12.21) $f = (I\alpha + Fr)/R$ where α is given by (12.23). The condition to be satisfied by μ is therefore

$$\mu \geq \frac{[IF/(I + MR^2)]\,[R\sin(\theta) - r] + Fr}{R\,[Mg - F\cos(\theta)]}\,.$$

Question. For $F\cos(\theta) = Mg$ the coefficient of friction must be infinite. How do you interpret this result? What happens for $F\cos(\theta) > Mg$? △

Example 12.7. Rolling Over an Edge. A homogeneous circular disk with mass M and radius R rolls with angular velocity ω on a horizontal floor. The disk collides with the edge of a low, fixed table B. The edge of the table is at a height h above the floor. We consider the case $h < R$. The collision is regarded as completely inelastic and of short duration. Assume that the disk does not slip on the edge.

Determine the least angular velocity which the disk must have if it is to "roll up onto the table".

Solution. The physical content of this example is very close to the physical content of Example 12.1.

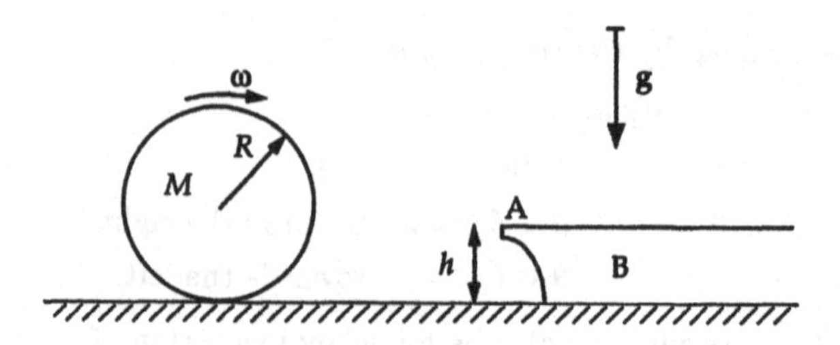

When the disk collides with the edge A, the disk begins a rotation about an axis through A, perpendicular to the plane of the paper. When the collision time is so short that the disk has not turned appreciably during the collision we can use the theorem of conservation of angular momentum about the axis through A during the collision (the collision approximation). This can be used to calculate the angular velocity ω_a with which the disk begins to turn about A. When the collision is over, the forces that act on the disk are conservative, and we may use the theorem of conservation of the mechanical energy to determine the critical angular velocity ω_0.

We have

$$\mathbf{L}_A = \mathbf{L}_{CM} + \mathbf{R}_{CM} \times M\mathbf{v}_{CM} \,.$$

Assume that the disk before the collision has just the critical angular velocity ω_0. Then conservation of angular momentum about the axis through A means that

$$I_{CM}\omega_0 + Mv_{CM}(R-h) = I_A\omega_a \,.$$

Using

$$I_{CM} = \frac{1}{2}MR^2 \,, \quad I_A = \frac{3}{2}MR^2 \,,$$

and $v_{CM} = R\omega_0$, we find

$$\omega_a = \omega_0 \left(1 - \frac{2}{3}\frac{h}{R}\right) .$$

We can now determine ω_0 by using the energy theorem:

$$\frac{1}{2}\left(\frac{3}{2}MR^2\right)\omega_a^2 = Mgh \,,$$

$$\omega_0 = \frac{2\sqrt{(1/3)gh}}{R(1-2h/3R)} \,.$$

In the limits

(a) $h \to 0 \Rightarrow \omega_0 \to 0$ (which makes sense!),
(b) $h \to \frac{3}{2}R \Rightarrow \omega_0 \to \infty$ (so this is a critical height of the table).

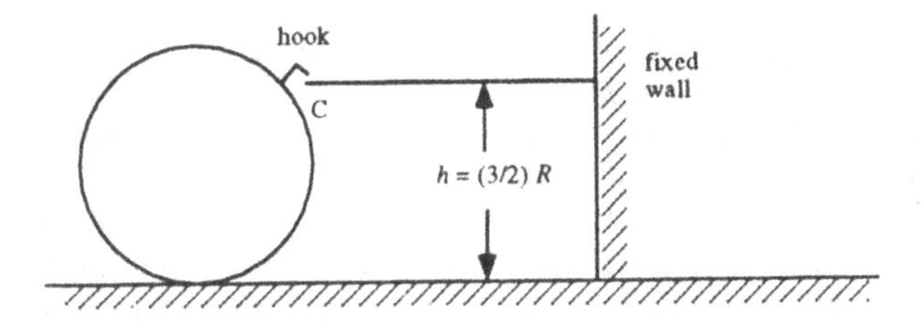

The cases $h > R$ can only be realized by a special setup (in the example we assumed $h < R$ – see the sketch).

Questions.

(1) A hole C in a boom is at a height $h = (3/2)R$ above the floor (see the sketch). Find the total angular momentum of the disk about the point C on the axis before the collision. Answer: $L_C = 0$. For $h > R$, the spin angular momentum and the orbital angular momentum are oppositely directed. At $h = (3/2)R$ the two contributions to the total angular momentum cancel.

(2) Consider again the original collision with the edge of the table. What fraction of the original mechanical energy of the disk is "lost" as heat in the collision? Answer:

$$f = \frac{4}{3}\frac{h}{R}\left(1 - \frac{1}{3}\frac{h}{R}\right),$$

for $h \rightarrow \frac{3}{2}R$ we have $f \rightarrow 1$.

△

Example 12.8. Determinism and Predictability. Newton's equations of motion have the mathematical property that, given a specific initial state of the system (initial positions and velocities), the solution at any finite future (or past) time is determined uniquely. A system of differential equations with this property is said to be *deterministic.*

It is important to realize, however, that this does *not* necessarily mean that almost identical initial states have almost identical solutions. Indeed, many conceptually simple systems have the remarkable property that almost identical initial states can have very different solutions.

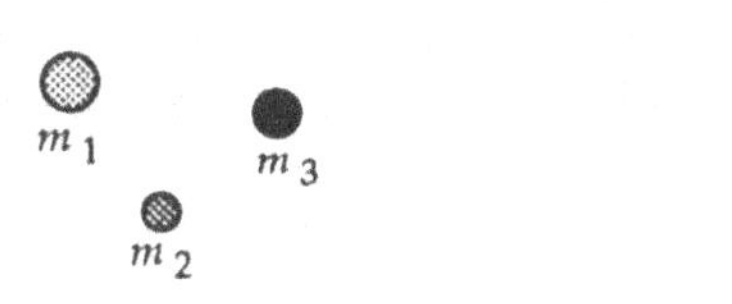

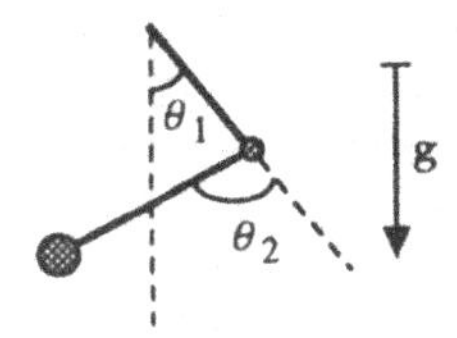

Three mutually gravitationally interacting masses, or a pendulum attached to a pendulum (the double pendulum) in a gravitational field, are examples of systems of this type, the complete solution of which cannot be provided by present-day mathematics.

Thus, while *deterministic* in the above sense, systems can exhibit behavior ("chaos") making it impossible to *predict* the evolution of the system without infinitely precise knowledge of the initial state. Such systems are the subject of much current investigation. See also Chapter 16. △

12.4 Problems

Problem 12.1.

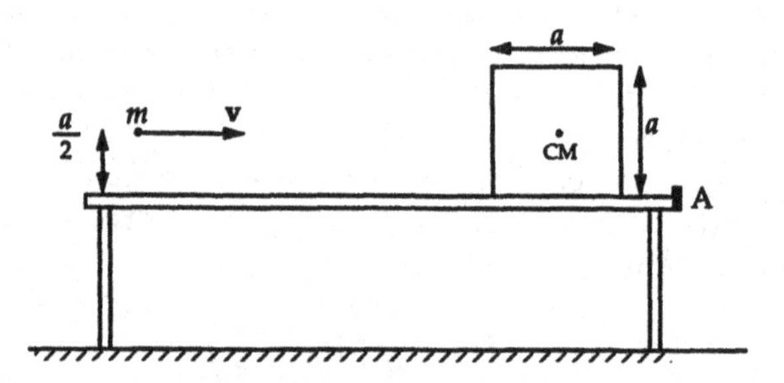

A homogeneous, cubic wooden block of mass M is at rest on a horizontal, frictionless table in the laboratory. The block has sides a. A pistol bullet of mass m and horizontal velocity $\mathbf{v}$ is shot into the block at a height $h = a/2$ over the table. The velocity of the bullet is perpendicular to the side of the block (see the figure). The bullet is assumed to come to rest in the center of mass of the block. The block now begins to slide along the frictionless table until the block is suddenly stopped in a completely inelastic collision with a lip, A, at the edge of the table. From this point the motion is pure rotation about A. The lip A is assumed to be so small that it can be considered an axis, and in particular it provides no torque on the block while it is in rotation about A. It is assumed that the collision time Δt is so short that the block does not turn appreciably about A during Δt.

(1) Determine the velocity (magnitude and direction) of the center of mass CM just after the block has been stopped at A.
(2) Determine the magnitude of the maximum velocity $v_{\mathrm{m}} = |\mathbf{v}_{\mathrm{m}}|$ that the pistol bullet may have if the block is *not* to fall down from the table.

Problem 12.2. A rod of length 1 m stands upright on a horizontal floor in the gravitational field of the Earth. The rod now begins to fall (from rest), rotating around a horizontal axis. (The end of the rod in contact with the floor does not slip on the floor.)

(1) Determine the velocity relative to the floor of the free end of the rod, just before it hits the floor.
(2) Determine the velocity of a particle falling freely from rest through 1 m.

Problem 12.3. A ring of mass M and radius r turns in a vertical plane without friction about a horizontal axis through the center of the ring because a mouse with mass m runs up the ring. The mouse remains on the same point B in space (!). B is the end point of a radius vector that forms the angle θ_0 with the vertical.

Determine the angular acceleration α of the ring.

Problem 12.4.

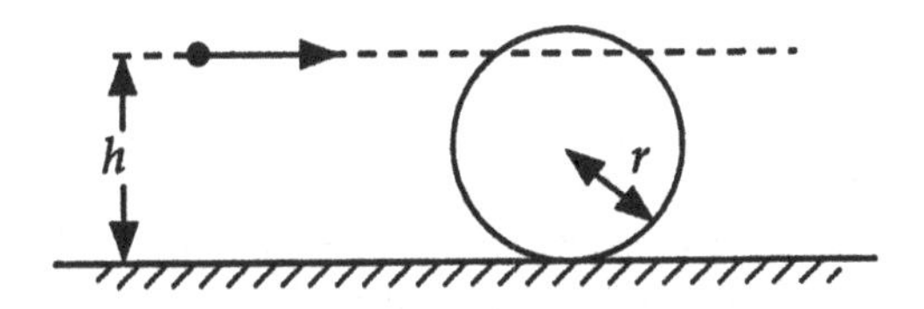

A homogeneous cylinder with circular cross section of radius r is at rest on a frictionless, horizontal surface. A small bullet is shot horizontally into the cylinder at the height h above the surface, and perpendicularly to a vertical plane through the cylinder axis. The bullet passes along a straight line through the cylinder in a very short time, and vertically above the CM of the cylinder. The bullet looses the momentum P to the cylinder.

Find h so that the motion of the cylinder after the "collision" is pure rolling, i.e., rolling without sliding.

Problem 12.5. A narrow straight tunnel has been drilled through the Earth between Copenhagen and Los Angeles (see Problem 8.4). How long will it take for a ball to roll through the tunnel?

Problem 12.6. A homogeneous ball with mass M and radius R is set in motion on a horizontal floor in such a way that the ball initially (i.e., at time $t = 0$) has the translational velocity V_0. We assume that the ball begins its motion as pure sliding. The coefficient of friction between the ball and the floor is μ. The motion of the ball between the time $t = 0$ and $t = t_r$ is a combination of sliding and rolling. At time $t = t_r$ the ball begins to roll without sliding. In the time between $t = 0$ and $t = t_r$ the ball has moved a distance D along the floor.

(1) Determine t_r.
(2) Determine D.
(3) Determine the work W that the forces of friction perform in the time from $t = 0$ to $t = t_r$, i.e. in the time where the ball was both rolling and sliding.
(4) Show that $W = \int_0^{t_r} \mu M g V_c(t) dt$, where V_c is the velocity of the contact point relative to the supporting surface (the floor).

Problem 12.7.

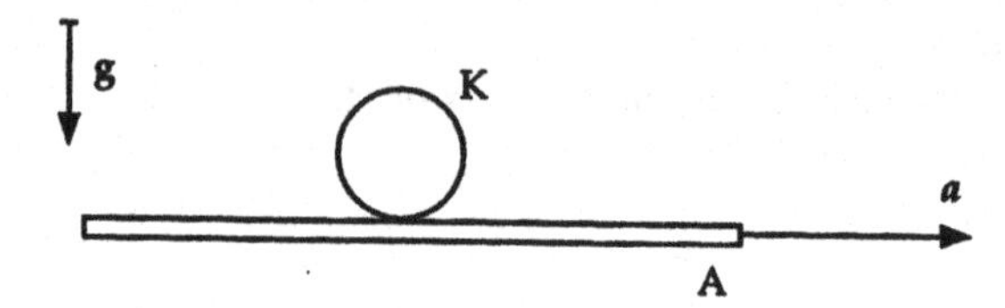

A homogeneous sphere K is at rest on a horizontal plate A. Initially, A is at rest in the laboratory. At $t = 0$ the plate A begins to accelerate with constant acceleration **a** in the horizontal direction (see the figure). The acceleration of gravity is g. The coefficient of friction between K and A is μ.

Find the smallest value $\mu = \mu_{\text{min}}$ the coefficient of friction may have if we demand that the sphere K will begin to roll without sliding relative to A, when we give A the acceleration **a**.

Problem 12.8.

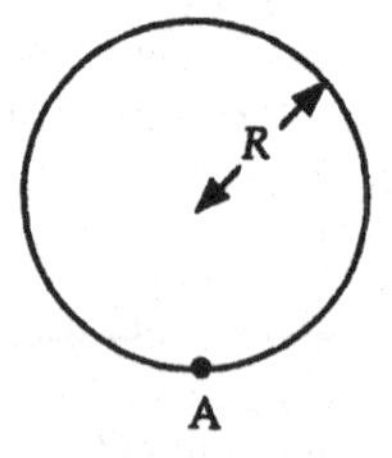

A thin homogeneous circular ring with mass M and radius R lies on a frictionless horizontal table in the laboratory. The ring may turn without friction around a vertical axis through the point A on the perimeter of the ring. The axis is fixed relative to the table (see the sketch).

A small beetle with mass m crawls around the ring, moving with constant speed v_0 relative to the ring. The ring is initially at rest, and the beetle starts out from the point A. As the beetle moves around the ring, the ring rotates around the axis through A.

Determine the angular velocity of the ring in its rotation around A when the beetle reaches the point half way around the ring. The angular velocity ω is measured relative to the lab frame, and ω should be expressed as a function of m, M, v_0, and R.

Problem 12.9.

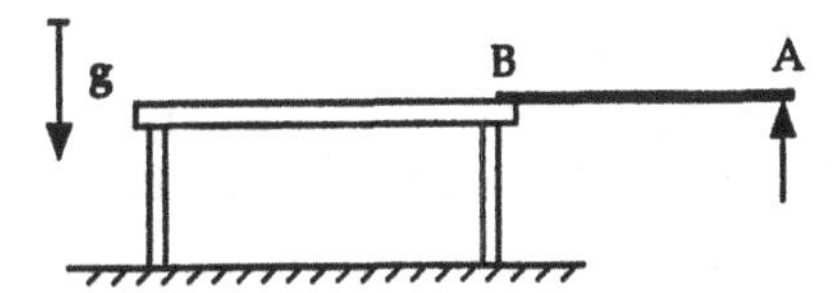

A thin homogeneous rod of mass M and length L is at rest and horizontal in the gravitational field of the Earth. One end, B, of the rod rests on the edge of a table. The other end A is supported by a vertical force (from a hand). The hand is suddenly removed, so that the end point A is no longer supported.

Determine the magnitude and direction of the force **F** by which the table acts on the rod just after the hand is removed.

Problem 12.10.

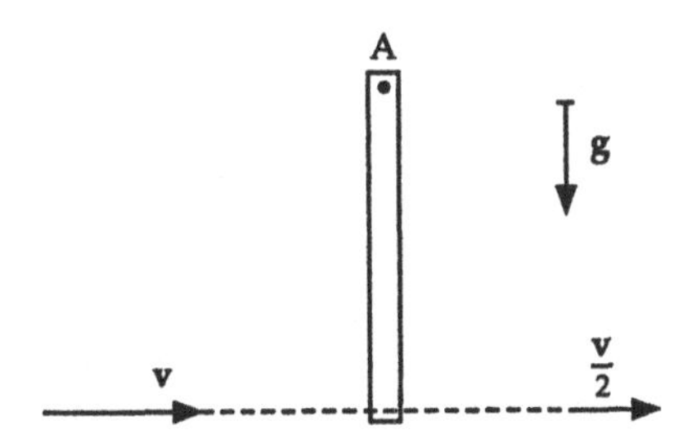

A heavy homogeneous rod has mass M and length L. The rod can turn without friction about a fixed horizontal axis A which is perpendicular to the rod (see the sketch). Initially the rod hangs at rest, vertically, in the gravitational field of the Earth. The acceleration of gravity is g.

A bullet with mass m passes in a very short time Δt through the rod near the lower end (so the distance from A to the point of impact can be taken to be L).

The direction of the velocity of the bullet – both before and after passing through the rod – is horizontal, perpendicular to the rod and to the axis A. Before the collision the velocity of the bullet is **v**, and after the collision it is **v**/2 (see the sketch).

We assume that Δt is so small that the rod does not move as the bullet passes the rod. Ignore air resistance.

(1) Determine the angular velocity ω_0 of the rod just after the bullet has passed through the rod. The angular velocity should be expressed as a function of m, M, v, and L.
(2) Determine what the speed v of the bullet should be such that the rod swings exactly 90° from its initial equilibrium position. Express v in terms of m, M, g, and L.

From its horizontal position the rod now swings back. As it does so, the axis through A acts on the rod with a reaction force $\mathbf{F}_1$.

(3) Determine $\mathbf{F}_1$ at the moment when the rod passes through its initial vertical position. Express $\mathbf{F}_1$ as a function of M and g.

We now return to the original situation and consider the rod as the bullet is passing through the rod. During this time, Δt, there is a reaction force from the axis A on the rod. The horizontal component is called $\mathbf{F}_2$. Assume that $\mathbf{F}_2$ is constant during the time Δt.

(4) Find $F_2 \equiv | \mathbf{F}_2 |$ for the situation described in question (2). Express F_2 in terms of Δt, M, g, and L.
(5) State the condition that Δt must satisfy for the approximation used to be valid.

Problem 12.11.

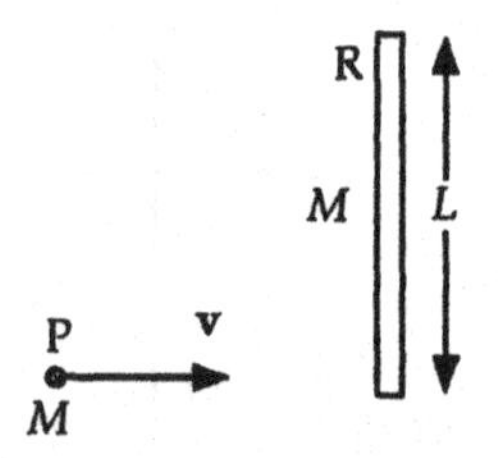

A thin homogeneous rod R has length L and mass M. The rod is at rest on a frictionless horizontal table. A small piece of sticky clay (a particle), of the same mass M strikes the rod, near the end, with velocity $\mathbf{v}$. The velocity $\mathbf{v}$ is perpendicular to the rod. The collision is completely inelastic, and after the collision the particle P sticks to the rod. After the collision the particle P and the rod move together as one body.

(1) Determine the velocity of the CM of the system before and after the collision (take the system to be both P and R).
(2) Determine the angular velocity ω in the rotation around the CM after the collision. The angular velocity should be given as a function of v and L.

A certain amount of mechanical kinetic energy ΔT is lost in the collision (converted into heat).

(3) Determine ΔT as a function of M and v.
(4) Assume now that the rod is fastened in such a way that it can turn without friction in a horizontal plane around a vertical axis through the point A. This axis is at a distance d from the point where P hits the rod (see the sketch). We assume that the duration of the collision is so short that the rod does not turn during the collision. Determine d in such a way that there are no reaction forces from A during the collision (the point A is then called the *center of percussion*).

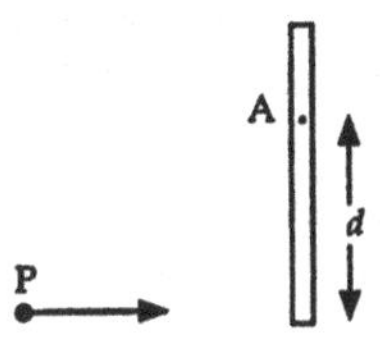

Problem 12.12. A rod of mass M and length L is kept at rest with one end on a frictionless horizontal floor and the other on a frictionless vertical wall. The rod is in a vertical plane perpendicular to the wall. The inclination with respect to the horizontal is initially θ_0. At a certain time the rod is left free to move.

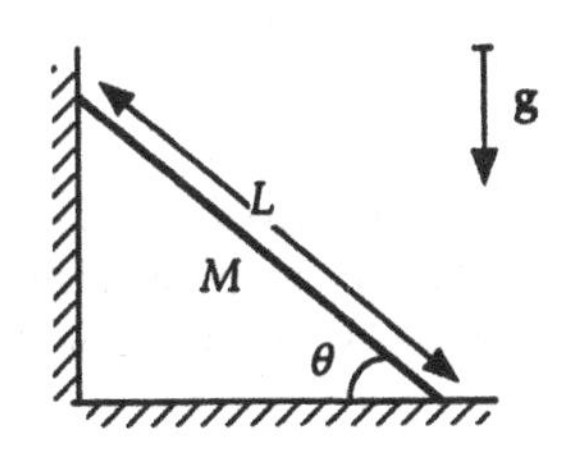

(1) Find $\dot{\theta}$ as a function of θ.
(2) Find the critical angle θ_1 for which the rod leaves the wall.
(3) Assume that $\theta_0 = 90°$ (we then have to give the rod a slight push to set it in motion). Find the horizontal component $\dot{X}_{\mathrm{CM}}$ of the velocity of the center of mass just before the center of mass reaches the floor.

Problem 12.13. A thin homogeneous rigid rod has length L and mass M. One endpoint of the rod is fastened to a light, rigid steel wire, also of length L.

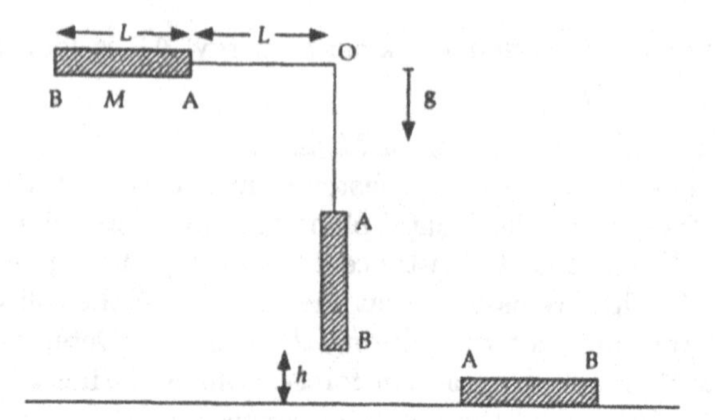

The other end of the wire is fastened to a horizontal axis O in the laboratory. The system can turn in a vertical plane. The steel wire cannot bend, and its mass can be ignored relative to M.

Initially the system is held at rest in a horizontal position. It is then set free to move under the influence of gravity. The acceleration of gravity is g. Ignore air resistance.

(1) At the moment the system begins to move, i.e., while the rod is still horizontal, the wire acts on the rod with a force $\mathbf{F}_1$, at the endpoint A. Determine $\mathbf{F}_1$.
(2) At the moment when the rod is vertical, the wire acts on the rod with a force $\mathbf{F}_2$. Determine $\mathbf{F}_2$.
(3) Assume now that the steel wire breaks at the moment when the rod is vertical. The lower point B of the rod is then at some height h above the laboratory floor. When the rod hits the floor is has turned so that it is exactly parallel to the floor. Determine h.

Problem 12.14. Find the angular velocity $\omega(t)$ for the sprinkler in Example 12.4 by considering the problem from a reference frame in which the tube is at rest. Use Newton's second law in an accelerated reference frame. Don't forget the term $m(\mathbf{r} \times \dot{\omega})$.

13. The General Motion of a Rigid Body

In the previous chapters we have considered rotations of rigid bodies possessing a high degree of symmetry. We have seen that the rotational motion of such a body around an axis fixed in inertial space and fixed relative to the body is as simple to describe as the motion of a single particle moving along a straight line. The same equations govern the motion in these two cases, when $\mathbf{p}$ is substituted for $\mathbf{L}$, and $\mathbf{v}$ is substituted for $\boldsymbol{\omega}$, etc.

Nevertheless we shall see in the following that, for certain types of rotational motion around an axis fixed in inertial space, new and somewhat surprising features may arise. As soon as we leave the case of rotation about an axis at rest both relative to an inertial frame and relative to the rotating body, the similarity between translational and rotational motions comes to an end.

13.1 Inertia in Rotational Motion

The differences between the description of translational motion and the general description of rotational motion, stem from the following facts.

For translational motion, the mass m (the "inertia") is a scalar quantity. For instance,

$$\mathbf{p} = m\mathbf{v} .$$

This means that the velocity $\mathbf{v}$ and the momentum $\mathbf{p}$ are always parallel vectors.

The corresponding equation for rotational motion, i.e., the connection between $\mathbf{L}$ and $\boldsymbol{\omega}$, is more complicated. The inertial properties of a rotating body cannot be expressed as a scalar quantity alone, it must be described by a quantity called a tensor. This tensor – called the inertia tensor – act in the equations as a scalar only in the simple case of rotation about an axis fixed in inertial space and fixed in the rotating body. For this special case, $L_{\mathrm{A}} = I_{\mathrm{A}}\omega$ in analogy to, say, $p_x = mv_x$.

Since the connection between the angular momentum vector and the angular velocity vector is in general expressed by means of a tensor, the angular momentum vector will not necessarily be parallel to the rotation vector.

Before proceeding to the description of the inertia tensor we wish to illuminate the essential issue by briefly describing a few particularly transparent cases of rotation about a fixed axis where **L** and ω are not parallel, but where the problem nevertheless can be solved without introducing the formal inertia tensor. The following three examples are of this nature.

Example 13.1. The Dumbbell

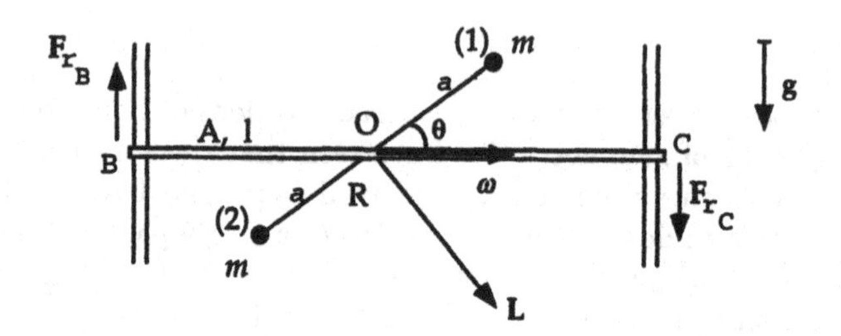

We regard a system consisting of the following parts: two, identical, small, heavy balls, each of mass m are connected with a thin rod R with total length $2a$ (a dumbbell). We disregard the mass of the rod, and the balls are considered so small that they can be treated as point masses.

The mid-point of the rod R is attached to the mid-point O of a horizontal axis A which can rotate without friction in two bearings B and C. The rod R forms the constant angle θ with the axis A. The axis A has total length $2\,l$.

The system is now set in rotation. The axis A rotates in the bearings B and C with the angular velocity ω. The rotation vector ω is thus directed along A (see the sketch). Note: the axis of rotation is fixed in the inertial system, but not relative to the body. The rod R moves on the surface of a cone.

As our system we consider the axis A plus the rod R and the two balls. As long as we disregard friction in the two bearings B and C (and air resistance), the system will continue its rotation with angular velocity ω. There are no external torques with components along ω.

Let us now analyse the motion. We first calculate the angular momentum with respect to O. The sketch shows an instant where the rod R is in the plane of the paper. The topmost ball (1) is moving out of the paper and the lower ball (2) is moving into the paper. The point masses give the same contribution to the angular momentum. The total angular momentum vector is in the plane of the paper and has the direction indicated on the sketch. During the entire motion **L** will lie in the plane spanned by ω and the rod R. The magnitude of the velocity relative to the lab frame of each of the two masses is

$$v = [a\sin(\theta)]\omega\,.$$

The magnitude of **L** is

$$L = 2amv = 2am[a\sin(\theta)]\omega = 2a^2m\omega\sin(\theta)\,.$$

The angular momentum vector moves around on the surface of a cone with its apex at O. The tip of **L** is on the above sketch moving into the plane of the paper. The direction of d**L**/dt is thus into the plane of the paper.

We now calculate the magnitude of d**L**/dt.

Note. In this example the motion is known and from this we determine the forces and torques necessary to maintain the motion.

$$\left|\frac{\mathrm{d}\mathbf{L}}{\mathrm{d}t}\right| = |\mathbf{L}|\cos(\theta)\omega = 2ma^2\omega^2\cos(\theta)\sin(\theta)\,.$$

In order to maintain this rate of change of the angular momentum vector, it is necessary to have an external torque on the system. This torque comes from the bearings. In each bearing there must be, because of the rotation, a reaction force, $\mathbf{F}_r$. The instantaneous directions are indicated on the sketch.

The torque $\mathbf{N}_O$ at O must have the same direction as d**L**/dt, that is, perpendicular to the axis A and into the paper. $\mathbf{N}_O$ has the magnitude:

$$N_O = 2F_r l = 2ma^2\omega^2\sin(\theta)\cos(\theta)\,,$$

from which

$$F_r = \frac{ma^2\omega^2}{l}\sin(\theta)\cos(\theta)\,.$$

Since the center of mass for the system is at rest, we know from the center of mass theorem that the sum of external forces on the system vanishes.

The total reaction force from the bearings thus becomes, counting all reaction forces positive upwards:

$F_r + mg$ in the left side, i.e., bearing B,

$-F_r + mg$ in the right side, i.e., bearing C.

Apart from the reaction forces in the bearings, the force of gravity acts on the balls. The total external force on the system is therefore

$$F_r + mg - F_r + mg - 2mg = 0.$$

The reaction forces caused by the rotation change direction as they rotate with the system. Note the following.

(1) There is no external torque with components along the rotational axis. Thus, ω remains constant. A possible friction in the bearings may provide a torque about the axis A and thus brake the rotation.

(2) for $\theta = \pi/2$ the angular momentum vector **L** is parallel to A and d**L**/dt vanishes. This case gives the smallest wear on the bearings; the system is said to be dynamically balanced.

△

Example 13.2. Flywheel on an Axis

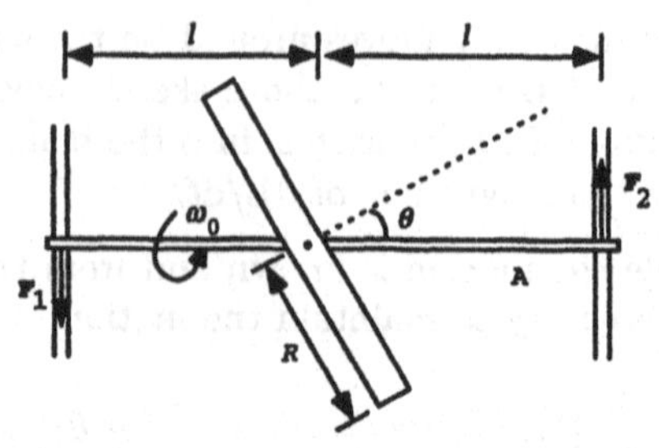

A heavy flywheel in the shape of a homogeneous circular disk has mass M and radius R. The wheel is fastened to a horizontal axis A passing through the center of mass of the wheel. The axis of symmetry for the wheel forms a small angle θ with the axis A. The flywheel is thus mounted out of alignment. The rotation axis A is assumed to be turning without friction in two bearings, one on each side of the wheel, both at a distance l from the center of mass of the wheel. The angular velocity of the axis – and consequently of the wheel – is ω_0.

Since we disregard friction in the bearings the wheel will continue to rotate with angular velocity ω_0. No external torques have components along ω_0. We shall see that the misalignment of the wheel causes a stress on the bearings due to the rotation.

As our system we take the wheel and the axis A which is considered massless. The reaction forces on the axis from the bearings are thus external forces for the system.

We begin by determining $\mathbf{L}_{CM}$, i.e., the angular momentum with respect to the center of mass (see the figure below).

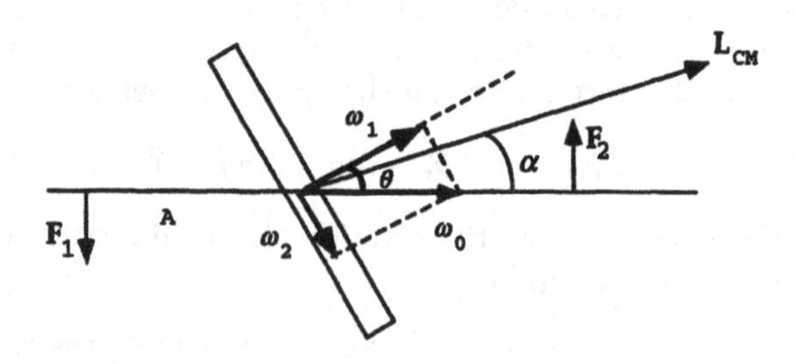

We split ω_0 into two components:

ω_1 (along the normal to the plane of the flywheel); and

ω_2 (the remainder, in the plane of the flywheel).

The angular momentum vector $\mathbf{L}_{CM}$ can also be split into two components, corresponding to the two components of ω_0.

$$\mathbf{L}_{CM} = \mathbf{L}_1 + \mathbf{L}_2 , \tag{13.1}$$

where

$$\begin{aligned} L_1 &\equiv |\mathbf{L}_1| = I_1\omega_0\cos(\theta) = \frac{1}{2}MR^2\omega_0\cos(\theta)\,, &(13.2)\\ L_2 &\equiv |\mathbf{L}_2| = I_2\omega_0\sin(\theta) = \frac{1}{4}MR^2\omega_0\sin(\theta)\,, &(13.3)\\ L_{\mathrm{CM}} &= \frac{1}{2}MR^2\omega_0\sqrt{\cos^2(\theta)+\frac{1}{4}\sin^2(\theta)}\,. &(13.4)\end{aligned}$$

The moment of inertia about the normal to the disk of the flywheel is different from the moment of inertia about a diameter of the disk. This is the cause of $\mathbf{L}_{\mathrm{CM}}$ being not parallel to ω_0. Had the two moments of inertia been equal, $\mathbf{L}_{\mathrm{CM}}$ would have been parallel to ω_0. Since the moment of inertia about an axis perpendicular to the disk is the greater, $\mathbf{L}_{\mathrm{CM}}$ lies as shown on the figure. The angle between ω_0 and $\mathbf{L}_{\mathrm{CM}}$ is denoted α. It may be determined from the relation

$$\tan(\theta-\alpha) = \frac{L_2}{L_1} = \frac{1}{2}\tan(\theta)\,.$$

In order to determine the torques at the bearings, i.e., the external torques on the system, we must first compute $\mathrm{d}\mathbf{L}_{\mathrm{CM}}/\mathrm{d}t$. The motion is known, and we are determining the forces.

The endpoint of $\mathbf{L}_{\mathrm{CM}}$ describes a uniform circular motion about the axis A. At the instant shown on the above sketch, the tip of $\mathbf{L}_{\mathrm{CM}}$ is moving out of the paper. The vector $\mathrm{d}\mathbf{L}_{\mathrm{CM}}/\mathrm{d}t$ is thus perpendicular to the paper, pointing out of the paper.

In order to determine the magnitude of $\mathrm{d}\mathbf{L}_{\mathrm{CM}}/\mathrm{d}t$ we need to first determine the component $L_\perp$ of $\mathbf{L}_{\mathrm{CM}}$ that is perpendicular to the axis A; we will then have

$$\left|\frac{\mathrm{d}\mathbf{L}_{\mathrm{CM}}}{\mathrm{d}t}\right| = L_\perp\omega_0\,.$$

From the sketch we can deduce the relation

$$L_\perp = L_1\sin(\theta) - L_2\cos(\theta)\,.$$

Inserting the previously determined L_1 and L_2, we find

$$\left|\frac{\mathrm{d}\mathbf{L}_{\mathrm{CM}}}{\mathrm{d}t}\right| = \frac{1}{4}MR^2\omega_0^2\cos(\theta)\sin(\theta)\,.$$

In order to keep the wheel rotating, there must be an external torque $\mathbf{N}$. This torque stems from the reaction forces $\mathbf{F}_1$ and $\mathbf{F}_2$ at the bearings. The directions are as shown on the figure above. The two forces have equal magnitude (why?) but the directions are different ("right-hand rule"). The magnitude F of the two forces is given by

$$\begin{aligned} 2Fl &= \frac{1}{4}MR^2\omega_0^2\cos(\theta)\sin(\theta)\,,\\ F &= \frac{1}{8}\frac{MR^2\omega_0^2\cos(\theta)\sin(\theta)}{l}\,.\end{aligned}$$

Because of the weight of the system there will also be reaction forces to counteract the force of gravity. At the instant shown on the sketch, the total reaction force on the bearings, with the upward direction as positive, is given by

Right bearing: $F + Mg/2$,
Left bearing: $-F + Mg/2$.
The total force in the vertical direction is thus

$$F + \frac{Mg}{2} - F + \frac{Mg}{2} - Mg = 0 .$$

This assures us that the center of mass has no net acceleration.

Note: the reaction forces due to the rotation change direction continuously, as the axis rotates. The reaction forces caused by gravity are always in the upwards direction.

When a flywheel is misaligned on its axis of rotation, the bearings will wear faster than if the wheel is accurately mounted perpendicular to the axis of rotation. A misaligned wheel is said to be dynamically imbalanced.

Question. Let us assume that the wheel rotates at 300 revolutions per minute. Calculate F, when $M = 1000$ kg, $R = 1$ m, $\theta = 1°$, $l = 0.2$ m .

Answer. $F \approx 1.08 \times 10^4$ N, the weight of the wheel is 9.8×10^3 N. △

Example 13.3. Precession of a Gyroscope. In this example we consider the physics behind the phenomenon known as the precession of a gyroscope.

A gyroscope (also known as a *top*) is, in its simplest version, a wheel or disk able to rotate about an axis, called the gyro-axis or the figure-axis, that passes through the center of the disk (wheel).

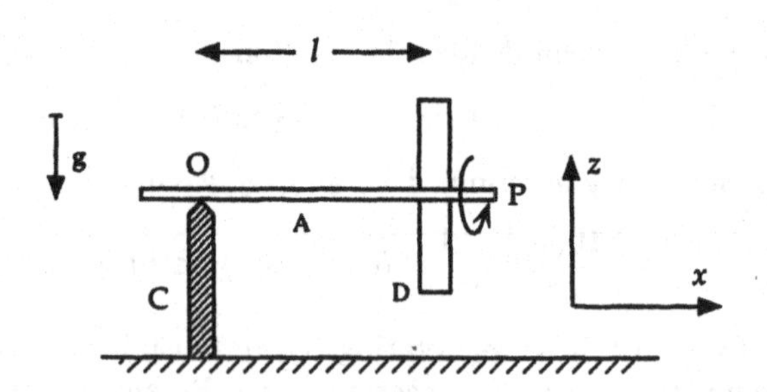

The figure shows a disk of mass M that can rotate about the axis A. We ignore the mass of A. This axis is suspended so that it can pivot about the origin O of the coordinate frame, assumed to be fixed (in the lab frame). Let us initially assume that the disk is *not* rotating about the axis A, and that the axis is also supported in the other end point, P, so that the axis A is along the (horizontal) x-axis. Then we remove the support at P.

The force of gravity, Mg, acting at the center of mass, has a torque N about O. This torque points in the y-direction (i.e., into the paper). The disk will begin to fall, initiating a rotation about the y-axis.

We now change the procedure. Before removing the support we set the disk spinning with a large angular velocity around the axis A. We again remove the support at P. One may then observe a peculiar phenomenon: The disk will not fall as in the case described above, instead the axis A will remain almost horizontal initiating a slow rotation (one says the gyroscope is precessing) around the z-axis. On the figure the disk will precess around the z-axis in the positive direction (the disk goes into the paper).

We wish to show that this behaviour is in accordance with the angular momentum theorem. The disk D of radius r is rotating with a large angular

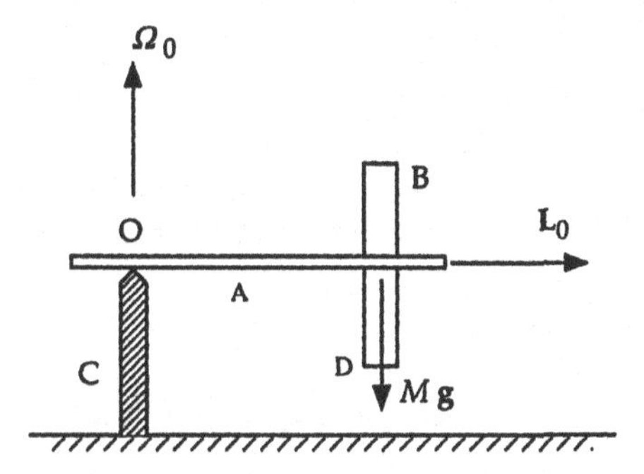

velocity ω_0 about the axis A. The rotation vector ω_0 is directed such that the rotation carries the point B out of the page. The distance from O to the center of mass of D is l. We assume that we start the gyroscope off in such a way that it carries out a so-called *regular precession.* We now show that such a motion is in accordance with the angular momentum theorem:

$$\frac{d\mathbf{L}}{dt} = \mathbf{N}_0 .$$

Note. We do not integrate this equation. We simply verify that regular precession is a possible motion.

The magnitude of the torque of gravity about O is

$$N = lMg ,$$

where $\mathbf{N}_0$ is directed into the plane of the paper.

When the gyroscope is in precession, the axis A continuously changes its direction and this motion carries the center of mass for the disk D around in a circular motion about the point O. This motion will add to the angular momentum about O. If, however, the angular velocity of the disk is quite large, the component of the angular momentum along A will be much greater than the component due to the precession about O. In that case we can assume as an approximation that $\mathbf{L}_0$ is parallel to A. The figure below shows the precessing gyroscope as seen from above. The end point of $\mathbf{L}_0$ moves in

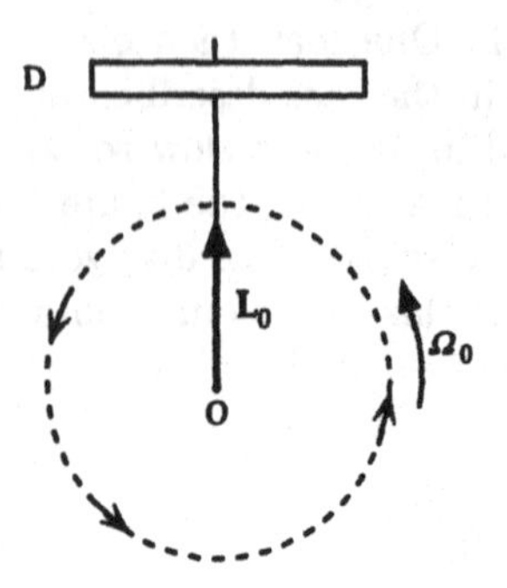

a uniform circular motion about O. The angular velocity in this motion is denoted Ω_0, and we wish to determine Ω_0 in accordance with the angular momentum theorem.

For the magnitude of $\mathbf{L}_0$ we have

$$L_0 = |\mathbf{L}_0| = I\omega_0 = \frac{1}{2}Mr^2\omega_0 ,$$

and consequently the magnitude of $\mathrm{d}\mathbf{L}_0/\mathrm{d}t$ is

$$\left|\frac{\mathrm{d}\mathbf{L}_0}{\mathrm{d}t}\right| = |\mathbf{L}_0|\,\Omega_0 = \frac{1}{2}Mr^2\omega_0\Omega_0 .$$

The angular momentum theorem demands that

$$\frac{1}{2}Mr^2\omega_0\Omega_0 = Mgl ,$$

$$\Omega_0 = \frac{2gl}{r^2\omega_0} .$$

The center of mass of the disk D moves in uniform circular motion about the point O. The centripetal force necessary to maintain this motion is the reaction force from the axis C. Furthermore, the center of mass has no acceleration in the vertical direction. From these two results we can determine the total reaction force from the axis C (see the figure below).

For the reaction force from C one gets

$$F_1 = Ml\Omega_0^2 = \frac{4Mg^2l^3}{r^4\omega_0^2}$$

$$F_2 = Mg$$

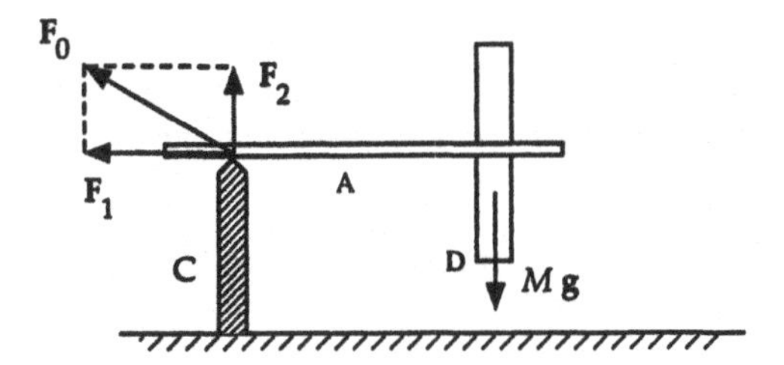

As a numerical example let us assume that we have a rapidly spinning top, e.g., $\omega_0 = 200\ \mathrm{s}^{-1}$. Furthermore: $l = 0.30$ m, $r = 0.15$ m, $M = 2$ kg. One then finds

$$\begin{aligned} \Omega_0 &= 1.3\ \mathrm{s}^{-1}\ (\text{note: } \Omega_0 \ll \omega_0\ ), \\ F_1 &= Ml\Omega_0^2 = 1\ \mathrm{N}, \\ F_2 &= Mg = 19.6\ \mathrm{N}. \end{aligned}$$

△

13.2 The Inertia Tensor

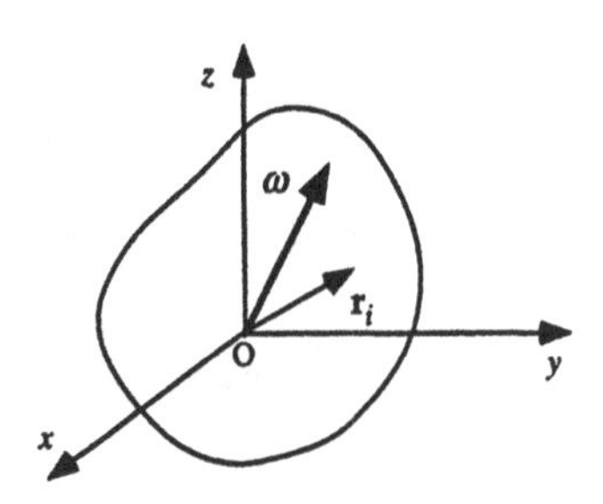

We shall now examine the connection between the general state of motion of a rigid body, and the quantities **L**, $\boldsymbol{\omega}$, and the kinetic energy T – all measured relative to an inertial frame. We shall restrict ourselves to the cases where one point O of the body is fixed in the inertial frame, so that the instantaneous motion of the body can be regarded as pure rotation around an axis through O.

This assumption is not excessively restrictive since the general motion of the body relative to an inertial frame can always be decomposed into translation of the point O and a rotation about an axis through O.

If the decomposition of the motion is performed such that the translation represents the motion of the center of mass, the rotation will represent the motion of the body in the center of mass frame.

We thus consider a rigid body moving in such a way that the point O is fixed in an inertial frame. The instantaneous motion of the body is characterised by the rotation vector $\boldsymbol{\omega}$. The total angular momentum with respect to O is

$$\mathbf{L}_O = \sum_i \mathbf{r}_i \times m_i \mathbf{v}_i = \sum_i \mathbf{r}_i \times m_i (\boldsymbol{\omega} \times \mathbf{r}_i) . \tag{13.5}$$

For the vector product of three vectors we have, in general,

$$\mathbf{A} \times (\mathbf{B} \times \mathbf{C}) = \mathbf{B} (\mathbf{A} \cdot \mathbf{C}) - \mathbf{C} (\mathbf{A} \cdot \mathbf{B}) .$$

Using this expression in (13.5) we obtain

$$\mathbf{L}_O = \sum_i \left[r_i^2 \boldsymbol{\omega} - \mathbf{r}_i (\mathbf{r}_i \cdot \boldsymbol{\omega}) \right] m_i . \tag{13.6}$$

Note: $\mathbf{L}_O$ is not necessarily parallel to $\boldsymbol{\omega}$. Using $\mathbf{r}_i = (x_i, y_i, z_i)$ and $\boldsymbol{\omega} = (\omega_x, \omega_y, \omega_z)$ we may write the component L_x along the x-axis

$$L_x = \sum_i \left[r_i^2 \omega_x - x_i (x_i \omega_x + y_i \omega_y + z_i \omega_z) \right] m_i ,$$

or, with $r_i^2 = x_i^2 + y_i^2 + z_i^2$,

$$L_x = \sum_i m_i (y_i^2 + z_i^2) \omega_x - \sum_i m_i x_i y_i \omega_y - \sum_i m_i x_i z_i \omega_z . \tag{13.7}$$

Defining

$$\begin{aligned} I_{xx} &= \sum_i m_i (y_i^2 + z_i^2) , \\ I_{xy} &= -\sum_i m_i x_i y_i , \\ I_{xz} &= -\sum_i m_i x_i z_i , \end{aligned}$$

we write (13.7) as follows:

$$L_x = I_{xx} \omega_x + I_{xy} \omega_y + I_{xz} \omega_z , \tag{13.8}$$

i.e., L_x depends linearly on all three components of $\boldsymbol{\omega}$.

Using the same procedure to write the components of $\mathbf{L}$ along the y and z axes we obtain the following important connection between the vectors $\mathbf{L} = (L_x, L_y, L_z)$ and $\boldsymbol{\omega} = (\omega_x, \omega_y, \omega_z)$:

$$\begin{aligned} L_x &= I_{xx} \omega_x + I_{xy} \omega_y + I_{xz} \omega_z , \\ L_y &= I_{yx} \omega_x + I_{yy} \omega_y + I_{yz} \omega_z , \\ L_z &= I_{zx} \omega_x + I_{zy} \omega_y + I_{zz} \omega_z . \end{aligned} \tag{13.9}$$

For instance:

$$I_{yy} = \sum_i m_i \left(x_i^2 + z_i^2\right) \ , \text{ moment of inertia around the } y\text{-axis},$$

$$I_{yx} = -\sum_i m_i y_i x_i \, , \text{ product of inertia. Note that } I_{yx} = I_{xy} .$$

A rigid body is composed of a continuous distribution of mass. If, instead of summation, we had used integration, the equations would have taken the following form:

$$\mathbf{L}_0 = \int \mathbf{r} \times (\boldsymbol{\omega} \times \mathbf{r}) \, \mathrm{d}m = \int \mathbf{r} \times (\boldsymbol{\omega} \times \mathbf{r}) \, \rho \mathrm{d}V \, ,$$

where ρ is the density.

$$\begin{aligned} \mathbf{L}_0 &= \int \left[r^2 \boldsymbol{\omega} - \mathbf{r}\,(\mathbf{r} \cdot \boldsymbol{\omega})\right] \rho \mathrm{d}V \, , \\ I_{xx} &= \int \left(y^2 + z^2\right) \rho \mathrm{d}V \, , \\ I_{xy} &= -\int xy \; \rho \mathrm{d}V \, , \end{aligned}$$

etc.

Consider again (13.9). From the definitions of the products of inertia: $I_{ij} = I_{ji}$. Of the nine quantities $I_{xx}, I_{xy}, \ldots$ only six are independent.

The nine quantities I_{ij} are the components of a (symmetric) tensor of rank 2. This particular tensor is called the tensor of inertia.

This description has the same geometric content as the following: The three quantities (L_x, L_y, L_z) are the components of a vector ($\mathbf{L}$).

The prescription for calculating the nine (six) components of the tensor of inertia are given above. The connection between $\mathbf{L}$ and $\boldsymbol{\omega}$ may – in the coordinate system used – be written by means of the matrix equation:

$$\begin{bmatrix} L_x \\ L_y \\ L_z \end{bmatrix} = \begin{bmatrix} I_{xx} & I_{xy} & I_{xz} \\ I_{yx} & I_{yy} & I_{yz} \\ I_{zx} & I_{zy} & I_{zz} \end{bmatrix} \begin{bmatrix} \omega_x \\ \omega_y \\ \omega_z \end{bmatrix} . \tag{13.10}$$

The nine (six) numbers in the above matrix represents – in the given coordinate system – a physical quantity, which exists in its own right, independent of the specific coordinate system chosen to represent the tensor (i.e, the physical quantity). The nine (six) numbers describe the instantaneous inertial properties of the body in rotation around a fixed point 0 $[\boldsymbol{\omega} = \boldsymbol{\omega}(t)]$. Such inertial properties exist of course independently of any particular coordinate system we choose to describe the inertial properties.

For a given coordinate system and a given orientation of the rigid body being studied there is a clear prescription for calculating the nine (six) components of the matrix representing the tensor in that particular coordinate system.

When we go from one coordinate system to another, the physical quantity called the inertia tensor will not change, but the components of the matrix representing the tensor will change. Compare with the three components representing a given vector in a chosen coordinate system.

Equation (13.10) may be written:

$$\mathbf{L} = \mathsf{I}\boldsymbol{\omega}$$

where I is a *linear operator* connecting two physically different vectors $\mathbf{L}$ and $\boldsymbol{\omega}$. The components of I in a given coordinate system has the dimensions of mass times length squared.

If we write the angular momentum theorem ("the equation of motion") it immediately becomes clear that the problem of motion is complicated:

$$\frac{\mathrm{d}}{\mathrm{d}t}(\mathsf{I}\boldsymbol{\omega}) = \mathbf{N}_0^{\mathrm{ext}} . \tag{13.11}$$

For instance, the x-component of (13.11) is

$$\frac{\mathrm{d}L_x}{\mathrm{d}t} = N_{0x} ,$$

or

$$I_{xx}\dot{\omega}_x + I_{xy}\dot{\omega}_y + I_{xz}\dot{\omega}_z + \dot{I}_{xx}\omega_x + \dot{I}_{xy}\omega_y + \dot{I}_{xz}\omega_z = (N_0)_x . \tag{13.12}$$

As (13.12) shows, generally both the components of the vector $\boldsymbol{\omega}$ and the components of the tensor I are time dependent, and makes the solution of the problem quite difficult.

There exist various methods for avoiding some of the difficulties, and in the following sections we shall demonstrate a few of these methods. First, however, we shall show directly how the idea of the tensor of inertia may be used to solve the problem of the rotating dumbbell of Example 13.1.

Example 13.4. The Dumbbell Revisited. See Example 13.1. We shall calculate here the angular momentum $\mathbf{L}$ for a rotating dumbbell using the tensor of inertia.

The coordinate frame (xyz) is fixed in the laboratory (assumed to be an inertial frame). The z-axis is along the vector $\boldsymbol{\omega}$, while the x-axis is perpendicular to the axis of rotation and in the plane of the paper. The y-axis is then perpendicular to the paper and pointing outwards.

It is assumed that the rod R is in the plane of the paper at the time $t = 0$, i.e., in the xz-plane. The coordinates of particle (1) and particle (2) at any time $t \geq 0$ are then:

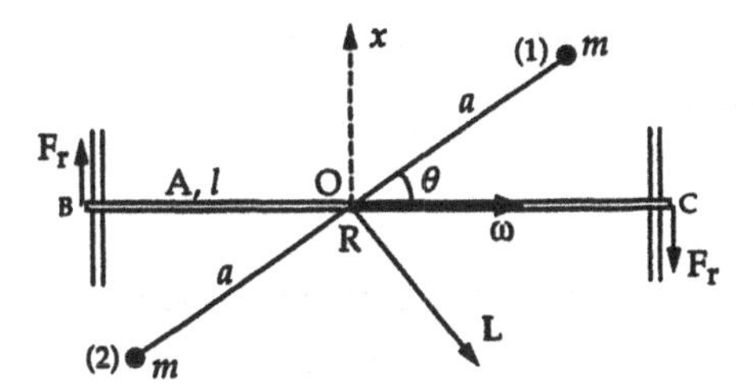

(1)

$$\begin{aligned} x_1 &= a\sin(\theta)\cos(\omega t) \equiv r\cos(\omega t)\,, \\ y_1 &= a\sin(\theta)\sin(\omega t) \equiv r\sin(\omega t)\,, \\ z_1 &= a\cos(\theta) \equiv d\,. \end{aligned}$$

(2)

$$\begin{aligned} x_2 &= -a\sin(\theta)\cos(\omega t) \equiv -r\cos(\omega t)\,, \\ y_2 &= -a\sin(\theta)\sin(\omega t) \equiv -r\sin(\omega t)\,, \\ z_2 &= -a\cos(\theta) \equiv -d\,. \end{aligned}$$

We have used the abbreviations

$$r \equiv a\sin(\theta)\,, \text{ and } d \equiv a\cos(\theta).$$

We now proceed to calculate the components of the tensor of inertia:

$$\begin{aligned} I_{xx} &= m\left(y_1^2 + z_1^2\right) + m(y_2^2 + z_2^2) \\ &= 2m\left(r^2\sin^2(\omega t) + d^2\right)\,, \end{aligned}$$

$$\begin{aligned} I_{xy} &= I_{yx} = -mx_1y_1 - mx_2y_2 \\ &= -2mr^2\sin(\omega t)\cos(\omega t)\,, \end{aligned}$$

$$I_{xz} = -2mdr\cos(\omega t)\,.$$

The process is analogous for the other components. We obtain for the tensor of inertia at the time t:

$$\mathsf{I} = 2m\begin{bmatrix} r^2\sin^2(\omega t) + d^2 & -r^2\sin(\omega t)\cos(\omega t) & -rd\cos(\omega t) \\ -r^2\sin(\omega t)\cos(\omega t) & r^2\cos^2(\omega t) + d^2 & -rd\sin(\omega t) \\ -rd\cos(\omega t) & -rd\sin(\omega t) & r^2 \end{bmatrix}.$$

The components of the matrix representation for the tensor of inertia are thus time dependent. The rotation vector in the (xyz) system has the components

$$\boldsymbol{\omega} = (0, 0, \omega)\,.$$

The angular momentum vector is now determined:

$$\mathbf{L} = \mathsf{I}\boldsymbol{\omega}\,.$$

We find

$$\begin{aligned} L_x &= -2mdr\cos(\omega t)\omega\,, \\ L_y &= -2mdr\sin(\omega t)\omega\,, \\ L_z &= 2mr^2\omega\,. \end{aligned}$$

From the angular momentum theorem applied in the inertial frame (the laboratory frame) we can determine the torque that the bearing has to supply in order to maintain the rotational motion:

$$\begin{aligned} N_x &= \frac{\mathrm{d}L_x}{\mathrm{d}t} = 2mdr\omega^2\sin(\omega t)\,, \\ N_y &= \frac{\mathrm{d}L_x}{\mathrm{d}t} = -2mdr\omega^2\cos(\omega t)\,, \\ N_z &= \frac{\mathrm{d}L_z}{\mathrm{d}t} = 0\,. \end{aligned}$$

Let us compare these with the results from Example 13.1.

The magnitude of the angular momentum vector is

$$\begin{aligned} L &\equiv |\mathbf{L}| = \sqrt{4m^2d^2r^2\omega^2 + 4m^2r^4\omega^2}\,, \\ L &= 2ma^2\omega\sin(\theta)\,. \end{aligned}$$

The magnitude of $\mathrm{d}\mathbf{L}/\mathrm{d}t$ is

$$\left|\frac{\mathrm{d}\mathbf{L}}{\mathrm{d}t}\right| = \sqrt{4m^2d^2r^2\omega^4} = 2m\omega^2a^2\sin(\theta)\cos(\theta)\,.$$

This is exactly what we found in Example 13.1. △

13.3 Euler's Equations

The inertia tensor for a rigid body rotating about a point fixed in an inertial frame can be written in matrix representation:

$$\mathsf{I} = \begin{bmatrix} I_{xx} & I_{xy} & I_{xz} \\ I_{yx} & I_{yy} & I_{yz} \\ I_{zx} & I_{zy} & I_{zz} \end{bmatrix}\,,$$

where I is a symmetric tensor, i.e., $I_{xy} = I_{yx}, I_{xz} = I_{zx}$, and $I_{yz} = I_{zy}$.

In books on geometry it is shown that for any symmetric tensor it is possible to find a coordinate system such that in this coordinate system the matrix representing the tensor has elements only along the diagonal. For the case of the inertia tensor this means that regardless of what shape a given body has it is always possible to find a coordinate system in which the products of inertia are zero. The inertia tensor is then said to be in diagonal form.

A coordinate system in which the tensor of inertia is in diagonal form is called a *principal* coordinate system for the body in question and with respect to the fixed point O. The coordinate axes for the principal coordinate system are called the *principal axes.* The three diagonal elements of the matrix representing the inertia tensor are the so-called *eigenvalues* of the inertia tensor.

If a principal coordinate system has been found, the inertia tensor expressed in that system is in diagonal form:

$$\mathsf{I} = \begin{bmatrix} I_1 & 0 & 0 \\ 0 & I_2 & 0 \\ 0 & 0 & I_3 \end{bmatrix} .$$

The eigenvalues I_1, I_2, and I_3 are the moments of inertia around the three principal axes. The quantities I_1, I_2, and I_3 are called the principal moments of inertia for the body, with respect to the point O.

In a principal coordinate system we have the following connection between **L** and $\boldsymbol{\omega}$:

$$\begin{bmatrix} L_x \\ L_y \\ L_z \end{bmatrix} = \begin{bmatrix} I_1 & 0 & 0 \\ 0 & I_2 & 0 \\ 0 & 0 & I_3 \end{bmatrix} \begin{bmatrix} \omega_x \\ \omega_y \\ \omega_z \end{bmatrix} .$$

In components this is

$$\begin{aligned} L_x &= I_1\omega_x , \\ L_y &= I_2\omega_y , \\ L_z &= I_3\omega_z . \end{aligned}$$

These results show that when the three principal moments of inertia are identical, the vectors **L** and $\boldsymbol{\omega}$ are parallel.

The principal coordinate system reflects the mass-geometric properties of the body. It is evident that a principal coordinate system offers advantages in the calculations. But, the advantages have a price. A principal coordinate system is fixed relative to the rotating body. Therefore a principal coordinate system will in general *not* be an inertial system.

L. Euler (1707–1783) showed that it is possible to re-formulate the equations of motion for a rigid body rotating around a fixed point, in such a way that we obtain certain advantages by using a principal coordinate system to describe the motion of the body.

13.3.1 Derivation of Euler's Equations

Consider a rigid body rotating about the origin O of an inertial frame I (the lab frame). In the figure below, the axes in I have been named X, Y, and Z.

The instantaneous rotation vector is $\boldsymbol{\omega}$. Let us emphasise: the symbol $\boldsymbol{\omega}$ always represents the rotation vector relative to the inertial (the laboratory) frame. The symbol $\mathbf{L}$ is always used to describe the angular momentum about the point O and relative to the inertial frame.

We now introduce a coordinate frame S fixed relative to the rigid body and we choose S in such a way that S is a principal coordinate frame for the body. In the figure the axes of S are labelled x, y and z, and are marked with dashed lines.

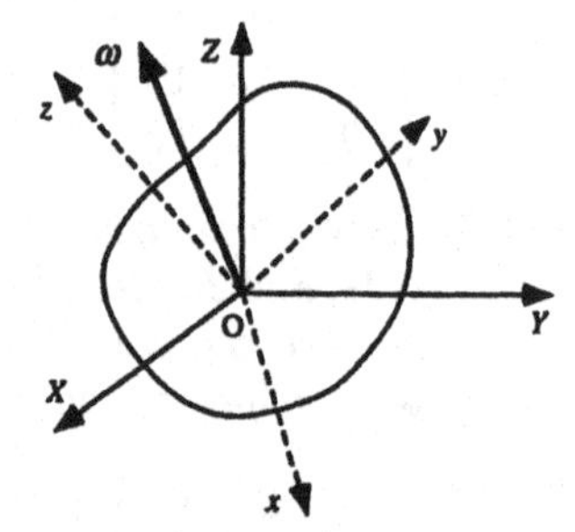

Consider an arbitrary particle (a mass point) which is part of the rotating body. Let the particle have position vector $\mathbf{r}$. Of course the particle will always have the velocity zero measured relative to S. Relative to the laboratory frame I, however, the particle has the velocity $\dot{\mathbf{r}} = \boldsymbol{\omega} \times \mathbf{r}$.

Consider now an arbitrary vector $\mathbf{Q}$. The time derivative of $\mathbf{Q}$ measured relative to the inertial frame XYZ is denoted

$$\left(\frac{\mathrm{d}\mathbf{Q}}{\mathrm{d}t}\right)_{\mathrm{I}} .$$

The time derivative of $\mathbf{Q}$ measured relative to the rotating coordinate frame (the principal frame) S is denoted

$$\left(\frac{\mathrm{d}\mathbf{Q}}{\mathrm{d}t}\right)_{\mathrm{S}} .$$

Suppose $\mathbf{Q}$ is constant in the principal system. This is for instance the case just considered, $\mathbf{Q} = \mathbf{r}$. We find

$$\left(\frac{\mathrm{d}\mathbf{Q}}{\mathrm{d}t}\right)_{\mathrm{I}} = \boldsymbol{\omega} \times \mathbf{Q} .$$

If $\mathbf{Q}$ varies with time, even relative to the principal frame S, we evidently have

$$\left(\frac{d\mathbf{Q}}{dt}\right)_I = \left(\frac{d\mathbf{Q}}{dt}\right)_S + \boldsymbol{\omega} \times \mathbf{Q}.$$

In operator form we thus have the following connection between differentiation with respect to time relative to the two coordinate frames I and S:

$$\left(\frac{d}{dt}\right)_I = \left(\frac{d}{dt}\right)_S + \boldsymbol{\omega} \times$$

Note: for the rotation vector $\boldsymbol{\omega}$ we have

$$\left(\frac{d\boldsymbol{\omega}}{dt}\right)_I = \left(\frac{d\boldsymbol{\omega}}{dt}\right)_S.$$

The same result is true also for position vectors for points on the line containing the instantaneous rotation vector $\boldsymbol{\omega}$.

We are now able to derive the Euler equations for a rigid body rotating about a point fixed in an inertial frame. We shall use the principal coordinate frame S, rotating with the body. In the coordinate frame S the components of the inertia tensor are constant (i.e., independent of time).

The angular momentum theorem relative to the inertial frame I (the laboratory) is

$$\left(\frac{d\mathbf{L}_0}{dt}\right)_I = \mathbf{N}_0.$$

From the operator equation given above the angular momentum theorem can be written in the form

$$\left(\frac{d\mathbf{L}_0}{dt}\right)_S + \boldsymbol{\omega} \times \mathbf{L}_0 = \mathbf{N}_0,$$

where the vectors are taken to be represented by their components in the principal coordinate frame S.

The advantage of this method is that in the coordinate frame S the components of the inertia tensor are independent of time. This means that

$$\left(\frac{d\mathbf{L}_0}{dt}\right)_S = \mathsf{I}\left(\frac{d\boldsymbol{\omega}}{dt}\right)_S \equiv \mathsf{I}\dot{\boldsymbol{\omega}}.$$

The vector equation of motion, relative to the frame S, becomes

$$\mathsf{I}\dot{\boldsymbol{\omega}} + (\boldsymbol{\omega} \times \mathbf{L}) = \mathbf{N},$$

with $\mathbf{L} = (I_x\omega_x, I_y\omega_y, I_z\omega_z)$, where I_x is the moment of inertia about the x-axis in the principal coordinate frame S, and analogously for I_y and I_z.

Written in components along the axis in the principal frame S the equations are

$$\begin{aligned} I_x\dot{\omega}_x - (I_y - I_z)\omega_y\omega_z &= N_x, \\ I_y\dot{\omega}_y - (I_z - I_x)\omega_z\omega_x &= N_y, \\ I_z\dot{\omega}_z - (I_x - I_y)\omega_x\omega_y &= N_z. \end{aligned}$$

In this form the equations of motion are known as the *Euler equations.*

It is important to be completely aware of the following point: Euler's equations are a set of coupled differential equations that describe how the rotation vector moves relative to the principal frame S. The rotation vector describes the instantaneous motion of the body relative to the inertial frame (the lab frame). Determining $\boldsymbol{\omega}(t)$ for a rotating body corresponds to determining $\mathbf{v} = \mathbf{v}(t)$ for the motion of a particle (CM).

At the risk of repeating ourselves, but to avoid misunderstandings:

Euler equations describe the motion of the body relative to the inertial frame I. The vectors that describe the motion in I are represented by their components in the principal (body) frame.

Note that Euler's equations for a homogeneous *sphere* rotating about CM has the simple form:

$$\begin{aligned} I\frac{\mathrm{d}\omega_x}{\mathrm{d}t} &= N_x\,, \\ I\frac{\mathrm{d}\omega_y}{\mathrm{d}t} &= N_y\,, \\ I\frac{\mathrm{d}\omega_z}{\mathrm{d}t} &= N_z\,. \end{aligned}$$

In this very special case the equations have the same form in the inertial frame as in any coordinate frame fixed in the body. In the description of the rotation of a homogeneous sphere there is no need to introduce a coordinate frame fixed relative to the sphere, since the inertia tensor is isotropic ($I_x = I_y = I_z = (2/5)MR^2$) in both cases.

13.4 Kinetic Energy

Sometimes it is useful to use a principal coordinate system when the problem is to calculate the kinetic energy T of a rotating rigid body, measured relative to the laboratory frame. The kinetic energy is:

$$\begin{aligned} T &= \sum_i \frac{1}{2} m_i v_i^2 = \sum_i \frac{1}{2} m_i \mathbf{v}_i \cdot (\boldsymbol{\omega} \times \mathbf{r}_i) \\ &= \frac{1}{2}\boldsymbol{\omega} \cdot \sum_i \mathbf{r}_i \times m_i \mathbf{v}_i \\ &= \frac{1}{2}\boldsymbol{\omega} \cdot \mathbf{L} \\ &= \frac{1}{2}\boldsymbol{\omega} \cdot \mathsf{I}\boldsymbol{\omega}\,. \end{aligned}$$

Vectors and tensors may be represented by their components in an arbitrary coordinate system.

Let us choose to express both $\boldsymbol{\omega}$ and I in a principal coordinate system:

$$T = \frac{1}{2}\left[\omega_x, \omega_y, \omega_z\right] \begin{bmatrix} I_x & 0 & 0 \\ 0 & I_y & 0 \\ 0 & 0 & I_z \end{bmatrix} \begin{bmatrix} \omega_x \\ \omega_y \\ \omega_z \end{bmatrix},$$

or

$$T = \frac{1}{2}\left(I_x\omega_x^2 + I_y\omega_y^2 + I_z\omega_z^2\right).$$

Again, T is the kinetic energy measured relative to the inertial system (the laboratory), but expressed through components of $\boldsymbol{\omega}$ and components of I relative to the principal coordinate system (which is fixed in the body).

13.5 Determination of the Principal Coordinate System

There is a systematic way to determine the principal coordinate system for a body with a given point O as its origin. This procedure may be found in books on geometry or in textbooks on analytical mechanics. In many cases, however, it is possible to avoid the rather complicated formal calculations, by making use of the symmetry properties of the bodies.

For example, if the fixed point O is on a symmetry axis of the homogeneous body, then that axis of symmetry is a principal axis for the point O. Suppose the z-axis coincides with the symmetry axis. For the mass density we then know that

$$\rho(x, y, z) = \rho(-x, -y, z).$$

This leads to

$$\begin{aligned} I_{xz} = I_{zx} &= -\int xz\rho \mathrm{d}V = 0, \\ I_{yz} = I_{zy} &= -\int yz\rho \mathrm{d}V = 0. \end{aligned}$$

This shows that the z-axis is principal.

Consider a homogeneous cylinder of radius R. The length of the cylinder is L. Let the fixed point O be the CM of the cylinder (see the sketch below). The symmetry axis (the length axis) is chosen as the z-axis. The other two axes lie in a plane perpendicular to the z-axis, and pass through O. This coordinate system is a principal system:

$$\begin{aligned} I_{xy} &= -\int xy\rho \mathrm{d}V = 0, \\ I_{xz} &= I_{yz} = 0. \end{aligned}$$

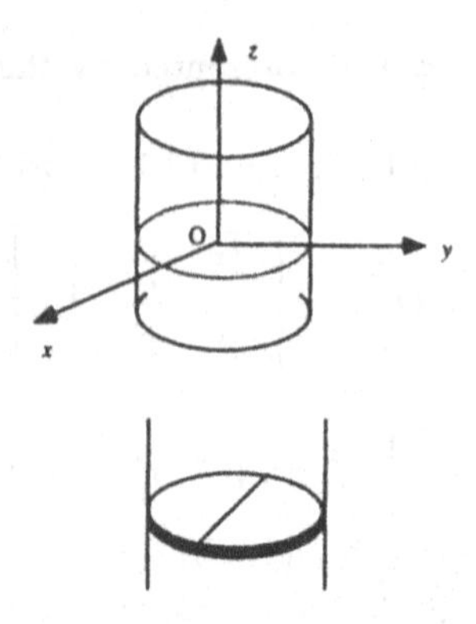

The moment of inertia around the x-axis is obtained as follows. Consider a thin disk as suggested on the sketch. The moment of inertia about a diameter of this disk is $(1/4)mR^2$, where m is the mass of the disk. The cylinder may be perceived as a rod made up of many such disks. The total moment of inertia about the x-axis becomes

$$I_{xx} = \frac{1}{4}MR^2 + \frac{1}{12}ML^2 \,.$$

We have used the parallel axis theorem for calculating the moment of inertia about the x-axis. Furthermore,

$$\begin{aligned} I_{yy} &= \frac{1}{4}MR^2 + \frac{1}{12}ML^2 \,, \\ I_{zz} &= \frac{1}{2}MR^2 \,. \end{aligned}$$

The tensor of inertia is diagonal with the three given principal moments of inertia. Note $I_{xx} = I_{yy}$; we may choose the principal directions as any two mutually perpendicular directions in the plane that is perpendicular to the z-axis through O.

Example 13.5. Rotating Dumbbell

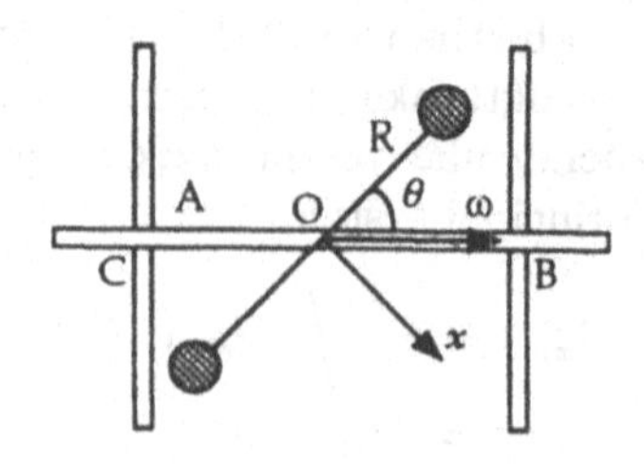

We return to the study of a dumbbell rotating about a fixed axis A (see Examples 13.1 and 13.4.). The system now consists of two homogeneous spheres, both of mass m and radius r. This means that we have to take into account the finite size of the spheres. The spheres are connected by a massless rod R. The distance between the spheres is $2a$. The length of the axis A is $2l$, and the axis rotates without friction in the bearings B and C. The angular velocity is ω.

The motion is thus given, and the goal is to determine the torques necessary to sustain the prescribed motion. In this example we shall solve the problem using a principal coordinate system. The y-axis coincides with the rod R and has its origin in the CM. The x-axis is perpendicular to R and is in the plane spanned by R and ω. In the instantaneous picture displayed in the above figure, the z-axis is pointing out of the paper. The described coordinate system is fixed relative to, and rotates together with, the body. The x-axis describes a cone with opening angle $90° - \theta$, the y-axis describes a cone with opening angle θ, and the z-axis describes a plane perpendicular to ω (i.e., a cone with opening angle 90°).

We express the inertia tensor I by its components in the chosen principal coordinate system:

$$\begin{aligned} I_x &= 2ma^2 + 2 \times \frac{2}{5}mr^2 , \\ I_y &= 2 \times \frac{2}{5}mr^2 , \\ I_z &= 2ma^2 + 2 \times \frac{2}{5}mr^2 . \end{aligned}$$

All the products of inertia are zero. The rotation vector ω, has the following components in the principal frame: $\omega = (\omega \sin\theta, \omega\cos\theta, 0)$. The components of the inertia tensor are of course time independent in the principal frame. In the specific case we consider here, the components of the rotation vector ω are also time independent in the principal frame.

The components in the principal frame of the angular momentum vector become

$$\begin{bmatrix} L_x \\ L_y \\ L_z \end{bmatrix} = \begin{bmatrix} 2ma^2 + \frac{4}{5}mr^2 & 0 & 0 \\ 0 & \frac{4}{5}mr^2 & 0 \\ 0 & 0 & 2ma^2 + \frac{4}{5}mr^2 \end{bmatrix} \begin{bmatrix} \omega\sin\theta \\ \omega\cos\theta \\ 0 \end{bmatrix} ,$$

or

$$\begin{aligned} L_x &= \left(2ma^2 + \frac{4}{5}mr^2\right) \omega \sin\theta , \\ L_y &= \left(\frac{4}{5}mr^2\right) \omega\cos\theta \\ L_z &= 0 . \end{aligned}$$

Note: **L** is not parallel to ω. The angular momentum vector **L** is constant in the principal system, i.e., **L** rotates with angular frequency ω relative to the laboratory frame. By employing Euler's equations we can determine the torques necessary to maintain the prescribed motion:

$$\begin{aligned} I_x\dot{\omega}_x - (I_y - I_z)\,\omega_y\omega_z &= N_x\,, \\ I_y\dot{\omega}_y - (I_z - I_x)\,\omega_z\omega_x &= N_y\,, \\ I_z\dot{\omega}_z - (I_x - I_y)\,\omega_x\omega_y &= N_z\,. \end{aligned}$$

Remember, the Euler equations describe how the rotation vector wanders relative to the principal frame, which is at rest relative to the body. In our case, ω is time independent and $\omega_z = 0$. Therefore, $N_x = N_y = 0$ and $N_z = -2ma^2\omega^2 \sin\theta\cos\theta$.

The torque derives from the reaction forces in the bearings. Note: N_z always points in the negative z-direction, and the torque thus turns along with the principal coordinate system.

Note: the CM of the system (located in the point O) remains at rest. Describe all the external forces acting on the system. △

Example 13.6. Flywheel. Consider again a flywheel mounted out of alignment (see Example 13.2). The flywheel is assumed to be rotating without friction and the goal is to determine the torques necessary to maintain the motion.

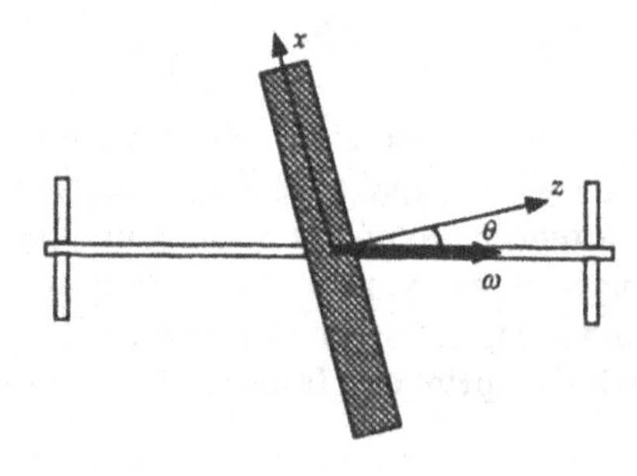

We want to determine the angular momentum of the wheel relative to the laboratory frame. We represent the angular momentum vector by its components in a principal coordinate frame. The z-axis is always the symmetry axis of the wheel (a disk). The origin is in the center of the wheel (see the figure). The other two axes lie in the plane of the disk, e.g., in such a way that the x-axis in the instant depicted is in the plane of the paper and pointing upwards. The x-axis is in the plane spanned by ω and the z-axis. The y-axis is then pointing out of the plane of the paper. Make sure that you understand how this coordinate frame, at rest relative to the moving wheel, moves relative to the laboratory frame.

The components of the tensor of inertia in the principal system are

$$I_x = \frac{1}{4}MR^2, \quad I_y = \frac{1}{4}MR^2, \quad I_z = \frac{1}{2}MR^2.$$

The products of inertia are all zero in this frame.

The components of the angular velocity are $\boldsymbol{\omega} = (-\omega\sin\theta, 0, \omega\cos\theta)$. We obtain:

$$\begin{bmatrix} L_x \\ L_y \\ L_z \end{bmatrix} = \begin{bmatrix} \frac{1}{4}MR^2 & 0 & \\ 0 & \frac{1}{4}MR^2 & 0 \\ 0 & 0 & \frac{1}{2}MR^2 \end{bmatrix} \begin{bmatrix} -\omega\sin\theta \\ 0 \\ \omega\cos\theta \end{bmatrix}.$$

The vectors $\mathbf{L}$ and $\boldsymbol{\omega}$ do not have the same direction. The angle between $\mathbf{L}$ and $\boldsymbol{\omega}$ is

$$\cos(\alpha) = \frac{\mathbf{L}\cdot\boldsymbol{\omega}}{|\mathbf{L}||\boldsymbol{\omega}|} = \frac{1+\cos^2\theta}{\sqrt{1+3\cos^2\theta}}.$$

Question. Show by means of Euler's equations that the torque necessary to maintain the prescribed motion of the wheel is

$$\begin{aligned} N_x &= 0, \\ N_y &= \frac{1}{4}MR^2\omega^2\cos\theta\sin\theta, \\ N_z &= 0. \end{aligned}$$

△

Example 13.7. The Gyroscope. In Example 13.3 we discussed the regular precession of a simple gyroscope: a rotating circular disk of radius r at the end of a light rod.

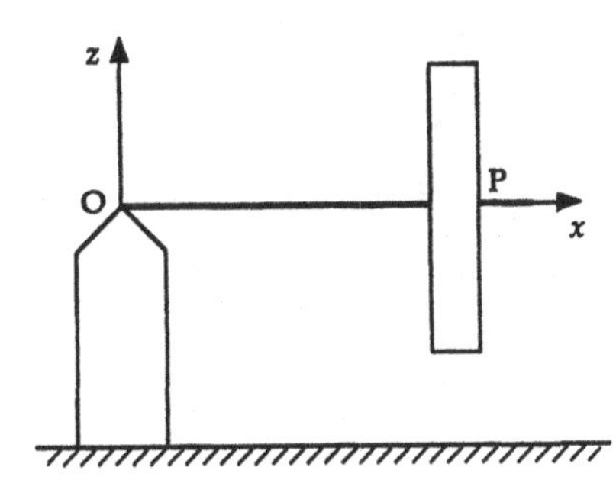

Assume that the disk is rotating with a high angular velocity ω_0. We say that the disk has a high spin. Assume furthermore that the point P is supported by a hand. By means of the hand we now lead P around in a uniform circular motion about O (into the paper). The angular velocity in this circular motion is chosen to be $\Omega_0 = (2gl)/(r^2\omega_0)$.

If P is led with this angular velocity we may remove the supporting hand, and the gyroscope will continue with the regular precession. The initial conditions are just right for the regular precession. We have, so to speak, by means of the hand given the gyroscope a vertical component of angular momentum. There are no external torques in this direction, and the vertical component of angular momentum therefore remains constant. The vertical component of the angular momentum that we give the system matches exactly the value required for regular precession under the influence of gravity.

If we start the gyroscope with other initial conditions the motion will be more complicated.

Let us assume that we keep the figure axis horizontal with a high spin (ω) of the disk. Suddenly we release the figure axis, without having given it any initial velocity in the lab frame. The CM of the system then begins to descend, i.e., $d\mathbf{L}/dt$ is directed downwards. The angular momentum theorem with respect to the fixed point O:

$$\frac{d\mathbf{L}}{dt} = \mathbf{N}_O ,$$

must be satisfied at any moment. A downward pointing $d\mathbf{L}/dt$ would demand a torque in O with a component pointing down. Such a torque does not exist. The only forces acting on the gyroscope are gravity and the reaction in O. If the suspension is really at a point (O) the reaction in O cannot have a torque about O.

As soon as we release the figure axis, the gyroscope starts its "fall", with a motion in the opposite direction of the "missing force", i.e., in a direction corresponding to the direction of the regular precession. The endpoint of the figure axis will – as long as we may neglect friction – describe a cycloid.

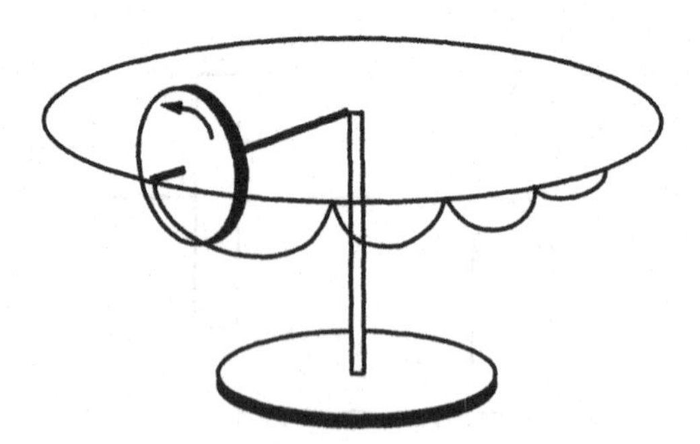

The figure axis starts a complicated "nodding" motion superimposed on a regular precession about O. The "nodding" motion is called *nutation*. If the spin rotation velocity is very large, the nutation will die out quickly due to friction in O. The regular precession remains.

When the nutation motions have been damped out, the figure axis will not be exactly horizontal. The angular momentum from the uniform circular motion will point vertically upwards. Therefore the angular momentum for the spin-motion will point a little downwards, in such a way that the vertical component of the angular momentum in O remains zero.

This component was zero to begin with, and it cannot change as there are no external torques with components in the vertical direction.

We proceed to determine the general expression for the total angular momentum of the gyroscope. The spin velocity is ω. The radius and mass of the disk, respectively, are r and M. The distance from the fixed point O is l. The point O is thus the fixed point around which the rigid body moves (see the sketch below).

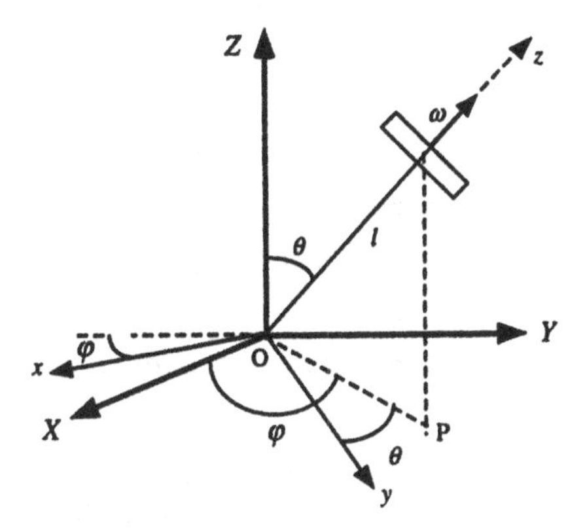

Let the XYZ system be the laboratory frame (an inertial frame). We might have chosen a "genuine" principal coordinate system, rotating with the disk, in such a way that the body was at rest in that frame. We shall, however, choose another coordinate system in which – due to the high degree of symmetry of the body – the tensor of inertia is also diagonal with time-independent components.

The coordinate axis xyz are chosen sò that this coordinate frame follows the gyroscope in its precession, rather than in the spin of the disk. The z-axis lies along the figure axis of the gyroscope. The figure axis forms the angle θ with the vertical, i.e., with the Z-axis.

The x-axis lies in the XY plane (always) and the y-axis is tilted the angle θ beneath the XY-plane. The projection of the CM of the disk on the XY-plane is called P. The angle from the X-axis to OP is called φ, the same angle that the x-axis forms with the negative direction of the Y-axis.

The orientation of the figure axes of the gyroscope is given by the two angles θ and φ. One can determine the placement of the xyz system starting from the XYZ system, as follows:

1: place the z-axis along the Z-axis, the x-axis along the $(-Y)$-axis and the y-axis along the X-axis;
2: turn the xyz system the angle φ about the Z-axis and then an angle θ about the new x-axis.

We shall now determine (1) the components of the total rotation vector $\boldsymbol{\Omega}$ and (2) the components of the tensor of inertia, with all components in the xyz-system.

First, $\boldsymbol{\Omega}$, the total rotation vector. The disk rotates with an angular rotation vector $\boldsymbol{\omega}$ about the z-axis. The total rotation vector also includes, however, contributions from θ and φ.

Let $\mathbf{i}$, $\mathbf{j}$, and $\mathbf{k}$ be unit vectors along the axes of the xyz system. A unit vector along the Z-axis is named $\mathbf{e}_z$. We then have

$$\boldsymbol{\Omega} = -\dot{\theta}\mathbf{i} + \dot{\varphi}\mathbf{e}_z + \omega\mathbf{k} .$$

For $\mathbf{e}_z$ we have

$$\mathbf{e}_z = -\sin\theta\mathbf{j} + \cos\theta\mathbf{k} .$$

The rotation vector in the xyz system becomes

$$\boldsymbol{\Omega} = -\dot{\theta}\mathbf{i} - \dot{\varphi}\sin\theta\mathbf{j} + (\dot{\varphi}\cos\theta + \omega)\,\mathbf{k} ,$$

or

$$\boldsymbol{\Omega} = \left(-\dot{\theta}, -\dot{\varphi}\sin\theta, \dot{\varphi}\cos\theta + \omega\right) .$$

Note. Even if the coordinate frame used is not fixed relative to the body, the components of the tensor of inertia are time independent. The reason for this is the high degree of symmetry of the body. The disk is at all times parallel to the xy-plane, and the moment of inertia about any diameter of the disk is $(1/4)Mr^2$.

The components of the angular momentum in the xyz coordinate system are now determined:

$$\begin{bmatrix} L_x \\ L_y \\ L_z \end{bmatrix} = \begin{bmatrix} \frac{1}{4}Mr^2 + Ml^2 & 0 & 0 \\ 0 & \frac{1}{4}Mr^2 + Ml^2 & 0 \\ 0 & 0 & \frac{1}{2}Mr^2 \end{bmatrix} \begin{bmatrix} -\dot{\theta} \\ -\dot{\varphi}\sin\theta \\ \dot{\varphi}\cos\theta + \omega \end{bmatrix} ,$$

or

$$\begin{aligned} L_x &= \left(\frac{1}{4}Mr^2 + Ml^2\right)(-\dot{\theta}) , \\ L_y &= \left(\frac{1}{4}Mr^2 + Ml^2\right)(-\dot{\varphi}\sin\theta) , \\ L_z &= \left(\frac{1}{2}Mr^2\right)(\dot{\varphi}\cos\theta + \omega) . \end{aligned}$$

We cannot use the angular momentum theorem in its simple form in the xyz frame because this frame is not inertial. We cannot use the Euler equations directly, because the xyz frame is not fixed relative to the rotating body, and the xyz system is thus not a "genuine" principal coordinate frame.

The problem of the general treatment of the motion of the gyroscope is fairly complicated. For a thorough treatment of the subject you should consult the book by Goldstein (see the literature list).

Here we shall limit ourselves to pointing out that the expressions for $\boldsymbol{\Omega}$ and $\mathbf{L}$ that we have derived contain as a special case the regular precession. We shall study a motion where θ is constant, but not necessarily equal to $\pi/2$.

There are no torques with components along the Z-axis; thus $\dot{\varphi}$ is constant. There are no torques with components along the z-axis; thus ω is constant.

Since θ is assumed constant, i.e., $\dot{\theta} = 0$, there is no nutation in the motion. Even with this assumption we see from the general expression for $\mathbf{L}$ that the expression for the angular momentum is still complicated:

$$\mathbf{L} = \left(\frac{1}{4}MR^2 + Ml^2\right)(-\dot{\varphi}\sin\theta\mathbf{j}) + \frac{1}{2}Mr^2\left(\dot{\varphi}\cos\theta + \omega\right)(\mathbf{k}) .$$

We now make the assumption that $\dot{\varphi} \ll \omega$, i.e., that the spin-rotation of the disk is very much faster that the precessional motion.

With these - very simplifying - assumptions, the angular momentum becomes

$$\mathbf{L} \approx \frac{1}{2}Mr^2(\omega\mathbf{k}) .$$

The angular velocity in the precession is

$$\boldsymbol{\Omega} = \dot{\varphi}\mathbf{e}_z .$$

Here $\mathbf{L}$ is a constant in the precessing coordinate frame. We evaluate the time derivative of $\mathbf{L}$, measured in the inertial frame:

$$\begin{aligned} \frac{\mathrm{d}\mathbf{L}}{\mathrm{d}t} &= \boldsymbol{\Omega} \times \mathbf{L} = \dot{\varphi}\mathbf{e}_z \times \frac{1}{2}Mr^2\omega\mathbf{k} \\ &= \frac{1}{2}Mr^2\omega\dot{\varphi}\sin\theta(-\mathbf{i}) . \end{aligned}$$

Gravity, acting on the CM, provides the torque around O necessary to sustain the precessional motion. This torque is likewise pointing in the direction of the negative x-axis. The equation of motion in the inertial frame becomes

$$\frac{1}{2}Mr^2\omega\dot{\varphi}\sin\theta = Mgl\sin\theta .$$

The angular velocity in the precession is

$$\dot{\varphi} = \frac{2Mgl}{Mr^2\omega} = \frac{2gl}{r^2\omega} .$$

The same results are obtained in Example 13.3. The angular precession velocity is independent of θ.

The general treatment of the gyroscope is of significance both in theoretical physics and in the applied sciences. Electrons and atomic nuclei with spin and charge perform precessional motion in an external magnetic field. This is the basis for important techniques in the study of the structure of matter. Inertial navigation is also based on a detailed understanding of the motion of gyroscopes. △

Example 13.8. Gyroscope Supported at the Center of Mass. A gyroscope is provided with a counterweight W such that the gyroscope is supported in its center of mass. The disk D with the mass M and radius r is spinning around the axis A with a large angular velocity . Since the gyroscope is supported in its center of mass there is no external torque on the system and the gyroscope will not precess.

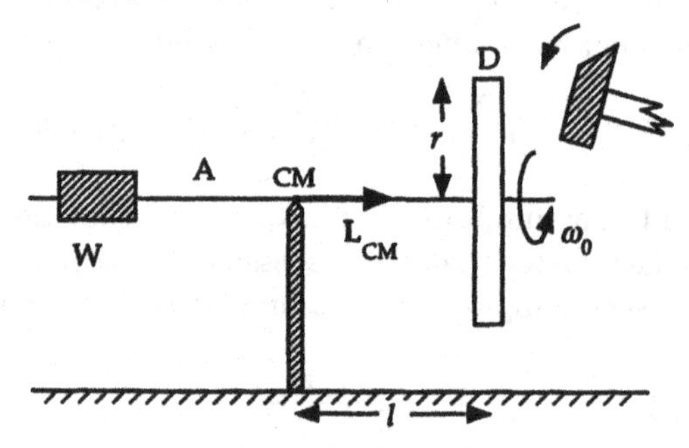

We now give the axis a tap (a light blow) with a hammer, at distance l from the CM. The tap may be described as an impulse $\mathbf{P} = \int_0^{\Delta t} \mathbf{F}(t)dt$. We assume that the force $\mathbf{F}(t)$ that the hammer exerts on A during this blow is parallel to the gravitational field.

What effect does the tap have on the gyroscope axis A? First, note the following. If the spin of the gyroscope is very high, the axis A (the figure axis) will hardly move during the blow and the small motion observed will be perpendicular to the direction of the blow. On the sketch above the motion will be into the paper.

In reality the axis A moves slightly. downwards in the direction of the blow. The axis moves on a circular cone the axis of which is nearly parallel to the axis shown in the sketch, but points slightly into the paper (see sketch). If no friction was present in the support (CM) the axis A would continue to describe a cone with a small opening angle. Due to friction in the bearings, however, the opening angle will become smaller and smaller and after a few revolutions the figure axis A will come to rest along the symmetry axis of the cone. The figure axis has thus experienced a slight net turn into the paper, i.e., in the direction perpendicular to the direction of the blow. This is the permanent effect of a light blow on a supported gyroscope with a high value of the spin rotation.

During the blow the gyroscope is subject to an external torque $\mathbf{N}_{\mathrm{CM}} = l\mathbf{e} \times \mathbf{F}(t)$ with respect to the point of support. Here $\mathbf{e}$ is a unit vector along A. $\mathbf{N}_{\mathrm{CM}}(t)$ is horizontal and pointing into the paper. During the blow we thus have

$$\frac{\mathrm{d}\mathbf{L}_{\mathrm{CM}}}{\mathrm{d}t} = \mathbf{N}_{\mathrm{CM}}(t) .$$

For the change $\Delta\mathbf{L}_{\mathrm{CM}}$ during the tap we find

$$\Delta\mathbf{L}_{\mathrm{CM}} = \int_0^{\Delta t} \mathbf{N}_{\mathrm{CM}}(t)\mathrm{d}t .$$

As $\mathbf{N}_{\mathrm{CM}}$ points into the paper, $\Delta\mathbf{L}_{\mathrm{CM}}$ is in this direction too. The value of $\Delta\mathbf{L}_{\mathrm{CM}}$ is

$$|\Delta\mathbf{L}_{\mathrm{CM}}| = \int_0^{\Delta t} l\,|\mathbf{F}(t)|\,\mathrm{d}t = Pl .$$

The axis moves perpendicular to the direction of the tap.

Note. A gyroscope mounted in such a way that it can turn about its CM is useful as a directional stabiliser. This is not because it does not turn when a torque is applied to it, but because it stops turning when the torque ceases. If a large nonrotating mass is mounted so that it is free to turn around its center of mass, it would acquire and maintain a small angular velocity if subjected to an external torque of a short duration. △

Example 13.9. The Earth as a Gyroscope

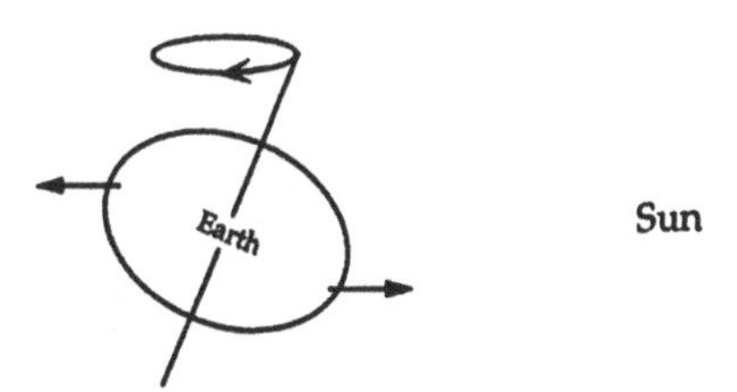

The sketch shows the Earth at the northern summer solstice (≈ 21st of June). In its orbit around the Sun, the Earth is – at the instant shown – on its way out of the plane of the paper. The axis of rotation points towards the North Star. It may be useful to emphasize that the North Star is so far away that no change in the direction of the axis of rotation relative to the North Star can be detected throughout a year.

As a result of the fact that the Earth is an ellipsoid of revolution, combined with the fact that the axis of rotation is tilted relative to the plane of the ecliptica, there is an external torque acting on the Earth. The torque is due to the tidal field

Consider first the torque **N** from the Sun. **N** is perpendicular to the plane of the paper and points out of the paper. Due to the fact that the Earth has an angular momentum relative to the heliocentric reference frame (inertial space!!), the rotation axis (i.e., **L**) will execute a precessional motion in a direction opposite to the direction in which the Earth moves around the Sun. We say that the rotation axis has *retrograde* motion.

The torque from the Sun varies according to the position of the Earth in its orbit. The torque is zero at each equinox (March and September) and maximum at solstice (June, December). The precession of the rotation is then – as mentioned – due to the tidal fields at the position of the Earth. The tides from the Moon is about $2\frac{1}{2}$ times stronger than the tidal fields from the Sun; the Moon therefore has the strongest influence on the precession of the axis of the Earth. Almost 35″, of the yearly precession of 50″, is due to the Moon.

The axis of the Earth describes a cone with an opening angle of $23\frac{1}{2}°$. A complete revolution takes about 25 800 years. Today, the axis of the Earth points towards the North Star. In about 12 000 years the rotation axis will point towards the star Vega in the constellation The Lyre.

The points where the instantaneous axis of rotation cuts the celestial sphère is called the pole of the heavens. That these poles move relative to the fixed stars was known to Hipparchus in about the year 120 BC. The precession of the rotation axis of the Earth was first explained by Newton.

△

13.6 Problems

Problem 13.1.

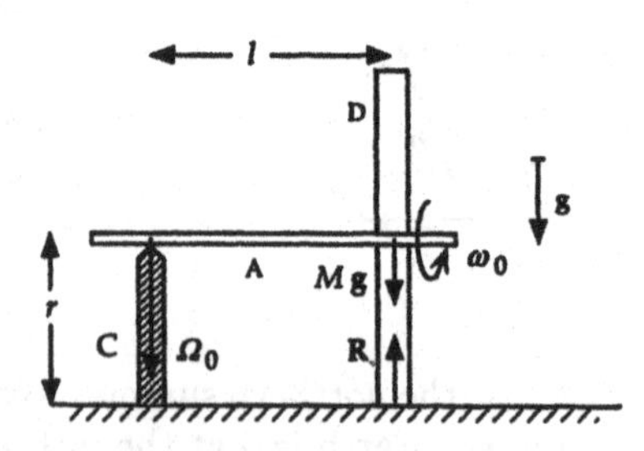

Let us assume that the support C in Example 13.3 is designed in such a way that it has exactly the height r. Let us furthermore assume that the disk D can roll without sliding on the horizontal surface. We still denote by ω_0 the (now much smaller) angular velocity about A. Since the disk rolls without sliding on the horizontal surface, this now determines the "angular velocity of precession" Ω_0, in such a way that $\Omega_0 = r(\omega_0/l)$.

Find the force R by which the surface must act on the disk D for this motion to be maintained.

Problem 13.2. Consider again the gyroscope supported in its CM (see Example 13.8).

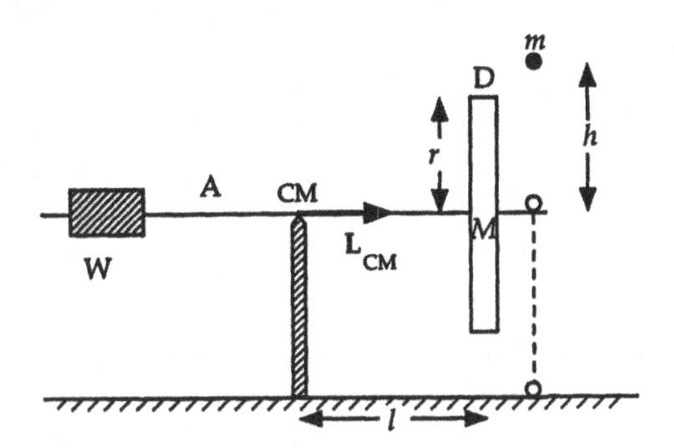

Let us assume that a particle with mass m falls freely through the height h and hits the axis A in a completely inelastic collision, but in such a way that the particle "falls off A when the collision is over".

The particle hits the axis A at a distance l from the CM. We neglect the effects of gravity during the collision.

(1) Find the angle $\Delta\theta$ that the axis A turns (into the paper) if: $M = 2.4$ kg, $l = 0.30$ m, $\omega_0 = 200\ \text{s}^{-1}$, $m = 0.10$ kg, $h = 0.50$ m, $r = 0.10$ m.
(2) What condition should be satisfied by the collision time Δt if it may be considered a good approximation to neglect gravity during the collision?

Problem 13.3.

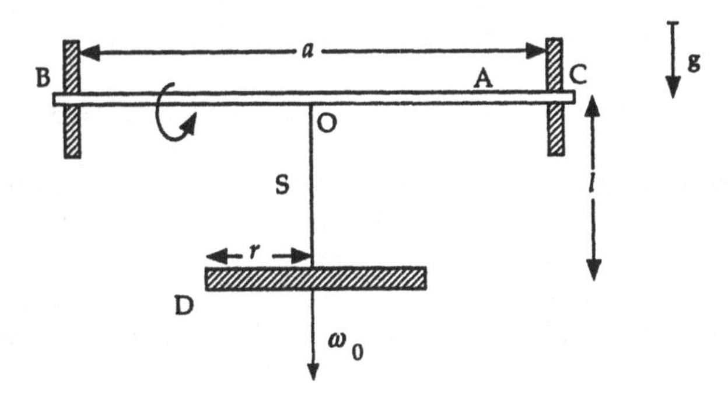

A gyro-wheel D (a disk) is fastened to the end of a stiff rod S. The wheel D is perpendicular to S. The rod S has length l and we neglect the mass of the rod. The wheel D rotates with constant angular velocity ω_0 (see the figure). The wheel D is a homogeneous thin circular disk of mass m and radius r. The rod S goes through the center of D.

The other end of S is fastened to the mid-point O of a thin, rigid horizontal axis A, which can turn without friction in two bearings, B and C, that are fixed in the laboratory. The axle A has length a, and its mass can be ignored in this problem.

The rod S may thus turn in a vertical plane perpendicular to the horizontal axis A. Initially, S is held horizontal, and at rest. At a certain time the rod is released and S begins to turn in the vertical plane through O. The initial angular velocity of the rod is thus zero.

Neglect all frictional effects (in particular, the disk D maintains its angular velocity ω_0).

(1) Find the angular velocity Ω of the rod S at the moment the disk D passes through its lowest position (where S is vertical).
(2) Find the forces $\mathbf{F}_B$ and $\mathbf{F}_C$ by which the axle bearing B and C respectively act on the axle A, when S passes through its lowest position (the angular velocity vector Ω of the rod is in the direction OC) .

Problem 13.4.

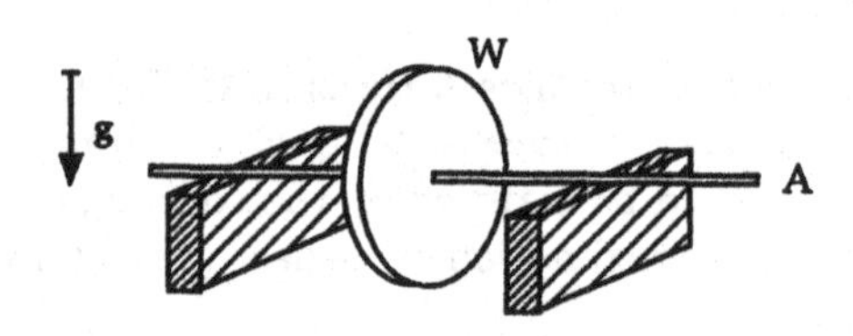

A wheel W is shaped as a homogeneous, flat, circular disk. The wheel is fastened to a thin axle A that passes through the center of the wheel. We ignore the mass of A.

The axle A rests in two bearings in such a way that the system can rotate without friction around a horizontal axis. Note that A is supported only from below! The radius R of the disk is $R = 50$ cm, and the distance between the two bearings is 1 m. The CM of the disk is exactly at the mid-point between the two bearings.

The wheel is fastened to the axle A in such a way that the normal to the disk of the wheel forms an angle of $\theta = 1°$ with the axle A. The acceleration of gravity is $g = 9.8\ \mathrm{m\,s^{-2}}$. The wheel is now set in motion, performing ν revolutions per second.

Determine the maximum value $\nu = \nu_0$ for the rotation frequency if the axle A must not leave the supporting bearings.

14. The Motion of the Planets

We shall now apply Newtonian mechanics to the study of the motion of planets.

Look at the sky on a dark, clear night. Seen from the Earth, the stars and the planets appear to be fastened inside a huge sphere, called the celestial sphere. The celestial sphere appears to rotate, once per day, around an axis passing through the north pole and the south pole of the Earth.

The stars are far away from the Earth. Therefore the stars seem to have fixed positions on the celestial sphere. The stars do not change positions relative to one another over periods of time comparable to a human life time.

The planets, on the other hand, are rather close to us. Even though from one night to the next, a planet appears to be fastened on the celestial sphere, over periods of weeks the line of sight, even to an outer planet, does change noticeably relative to the stars.

The roots of modern science are found in the study of the motion of the planets. Therefore we begin with some remarks of a historical nature.

14.1 Tycho Brahe

On the island of Hven, between Sweden and Denmark, Brahe built, in the years 1576–1597, an astronomical research institution of historical significance. Brahe understood the significance of precise astronomical observations, and he had the means to construct the necessary instruments. Before the time of Brahe, the positions of the heavenly bodies were known with the precision of 10 minutes of arc ($10'$). Brahe improved the precision significantly and reached an accuracy of 1 to 2 minutes of arc, which is close to the limit of precision obtainable with the unaided eye.

The years 1576–1597, in which Brahe performed his measurements, will be remembered as one of the most decisive periods in the history of science, and indeed in the history of man. This is impressive, especially when we recall that the naked-eye observational methods of Brahe were deemed to become obsolete after half a score of years. In 1609, Galileo pointed a telescope towards the sky, and science was changed forever.

Nevertheless Brahe's results, particularly from the observation of the planet Mars, became of fundamental significance. As a kind of afterthought one might add that if the observations of the planets made by Brahe had been even more precise, if these observations had disclosed the "irregularities" in for instance the motion of Mars (irregularities caused by gravitational fields from the other planets), Kepler might not have uncovered his three simple laws for the orbits of the planets. The gravitational law of Newton might have been more difficult to find, and the history of man had changed. Such speculations may be considered entertaining, but are not particularly useful. Seen from the viewpoint of physics the remarks merely illustrate that it is important to find the essential aspects of experimental or observational material.

14.2 Kepler and the Orbit of Mars

Contrary to Brahe, Kepler initially accepted the heliocentric model of the solar system, as described by N. Copernicus. The Sun is at rest in the center of the system, with the planets moving in circular orbits around the Sun.

Originally Kepler looked for a connection between forces and the structure of the solar system. He realized very soon that the periods of the planets increase with the distance of the planet from the Sun. It was Kepler's belief that the increase in the periods of the planets was connected to a force from the Sun, a force that decreased with distance.

Kepler did not succeed in connecting the motions in the solar system to the concept of force. According to historians of science, he had a quite clear understanding of the importance of this task. The title of Kepler's principal book (published in the year 1609) was:

A New Astronomy Based on Causation
or
A Physics of the Sky
Derived from Investigations of the Motions of the Star Mars.
Founded on Observations made by the Nobleman Tycho Brahe.

If Kepler did not succeed in explaining the dynamics of planetary motions, he did succeed in using the empirical material of Brahe in a masterly way. Kepler condensed this material into an elegant form, a form that became decisive for the work of Newton. Through a nearly superhuman effort of calculation, Kepler succeeded in showing that planets move around the Sun, not in circular orbits with the Sun in the center, but in elliptical orbits with the Sun in one focus.

The difficult problem confronting Kepler was to determine the orbit of a planet relative to the Sun, based solely on observed *directions* to the planet, as seen from a rotating Earth, itself following an unknown curve on its way around the Sun. Kepler set out to determine the orbit of Mars.

14.2.1 The Length of a Martian Year

The time needed for a planet to complete one revolution around the Sun is called the *sidereal* period. The sidereal period for a planet cannot be directly observed, but it can be determined as shown below. Assume that the orbits of the Earth (E) and Mars (M) are circles with the Sun (S) in the center. The planet Mars is said to be in opposition, when the Sun, the Earth, and

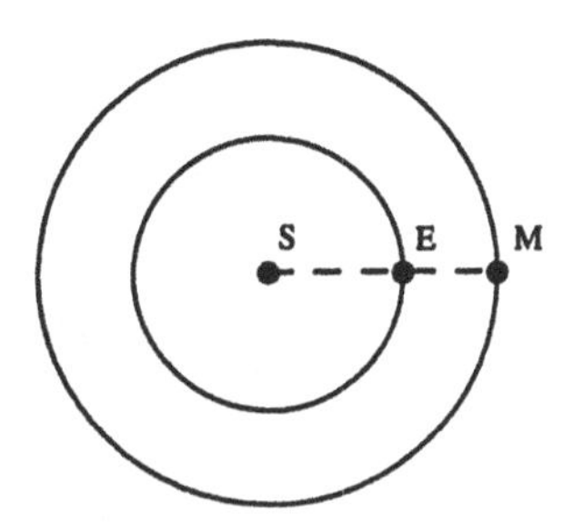

Mars lie on a straight line, and Mars is closer to the Earth than to the Sun (see the figure).

The time between two successive oppositions is called the *synodic* period of the planet. The synodic period can be determined by observation. When the synodic period is known the sidereal period may be calculated as follows.

Let the sidereal period of Mars be T and the synodic period S. The sidereal period of the Earth – i.e., one year – is called A. The quantities T, S, and A are the respective periods measured in days (1 day = 24 h).

Mars is an outer planet relative to Earth. In the time between two successive oppositions, i.e., during one synodic period, Mars moves 360^0 less relative to the Sun than the Earth moves in the same time.

In one sidereal period Mars moves 360° relative to the Sun. In one day Mars moves $360°/T$ relative to the Sun.

In one synodic period Mars moves $(S/T)360°$ relative to the Sun.

During one synodic period for Mars, the Earth moves $(S/A)360°$ relative to the Sun.

The equation to determine T is thus:

$$\frac{S}{T}360° = \frac{S}{A}360° - 360° ,$$

or

$$\frac{1}{T} = \frac{1}{A} - \frac{1}{S},$$

From Brahes measurements Kepler knew that the synodic period for Mars – i.e., the time between two successive oppositions – was S = 779.8 days. Consequently

$$\frac{1}{T} = \frac{1}{365.24} - \frac{1}{779.8} = 0.0014555 \text{ days}^{-1},$$

$$T = 687 \text{ days.}$$

To trace its orbit around the Sun, Mars thus needs 687 days = 1.88 years.

Kepler actually started by determining the orbit of the Earth. To do this he used his knowledge about the sidereal period of Mars (687 days) to identify the dates on which Mars was back in a given point in its orbit.

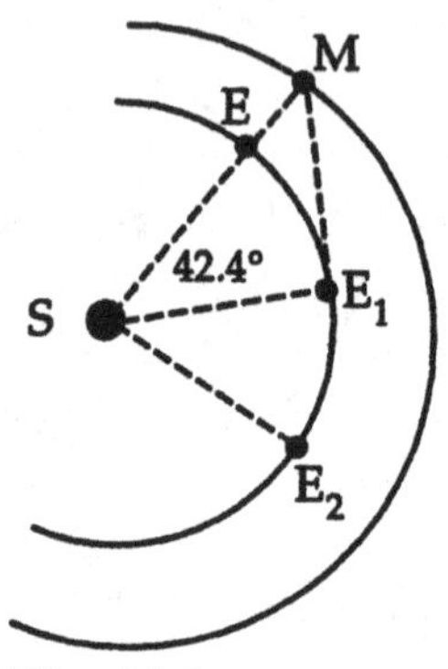

Fig. 14.1.

We shall show only the principles used by Kepler. Consider Figure 14.1. Let the point M mark an opposition of the planet Mars. It will take Mars 687 days to return to the point M in the orbit. During 687 days the Earth has completed 687/365 ≈ 1.88 revolutions around the Sun. The Earth has thus moved 1.882 × 360° = 677.6° in its orbit, or 42.4° less than two complete revolutions. The Earth will then be located in the point E_1, as shown on Figure 14.1, i.e., 42.4° "behind" Mars. After an additional 687 days Mars will again be in the point M, while the Earth will be in the point marked E_2 on Figure 14.1 (assuming circular orbits).

From each successive complete revolutions of the planet Mars, Kepler was able to find one point of the orbit of the Earth.

Brahe had observed Mars for more than 20 years. The observations included ten oppositions of Mars.

Using the method outlined above, Kepler was able to construct the orbit of the Earth relative to the Sun (relative to the heliocentric reference frame!). Kepler found that, within the precision of observation, the orbit of the Earth was a circle, but with the essential feature that the Sun was *not* located at the center of the circle.

By plotting the position of the Earth at various dates, Kepler discovered that the Earth does not move with the same speed all year round. The Earth moves faster when it is closer to the Sun. Here is the beginning of the discovery that later became known as Kepler's second law:

The radius vector from the Sun to the planet sweeps out equal areas in equal amounts of time.

As we have already seen (Chapter 10), this law is a direct consequence of the conservation of angular momentum in a central field of force.

Now Kepler knew the orbit of the Earth. The next problem was – based on the observations of Brahe and on the knowledge of the orbit of the Earth – to find the orbit of Mars relative to the Sun.

14.2.2 The Orbit of the Planet Mars

Kepler utilized the fact that he knew the length of a Martian year (sidereal period, 687 days).

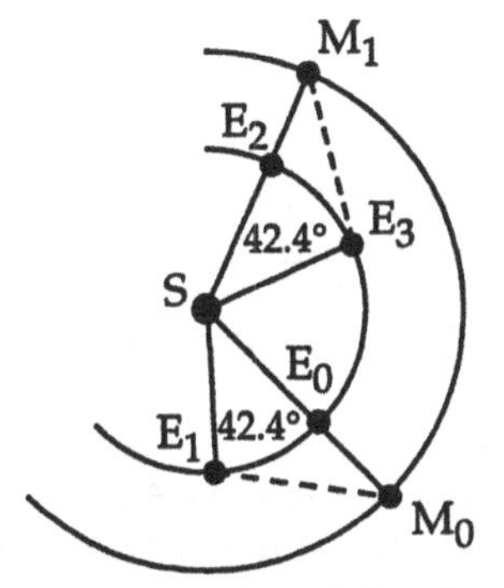

Fig. 14.2. Kepler's determination of the orbit of the planet Mars. (SE_0M_0) is one opposition of Mars and (SE_2M_1) is another

Kepler again used the oppositions of Mars. Consider the opposition marked by the line SE_0M_0 in Figure 14.2. The line of sight to Mars relative to the stars was known from the measurements of Brahe. When Mars, 687 days later, again is in the point marked M_0, the planet is *not* in opposition, because the Earth will now be in the point marked E_1, i.e., 42.4° "behind" Mars. Kepler knew the orbit of the Earth, and the position of the Earth at any given date was now also known. That means: Kepler knew the date on which the Earth was in the point of its orbit marked E_1. The position of Mars, i.e. the line of sight to Mars on the day the Earth was in E_1 could be found from the tables of Brahe. Kepler was able to calculate the angle SE_1M_0. Using the distance from the Sun to the Earth as unit of length, Kepler found one point in the orbit of Mars by triangulation.

Using the ten oppositions studied by Brahe, Kepler was able to determine ten points in the orbit of Mars. Kepler tried to find a circle passing through those ten points. He found for some of the points the now famous deviation of 8′ (minutes of arc) from the circle which gave the best fit (see Chapter 1).

It is interesting to note that before Brahe the positions of the planets were known with an accuracy of 10′. The improvement of the precision of observation to about 2′ was therefore decisive. If – before Brahe – one had tried to fit a circle to the observation points it would have been considered successful.

Kepler had confidence in the data of Brahe, and he took the decisive step. The hypothesis of circular orbits must be rejected. Instead of conserving "the ideal circular orbits" Kepler chose to believe in observations.

The author Stefan Zweig has written a book with the title *Sternstunden der Menschheit*. Here is a *Sternstunde* in our history, far more important than any described by Stefan Zweig. A description of Kepler's achievements cannot be found in the book by Stefan Zweig.

Kepler found that the ten points of the orbit of Mars could be fitted to a curve known to the mathematicians for a long time: an ellipse. Johannes Kepler had thus reached the law that has become known as Kepler's first law:

> The orbit of a planet relative to the Sun lies in a fixed plane containing the Sun, and each planet moves around the Sun in an elliptical orbit with the Sun in one focus.

Through the work of Brahe and Kepler the solar system had disclosed one of its deepest secrets.

The study of the solar system has resulted in several other decisive advances in physics: the law of gravity, the finite velocity of light (the "lingering of light", Ole Rømer), and the rotation of the perihelion of the elliptical orbit of the planet Mercury. Somewhere in the solar system – perhaps on Mars – we may find the key to the greatest riddle of the natural sciences: the origin of life itself.

The reason why it was not possible to fit the points of observation of Mars into a circular orbit around the Sun is the substantial eccentricity ("flatness") of the Martian ellipse. For the precise definition of eccentricity, see below. The eccentricity of the elliptical orbit of Mars is $e = 0.09$, which is five times larger than the eccentricity of the elliptical orbit of the Earth, and more than twelve times the eccentricity of the orbit of Venus. It is, however, important to note that even for Mars the deviation from the circular form is small.

Kepler's third law was published – among several more obscure results – in the year 1619:

> The square of the period of revolution of a planet is proportional to the third power of the greatest semi axis of the ellipse.

Specifically, if T denotes the period and a the major semi axis, $T^2/a^3 = C$, where C is a constant, the same for all planets.

14.2.3 Determination of Absolute Distance in the Solar System

The sidereal period of revolution T_p for a planet may be determined via the observation of the synodic period. From Kepler's third law, the semi-major axis a_p for the planetary orbit (ellipse, see below) may then be found using the astronomical unit as the basic measure of distance. One astronomical unit (1 AU) is defined as the mean value of the distance of the Earth from the Sun. Measuring T_p in years we have

$$\frac{T_{\mathrm{P}}^2}{a_{\mathrm{P}}^3} = \frac{T_{\mathrm{E}}^2}{a_{\mathrm{E}}^3} = \frac{1^2}{1^3} = 1 \,.$$

The absolute distances in the solar system, i.e., distances measured in meters, can be found only when one distance – for instance the distance from Earth to Venus, or from Earth to Mars – has been determined.

The problem has not been simple to solve. Today one can measure the distance say, from the Earth to Venus, with high precision by means of radar signals reflected from the surface of Venus.

Historically the problem was first solved by triangulation. From two points on the Earth, a large distance apart, the direction of the line of sight to a planet is measured. The two directions of the line of sight will then form a certain angle, which is larger the closer the planet is to the Earth. The difficulties with this measurement is obviously the small value of the angle between the lines of sight. The angle between two lines of sight from the Earth to the Moon may be about 1^0. The angle between two lines of sight from the Earth to even the nearest planets will never be more than $1'$ (1 arc minute).

Mars is closest to the Earth when in opposition. In the most favorable oppositions Mars is 0.37 AU from the Earth. Venus may come even closer (0.26 AU). When Mars is in opposition the illuminated half sphere of Mars is facing the Earth. Therefore Mars is easy to observe during an opposition. The orbit of Venus lies within the orbit of the Earth. Therefore when Venus is closest to the Earth, Venus will have its dark side facing the Earth. Venus is therefore impossible to observe when it is closest to Earth unless the planet passes in front of the solar disk. Venus will then be observable as a small dark spot against the large luminous disk of the Sun. This phenomenon is called a transit of Venus. The orbits of the Earth and the orbit of Venus lie nearly in the same plane, but not exactly so. As a rule Venus will bypass the Sun. A Venus transit is a rare phenomenon. They come in pairs. There were two in the 19th century (1874 and 1882), and the next pair is 2004 and 2012.

From the first two of the mentioned Venus transits a triangulation measurement was made. Based on this the AU was estimated to be between 147×10^6 km and 149×10^6 km.

Even better than Venus for the determination of the astronomical unit is the asteroid Eros, which was discovered in 1898. The distance from Earth to

Eros may be as small as 0.15 AU. A favorable opposition of Eros took place in 1930. At this opposition the astronomical unit was determined as 149.7×10^6 km. The present value is 1 AU = 149.598×10^6 km.

Kepler published his laws as unexplained facts. The full dynamical consequences of these laws were recognized by Newton, after he had formulated his general laws of motion. We shall show that gravitational attraction, i.e., Newton's law of gravity is implied by Kepler's laws. After this we shall demonstrate the converse: Kepler's three laws are consequences of Newtonian mechanics and the law of gravity.

Before proceeding, we shall briefly review some results related to the geometry of conic sections.

14.3 Conic Sections

Detailed descriptions of conic sections may be found in books on geometry or calculus. Here we give a rudimentary introduction.

The curves obtained by intersecting a cone with a plane which does not pass through the vertex of the cone, are called conic sections. If the plane

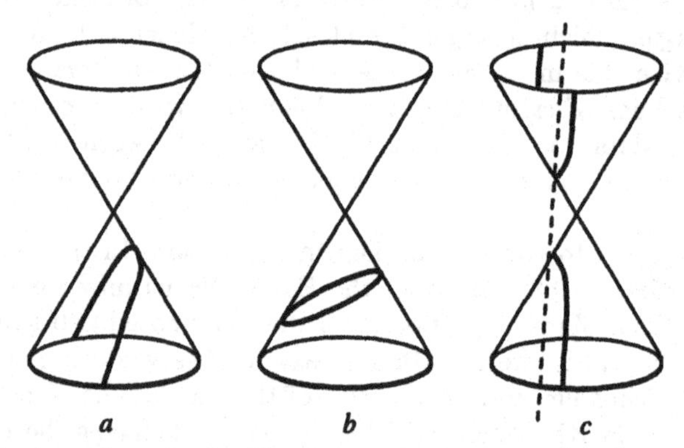

Fig. 14.3. Conic sections obtained by intersecting a cone with a plane: (a) parabola, (b) ellipse, (c) hyperbola

intersecting the cone is parallel to a generator of the cone (Figure 14.3a), the conic section becomes a parabola. Otherwise the curve produced is called an ellipse or a hyperbola, depending on whether the plane intersects one portion of the cone (Figure 14.3b) or both portions (Figure 14.3c). A circle is a special case of an ellipse.

The three types of nondegenerate conic sections may be characterized within the plane, in the following manner. A conic section is the set of all points P for which the distance from a fixed point F (the *focus*) and a fixed line l, has a constant ratio e, called the *eccentricity*:

$$\frac{\mathrm{PF}}{\mathrm{Pl}} = e\,, \tag{14.1}$$

where PF is the distance from P to F and Pl is the distance from P to l. The line l, which is called the directrix, does not pass through F.

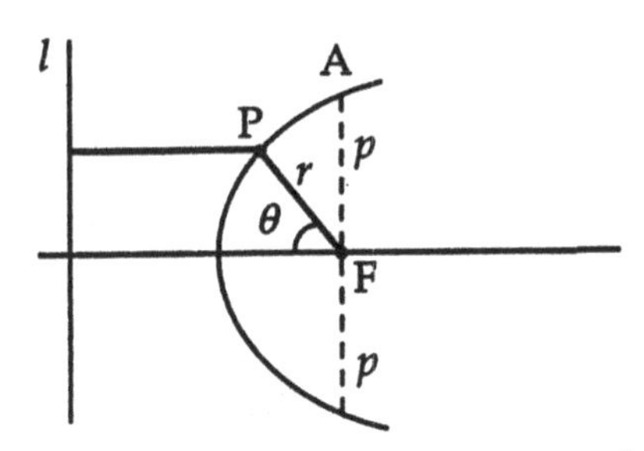

Fig. 14.4. The equation for conic sections using polar coordinates

Let $2p$ be the length of the chord perpendicular to the axis of the conic section and passing through the focus F. By choosing the point P at the endpoint of the chord, e.g., at the point A (see Figure 14.4) the defining equation becomes

$$p = e(\mathrm{Fl})\,. \tag{14.2}$$

The quantity p is called the parameter of the conic section.

Measuring the angle from the symmetry axis as in the figure and using (14.1) and (14.2) we find

$$r = \mathrm{FP} = e(\mathrm{Pl}) = e(\mathrm{Fl} - r\cos\theta)\,,$$

$$r = e\left(\frac{p}{e} - r\cos\theta\right) = p - er\cos\theta\,.$$

Finally we get

$$r = \frac{p}{1 + e\cos\theta}\,. \tag{14.3}$$

The expression (14.3) is the equation for a conic section for all three cases.

We get an ellipse for $0 \le e < 1$, a hyperbola for $1 < e$, and a parabola for $e = 1$. In Figure 14.4 only a part of the curve close to F has been shown. In this way all three cases may be said to be included in the figure.

For the ellipse ($e < 1$) the angle θ may take all values from 0 to 2π. Since $e < 1$ the denominator can never become zero. For the parabola ($e = 1$) we have: $r \to \infty$ for $\theta \to \pi$ (or $-\pi$). For the hyperbola one obtains all points on the branch considered, when θ is limited to $|\,\theta\,| < \theta_0$, where $\cos\theta_0 \equiv -1/e$; $\pi/2 < \theta_0 < \pi$. One finds $r \to \infty$ for $|\,\theta\,| \to \theta_0$, which give the directions of the asymptotes. (If θ runs through the intervals $\theta_0 < |\,\theta\,| \le \pi$ one gets negative values of r, which might be considered to correspond to the other branch of the hyperbola.)

Concluding: for any value of $e \geq 0$ and $p > 0$, (14.3) describes a conic section. The result (14.3) also includes the circle, which is the special case $e = 0$. In general, e determines the *shape* of the conic section, and p determines the *size*.

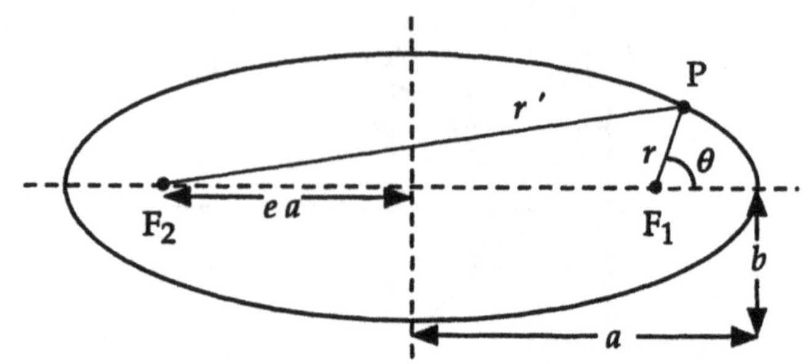

Fig. 14.5. The ellipse is the set of points where the distances to F_1 and F_2 have a constant sum

The ellipse can also be defined as the set of points P where the distances from two fixed points (the *foci*) have a constant sum (see Figure 14.5).

The major axis has length $2a$, the minor axis $2b$. The distance between the foci is $e \cdot 2a$, where – as we shall see below – e is the eccentricity. From the definition we obtain

$$r + r' = 2a\,.$$

Furthermore (see Figure 14.5),

$$(r')^2 = (2ea + r\cos\theta)^2 + (r\sin\theta)^2\,.$$

Using these two equations we obtain

$$r = \frac{a(1-e^2)}{1+e\cos\theta}\,. \tag{14.4}$$

By comparing (14.4) and (14.3) we find $p = a(1-e^2)$.

The perihelion (smallest value of r) and the aphelion (largest value of r) are determined by

$$\theta = 0 \text{ (perihelion)} \Rightarrow r_{\text{min}} = \frac{p}{1+e} = a(1-e)\,,$$

$$\theta = \pi \text{ (aphelion)} \Rightarrow r_{\text{max}} = \frac{p}{1-e} = a(1+e)\,.$$

From this

$$\frac{r_{\text{max}}}{r_{\text{min}}} = \frac{1+e}{1-e}\,, \tag{14.5}$$

or

$$e = \frac{r_{\text{max}} - r_{\text{min}}}{r_{\text{max}} + r_{\text{min}}}\,.$$

The connection between semi-major axis a, semi-minor axis b, and the eccentricity is

$$b = a\sqrt{1-e^2}\,. \tag{14.6}$$

If we – instead of the angle θ – use the angle $\varphi \equiv \pi - \theta$ as the polar angle we will have

$$r = \frac{a(1-e^2)}{1-e\cos\varphi} = \frac{p}{1-e\cos\varphi}\,. \tag{14.7}$$

The perihelion is at $\varphi = \pi$.

14.4 Newton's Law of Gravity Derived from Kepler's Laws

Newton's gravitational law is contained within Kepler's three laws. In Chapter 1 we demonstrated this for the special case of uniform circular motion. Below we present the general calculations for elliptic orbits.

Kepler's first law was formulated in two steps. Kepler first showed that the orbit of a given planet lies in a fixed plane containing the center of the Sun; then that the planet moves in an ellipse with the Sun at one focus.

We consider motion in a fixed plane. The problem is to deduce the $1/r^2$ dependence of gravitational attraction only from the observed motion of the planets.

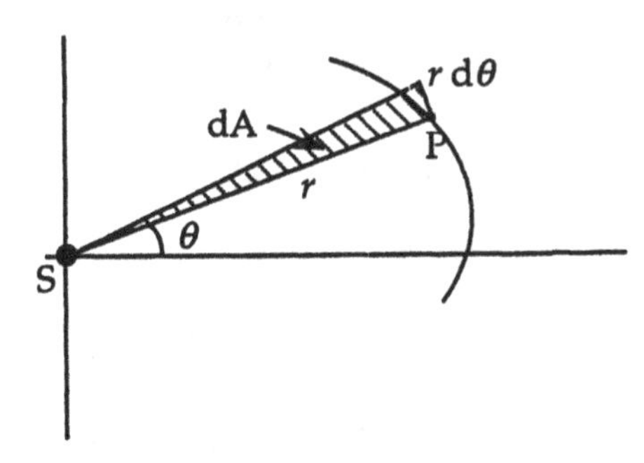

Fig. 14.6. The orbit of a planet, P, around the Sun, S

We start by calculating the acceleration of a planet moving in an elliptical orbit with the Sun in one focus. The element of area $\mathrm{d}A$ (see Figure 14.6) in polar coordinates (with origin in the Sun) is

$$\mathrm{d}A = \frac{1}{2}r r\mathrm{d}\theta = \frac{1}{2}r^2\mathrm{d}\theta\,. \tag{14.8}$$

Kepler's second law states that the area velocity is constant and we denote the constant by $h/2$. We have

$$\mathrm{d}A = \frac{1}{2}h\mathrm{d}t\,, \quad \text{or} \quad \dot{A} = \frac{1}{2}h\,.$$

From Kepler's second law it thus follows that

$$2\dot{A} = r^2\dot{\theta} = h\,. \tag{14.9}$$

We seek the acceleration vector for a Kepler orbit. In polar coordinates the two components of the acceleration are as follows (see the Appendix):

Radial component:

$$a_r = \frac{\mathrm{d}^2 r}{\mathrm{d}t^2} - r\left(\frac{\mathrm{d}\theta}{\mathrm{d}t}\right)^2 = \ddot{r} - r\dot{\theta}^2\,. \tag{14.10}$$

Angular component:

$$a_\theta = \frac{1}{r}\left[\frac{\mathrm{d}}{\mathrm{d}t}\left(r^2\frac{\mathrm{d}\theta}{\mathrm{d}t}\right)\right] = 2\dot{r}\dot{\theta} + r\ddot{\theta}\,. \tag{14.11}$$

It will be useful to introduce a substitution. Instead of r we shall use $u \equiv 1/r$ as a new variable.

From Kepler's first law we know the shape of the curve $r = r(\theta)$ or $u = u(\theta)$. The goal is to find the two components of the acceleration, a_r and a_θ, through $u = u(\theta)$, i.e., through the equation for the orbit. With this purpose in mind we eliminate the time from (14.10) and (14.11), i.e., we express $\dot{r}, \ddot{r}, \dot{\theta}$, and $\ddot{\theta}$ by $u, (\mathrm{d}u/\mathrm{d}\theta)$ and $(\mathrm{d}^2u/\mathrm{d}\theta^2)$.

From (14.9)

$$\dot{\theta} = \frac{h}{r^2} = hu^2\,. \tag{14.12}$$

Differentiating (14.12) with respect to time:

$$\ddot{\theta} = 2hu\dot{u} = 2hu\frac{\mathrm{d}u}{\mathrm{d}\theta}\frac{\mathrm{d}\theta}{\mathrm{d}t} = 2h^2u^3\frac{\mathrm{d}u}{\mathrm{d}\theta}\,. \tag{14.13}$$

Moreover, because $(\mathrm{d}u/\mathrm{d}\theta) = -(1/r^2)(\mathrm{d}r/\mathrm{d}\theta)$, we have

$$\dot{r} = \frac{\mathrm{d}r}{\mathrm{d}\theta}\frac{\mathrm{d}\theta}{\mathrm{d}t} = -\frac{1}{u^2}\frac{\mathrm{d}u}{\mathrm{d}\theta}\dot{\theta} = -h\frac{\mathrm{d}u}{\mathrm{d}\theta}\,. \tag{14.14}$$

The quantities $\dot{\theta}$, $\ddot{\theta}$, and $\dot{r}$ are now expressed by $u = u(\theta)$.

We can then determine a_θ:

$$a_\theta = 2\dot{r}\dot{\theta} + r\ddot{\theta} = 2\left(-h\frac{\mathrm{d}u}{\mathrm{d}\theta}\right)hu^2 + \frac{1}{u}2h^2u^3\frac{\mathrm{d}u}{\mathrm{d}\theta}\,,$$

or

$$a_\theta = 0.$$

Based on Kepler's second law we have shown that the acceleration of the planet is directed in a radial direction (i.e., towards the Sun). This result, well known from circular motion, is thus also true for a general elliptical orbit.

We now seek a_r. From (14.10) we see that we have to find $\ddot{r}$ expressed by $u = u(\theta)$.

$$\begin{aligned} \ddot{r} &= \frac{\mathrm{d}\dot{r}}{\mathrm{d}t} = \frac{\mathrm{d}}{\mathrm{d}t}\left(-h\frac{\mathrm{d}u}{\mathrm{d}\theta}\right) \\ &= \frac{\mathrm{d}}{\mathrm{d}\theta}\left(-h\frac{\mathrm{d}u}{\mathrm{d}\theta}\right)\frac{\mathrm{d}\theta}{\mathrm{d}t} = -h^2u^2\frac{\mathrm{d}^2u}{\mathrm{d}\theta^2}\,. \end{aligned} \tag{14.15}$$

By inserting (14.15) and (14.12) into (14.10) we obtain

$$a_r = -h^2u^2\left[\frac{\mathrm{d}^2u}{\mathrm{d}\theta^2} + u\right]. \tag{14.16}$$

From Kepler's second law we have found the radial acceleration, expressed by the area velocity constant h, and the equation of the orbit $u = u(\theta)$.

From Kepler's first law

$$r = \frac{p}{1 + e\cos\theta} \quad \text{(an ellipse)},$$

or

$$u = u(\theta) = \frac{1}{p}\left[1 + e\cos\theta\right].$$

Thus

$$\frac{\mathrm{d}^2u}{\mathrm{d}\theta^2} = -\frac{e}{p}\cos\theta\,.$$

Inserting $(\mathrm{d}^2u/\mathrm{d}\theta^2)$ and u into (14.16) we finally obtain

$$a_r = -\frac{h^2}{p}\frac{1}{r^2}\,. \tag{14.17}$$

Conclusion. The acceleration of the planet is directed towards the Sun (the minus sign), and the acceleration is inversely proportional to the square of the distance from the Sun.

We proceed to apply Kepler's third law, in order to demonstrate that the constant h^2/p can depend only on the physical nature of the Sun, i.e., the value for h^2/p is the same for all planets.

Kepler's third law may be written

$$\frac{a^3}{T^2} = C\,,$$

where C is a constant, i.e., the same for all planets. We introduce the sidereal period of revolution T into (14.9).

By integration over a complete revolution (14.9) becomes

$$2A = hT. \tag{14.18}$$

The area A of an ellipse is $A = \pi ab$, where a and b are the semi-major and semi-minor axes respectively. Furthermore,

$$p = a(1 - e^2) \quad \text{and} \quad b = a\sqrt{1 - e^2} = \frac{p}{\sqrt{1 - e^2}}.$$

From (14.18) we therefore get, using $b^2 = ap$,

$$T^2 = \left(\frac{2}{h}A\right)^2 = \left(\frac{2\pi}{h}\right)^2 a^2 b^2 = 4\pi^2 a^3 \frac{p}{h^2}. \tag{14.19}$$

From (14.19) – and applying Kepler's third law, $a^3/T^2 = C$ – we find that

$$\frac{h^2}{p} = 4\pi^2 \frac{a^3}{T^2} = 4\pi^2 C.$$

We may then finally write (see 14.17)

$$a_r = -\frac{4\pi^2 C}{r^2}, \tag{14.20}$$

where C is the same for all planets, i.e., C depends (at most) on properties of the Sun only.

From Kepler's three laws we have computed the acceleration of the planet, and seen that it depends only on the distance of the planet from the Sun: The acceleration is directed towards the Sun, and the acceleration is inversely proportional to the square of the distance from the Sun.

Newton added a decisive new feature to these results in the form of a theoretical interpretation of the derived formula. Newton introduced the Sun as the *cause* of the acceleration of the planets, and this guided him to the fundamentally new idea about *universal gravitation* (see Chapter 1).

This most surprising step, rightfully admired by both Newton's contemporaries and by later generations, was Newton's linking of the fall of bodies towards the Earth with the motion of celestial bodies.

The interaction that makes an apple fall to the ground also holds the Moon in its orbit around the Earth.

In Chapter 8 we proved that the Earth acts gravitationally as if all of its mass was concentrated in the center. From (14.20) we know that the acceleration near the surface of the Earth is (ρ = radius of the Earth)

$$g = \frac{4\pi^2 C'}{\rho^2}.$$

The constant C' is the same for *all* bodies moving in the gravitational field of the Earth. The constant may therefore be calculated from data of the orbit of the Moon: $C' = r^3/T^2$, where r = radius of the lunar orbit and T is the sidereal period of revolution of the Moon. Introducing numerical values, Newton found the gravitational acceleration g near the surface of the Earth:

$$g = \frac{4\pi^2 r^3}{\rho^2 T^2} = 9.8\,\mathrm{m\,s^{-2}}\,,$$

which is in accordance with the observed value. The greatest achievement in the history of man was completed.

14.5 The Kepler Problem

The derivation of Kepler's three laws, setting out from Newtonian mechanics and the law of gravitational attraction, is called the Kepler problem. The solution of this problem is one of the jewels of theoretical physics.

We start by deriving Kepler's first law. We first solve the so-called one-

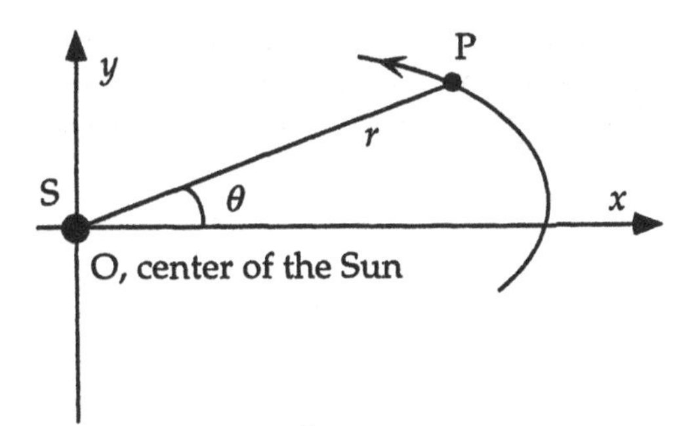

Fig. 14.7. A planet moving around the Sun

body problem. It is assumed that the Sun is fixed in the origin of the coordinate system. We furthermore neglect the gravitational interactions between the planets. We thus consider one planet moving in the gravitational field of the Sun, which is at rest in the origin of an inertial system (the heliocentric reference frame).

The angular momentum $\mathbf{L}_O$ of the planet around O is

$$\mathbf{L}_O = \mathbf{r} \times m\dot{\mathbf{r}} = \mathbf{r} \times m\mathbf{v}.$$

In a central force field the angular momentum is a constant of the motion. The plane spanned by $\mathbf{r}$ and $\mathbf{v}$ is the plane of motion for the planet. The plane of motion is fixed perpendicular to $\mathbf{L}_O$, and passes through the center of the Sun.

We seek the orbit that the planet will traverse, when it is started with a given velocity $\dot{\mathbf{r}}$.

We use polar coordinates. The mass of the planet is m. We write the equation of motion of the planet:

$$m\frac{\mathrm{d}^2\mathbf{r}}{\mathrm{d}t^2} = -\frac{GMm}{r^2}\frac{\mathbf{r}}{r}\,. \tag{14.21}$$

The term on the right is the force on the planet. The equation of motion may be written as:

$$m\frac{\mathrm{d}^2\mathbf{r}}{\mathrm{d}t^2} = \frac{C}{r^2}\frac{\mathbf{r}}{r}\,, \tag{14.22}$$

where $C \equiv -GMm$.

In the form (14.22) the equation is more general: for attractive forces C is negative (gravitational forces, electron in the Coulomb field of a proton). For repulsive forces C is positive (two electrically charged particles with the same sign of the charge).

In polar coordinates the acceleration is

$$\frac{\mathrm{d}^2\mathbf{r}}{\mathrm{d}t^2} \equiv \mathbf{a} = (\ddot{r} - r\dot{\theta}^2)\mathbf{e}_r + \frac{1}{r}\frac{\mathrm{d}}{\mathrm{d}t}(r^2\dot{\theta})\mathbf{e}_\theta\,,$$

with $\mathbf{e}_r$ being the unit vector along the radius vector, and $\mathbf{e}_\theta$ the unit vector perpendicular to radius vector. The equation of motion (14.22) written in two components, one along $\mathbf{e}_r$ and one along $\mathbf{e}_\theta$, becomes

$$m(\ddot{r} - r\dot{\theta}^2) = \frac{C}{r^2}\,, \tag{14.23}$$

$$m\frac{1}{r}\frac{\mathrm{d}}{\mathrm{d}t}(r^2\dot{\theta}) = 0\,. \tag{14.24}$$

From (14.24)

$$\frac{\mathrm{d}}{\mathrm{d}t}(mr^2\dot{\theta}) = 0\,,$$

or, after integration,

$$mr^2\dot{\theta} = L\,, \tag{14.25}$$

where L is the magnitude of the angular momentum of the planet relative to O. The magnitude of the angular momentum, $L = |\,\mathbf{L}\,|$, is constant and determined by the initial conditions. We have thus introduced a constant of motion into the process of integration.

From (14.25) we find

$$\dot{\theta} = \frac{L}{mr^2}\,. \tag{14.26}$$

Introducing $\dot{\theta}$ into (14.23) gives

$$\ddot{r} - \frac{L^2}{m^2r^3} = \frac{C}{mr^2}\,. \tag{14.27}$$

The equation (14.27) is a differential equation for r as function of t. In order to derive Kepler's first law we are interested, not in an instantaneous location of the planet, but in the shape of the orbit. In other words: We are interested in determining r as a function of θ, not r as a function of t.

We eliminate t from (14.27) by using (14.26). First we determine

$$\dot{r} \equiv \frac{\mathrm{d}r}{\mathrm{d}t}\,, \quad \text{then} \quad \ddot{r} \equiv \frac{\mathrm{d}^2 r}{\mathrm{d}t^2}\,,$$

both expressed by θ instead of t.

$$\frac{\mathrm{d}r}{\mathrm{d}t} = \frac{\mathrm{d}r}{\mathrm{d}\theta}\frac{\mathrm{d}\theta}{\mathrm{d}t} = \frac{\mathrm{d}r}{\mathrm{d}\theta}\frac{L}{mr^2}\,, \tag{14.28}$$

$$\frac{\mathrm{d}^2 r}{\mathrm{d}t^2} = \frac{\mathrm{d}}{\mathrm{d}t}\left(\frac{\mathrm{d}r}{\mathrm{d}\theta}\frac{L}{mr^2}\right)\,,$$

or

$$\frac{\mathrm{d}^2 r}{\mathrm{d}t^2} = \frac{L^2}{m^2 r^4}\left[\frac{\mathrm{d}^2 r}{\mathrm{d}\theta^2} - \frac{2}{r}\left(\frac{\mathrm{d}r}{\mathrm{d}\theta}\right)^2\right]\,. \tag{14.29}$$

We introduce a new variable $u(\theta) = 1/r(\theta)$. The reason for introducing u as variable instead of r is that the parenthesis in (14.29) is close to being equal to $(\mathrm{d}^2 u/\mathrm{d}\theta^2)$. We find

$$\frac{\mathrm{d}u}{\mathrm{d}\theta} = -\frac{1}{r^2}\frac{\mathrm{d}r}{\mathrm{d}\theta}\,,$$

$$\frac{\mathrm{d}^2 u}{\mathrm{d}\theta^2} = -\frac{1}{r^2}\left[\frac{\mathrm{d}^2 r}{\mathrm{d}\theta^2} - \frac{2}{r}\left(\frac{\mathrm{d}r}{\mathrm{d}\theta}\right)^2\right]\,.$$

Using these results we obtain

$$\frac{\mathrm{d}^2 r}{\mathrm{d}t^2} \equiv \ddot{r} = -\frac{L^2}{m^2 r^2}\frac{\mathrm{d}^2 u}{\mathrm{d}\theta^2}\,. \tag{14.30}$$

Introducing (14.30) into (14.27) and using $1/r = u$ we obtain the following differential equation for $u = u(\theta)$:

$$\frac{\mathrm{d}^2 u}{\mathrm{d}\theta^2} + u = -\frac{Cm}{L^2}\,. \tag{14.31}$$

This differential equation has the same form as the equation describing the oscillation of a mass at the end of a spring, hanging in the gravitational field of the Earth (see Example 2.3).

The solution of (14.31) is

$$u = A\cos(\theta + \varphi_0) - \frac{Cm}{L^2}\,. \tag{14.32}$$

The quantities A and φ_0 are constants of integration. By the substitution of (14.32) into (14.31) it may be verified that the given expression for u is a solution of the differential equation.

The integration constant φ_0 describes the orientation of the orbit in the plane. By choosing the polar axis in a suitable way we can obtain $\varphi_0 = 0$. The result (14.32) may thus be written as follows:

$$\frac{1}{r} = A\cos\theta - \frac{Cm}{L^2}\,. \tag{14.33}$$

We proceed by introducing another constant of integration: The mechanical energy E of the planet. The planet moves in a conservative field of force. The energy is therefore conserved.

We express the integration constant A through E.

$$E \equiv \frac{1}{2}mv^2 + \frac{C}{r} = \frac{1}{2}m(\dot{r}^2 + r^2\dot{\theta}^2) + \frac{C}{r}\,. \tag{14.34}$$

Using the expressions for $\dot{r}$ (14.28) and $\dot{\theta}$ (14.26) we obtain

$$E = \frac{1}{2}m\left(\frac{L^2}{m^2r^4}\right)\left[\left(\frac{\mathrm{d}r}{\mathrm{d}\theta}\right)^2 + r^2\right] + \frac{C}{r}\,. \tag{14.35}$$

The total mechanical energy E is – in (14.35) – expressed in terms of the parameters of the orbit. Using (14.33) we find an equation connecting E and A. From (14.33)

$$\frac{\mathrm{d}r}{\mathrm{d}\theta} = r^2 A\sin\theta\,.$$

Introducing $\mathrm{d}r/\mathrm{d}\theta$ in (14.35) gives

$$E = \frac{1}{2}m\left(\frac{L^2}{m^2r^4}\right)(r^4A^2\sin^2\theta + r^2) + \frac{C}{r}\,.$$

By using (14.33) again we find

$$E = \frac{1}{2}\frac{L^2}{m}A^2 - \frac{C^2m}{2L^2}\,,$$

or

$$A = \frac{Cm}{L^2}\left(1 + \frac{2EL^2}{C^2m}\right)^{1/2}\,. \tag{14.36}$$

Consider again (14.33). Introducing (14.36) and using $C \equiv -GMm$ we obtain the expression for the orbit of the planet in polar coordinates and expressed by two constants of the motion, L and E:

$$\frac{1}{r} = \frac{Gm^2M}{L^2}\left[1 - \left(1 + \frac{2EL^2}{G^2m^3M^2}\right)^{1/2}\cos\theta\right]\,. \tag{14.37}$$

Equation (14.37) describes an ellipse with perihelion for $\theta = \pi$ (see Section 14.3, Equations (14.7) and (14.3)). By choosing $\varphi_0 = \pi$ in (14.32), instead of $\varphi_0 = 0$, (14.37) becomes

$$\frac{1}{r} = \frac{Gm^2M}{L^2}\left[1 - \left(1 + \frac{2EL^2}{G^2m^3M^2}\right)^{1/2} \cos(\theta + \pi)\right] .$$

We prefer this choice of φ_0 and write our final result as

$$\frac{1}{r} = \frac{Gm^2M}{L^2}\left[1 + \left(1 + \frac{2EL^2}{G^2m^3M^2}\right)^{1/2} \cos\theta\right] \tag{14.38}$$

$$\frac{1}{r} = \frac{1}{p}\left[1 + e\cos\theta\right] . \tag{14.39}$$

Equation (14.39) is the equation, in polar coordinates, for a conic section.

Our result, (14.38), describes a conic section with parameter

$$p = \frac{L^2}{Gm^2M} ,$$

and eccentricity

$$e = \sqrt{1 + \frac{2EL^2}{G^2m^3M^2}} .$$

The total energy,

$$E = \frac{1}{2}mv^2 - \frac{GMm}{r} ,$$

may be either negative, positive, or zero.

From (14.38) we conclude:

1. For $E < 0$, $e < 1$, we have an ellipse, $e = 0$ corresponds to a circular motion
2. For $E > 0$, $e > 1$, we have a hyperbola
3. For $E = 0$, $e = 1$, we have a parabola

> We have shown that Kepler's first law follows from Newton's second law, in combination with the law of gravitational attraction.

Many other important consequences concerning the motion of celestial bodies may be read from (14.38). Let us conclude.

A planet, a comet, an asteroid, or any heavenly body whatsoever, governed by the gravitational field of the Sun, will traverse an orbit that is a conic section. The form of the conic section is determined solely by the total mechanical energy E, given by the initial conditions. If $E \geq 0$ the body is not bound to the solar system. The orbit is a hyperbola, or for $E = 0$, a parabola. If a body with $E \geq 0$ passes "close to the Sun", i.e., if such a body appears in the solar system at all, it will happen only once. No comet with $E > 0$ has been observed until now. For $E < 0$ we have a "bound state" of the planet.

The angular momentum $\mathbf{L}$ is likewise a constant of the motion, given by the initial conditions. The plane of the orbit is perpendicular to $\mathbf{L}$. The magnitude L of $\mathbf{L}$ determines the parameter $p = L^2/Gm^2M$.

14.5.1 Derivation of Kepler's 3rd Law from Newton's Law of Gravity

In Chapter 10 we proved that Kepler's second law is a consequence of conservation of angular momentum in a central force field. We shall now show that Kepler's third law also follows from Newtonian mechanics.

Kepler's third law is

$$T^2 = \frac{1}{C}a^3 \equiv ka^3 .$$

The constant of proportionality k is the same for all planets.

Kepler's second law can be written as

$$\frac{\mathrm{d}A}{\mathrm{d}t} = \frac{L}{2m} . \tag{14.40}$$

To prove the third law we have to introduce T, the sidereal period, into (14.40). By integration over a complete revolution,

$$A = \frac{LT}{2m} , \tag{14.41}$$

where $A = \pi ab$ (the area of the ellipse).

From (14.41)

$$T^2 = \left(\frac{2m}{L}\right)^2 A^2 = \left(\frac{2m}{L}\right)^2 \pi^2 a^2 b^2 . \tag{14.42}$$

From Section 14.2

$$b^2 = a^2(1 - e^2) ,$$

and

$$p = a(1 - e^2) = \frac{L^2}{Gm^2 M} .$$

From (14.42) we obtain

$$T^2 = \frac{4\pi^2}{GM}a^3 = ka^3 . \tag{14.43}$$

The constant $k = 4\pi^2/GM$ depends only on the mass of the Sun, and is thus the same for all planets. Kepler's third law has been derived from Newtonian mechanics.

From (14.43)

$$T = (ka^3)^{1/2} ,$$

or

$$\log a = \frac{2}{3}\log T + B, \quad \text{where } B \text{ is a constant.}$$

Figure (14.8) illustrates Kepler's third law for the solar system. The figure shows that $\log a$ is a linear function of $\log T$, and the slope of the line is 2/3.

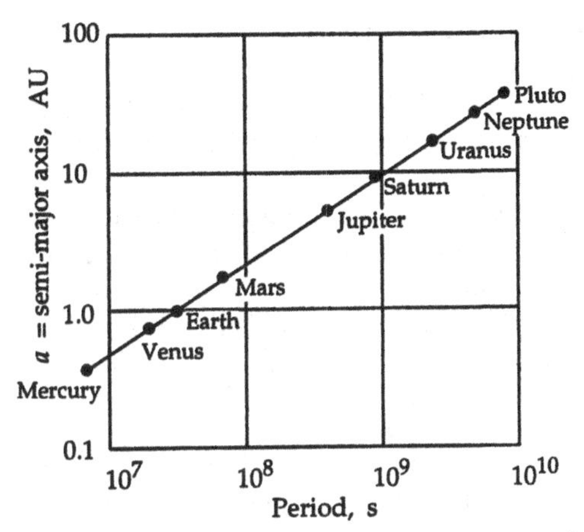

Fig. 14.8. Kepler's third law for the solar system, $\log a = \frac{2}{3}\log T + B$, $B \equiv$ constant. Based on *Berkeley Physics Course*

Kepler's third law is a consequence of the universal law of gravity and Newton's laws of motion. The law is valid also for elliptical orbits of moons moving around planets. The mass M of the Sun is then replaced by the mass of the given planet.

Newton tested the validity of Kepler's third law on the four Jupiter moons known to him. Newton knew the periods of revolution of the moons of Jupiter with fairly good accuracy.

The table below shows the radius ρ in the orbits of the moons; $\rho = r/R_j$ is measured in units of the radius R_j of Jupiter. The table furthermore gives the period of revolution T for the moons, and finally (ρ^3/T^2). Kepler's third law is seen to be valid to a high order of accuracy.

The radius in the orbits of the moons, as given in the table, is measured in units of the radius of Jupiter. The knowledge of the absolute distances in the solar system was limited at the time of Newton, and the size of Jupiter was consequently not known.

	r/R_j	T (s)	ρ^3/T^2 (s^{-2})
Io	5.58	1.53×10^5	7.4×10^{-9}
Europa	8.88	3.07×10^5	7.5×10^{-9}
Ganymede	14.16	6.19×10^5	7.5×10^{-9}
Callisto	24.90	1.45×10^6	7.4×10^{-9}

14.6 The Effective Potential

In this section we shall briefly describe another procedure for the integration of the equation of motion for a planet moving in the gravitational field of the Sun.

The angular momentum $\mathbf{L}$ of the planet is assumed to be different from zero ($L \equiv |\mathbf{L}| = 0$ corresponds to the planet moving along a radius vector, away from or into the Sun).

From Example 10.2 it is known that the total energy of the planet may be written as follows:

$$E = \frac{1}{2}m\dot{r}^2 + \frac{L^2}{2mr^2} - \frac{GMm}{r} . \tag{14.44}$$

The term $L^2/2mr^2$ is called the *centrifugal potential energy.*

We look for the differential equation of the orbit. The magnitude of the angular momentum is

$$L = mr^2\frac{\mathrm{d}\theta}{\mathrm{d}t} .$$

The energy E and the magnitude of the angular momentum L are known constants of motion.

We transform differentiation with respect to time into differentiation with respect to θ:

$$\frac{\mathrm{d}}{\mathrm{d}t} = \frac{\mathrm{d}\theta}{\mathrm{d}t} \cdot \frac{\mathrm{d}}{\mathrm{d}\theta} = \frac{L}{mr^2}\frac{\mathrm{d}}{\mathrm{d}\theta} ,$$

Furthermore, we apply the variable transformation $u = 1/r$. This means that

$$\frac{\mathrm{d}r}{\mathrm{d}t} = \frac{\mathrm{d}}{\mathrm{d}t}\left(\frac{1}{u}\right) = -\frac{1}{u^2}\frac{\mathrm{d}u}{\mathrm{d}t} = -\frac{L}{m}\frac{\mathrm{d}u}{\mathrm{d}\theta} .$$

Equation (14.44) may thus be rewritten as

$$\frac{1}{2}\frac{L^2}{m}\left(\frac{\mathrm{d}u}{\mathrm{d}\theta}\right)^2 + \frac{1}{2}\frac{L^2}{m}u^2 - GMmu = E . \tag{14.45}$$

This is a differential equation for $u = u(\theta)$. The equation may be simplified by differentiation with respect to θ and division by $(L^2/m)(\mathrm{d}u/\mathrm{d}\theta)$. We find

$$\frac{\mathrm{d}^2u}{\mathrm{d}\theta^2} + u = \frac{GMm^2}{L^2} .$$

This is the equation integrated in Section 14.5. The solution may be written – with a suitable choice of polar axes – as $u = A\cos\theta + 1/p$, where we have introduced $p \equiv L^2/GMm^2$.

Instead of the integration constant A we shall use $A \equiv e/p$, where e is a new integration constant. The possible orbits then have the form

$$r = \frac{p}{1 + e\cos\theta} ,$$

which is the well-known equation for a conic section.

It is convenient to express the integration constant e – which obviously is the eccentricity – by means of E. By differentiating $u = (1 + e\cos\theta)/p$ with respect to θ and inserting the result into equation (14.45) we find

$$E = \frac{G^2M^2m^3}{2L^2}(e^2 - 1),$$

or

$$e = \sqrt{1 + \frac{2EL^2}{G^2M^2m^3}},$$

which is identical to the result found previously.

14.7 The Two-Body Problem

We have solved the so-called one-body problem: a material particle moves in a central field of force of the type $1/r^2$. The model corresponds to a planet moving in the gravitational field of the Sun, where the Sun is assumed to be fixed at the origin.

Below we investigate the motion of two spherically symmetrical bodies moving in astronomical space. The two bodies are assumed to move exclusively under their mutual gravitational interaction. The problem is called the two-body problem. As a model one may think of the Sun of mass M and one of its planets, say Jupiter, of mass m. Consider Figure 14.9.

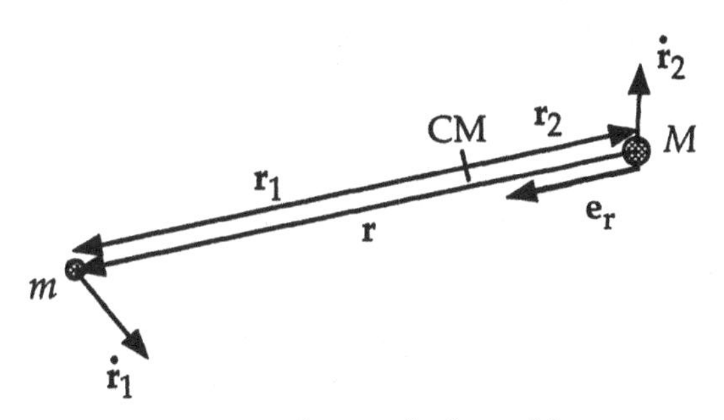

Fig. 14.9. The two-body problem

The center of mass, CM, of the system will be either "at rest" or move with constant velocity, because no external forces are acting on the system. Seen from the point of view of the original Newtonian mechanics this means that CM is either at rest in absolute space (astronomical space) or moves with a constant velocity relative to that space. The modified form of Newtonian mechanics are now in logical difficulties: we look for the motion of the Sun under the influence of the gravitational field of the planet (Jupiter). Therefore it has no meaning to use the heliocentric reference frame as the basis for the description. The astronomical two-body problem illustrates the profound difficulties connected with the choice of inertial systems.

We choose CM as the origin for an inertial system (see Figure 14.9):

$$M\mathbf{r}_2 + m\mathbf{r}_1 = 0\,.$$

The radius vector to m measured from M is denoted $\mathbf{r}$:

$$\mathbf{r} \equiv \mathbf{r}_1 - \mathbf{r}_2\,.$$

A unit vector in the direction of $\mathbf{r}$ is denoted $\mathbf{e}_r$:

$$\mathbf{e}_r \equiv \frac{\mathbf{r}}{r} = \frac{\mathbf{r}_1 - \mathbf{r}_2}{r}\,.$$

The equations of motion for each of the two bodies are

$$m\frac{\mathrm{d}^2\mathbf{r}_1}{\mathrm{d}t^2} = -\frac{GMm}{r^2}\mathbf{e}_r\,,$$

$$M\frac{\mathrm{d}^2\mathbf{r}_2}{\mathrm{d}t^2} = \frac{GMm}{r^2}\mathbf{e}_r\,.$$

If we add these two equations we reach the not surprising conclusion that the total momentum of the system is a constant.

Our real aim is to obtain a differential equation for $\mathbf{r} = \mathbf{r}_1 - \mathbf{r}_2$. Transferring the masses to the right side of the equations and subtracting the second equation from the first we get

$$\frac{\mathrm{d}^2\mathbf{r}_1}{\mathrm{d}t^2} - \frac{\mathrm{d}^2\mathbf{r}_2}{\mathrm{d}t^2} = -\frac{GMm}{r^2}\left[\frac{1}{m} + \frac{1}{M}\right]\mathbf{e}_r\,,$$

or

$$\frac{mM}{M+m}\frac{\mathrm{d}^2\mathbf{r}}{\mathrm{d}t^2} = -\frac{GMm}{r^2}\mathbf{e}_r\,.$$

The reduced mass μ is defined as

$$\mu \equiv \frac{mM}{m+M}\,,$$

or

$$\frac{1}{\mu} = \frac{1}{m} + \frac{1}{M}\,. \tag{14.46}$$

We obtain the following differential equation for $\mathbf{r}$, the radius vector from M to m:

$$\mu\frac{\mathrm{d}^2\mathbf{r}}{\mathrm{d}t^2} = -\frac{GMm}{r^2}\mathbf{e}_r = -\frac{G(M+m)\mu}{r^2}\mathbf{e}_r\,.$$

This differential equation has the same form as the equation we solved for the one-body problem. We have reached a fundamental result: the motion of m relative to M may be determined *as if* M were at rest at the origin of an inertial system, if we – instead of the inertial mass m – use the reduced mass μ as defined in (14.46). The two-body problem is reduced to a one-body problem for the motion of a mass μ in the gravitational field of a mass of magnitude $M+m$.

Note that m and M enter the problem in a completely symmetrical way. We could equally well have used $(-\mathbf{r})$ for a description of the motion.

We shall briefly show how the two constants of motion, the angular momentum $\mathbf{L}$ and the mechanical energy E may be expressed by the reduced mass μ.

See again Figure 14.9. Consider first the total angular momentum $\mathbf{L}$ with respect to CM:

$$\mathbf{L}_{\mathrm{CM}} = \mathbf{r}_2 \times M\dot{\mathbf{r}}_2 + \mathbf{r}_1 \times m\dot{\mathbf{r}}_1 \,.$$

Eliminate M by means of the definition of CM:

$$-M\mathbf{r}_2 = m\mathbf{r}_1 \,.$$

We get

$$\begin{aligned} \mathbf{L}_{\mathrm{CM}} &= -\mathbf{r}_2 \times m\dot{\mathbf{r}}_1 + \mathbf{r}_1 \times m\dot{\mathbf{r}}_1 \\ &= (\mathbf{r}_1 - \mathbf{r}_2) \times m\dot{\mathbf{r}}_1 \,. \end{aligned}$$

The term $m\dot{\mathbf{r}}_1$ may be rewritten:

$$\begin{aligned} m\dot{\mathbf{r}}_1 &= \frac{m(m+M)}{m+M}\dot{\mathbf{r}}_1 \\ &= \frac{m}{m+M}(m\dot{\mathbf{r}}_1 + M\dot{\mathbf{r}}_1) \\ &= \frac{m}{m+M}(-M\dot{\mathbf{r}}_2 + M\dot{\mathbf{r}}_1) \\ &= \mu\dot{\mathbf{r}} \,. \end{aligned}$$

The total angular momentum for the two masses in their motion around CM is

$$\mathbf{L}_{\mathrm{CM}} = \mathbf{r} \times \mu\dot{\mathbf{r}} \,.$$

We conclude: the angular momentum $\mathbf{L}_{\mathrm{CM}}$ may be calculated *as if* the mass μ moved around M. For this calculation the mass M may be taken to be at rest in an inertial frame.

Next we calculate the total mechanical energy:

$$\begin{aligned} E &= \frac{1}{2}m\dot{\mathbf{r}}_1^2 + \frac{1}{2}M\dot{\mathbf{r}}_2^2 - \frac{GMm}{r} \\ &= \frac{1}{2}m\dot{\mathbf{r}}_1 \cdot \dot{\mathbf{r}}_1 + \frac{1}{2}M\dot{\mathbf{r}}_2 \cdot \dot{\mathbf{r}}_2 - \frac{GMm}{r} \,, \end{aligned}$$

or, using $\dot{\mathbf{r}}_2 = -(m/M)\dot{\mathbf{r}}_1$,

$$E = \frac{1}{2}\left(m + \frac{m^2}{M}\right)\dot{\mathbf{r}}_1 \cdot \dot{\mathbf{r}}_1 - \frac{GMm}{r} \,.$$

Making use of

$$\dot{\mathbf{r}}_1 = \frac{M}{m+M}(\dot{\mathbf{r}}_1 - \dot{\mathbf{r}}_2) = \frac{M}{m+M}\dot{\mathbf{r}},$$

we find

$$E = \frac{1}{2}\mu\dot{\mathbf{r}}^2 - \frac{GMm}{r} = \frac{1}{2}\mu\dot{\mathbf{r}}^2 - \frac{G(M+m)\mu}{r}.$$

We conclude that the total energy E may be calculated *as if* the mass μ moved around M. For this calculation the mass M may be taken to be at rest in an inertial frame.

14.7.1 The Two-Body Problem and Kepler's 3rd Law

The differential equations describing the one-body problem and the two-body problem are of similar form. With an easily understandable notation the equations may be written as

$$\begin{aligned} \text{one-body:} \quad \frac{d^2\mathbf{r}}{dt^2} &= -\frac{GM}{r^2}\frac{\mathbf{r}}{r}, \\ \text{two-body:} \quad \frac{d^2\mathbf{r}}{dt^2} &= -\frac{G(M+m)}{r^2}\frac{\mathbf{r}}{r}. \end{aligned}$$

For the one-body problem we found Kepler's third law:

$$\frac{T^2}{a^3} = \frac{4\pi^2}{GM}.$$

For the two-body problem the corresponding expression becomes

$$\frac{T^2}{a^3} = \frac{4\pi^2}{G(M+m)}.$$

The ratio T^2/a^3 is thus not exactly the same for all planets, due to the fact that m varies from planet to planet. Due to the large mass of the Sun compared to planetary masses the deviations from planet to planet are small.

14.8 Double Stars: The Motion of the Heliocentric Reference Frame

Many stars are double stars, i.e., two neighboring stars moving under their mutual gravitational interaction. For simplicity we assume that the two stars move in circles around their common CM.

The distance from CM to mass M is called R, and the distance from CM to m is denoted r $(mr = MR)$.

We use the results from the study of the two-body problem. The motion can be described as the motion of a single particle with the reduced mass μ

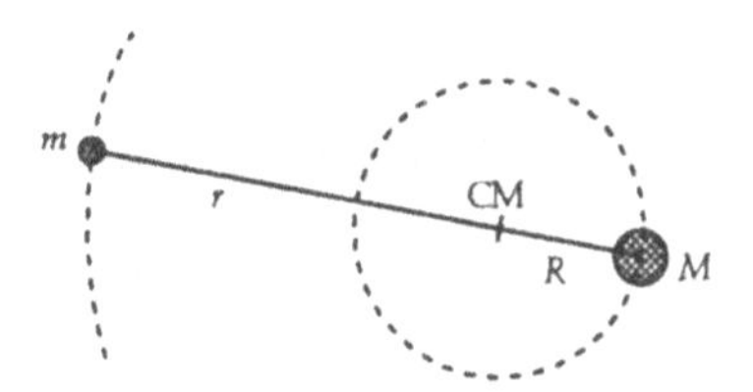

Fig. 14.10. Double stars

moving around M. The radius in that circular motion is called ρ. We have: $\rho = r + R$.

$$\mu\rho\omega^2 = G\frac{Mm}{\rho^2}, \quad \mu \equiv \frac{Mm}{M+m} .$$

For the angular frequency we find

$$\omega^2 = \frac{G(M+m)}{\rho^3} .$$

The period of revolution is determined by $\omega T = 2\pi$.

Relative to what do the two stars move? It would be absurd to use the heliocentric reference frame!

Our own solar system is "nearly a double star system". The mass of Jupiter dominates the planetary system. In units of the mass of the Earth the masses in the solar system are:

Sun: 332 946
Jupiter: 317.9
Saturn: 95.2
The rest of the planets together: 33.7

Neglecting the mass of all the planets except the mass of Jupiter, we can estimate the position of the CM of the solar system. The distance of Jupiter from the Sun is

$$5.2 \text{ AU} = 5.2 \times 1.5 \times 10^8 \text{ km},$$

$$R_{\mathrm{CM}} \approx \frac{m_{\mathrm{J}}}{M_{\mathrm{S}}} r_{\mathrm{J}} \approx 744\,750 \text{ km}.$$

The radius of the Sun is 700 000 km. The CM of the solar system is thus located about 50 000 km above the surface of the Sun.

A coordinate frame with its origin in the center of the Sun, i.e., the heliocentric reference frame, has an acceleration relative to the CM of the solar system. The CM of the solar system moves around the center of the galaxy. The Sun is about 30 000 light years or 3×10^{22} cm from the galactic center.

The time for one revolution around the galactic center has been estimated as about 250 million years, or

$$T \approx 8 \times 10^{15}\ \text{s}.$$

The acceleration of the CM relative to the galactic center is thus

$$a = \frac{v^2}{r} = \frac{4\pi^2 r}{T^2} \approx 1.9 \times 10^{-6}\ \text{cm}\,\text{s}^{-2}.$$

The tides in the freely falling heliocentric reference frame are so small that we have not been able to measure them (yet). Therefore we use the heliocentric reference frame as a local inertial reference frame.

From measurements on radioactive isotopes in minerals in meteorites (and in rocks from the Moon and the Earth) we know that the solar system formed 4.6×10^9 years ago. The solar system has completed

$$\frac{4.6 \times 10^9}{2.5 \times 10^8} \approx 18$$

revolutions around the center of the galaxy, since the system was born. What did we meet on this long journey? Supernova explosions? Interstellar clouds?

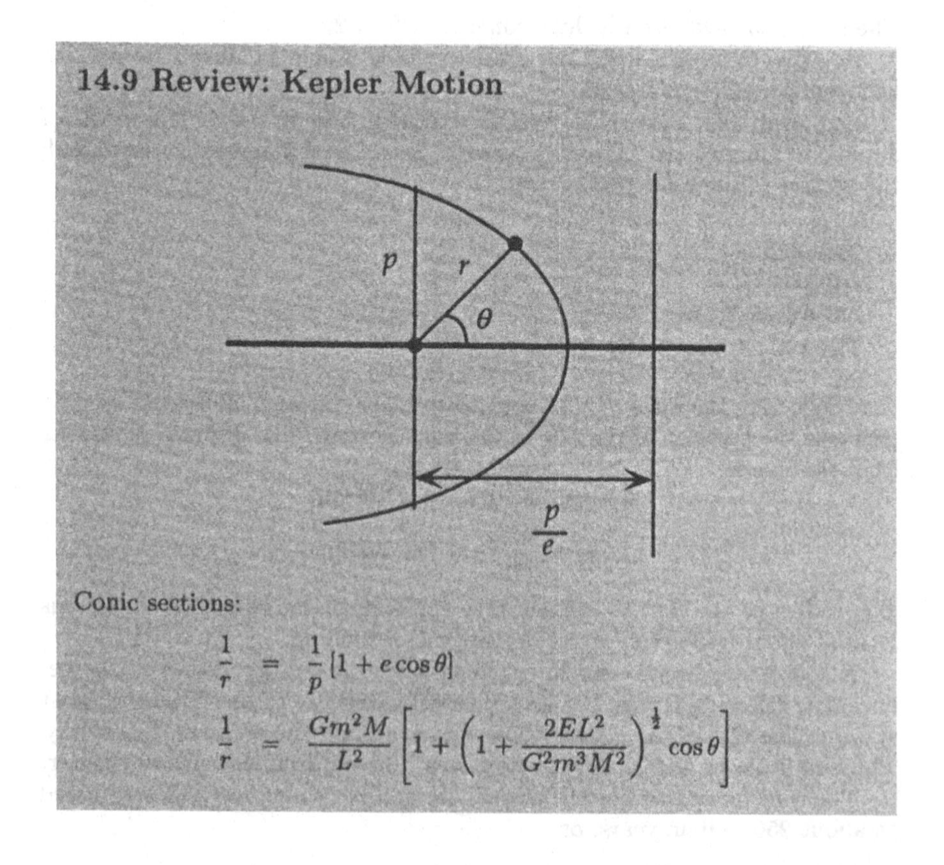

14.9 Review: Kepler Motion
p
r
θ
p/e
Conic sections:
$\frac{1}{r} = \frac{1}{p}[1 + e\cos\theta]$
$\frac{1}{r} = \frac{Gm^2M}{L^2}\left[1 + \left(1 + \frac{2EL^2}{G^2m^3M^2}\right)^{\frac{1}{2}}\cos\theta\right]$

Conservation laws:

$$L = mr^2\dot{\theta} = \text{constant}$$

$$E = \frac{1}{2}mv^2 - \frac{GmM}{r} = \text{constant}'$$

14.10 Examples

In this section we discuss a few examples of motion in the solar system.

Example 14.1. Planetary Orbits and Initial Conditions. We consider a family of possible orbits for a planet around the Sun, S. The Sun has the mass M. The planet, P, with mass m, is imagined to be started with a velocity always perpendicular to the line SP, but with various values of the magnitude of the initial velocity. The initial distance between S and P is r_0.

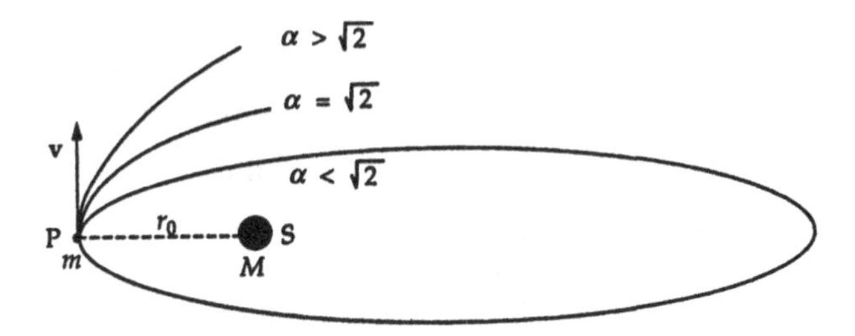

We begin by determining the magnitude of the velocity necessary for a uniform circular motion:

$$v_0 = \sqrt{\frac{GM}{r_0}} .$$

The planet is then imagined to be started with an arbitrary magnitude of the velocity v_P (still perpendicular to SP). We introduce the ratio

$$\alpha \equiv \frac{v_P}{v_0} .$$

The value $\alpha = 1$ thus corresponds to a circular orbit. Below we shall show that

$$\begin{aligned} \text{for } \alpha &= \sqrt{2} \text{ the orbit is a parabola,} \\ \text{for } \alpha &< \sqrt{2} \text{ the orbit is an ellipse,} \\ \text{for } \alpha &> \sqrt{2} \text{ the orbit is a hyperbola.} \end{aligned}$$

Calculate first the total energy in the orbit expressed in terms of α :

$$\begin{aligned} E &= \frac{1}{2}mv_p^2 - \frac{GMm}{r_0} \\ &= \frac{1}{2}mv_0^2\alpha^2 - \frac{GMm}{r_0} \\ &= \frac{1}{2}(\alpha^2 - 1)mv_0^2 + \frac{1}{2}mv_0^2 - \frac{GMm}{r_0} \,. \end{aligned}$$

The last two terms in this expression form the energy E_0 in a circular orbit.

$$E = E_0 + \frac{1}{2}(\alpha^2 - 1)mv_0^2 \,.$$

Question: Show that the total energy in a circular orbit may be written $E_0 = -\frac{1}{2}mv_0^2$.

We finally obtain

$$E = E_0(2 - \alpha^2) \,,$$

or, as E_0 is negative,

$$E = (\alpha^2 - 2) \mid E_0 \mid \,.$$

From this result we see that:

$$\begin{aligned} &\text{for } \alpha > \sqrt{2} \quad E > 0 \quad \text{(hyperbola)}, \\ &\text{for } \alpha = \sqrt{2} \quad E = 0 \quad \text{(parabola)}, \\ &\text{for } \alpha < \sqrt{2} \quad E < 0 \quad \text{(ellipse)}. \end{aligned}$$

The shape of the orbit is determined not only by Newton's laws but also by the initial conditions. This fact makes it possible to find out something about the origin of the solar system. The fact that the planetary orbits lie nearly in the same plane has something to do with the initial conditions of the system, and is not dictated by the laws of force and motion. See Example 10.2. △

Example 14.2. Shape and Size of Planetary Orbits. Consider the figure (see also Figure 14.4 for the definition of conic sections).

We draw the chord through the focal point F and perpendicular to the axis of the conic section. The points where the chord intersects the conic section are denoted A and B. All conic sections with the same value of the parameter p pass through A and B. The shape of the conic section is determined by the eccentricity, which again is determined by the distance $d = p/e$ to the directrix l.

We have (for a celestial body):

$$p = \frac{L^2}{GMm^2} \,, \quad e = \sqrt{1 + \frac{2EL^2}{G^2m^3M^2}} \,.$$

From this, the orbit for all celestial bodies, with the same magnitude of angular momentum L, has the same value of parameter p. All such orbits pass through the points A and B, if the conic sections have the same axis.

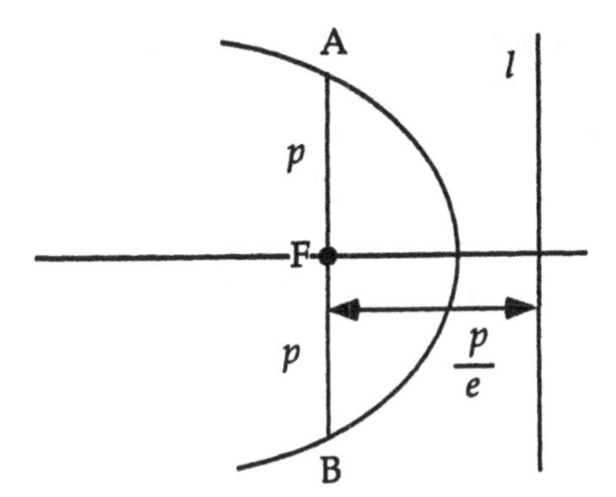

The eccentricity, and consequently the shape of the conic section, is also determined by the total energy E.

Question. Does the value of m influence the size and shape of the orbit?

Answer. No. We proceed to show that the semi-major axis of an elliptical orbit depends only on E.

For an ellipse $p = a(1 - e^2)$. From p and e as given above we get (note $E < 0$)

$$a = \frac{GMm}{2(-E)} ,$$

or

$$E = -\frac{GMm}{2a} .$$

For a circle, $a = \frac{1}{2}(r + r) = r$.

Consider the following figure showing elliptical orbits with the same semi-major axis a.

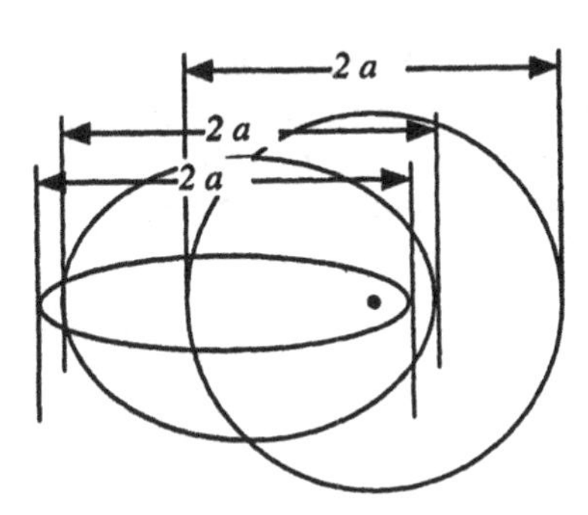

Planets moving in these elliptical orbits have the same energy, but not the same angular momentum.

Of all planetary orbits, with the same angular momentum, the circle has the lowest energy E. This is seen from

$$\frac{2EL^2}{G^2M^2m^3}+1=e^2\ ,$$

or

$$E=\frac{(e^2-1)G^2M^2m^3}{2L^2}\ .$$

For a circular orbit $e^2=0$.

△

Example 14.3. Motion Near the Surface of the Earth

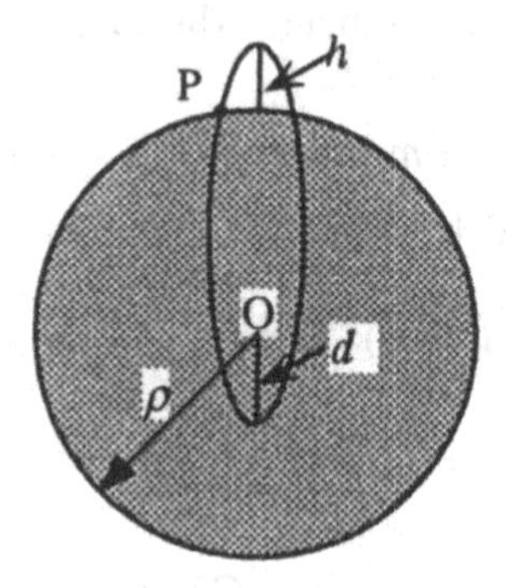

The trajectory of a cannonball near the surface of the Earth is – neglecting air resistance – a parabola. The approximation made is that the acceleration due to gravity is a constant vector.

The approximation is very good indeed, but strictly speaking we should consider the top point in the orbit as the aphelion (apogee) in an elongated ellipse, where the center 0 of the Earth is in one of the foci of the ellipse.

Assume that a cannonball is fired with the initial velocity v_0 and from the point P. Assume further more that the Earth is at rest in an inertial frame. The total energy in an elliptical orbit is determined exclusively by the semi-major axis a. The energy E is

$$E=\frac{1}{2}mv_0^2-\frac{GMm}{\rho}=-\frac{GMm}{2a}\ ,$$

where

$$\begin{aligned}\rho &\equiv \text{radius of the Earth,}\\ M &\equiv \text{mass of the Earth,}\\ m &\equiv \text{mass of the canonball.}\end{aligned}$$

The major axis $2a$ is only slightly different from the radius of the Earth, ρ. We write

$$2a = \rho + \Delta = \rho \left(1 + \frac{\Delta}{\rho}\right) .$$

Let us estimate the magnitude of Δ:

$$\frac{1}{2}mv_0^2 - \frac{GMm}{\rho} = -\frac{GMm}{\rho}\left(1 + \frac{\Delta}{\rho}\right)^{-1} ,$$

$$\frac{1}{2}mv_0^2 - \frac{GMm}{\rho} \approx -\frac{GMm}{\rho}\left(1 - \frac{\Delta}{\rho}\right) .$$

We thus obtain

$$\Delta \approx \frac{v_0^2}{2GM/\rho^2} = \frac{v_0^2}{2g} ,$$

where g is the acceleration of gravity at the surface of the Earth. We have that $\Delta = h + d$ (see the figure).

If the start velocity of the cannonball is $v = 1\ \mathrm{km\,s^{-1}}$ we find that

$$\Delta \cong \frac{v^2}{2g} \approx 51\ \mathrm{km}.$$

The shape of the elliptic orbit depends not only on the speed v_0 but also on the firing angle.

Note. Close to the perihelion it is difficult to distinguish a "long ellipse" from a parabola. For instance, many of the comets observed until now are in orbits with excentricities very close to 1, i.e., many comets are in orbits that are nearly parabolic. Δ

Example 14.4. Velocities in an Elliptical Orbit

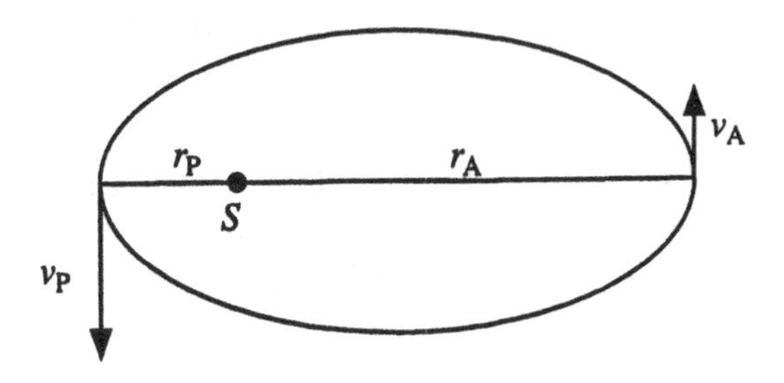

The velocity of a planet in the perihelion (perigæum) is v_P and the corresponding velocity in the aphelion (apogæum) is v_A. Both velocities are perpendicular to the axis of the ellipse, and measured relative to the heliocentric reference frame.

From conservation of angular momentum:

$$mr_\mathrm{A}v_\mathrm{A} = mr_\mathrm{P}v_\mathrm{P} ,$$

$$v_\mathrm{A} = \frac{r_\mathrm{P}}{r_\mathrm{A}}v_\mathrm{P} = \frac{1-e}{1+e}v_\mathrm{P} .$$

From conservation of energy:

$$\frac{1}{2}mv_\mathrm{A}^2 - \frac{GMm}{a(1+e)} = \frac{1}{2}mv_\mathrm{P}^2 - \frac{GMm}{a(1-e)} .$$

By using $v_\mathrm{A} = (1-e)v_\mathrm{P}/(1+e)$ we obtain

$$v_\mathrm{P} = \sqrt{\frac{GM}{a}\frac{1+e}{1-e}} ,$$

$$v_\mathrm{A} = \sqrt{\frac{GM}{a}\frac{1-e}{1+e}} .$$

For the planet Earth:

$$e = 0.0167, \qquad a = 1\ \mathrm{AU} = 149.6 \times 10^6\ \mathrm{km},$$

$$\sqrt{\frac{GM}{a}} = 29.78\ \mathrm{km\,s^{-1}}, \qquad \sqrt{\frac{1+e}{1-e}} = 1.0168,$$

$$v_\mathrm{P} = 30.3\ \mathrm{km\,s^{-1}}, \qquad v_\mathrm{A} = 29.3\ \mathrm{km\,s^{-1}}$$

For the planet Mars:

$$e = 0.0933, \qquad a = 1.524\ \mathrm{AU},$$

$$\sqrt{\frac{GM}{a}} = 24.24\ \mathrm{km\,s^{-1}}, \qquad \sqrt{\frac{1+e}{1-e}} = 1.098,$$

$$v_\mathrm{P} = 26.6\ \mathrm{km\,s^{-1}}, \qquad v_\mathrm{A} = 22.1\ \mathrm{km\,s^{-1}}.$$

$\triangle$

Example 14.5. Hohman Orbit to Mars. When a spaceship is sent to another planet, the ship is first placed in a so-called parking orbit around the Earth. To enter the transfer orbit, the spaceship must leave the parking orbit and escape the gravitational field of the Earth.

The rocket engine delivers the thrust necessary for placing the spaceship in the interplanetary orbit. Exactly when the rocket engines should be started depends on the relative position of the Earth and the planet of destination.

In calculating the transfer orbit from the Earth to another planet we shall make a series of simplifying assumptions. We ignore the binding energy of the spacecraft in the gravitational field of the Earth in the parking orbit.

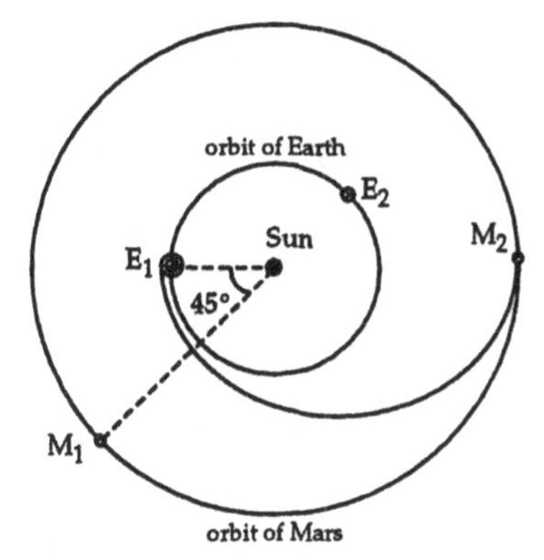

Fig. 14.11. Hohman orbit to Mars. Launch window: E_1 Earth at launch; E_2 Earth at arrival; M_1 Mars at launch; M_2 Mars at arrival

We shall briefly discuss a journey from the Earth to Mars along the so-called *Hohman orbit*, named after the astronomer who first calculated this transfer orbit.

The launch should take place as the spacecraft is on the dark side of the Earth. The velocity of the spacecraft in the parking orbit is then in the same direction as the velocity of the Earth in its orbit around the Sun.

Let us now assume that the spacecraft is nearly free of the gravitational field of the Earth, i.e., we neglect the gravitational field of the Earth. The velocity of the spacecraft in the heliocentric reference frame is assumed to be the same as the orbital velocity of the Earth in this frame ($\approx$ 30 km/s). As we shall demonstrate below only a rather modest increase in the velocity of the spacecraft relative to the heliocentric reference frame is necessary to bring the craft into an elliptical orbit towards Mars.

The Hohman orbit is tangential to the orbit of the Earth at launch, and tangential to the orbit of Mars at arrival. The Hohman orbit is thus a semi-elliptical orbit, whose perihelion coincides with the orbit of the Earth and whose aphelion coincides with the orbit of Mars.

The exact calculation of a Hohman orbit is involved, particularly due to the fact that the plane of the orbit of Mars is slightly tilted relative to ecliptica ($i = 1°51'$).

The essential aspects of the determination of the transfer orbit are nevertheless present in the calculations below, in terms of our simplified model.

Assume that the orbits of the Earth and Mars are in the same plane, the ecliptica. Furthermore, assume that the Earth and Mars perform uniform circular motions around the Sun, with the Sun located in the common center of the orbits.

The radius in the orbit of the Earth is 1 AU, and the radius in the orbit of Mars is 1.52 AU. With the assumption of a circular orbit, the velocity of the Earth is 29.8 $\mathrm{km\,s^{-1}}$ and that of Mars is 24.2 $\mathrm{km\,s^{-1}}$.

The semi-major axis for the Hohman orbit becomes

$$a_{\rm H} = \frac{1+1.52}{2} = 1.26 \ \ \text{AU}.$$

The trip to Mars along the Hohman orbit may be determined from Kepler's third law. The time for one complete revolution in a Hohman orbit is denoted $T_{\rm H}$. Let $T_{\rm H}$ be measured in years. The time for a complete revolution of the Earth (one year) is called $T_{\rm E} = 1$ year, Then, from Kepler's third law:

$$\frac{T_{\rm H}^2}{a_{\rm H}^3} = \frac{T_{\rm E}^2}{a_{\rm E}^3} = \frac{1^2}{1^3} = 1,$$

$$T_{\rm H} = a_{\rm H}^{3/2} = 1.414 \ \text{years}.$$

The travel time τ to Mars corresponds to one half revolution:

$$\tau = 258 \ \text{days}.$$

The sidereal time of revolution for Mars is 687 days. As the spaceship has moved along the Hohman orbit, Mars has moved

$$360^\circ \, \frac{258}{687} \cong 135^\circ \, .$$

If Mars is 45° ahead of the Earth at launch, the spaceship will meet Mars at the point where the Hohman orbit touches the orbit of Mars.

The velocity at launch. The spaceship is in the perihelion of the Hohman orbit at launch. The initial velocity should then be

$$v_{\rm P} = \sqrt{\frac{GM}{a_{\rm H}} \frac{1+e}{1-e}} = \sqrt{\frac{GM}{a_{\rm H}} \frac{r_{\rm A}}{r_{\rm P}}} \, ,$$

where $a_{\rm H} = 1.26$ AU.

For the Earth we know that

$$\sqrt{\frac{GM}{a_{\rm E}}} = 29.8 \ \text{km}\,\text{s}^{-1} \, .$$

For $v_{\rm P}$ we obtain

$$v_{\rm P} = 29.8 \sqrt{\frac{1}{1.26}} \sqrt{\frac{1.52}{1}} = 32.7 \ \text{km}\,\text{s}^{-1} \, .$$

When the rocket engine has released the spaceship from the gravitational field of the Earth, the ship has the same velocity as the Earth relative to the heliocentric frame, i.e., 29.8 $\text{km}\,\text{s}^{-1}$. By means of the rocket engine the spaceship should be given an increase in velocity of Δv, where

$$\Delta v = 32.7 - 29.8 = 2.9 \ \text{km}\,\text{s}^{-1}.$$

When this has taken place, the ship will "fall" along the Hohman orbit to Mars, guided by the gravitational field of the Sun.

At the arrival to Mars, the spaceship is in the apehelion of the Hohman orbit. The velocity of the ship is then

$$v_{\mathrm{A}} = v_{\mathrm{P}} \frac{r_{\mathrm{P}}}{r_{\mathrm{A}}} = 32.7 \frac{1}{1.52} \cong 21.5 \text{ km s}^{-1} .$$

If the spaceship is bound to enter an orbit around Mars the rockets must adjust the velocity of the ship to the velocity of the planet.

Even if the Hohman orbit is inexpensive from the point of view of fuel, it will not be used in the manned expedition to Mars, due to the long time of flight. △

14.11 Problems

Problem 14.1.

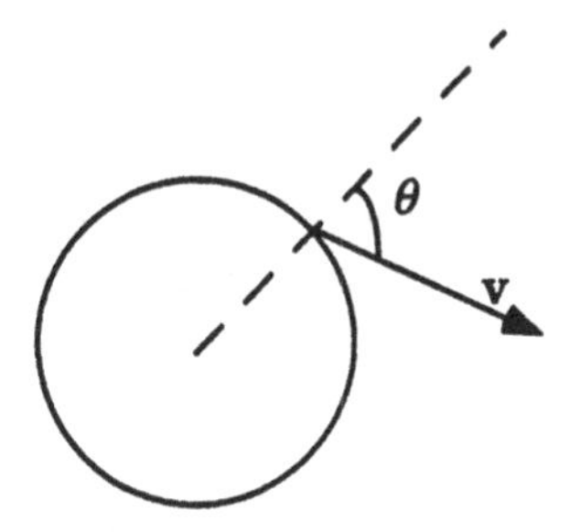

Assume that the Earth is at rest in an inertial frame. A rocket is started, not along the vertical, but in a direction forming an angle θ with the vertical.

(1) Calculate the magnitude of the start velocity v_0, when it is assumed that the rocket just escapes the gravitational field of the Earth.

Remark: the launch facilities of the European Space Agency are located in South America. Why not in, say, northern Norway?

(2) This question deals with the escape velocity from the solar system from a point in the orbit of the Earth. Assume that a rocket interacts only with the gravitational field of the Sun. Determine the smallest velocity v relative to the sun that a spacecraft should be given at the distance of 1 AU from the Sun, so that the spacecraft leaves the solar system.

Problem 14.2. Consider a planet moving in a uniform circular motion in the gravitational field of the Sun. Perform an explicit calculation of the total mechanical energy E and the magnitude L of the angular momentum for the planet, and show explicitly that the excentricity e is zero. Use

$$e^2 = 1 + \frac{2EL^2}{G^2 m^3 M^2},$$

M = mass of Sun,
m = mass of planet,
G = gravitational constant.

Problem 14.3. In a double star system (also called a binary star) one of the stars has the mass $m = 3 \times 10^{30}$ kg and the other has the mass $M = 4 \times 10^{30}$ kg.

Each of the stars performs a uniform circular motion around the center of mass (CM) of the system and relative to an inertial frame. The stars may be considered as mass points. The distance between the stars is 10^{13} m.

(1) Determine the angular velocity ω of the motion of the stars.
(2) Determine the magnitude of the total inner angular momentum of the system, i.e., determine $L_{\mathrm{CM}} = |\,\mathbf{L}_{\mathrm{CM}}\,|$, the angular momentum relative to the CM of the system.

Problem 14.4. Consider the Earth–Moon as an isolated two body system. The Earth and the Moon are assumed to move in circular orbits around the center of mass (CM) of the system.

(1) Determine the position of the center of mass (CM) for the Earth–Moon system.
(2) Determine the orbital speed and the orbital period of revolution of the Moon. (The CM is assumed to be at rest in an inertial system.)

Problem 14.5. Comet Halley orbits the Sun in an elliptical orbit. At perihelion, the distance of the comet from the Sun is 87.8 $\times 10^6$ km. At aphelion the distance from the Sun is 5280 $\times 10^6$ km.

(1) Calculate the period of the comet.
(2) Calculate the speed of the comet relative to the heliocentric reference system when the comet is in the perihelion (V_{P}) and when the comet is in the aphelion (V_{A}).

Problem 14.6. The first artificial satellite, the *Sputnik 1*, was launched on October 4, 1957. Sputnik 1 had a perihelion of 227 km above the surface of the Earth. The speed at perihelion was 8 km s^{-1}, measured relative to the geocentric absolute system (origin at the center of the Earth and axes that point towards the same three fixed stars).

(1) Determine the height above the surface of the Earth that Sputnik 1 had at aphelion.
(2) Determine the orbital period of revolution for Sputnik 1.

Problem 14.7. This problem deals with a Hohman transfer orbit to Venus.

Assume that the Earth and Venus move in the same plane (ecliptica) and in circular orbits around the Sun. The radius of the orbit of Venus is 0.72 AU. (1 AU $= 1.5 \times 10^8$ km). Compare the present problem with Example 14.5. The

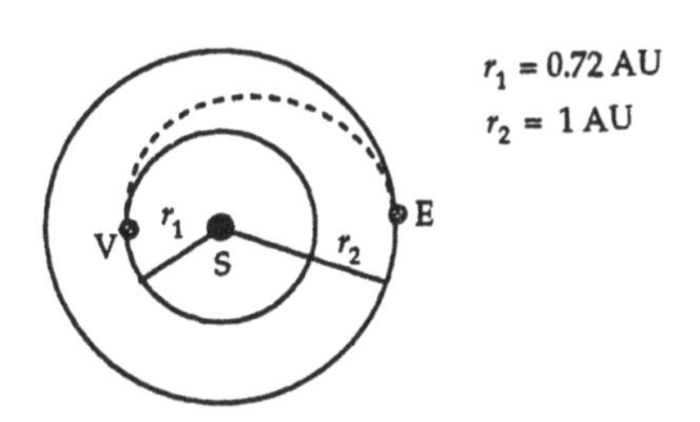

spacecraft is in a parking orbit around the Earth. The launch into a Hohman orbit to Venus (an inner planet) occurs when the spacecraft emerges onto the sunlit side of the Earth. The initial velocity of the spacecraft includes two contributions: the orbital velocity of the Earth about the Sun plus the orbital velocity of the spacecraft around the Earth. When the spacecraft is on the sunlit side of the Earth these contributions are in opposite directions. A rocket thrust in the direction of the orbital motion of the craft around the Earth will allow the spacecraft to escape from the gravitational field of the Earth.

Once the spacecraft is essentially free of the influence of the Earth, the spacecraft will move in an elliptical orbit around the Sun, with an initial speed v_0 relative to the heliocentric reference frame. Note: $2a = 1.72$ AU.

(1) Determine v_0 such that the spacecraft enters a Hohman transfer orbit to Venus (compare with Example 14.5). Show that $v_0 < v_E$, where v_E is the orbital speed of the Earth around the Sun.
(2) Determine the travel time τ to Venus.
(3) Determine the speed v_1 of the spacecraft when it reaches Venus. Show that $v_1 > v_V$, where v_V is the orbital speed of Venus in the heliocentric reference frame.
(4) Discuss the relative positions at launch of Earth and Venus necessary for a Hohman transfer orbit to be realized.

15. Harmonic Oscillators

15.1 Small Oscillations

A particle moving in response to an elastic force proportional to the distance of the particle from the equilibrium point is called a harmonic oscillator. The harmonic oscillator has been treated in the Examples 2.2, 2.3, and 2.7. In Example 2.7 we noted that harmonic motion is a good approximation to many types of oscillations around a point of stable equilibrium.

For a one dimensional harmonic oscillator with force constant k, the potential energy V is given by (motion along x-axis)

$$V(x) = \frac{1}{2}kx^2$$

The potential energy curve for a harmonic oscillator is a parabola. In Figure 15.1 $U(x)$ is a potential energy curve for a particle moving in a certain force field. The particle has an equilibrium point for $x = 0$. The parabola approximating the real potential energy curve around $x = 0$ is shown as a dashed line on the figure. For oscillations with small amplitudes around the equilibrium point, the motion of the particle may be approximated by a harmonic oscillation; we then speak of *small oscillations.*

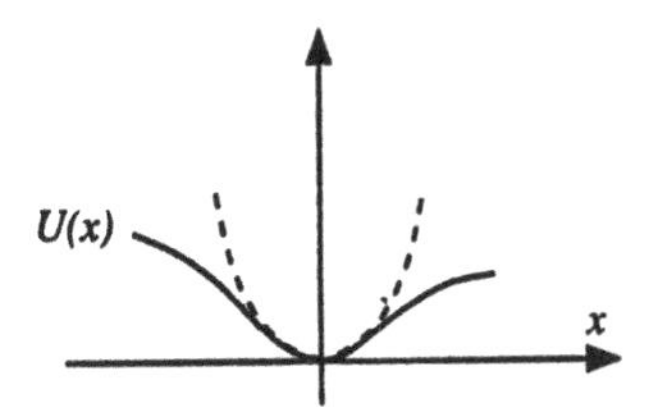

Fig. 15.1. The potential energy function $U(x)$ can be approximated by a quadratic function near the equilibrium point

In Chapter 3 we saw that the mathematical pendulum, i.e., a heavy mass fastened to a light string, will perform harmonic oscillations for small amplitudes (see Section 3.4). The potential energy for the mathematical pendulum as function of θ is:

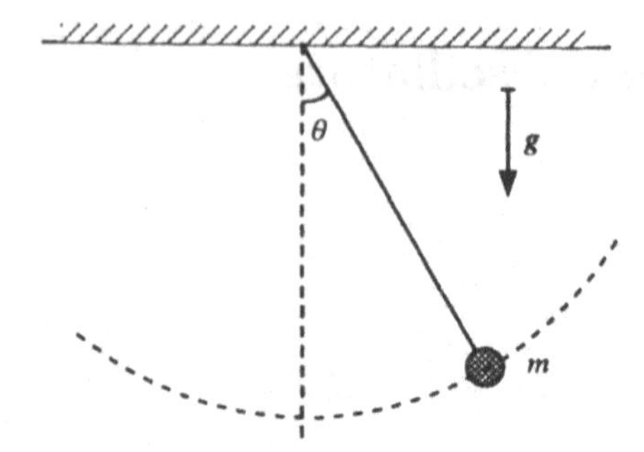

Fig. 15.2. Mathematical pendulum. Potential energy: $U(\theta) = mgL(1 - \cos\theta)$

$$\begin{aligned} U(\theta) &= mgL(1-\cos\theta) \\ &= mgL - mgL\left\{1 - \frac{\theta^2}{2} + \frac{1}{24}\theta^4 + ...\right\} \\ &\cong \frac{1}{2}mgL\theta^2 \end{aligned}$$

For angular amplitudes, where $\cos\theta$ may be approximated by $\cos\theta \cong 1 - (\theta^2/2)$ the pendulum will perform harmonic oscillations.

15.2 Energy in Harmonic Oscillators

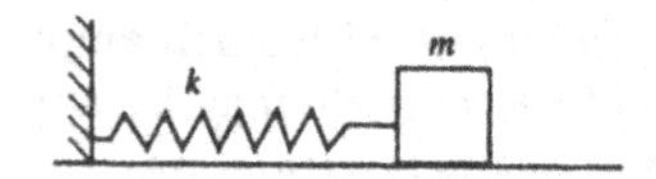

Fig. 15.3. Harmonic oscillator

The equation of motion for the oscillator sketched in Figure 15.3 is:

$$m\ddot{x} = -kx\,,$$

or, with $\omega_0^2 = k/m$,

$$\ddot{x} + \omega_0^2 x = 0\,. \tag{15.1}$$

The general solution may be written as follows:

$$x = x_0 \cos(\omega_0 t + \varphi)\,. \tag{15.2}$$

The amplitude x_0 and the phase angle φ are given by the initial conditions. The period of the oscillations is $T = 2\pi/\omega_0$. Let us assume we have the following initial conditions:

$$t = 0, \qquad x(0) = x_0, \qquad \text{and} \qquad \text{v}(0) = 0.$$

Equation (15.2) becomes

$$x(t) = x_0 \cos \omega_0 t.$$

The velocity is

$$v(t) = -\omega_0 x_0 \sin \omega_0 t.$$

We may now calculate the total mechanical energy E_0 in the oscillation:

$$E_0 = U + K,$$

where K is the kinetic energy and U is the potential energy.

$$E = \frac{1}{2} k x_0^2 \cos^2 \omega_0 t + \frac{1}{2} m \omega_0^2 x_0^2 \sin^2 \omega_0 t,$$

or, since $k \equiv m\omega_0^2$,

$$E = \frac{1}{2} m \omega_0^2 x_0^2.$$

The total energy is – of course – constant in time (conservative force). During the motion the energy oscillates between potential energy and kinetic energy.

The time average of the kinetic energy K over a period is

$$\langle K \rangle = \frac{1}{T} \int_0^T \frac{1}{2} m \dot{x}^2 \mathrm{d}t.$$

Using $\int_0^T \sin^2 \omega_0 t \mathrm{d}t = \frac{1}{2}$ we obtain

$$\langle K \rangle = \frac{1}{4} m \omega_0^2 x_0^2.$$

For the potential energy averaged over one period we get by similar calculations

$$\langle U \rangle = \frac{1}{4} m \omega_0^2 x_0^2.$$

The equality of the average value of the potential and the kinetic energy is a property of harmonic oscillations. The time average of the total energy is equal to the total energy in the system:

$$\langle E \rangle = \langle U \rangle + \langle K \rangle = E = \frac{1}{2} m \omega_0^2 x_0^2.$$

15.3 Free Damped Oscillations

Consider first an oscillator moving according to the equation of motion

$$m\ddot{x} = -kx\,. \tag{15.3}$$

This equation will have the solution given by (15.2):

$$x = x_0 \cos(\omega_0 t + \varphi)\,.$$

The oscillator would move forever, but this will never happen for any real oscillator. Friction, in some form or another, will always occur.

We have previously considered particles moving under frictional forces (see, for instance, Example 2.4). We now proceed to discuss an important case of the damped oscillator. We assume that the frictional force is proportional to the velocity of the particle. This will be the case, approximately, for a particle moving through a fluid (oil, air). Later in your studies you will find that a frictional force proportional to the velocity of the particle to a good approximation describes even the damping of an oscillating electric charge (an electron) as it emits electromagnetic waves.

There are thus several justifications for studying an oscillator that has the following equation of motion instead of (15.3):

$$m\ddot{x} = -kx - b\dot{x}\,, \tag{15.4}$$

where b is a constant. Rearranging the equation:

$$\ddot{x} + \frac{b}{m}\dot{x} + \frac{k}{m}x = 0\,.$$

We use the definitions

$$\frac{b}{m} \equiv \gamma\,, \qquad \text{and} \qquad \frac{k}{m} \equiv \omega_0^2\,,$$

and obtain

$$\ddot{x} + \gamma\dot{x} + \omega_0^2 x = 0\,. \tag{15.5}$$

Equation (15.5) is a homogeneous, linear, second-order differential equation with constant coefficients. The general solution of this type of differential equations may be found in books on calculus. We shall here simply discuss three cases.

15.3.1 Weakly Damped Oscillations

We assume that the damping is light, so that the oscillator performs nearly harmonic oscillations. More precisely, we assume that $\gamma \ll \omega_0$ (weak damping). In this case the solution of (15.5) may be written as follows:

$$x = x_0\, \mathrm{e}^{-\frac{\gamma t}{2}} \cos(\omega_\mathrm{d} t + \theta)\,, \tag{15.6}$$

where

$$\omega_\mathrm{d} = \omega_0 \sqrt{1 - (\gamma/2\omega_0)^2}\,.$$

The amplitude and the phase angle θ is determined by the initial conditions. Note: our assumption $\gamma \ll \omega_0$ implies that $1 - (\gamma/2\omega_0)^2 > 0$, which again means that the oscillation frequency ω_d is a real number.

Question. Show by substitution that (15.6) is a solution of (15.5).

The word 'frequency' is often used for ω. The correct designation of ω is *angular frequency*. The frequency ν denotes the number of oscillations per second.

$$\nu = \frac{1}{T}\,; \quad \omega = 2\pi\nu\,.$$

Figure 15.4 shows a graphical representation of the solution (15.6).

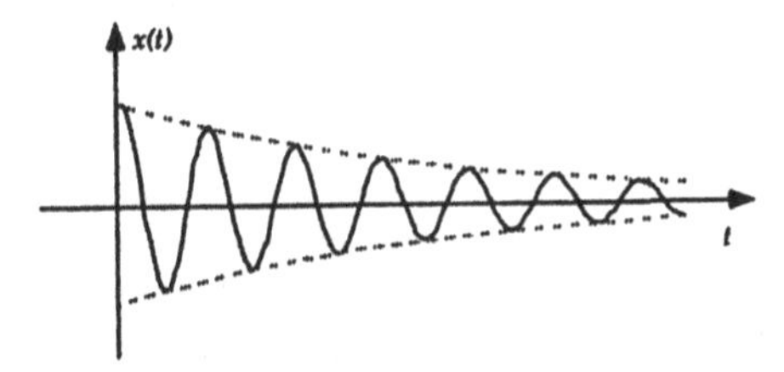

Fig. 15.4. Damped harmonic oscillations. The dashed line corresponds to $\mathrm{e}^{-\gamma t/2}$. See (15.6)

Note. The damping makes the oscillation frequency ω_d smaller than the corresponding frequency ω_0 for undamped harmonic oscillations. If $\gamma \ll \omega_0$ we see from (15.6) that $\omega_\mathrm{d} \approx \omega_0$. The case $\gamma \ll \omega_0$ is by far the most important case for harmonic oscillations.

When a hammer hits a church bell in a brief blow, the church bell starts to ring. The surface of the bell performs harmonic oscillations with several frequencies, and, due mainly to a coupling to the surrounding air, the oscillations of the bell are damped. The energy in the oscillations of the bell is slowly transferred to other degrees of freedom, in this case sound waves. The bell emits sound waves corresponding to the frequency of the oscillations. The bell will ring for some time. It is a weakly damped oscillator, $\gamma \ll \omega_0$. One might say that γ in this case describes the coupling between the oscillations of the bell and the sound field.

15.3.2 Strongly Damped Oscillations

For $\gamma/2 > \omega_0$ the oscillation frequency $\omega_d = \omega_0 \left[1 - (\gamma/2\omega_0)^2\right]^{1/2}$ would become imaginary. For $\gamma/2 > \omega_0$ one obtains the following solutions of (15.5):

$$x(t) = A \exp\left[-\frac{\gamma}{2} + \sqrt{\frac{\gamma^2}{4} - \omega_0^2}\right] t + B \exp\left[-\frac{\gamma}{2} - \sqrt{\frac{\gamma^2}{4} - \omega_0^2}\right] t . \quad (15.7)$$

This represent a nonoscillating motion of the particle, i.e., an aperiodic motion (see Figure 15.5).

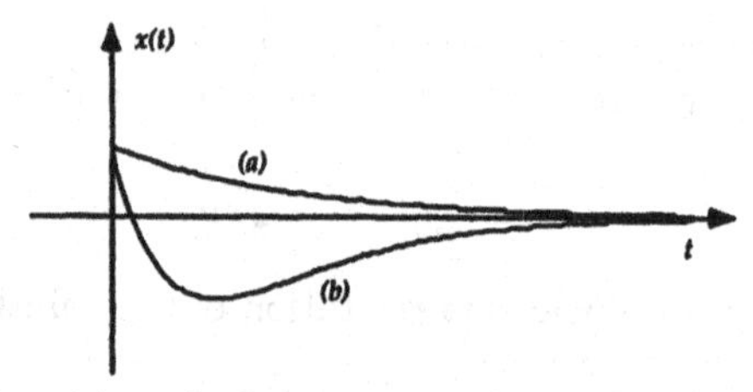

Fig. 15.5. Aperiodic motion of a harmonic oscillator. Graph marked (a) illustrates Equation (15.7). Graph marked (b) illustrates Equation (15.8)

Physically the solution (15.7) means, for instance, that the mass is pulled out at a given distance from the point of equilibrium and released with zero velocity. For $\gamma/2 > \omega_0$ the damping is so strong that the particle cannot oscillate at all. The particle slowly moves towards the point of equilibrium.

15.3.3 Critical Damping

For $\frac{\gamma}{2} = \omega_0$ the solution becomes

$$x(t) = A \exp\left(-\frac{\gamma t}{2}\right) + Bt \exp\left(\frac{-\gamma t}{2}\right) . \quad (15.8)$$

See the curve marked (*b*) in Figure 15.5. Equation (15.8) again describes a nonoscillating solution. The motion for $\gamma/2 = \omega_0$ is called critically damped motion.

We shall not discuss the nonoscillating solutions any further. Critical damping ($\gamma/2 = \omega_0$) is of interest in the design of some measuring instruments.

15.4 Energy in Free, Weakly Damped Oscillations

In the case of weak damping ($\gamma \ll \omega_0$) the solution of the differential Equation (15.5) is (15.6). We shall use the approximation

$$\omega_\mathrm{d} = \omega_0\sqrt{1 - (\gamma/2\omega_0)^2} \approx \omega_0 \,.$$

With this approximation the solution becomes

$$x(t) = x_0 \exp\left(-\frac{\gamma t}{2}\right)\cos(\omega_0 t + \theta)\,. \tag{15.9}$$

From (15.9) we could calculate the velocity of the mass and then the kinetic energy as functions of time for the weakly damped oscillator. We shall, however, make some further simplifying assumptions.

During one period ($T = 2\pi/\omega_0$) the amplitude will decrease by an amount $\exp(-\gamma T/2) = \exp(-\pi\gamma/\omega_0)$. Since $\gamma \ll \omega_0$, the amplitude will decrease very little during one oscillation. In the following calculations we assume that the amplitude is constant over one oscillation. The velocity then becomes (approximately)

$$v \approx -\omega_0 x_0 \exp\left(-\frac{\gamma t}{2}\right)\sin(\omega_0 t + \theta)\,. \tag{15.10}$$

For the energy in the oscillations we then get

$$E = K + U\,,$$

$$E = \frac{1}{2}m\omega_0^2 x_0^2 \exp\left(-\frac{\gamma}{t}\right)\sin^2(\omega_0 t + \theta) + \frac{1}{2}k x_0^2 \exp\left(\frac{-\gamma}{t}\right)\cos^2(\omega_0 t + \theta)\,.$$

Using

$$\langle \sin^2(\omega_0 t + \theta)\rangle = \langle \cos^2(\omega_0 t + \theta)\rangle = \frac{1}{2}\,.$$

we get the following averages:

$$\langle K\rangle = \frac{1}{4}m\omega_0^2 x_0^2 \exp(-\gamma t)\,, \tag{15.11}$$

$$\langle U\rangle = \frac{1}{4}m\omega_0^2 x_0^2 \exp(-\gamma t)\,, \tag{15.12}$$

and

$$\langle E\rangle = \frac{1}{2}m\omega_0^2 x_0^2 \exp(-\gamma t)\,. \tag{15.13}$$

That means: the energy in the oscillations decreases exponentially. For instance, pull the oscillator out to a given distance x_0, and release it with the velocity zero. The oscillator has received the (potential) energy

$$E_0 = \frac{1}{2}k x^2(0) = \frac{1}{2}m\omega_0^2 x_0^2\,.$$

We say that the oscillator is in an excited state, with the excitation energy E_0. In the ensuing oscillations the oscillator will loose energy due to the damping. The decrease of the (average) energy will follow the differential equation

$$\frac{dE}{dt} = -\gamma E. \tag{15.14}$$

This equation has the solution

$$E = E_0 \exp(-\gamma t). \tag{15.15}$$

With $\tau \equiv 1/\gamma$, (15.15) becomes

$$E = E_0 \exp\left(-\frac{t}{\tau}\right). \tag{15.16}$$

The parameter τ has the dimension of time and τ is called the lifetime of the excited state of the oscillator.

Consider again (15.11)–(15.13). In spite of the fact that these equations describe the time average over one period they contain the time variable t! These equations represents the average over one period at about the time t. As t increases, the average values decrease exponentially. Note: the energy decreases as $\exp -\gamma t$, while the amplitude decreases as $\exp -\gamma t/2$. The energy is proportional to the square of the amplitude.

The frictional force, $f = -b\dot{x} = -m\gamma\dot{x}$, is responsible for the dissipation of energy in the oscillations. Put differently, the frictional force describes the fact that during the oscillations part of the oscillation energy passes into other degrees of freedom (e.g., into heat as in the damping of a pendulum or into sound waves as in case of the vibrating church bell).

The power dissipated is equal to the negative of the (average) rate at which the frictional force performs work. Assuming $\gamma \ll \omega_0$ (weak damping) we obtain for the power dissipated:

$$\begin{aligned} P = \langle fv \rangle &\approx -\langle m\gamma\dot{x}v \rangle \\ &= -\gamma\frac{1}{2}m\omega_0^2 x_0^2 \exp(-\gamma t) \\ &= -\gamma E\,, \end{aligned}$$

in agreement with (15.14). We have used (15.10) and $\dot{x} = v$.

15.5 Forced Oscillations

Consider first an oscillator without damping on which an external force $\mathbf{F}(t)$ acts along the x-axis. The equation of motion is then

$$m\ddot{x} = -kx + F(t). \tag{15.17}$$

The external driving force may have any kind of functional dependence on time. We shall here consider the particular case of an oscillating force $F(t)$ of the form

$$F(t) = F_0 \cos \omega t \,. \tag{15.18}$$

Note. The frequency ω of the applied force is in general different from the so-called eigenfrequency $\omega_0 = (k/m)^{1/2}$ of the undamped oscillator.

If you want a concrete situation to think about, you may assume that the oscillating mass carries an electric charge q. If a monochromatic, electromagnetic wave of frequency ω (a light wave) passes over the mass (an electron) the electric field vector $\mathbf{E}_0 \cos \omega t$ acts on the charge with a periodic force, $\mathbf{F}(t) = q\mathbf{E}_0 \cos \omega t$ ($\mathbf{E}_0$ is in the x-direction).

The use of the expression (15.18) is not an unreasonable limitation. Any periodic function $F(t)$ may be approximated very accurately by a sum of sine or cosine functions, a so-called Fourier series. A nonperiodic function may be expressed in terms of a so-called Fourier integral.

With $F(t) = F_0 \cos \omega t$ the equation of motion may be written

$$\ddot{x} + \omega_0^2 x = \frac{F_0}{m} \cos \omega t \,. \tag{15.19}$$

We shall not discuss here the general solution of (15.19). We seek only one solution, the so-called steady state solution (see below).

It is natural to assume that the oscillator starts to oscillate with the frequency ω of the applied force. Our problem may then be formulated as follows.

Determine the value of the amplitude A for which the following expression is a solution of (15.19):

$$x = A \cos \omega t \,. \tag{15.20}$$

Insert into the equation of motion:

$$-\omega^2 A \cos \omega t + \omega_0^2 A \cos \omega t = \frac{F_0}{m} \cos \omega t \,,$$

We find that

$$A = \frac{F_0/m}{\omega_0^2 - \omega^2} \,.$$

The solution of (15.19) becomes

$$x(t) = \frac{F_0/m}{\omega_0^2 - \omega^2} \cos \omega t \,. \tag{15.21}$$

This means that the mass oscillates with the frequency ω of the applied force and with an amplitude that depends on the eigenfrequency of the oscillator. For $\omega < \omega_0$ the oscillations of the mass and the applied force are in phase.

For $\omega > \omega_0$ the oscillations of the mass and the applied force are in antiphase. For large values of ω the amplitude of the oscillations becomes small, $\omega \to \infty$, $A \to 0$. The oscillator cannot follow the rapidly oscillating impressed force $F(t) = F_0 \cos \omega t$.

According to the result (15.21) a remarkable thing happens when ω is close to ω_0. Actually, for $\omega \to \omega_0$, $A \to \infty$. The response of the oscillator to the applied force grows without bounds. An infinite amplitude is of course impossible in a real physical system. As we shall see below, when we include friction, we will get a large amplitude for $\omega = \omega_0$, but certainly not an infinite amplitude. The phenomenon appearing for $\omega = \omega_0$ is called *resonance*.

The phenomenon of resonance is well known to anyone who has pushed a child on a swing. If we choose the right time for pushing and the right frequency, the swing will eventually go very high, even with a slight push.

15.6 The Forced Damped Harmonic Oscillator

The solution of this general problem will contain as special cases the solution of the unforced oscillator (with and without damping) as well as the solution for the undamped forced oscillator.

The equation of motion is now

$$m\ddot{x} = -kx - m\gamma\dot{x} + F_0 \cos \omega t \,,$$

or

$$\ddot{x} + \gamma\dot{x} + \omega_0^2 x = \frac{F_0}{m} \cos \omega t \,. \tag{15.22}$$

From the theory of differential equations it is known that the solution of (15.22) may be written as a sum of two terms: (1) the complete solution, $x_1(t)$, of the corresponding homogeneous equation $\ddot{x} + \gamma\dot{x} + \omega_0^2 x = 0$, and (2) an arbitrary solution, $x_2(t)$, to the inhomogeneous equation (15.22).

We write the solution in the form

$$x(t) = x_1(t) + x_2(t) \,,$$

$$x(t) = x_0 \exp\left(-\frac{\gamma t}{2}\right) \cos(\omega_{\mathrm{d}} t + \varphi) + A \cos(\omega t - \theta) \tag{15.23}$$

The first term, $x_1 = x_0 \exp(-\gamma t/2) \cos(\omega_{\mathrm{d}} t + \varphi)$, is the solution obtained for the free (no external force) damped oscillator. As before

$$\omega_{\mathrm{d}} = \omega_0 \left[1 - (\gamma/2\omega_0)^2\right]^{1/2} .$$

The integration constants x_0 and φ are determined by the initial conditions. The initial conditions are, however, not interesting in the present case. The exponential factor $\exp(-\gamma t/2)$ makes the solution $x_1(t)$ a so-called transient which dies out, as t increases, leaving the solution as

$$x_2 = A\cos(\omega t - \theta)\,. \tag{15.24}$$

This so-called steady-state solution is reached when the transient has died out. In the steady state the mass oscillates with the frequency of the applied force and with a constant amplitude A. With the chosen sign for the phase angle θ in the term $\cos(\omega t - \theta)$, θ gives the lagging of the oscillations of the mass behind the phase of the applied force, $F_0 \cos \omega t$. A negative value of θ will mean that the oscillations of the mass are ahead of the applied force in phase.

We now determine the amplitude A and the phase angle θ. We rewrite (15.24) as follows:

$$x_2(t) = A\cos\omega t\cos\theta + A\sin\omega t\sin\theta\,. \tag{15.25}$$

To determine for which values of A and θ the expression (15.25) is a solution for the equation of motion (15.22), we differentiate twice:

$$\dot{x}_2 = A\omega(-\sin\omega t\cos\theta + \cos\omega t\sin\theta)\,,$$
$$\ddot{x}_2 = A\omega^2(-\cos\omega t\cos\theta - \sin\omega t\sin\theta)\,.$$

Substituting these results into (15.22) we obtain, after some rearrangements,

$$\left[-\omega^2 A\cos\theta + \gamma\omega A\sin\theta + \omega_0^2 A\cos\theta - \frac{F_0}{m}\right]\cos\omega t$$
$$+\left[-\omega^2 A\sin\theta - \gamma\omega A\cos\theta + \omega_0^2 A\sin\theta\right]\sin\omega t = 0$$

This equation must be true for all values of t. Consequently the expressions in the parenthesis must be *zero separately*. We therefore get a pair of equations in the two unknowns, θ and A:

$$\left.\begin{array}{rcl} \left(\omega_0^2 - \omega^2\right)\cos\theta + \gamma\omega\sin\theta & = & \frac{F_0}{mA} \\ \left(\omega_0^2 - \omega^2\right)\sin\theta - \gamma\omega\cos\theta & = & 0 \end{array}\right\}. \tag{15.26}$$

From the second equation we see that θ can be determined (up to multiples of π) as a function of ω. We fix θ as the $0 < \theta < \pi$ solution of

$$\cot\theta = \frac{\omega_0^2 - \omega^2}{\gamma\omega} \tag{15.27}$$

for $0 < \omega < \infty$.

Squaring the two equations (15.26), and adding, we get

$$\left(\frac{F_0}{mA}\right)^2 = (\omega_0^2 - \omega^2)^2 + (\gamma\omega)^2$$

from which we isolate A:

$$A = \frac{F_0/m}{\sqrt{(\omega_0^2 - \omega^2)^2 + \gamma^2\omega^2}}\,. \tag{15.28}$$

The expressions (15.27) for $\cot\theta$ and (15.28) for the amplitude A give, together with (15.23), the complete solution of the problem of the damped, forced harmonic oscillator. Note: $F_0/(mA)$ is always positive.

15.7 Frequency Characteristics

A harmonic oscillator moves in response to an impressed external force $F = F_0 \cos \omega t$. The transients die out exponentially with time. The steady state response is $x(t) = A\cos(\omega t - \theta)$, where

$$A = \frac{F_0/m}{\sqrt{(\omega_0^2 - \omega^2)^2 + \gamma^2\omega^2}},$$

$$\tan\theta = \frac{\gamma\omega}{\omega_0^2 - \omega^2}.$$

Figure (15.6) shows graphs of the amplitude function $A(\omega)$ and $\theta(\omega)$, where θ is the phase. The joint behavior of both amplitude and phase of a (harmonically) forced oscillator, often summarized graphically, is known as the *frequency characteristics* of the oscillator.

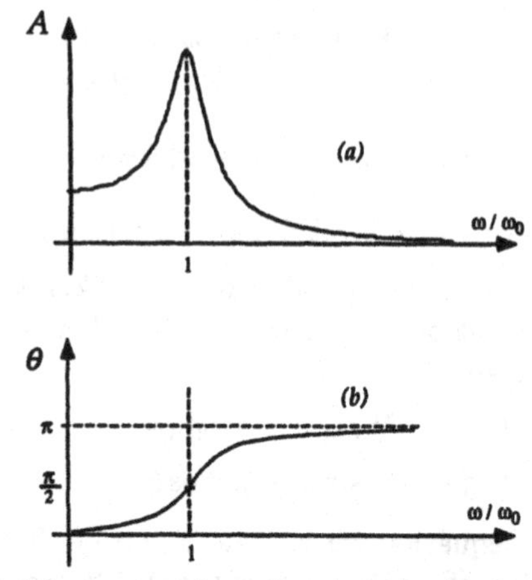

Fig. 15.6. (a) Amplitude, (b) phase angle as function of ω/ω_0 for a lightly damped ($\gamma \ll \omega_0$) forced harmonic oscillator; ω is the frequency of the external force $F = F_0 \cos \omega t$

We shall consider some limiting cases, assuming always the weakly damping case, $\gamma \ll \omega_0$, or $\omega_0\tau \gg 1$, where $\tau \equiv 1/\gamma$.

15.7.1 $\omega \ll \omega_0$: A Low Driving Frequency

For a very low frequency of the external driving force, i.e., for $\omega \to 0$ we obtain

$$A \to \frac{F_0}{m\omega_0^2}, \qquad \text{and} \qquad \theta \to 0.$$

The amplitude is the same as the amplitude for a static force and the oscillations are in phase with the impressed, low-frequency force.

15.7.2 $\omega \gg \omega_0$: A High Driving Frequency

$$A \to \frac{F_0}{m\omega^2}, \quad \text{and } \theta \to \pi$$

Note: for $\omega = \omega_0$, $\theta = \frac{\pi}{2}$.

15.7.3 $\omega \cong \omega_0$: Resonance

The case where the driving frequency ω is close to or equal to the eigenfrequency ω_0 is called resonance. For $\omega = \omega_0$ we get

$$A = \frac{F_0}{m\gamma\omega_0}, \quad \text{and } \theta = \frac{\pi}{2}.$$

The maximum response does not occur exactly for $\omega = \omega_0$. Let us determine the frequency ω_m for which the denominator of A has a minimum:

$$\frac{\mathrm{d}}{\mathrm{d}\omega}\left[\left(\omega_0^2 - \omega^2\right)^2 + \gamma^2\omega^2\right] = 0$$

$$2(\omega_0^2 - \omega^2)(-2\omega) + 2\gamma^2\omega = 0$$

$$\omega_\mathrm{m} = \omega_0\sqrt{1 - \frac{\gamma^2}{2\omega_0^2}}. \tag{15.29}$$

The exact oscillation frequency for the free, damped harmonic oscillator was found to be [see (15.6)]

$$\omega_\mathrm{d} = \omega_0\sqrt{1 - \frac{\gamma^2}{4\omega_0^2}}.$$

For weak damping, i.e., for $\gamma \ll \omega_0$, $\omega_\mathrm{m} \approx \omega_\mathrm{d} \approx \omega_0$. At resonance the phase angle θ is $\pi/2$. While the displacement is out of phase with the force in the case of resonance, we can easily prove that for $\theta = \pi/2$ the velocity is exactly in phase with the force:

$$x = A\cos(\omega t - \theta)$$

$$\dot{x} = -A\omega\sin(\omega t - \theta), \qquad \text{for } \theta = \frac{\pi}{2},$$

$$\dot{x} = A\omega\cos\omega t.$$

That means: in resonance, i.e., for $\theta = \pi/2$, the velocity of the mass is in phase with the impressed force, $F = F_0\cos\omega t$. At resonance the work performed by the applied force on the oscillating mass reaches a maximum. Due to the perfect phase matching the mass gets pushed at just the right times and at the right places, and the oscillator absorbs maximum energy per unit of time from the applied force.

In the next section we shall calculate the time average of the power absorbed as function of the impressed frequency ω.

15.8 Power Absorption

The time average of the work performed by the applied force on a driven oscillator and for a given driving frequency ω is

$$P = \langle F(t)\dot{x}(t)\rangle = \langle F_0 \cos\omega t(-1)A\omega \sin(\omega t - \theta)\rangle \,.$$

The calculation of P is somewhat tedious, but the result is interesting and important. Using the basic identities

$$\sin(\omega t - \theta) = \sin\omega t \cos\theta - \cos\omega t \sin\theta \,,$$

and

$$\langle \cos\omega t \sin\omega t\rangle = 0 \,, \quad \langle \cos^2\omega t\rangle = \frac{1}{2} \,,$$

we get

$$\begin{aligned} P &= -F_0 A\omega \left\{\langle \cos\omega t \sin\omega t \cos\theta\rangle - \langle \cos^2\omega t \sin\theta\rangle\right\} \\ &= \frac{1}{2} F_0 \omega A \sin\theta \,. \end{aligned} \tag{15.30}$$

The phase angle θ is decisive for the power absorption. To find the product $A\sin\theta$, we multiply the first equation in (15.26) by $\gamma\omega$, and the second equation by $(\omega_0^2 - \omega^2)$. Adding the two resulting equations, we find that

$$A\sin\theta = \gamma\omega \frac{F_0/m}{[(\gamma\omega)^2 + (\omega_0^2 - \omega^2)^2]} \,.$$

Substituting this into (15.30), we get

$$P = \frac{1}{2}\frac{F_0^2}{\gamma m}\frac{\gamma^2\omega^2}{[(\gamma\omega)^2 + (\omega_0^2 - \omega^2)^2]} \,. \tag{15.31}$$

Figure 15.7 shows the power absorption as a function of frequency for a weakly damped oscillator.

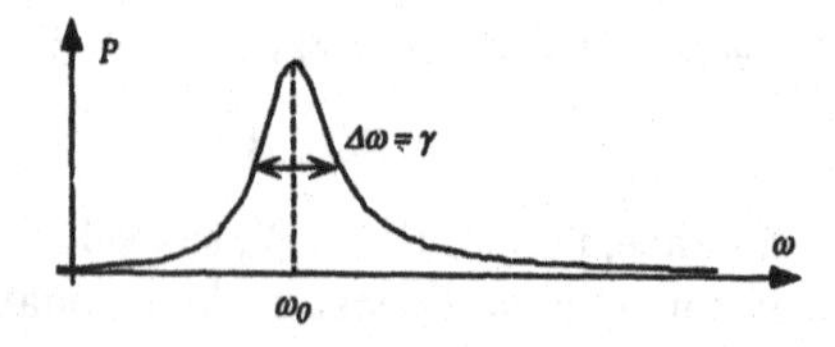

Fig. 15.7. Absorption of power for a weakly damped, driven harmonic oscillator. $F = F_0 \cos\omega t$

For $\omega = \omega_0$ the absorption of power becomes

$$P = \frac{1}{2}\frac{F_0^2}{\gamma m}\,.$$

The power absorption is reduced to one half the value at resonance for

$$\frac{(\gamma\omega)^2}{(\omega_0^2 - \omega^2)^2 + \gamma^2\omega^2} = \frac{1}{2}\,,$$

$$\pm\gamma\omega = \omega_0^2 - \omega^2\,.$$

We are interested only in frequencies ω close to the eigenfrequency ω_0. We write

$$\gamma\omega = (\omega_0 + \omega)(\omega_0 - \omega) \approx 2\omega_0(\omega_0 - \omega)\,,$$

or

$$\gamma\omega_0 \approx 2\omega_0(\omega_0 - \omega)\,,$$

$$\omega_0 - \omega = \frac{\gamma}{2}\,.$$

For frequencies $\omega_2 = \omega_0 - \gamma/2$ and $\omega_1 = \omega_0 + \gamma/2$ the absorption has decreased to half the value it has at resonance. The full width at half maximum (FWHM) of the absorption line (see Figure 15.7) is $\Delta\omega = \gamma$. The damping constant γ thus determines the width of the resonance line.

15.9 The Q-Value of a Weakly Damped Harmonic Oscillator

The resonance properties of a given weakly damped oscillator are often characterized by the so-called Q-value.

Consider first an unforced, weakly damped harmonic oscillator. The energy dissipated during one oscillation is small compared to the amount of energy stored in the oscillations. The Q-value for such an oscillator is defined as 2π times the mean energy stored, divided by the energy dissipated per period. (The factor 2π could have been avoided if the work done per radian had been used instead of the work per period.)

Q is a useful number only if $Q \gg 1$, i.e., if we consider weakly damped oscillations $(\gamma \ll \omega_0)$. Therefore we may assume that

$$\omega_d = \omega_0\sqrt{1 - (\frac{\gamma}{2\omega_0})^2} \approx \omega_0$$

For Q we obtain (remember $T = 2\pi/\omega$)

$$Q = 2\pi\frac{E}{PT} = \frac{\omega_0 E}{P}\,.$$

Q is a dimensionless number. Previously we found (see Section 15.4)

$$P \equiv \langle fv \rangle = -\gamma E = -\frac{E}{\tau}\,.$$

From this,

$$Q = \omega_0 \tau = \frac{\omega_0}{\gamma}\,.$$

Expressed by means of Q, the energy for a weakly damped oscillator decreases according to the equation

$$\frac{\mathrm{d}}{\mathrm{d}t}E = -\frac{\omega_0}{Q}E\,,$$

$$E(t) = E(0)\exp\left(-\frac{\omega_0}{Q}t\right)\,.$$

The amplitude response to an applied, static force F_0 is

$$A(\omega) = A(0) = \frac{F_0}{m\omega_0^2}\,.$$

The amplitude at resonance for a driven oscillator is

$$A(\omega) = A(\omega_0) = \frac{F}{\gamma m \omega_0}\,.$$

We thus have

$$Q = \frac{A(\omega_0)}{A(0)} = \frac{\omega_0}{\gamma} = \omega_0 \tau\,.$$

Sometimes Q is called "the resonance amplification" or "the Quality factor" of an oscillator. A high value of Q signifies high amplification, i.e., small damping.

Considered as an amplifying system the weakly damped oscillator reacts mainly in a frequency interval of width $\Delta\omega = \gamma$ around the resonance frequency ω_0. This frequency interval is often called the band width. Let the relative band width be called W :

$$W \equiv \frac{\Delta\omega}{\omega_0} = \frac{\gamma}{\omega_0}\,.$$

For the amplification we found

$$Q = \frac{\omega_0}{\gamma}\,.$$

The connection between the amplification Q and the relative bandwidth W is thus

$$Q \times W = 1\,.$$

This connection is fundamental for all oscillating systems. It may be a mass on a spring or an oscillator in a radio receiver. If we want a system to react in a broad band of frequencies we must pay the price of a decrease in amplification. The Q-value for mechanical oscillators can reach values of 10^3. This is obtained for, say, a vibrating violin string. A church bell will have a somewhat lower value of Q. For electric oscillators higher values are obtained. For vibrating electrons inside atom, i.e., "antennas" emitting light waves, Q is about 10^7. For nuclei emitting γ-rays, Q-values of about 10^{12} may be reached.

Q is sometimes also used to describe the shape of the resonance line. The width of the resonance line is

$$\gamma = \frac{\omega_0}{Q} .$$

The exact value of the frequency where maximum amplitude occurs is

$$\omega = \omega_0 \sqrt{1 - \frac{1}{2Q^2}} .$$

For $Q = 1000$

$$\omega = \omega_0 \sqrt{1 - \frac{1}{2} \times 10^{-6}} ,$$

$$\omega \cong \omega_0 \left\{ 1 - \frac{1}{4} \times 10^{-6} + \dots \right\} ,$$

where $\omega \cong \omega_0$ with a precision of about 1 in one million.

15.10 The Lorentz Curve

Consider again the result (15.31):

$$P = \frac{1}{2} \frac{F_0^2}{\gamma m} \frac{\gamma^2 \omega^2}{(\omega_0^2 - \omega^2)^2 + \gamma^2 \omega^2} .$$

We assume weak damping, $\gamma \ll \omega_0$. We are then normally interested only in frequencies close to ω_0. The often used approximation

$$\omega_0^2 - \omega^2 = (\omega_0 + \omega)(\omega_0 - \omega) \cong 2\omega_0(\omega_0 - \omega)$$

gives

$$P = \frac{1}{2} \frac{F_0^2}{\gamma m} \frac{\gamma^2/4}{(\omega_0 - \omega)^2 + \gamma^2/4} . \tag{15.32}$$

This is the so-called *Lorentz curve* of a resonance line for a weakly damped oscillator. You will meet this result often in deeper studies of physics, particularly in the study of emission and absorption of electromagnetic radiation. The result (15.32) describes for instance the shape of the so-called spectral lines emitted by atoms. The shape is similar to the one shown in Figure 15.7. From (15.32) it is easily demonstrated that the full width at half maximum (FWHM) of the Lorentz curve is $\Delta\omega = \gamma$.

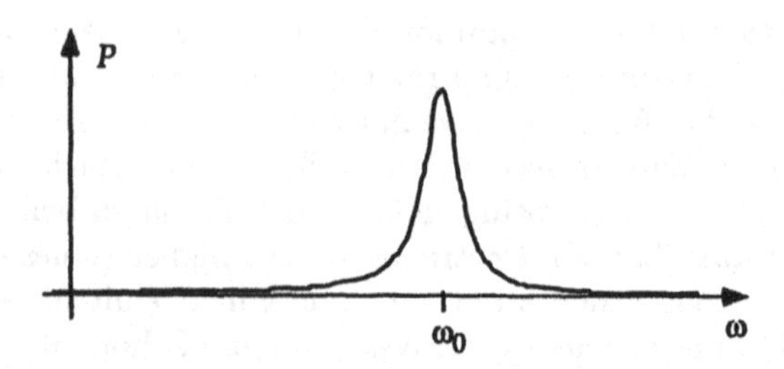

15.11 Complex Numbers

The solution of the equation of motion for the harmonic oscillator becomes more elegant if we employ complex numbers in the derivation. The equation of motion is

$$\ddot{x} + \gamma\dot{x} + \omega_0^2 x = \frac{F_0}{m}\cos\omega t\,. \tag{15.33}$$

We represent the applied force $F(t)$ as the real part of a complex number (we assume $\mathbf{F}$ is along the x-axis) as

$$F(t) = F_0 \exp \mathrm{i}\omega t\,, \tag{15.34}$$

where in principle the amplitude F_0 may itself be a complex number, $F_0 = a + ib$. The amplitude F_0 is independent of time.

Analogously, we assume a solution $x(t)$ of the form

$$x(t) = x_0 \exp \mathrm{i}\omega t\,, \tag{15.35}$$

where x_0 is the (complex) amplitude. Inserting $x(t)$ and $F(t)$ in the equation of motion we obtain

$$\left[(\mathrm{i}\omega)^2 x_0 + \mathrm{i}\omega\gamma x_0 + \omega_0^2 x_0\right] \exp \mathrm{i}\omega t = \frac{F_0}{m}\exp \mathrm{i}\omega t\,,$$

or

$$x_0 = \frac{F_0}{m(\omega_0^2 - \omega^2 + \mathrm{i}\gamma\omega)}\,. \tag{15.36}$$

The new complex amplitude x_0 is again proportional to the amplitude of the force, F_0. We choose the phase angle of the force such that $\mid \mathbf{F}_0 \mid = F_0$ is a real number.

That means $F(t) = F_0 \exp \mathrm{i}\omega t = F_0 \cos\omega t + \mathrm{i}F_0 \sin\omega t$. We write $x_0 \equiv CF_0$, where the complex factor C is

$$C = \frac{1}{m(\omega_0^2 - \omega^2 + \mathrm{i}\gamma\omega)}\,. \tag{15.37}$$

The factor C may be written as

$$C = \rho \exp \mathrm{i}\theta\,.$$

The solution then becomes

$$x(t) = \rho F_0 \exp \mathrm{i}(\omega t + \theta)\,. \tag{15.38}$$

The solution is the real part of (15.38), i.e.,

$$\mathrm{Re}\{x(t)\} = \rho F_0 \cos(\omega t + \theta)\,.$$

To determine ρ we return to (15.37):

$$CC^* = \rho^2$$

($C^* \equiv$ complex conjugate of C).

$$\begin{aligned}\rho^2 &= \frac{1}{m(\omega_0^2 - \omega^2 + \mathrm{i}\gamma\omega)}\frac{1}{m(\omega_0^2 - \omega^2 - \mathrm{i}\gamma\omega)} \\ &= \frac{1}{m^2\left[(\omega_0^2 - \omega^2)^2 + \gamma^2\omega^2\right]}\,.\end{aligned}$$

The phase angle θ may be found in the following way:

$$\frac{1}{C} = \frac{1}{\rho}\exp -\mathrm{i}\theta = m\left[\omega_0^2 - \omega^2 + \mathrm{i}\gamma\omega\right]\,,$$

$$\begin{aligned}\tan(-\theta) &= \frac{\gamma\omega}{\omega_0^2 - \omega^2}\,, \\ \tan\theta &= -\frac{\gamma\omega}{\omega_0^2 - \omega^2}\,.\end{aligned}$$

Note. Here we have used a sign convention for θ that is the opposite of the one used in Section 15.6.

The solution of the equation of motion becomes

$$x(t) = \mathrm{Re}\left\{\rho F_0 \exp \mathrm{i}(\omega t + \theta)\right\} = \frac{F_0}{m\sqrt{(\omega_0^2 - \omega^2)^2 + \gamma^2\omega^2}}\cos(\omega t + \theta)\,,$$

the same as given in Section 15.6, except for the sign convention on θ.

Clearly the use of complex numbers substantially shortens the calculations. Compare this Section with Section 15.6.

15.12 Problems

Problem 15.1. A homogeneous thin rod of length 1 m may turn without friction around a horizontal axis through one of its endpoints. The acceleration of gravity is g.

(1) Find the period T for small oscillations.

Assume now that the rod is rotating around a horizontal axis through a point at a distance x from one of the ends of the rod.

(2) Determine x such that the period T is as small as possible, and give the value of T.

Problem 15.2. Consider damped harmonic oscillations. Let the coefficient of friction γ be half the value of the one that just gives critical damping.

(1) How many times is the period T larger than it would be for $\gamma = 0$?
(2) Determine the ratio between two successive swings to the same side.

Problem 15.3. Consider an undamped harmonic oscillator ($\gamma = 0$) with eigenfrequency ω_0. An external driving force $F = F_0 \cos \omega t$ (where $\omega \neq \omega_0$) is applied. The equation of motion is

$$\ddot{x} + \omega_0^2 x = \frac{F_0}{m} \cos \omega t .$$

The general solution may be written as

$$x(t) = A \cos \omega_0 t + B \sin \omega_0 t + \frac{F_0}{m} \frac{\cos \omega t}{\omega_0^2 - \omega^2} .$$

(1) Determine the initial conditions such that the undamped oscillator will begin steady state motion immediately.
(2) Determine the value of x_0 and θ in

$$x = x_0 \cos(\omega t + \theta)$$

for the case described in question (1).

Problem 15.4. Consider a damped, forced harmonic oscillator.

(1) Calculate the work performed by the frictional force at resonance, and show that it is equal to the work performed by the impressed force at resonance.

(2) A body is suspended from a spring. The period of oscillations is $T = 0.5$ s. An impressed force $F = F_0 \cos \omega t$ acts vertically on the body.

The amplitude F_0 is $F_0 = 0.1$ N. A frictional force f (proportional to the velocity of the mass) acts on the body too. The amplitude at resonance, A_r, is 5 cm. Determine the damping constant $b \equiv m\gamma$.

Problem 15.5. A tuning fork vibrates with the frequency $\upsilon = 440$ cycles per second ($\omega = 2\pi\upsilon = 2\pi \times 440\ \text{s}^{-1}$). The tuning fork emits 1/10 of its stored energy in 1 second. Determine the Q-value of the tuning fork.

16. Remarks on Nonlinearity and Chaos

16.1 Determinism vs Predictability

In Newtonian mechanics, dynamical systems are described by differential equations which, if supplied with initial conditions, can be integrated to a unique solution at any future time t. A *unique* solution means that for each set of initial conditions there exists only one solution. A set of initial conditions determines exactly one solution. Dynamical systems having this property are called *deterministic*. See Example 12.8.

In simple systems such as occur in the examples and problems of the preceding chapters, the solution can be written explicitly in terms of elementary functions containing the initial conditions as parameters. The correspondence between the initial conditions and solutions is then quite transparent. One may come to suspect that the correspondence is always this 'regular', in particular that two sets of initial conditions that are almost identical will also correspond to two solutions that are almost identical for all times.

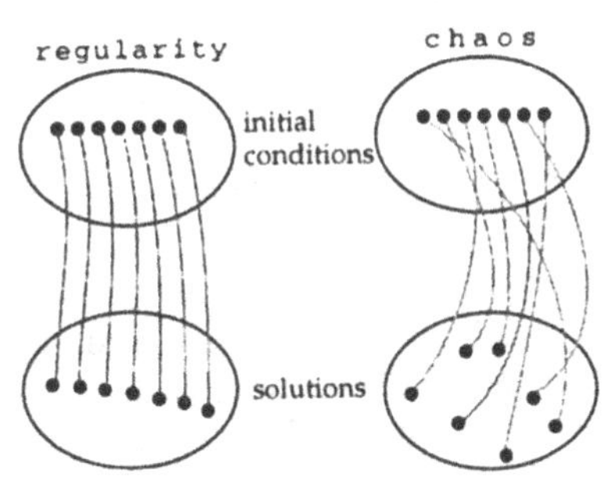

Fig. 16.1.

This, however, is *not* true in general. Many, even conceptually simple, dynamical systems have the property that their solutions depend in an extremely sensitive manner on the initial conditions, in such a way that nearly identical initial conditions quickly evolve to completely different solutions. The number of digits needed to specify an initial condition in order to make a meaningful prediction a time t ahead, can be an exponentially increasing function of t. If, say, one additional decimal place is required in the specification of the initial values for each second ahead that one wishes to predict, the exact solution quickly becomes impossible to predict from an integration of the equations of motion. Dynamical systems with this property are loosely termed 'chaotic', and have recently (not least spurred by the evolution of inexpensive, high-precision, high-speed computers) been the focus of much attention.

A complete understanding and classification of all such systems seems beyond the abilities of present-day mathematics, even though much progress has been made. The problems involved have forced mathematicians and physicists to re-evaluate the very concept of what we mean by the 'solution' to a dynamical system. In the study of such 'chaotic' systems, the emphasis has shifted towards obtaining qualitative, global information about the system, rather than seeking the type of detailed and local solutions discussed in the preceding chapters.

The mathematical techniques involved in gaining even a qualitative understanding of chaotic systems are fairly sophisticated. In this chapter we shall merely try to give the reader a glimpse of some of the aspects of the topic.

It it essential, however, to be aware of the following fact: *even the most 'chaotic' macroscopic system obeys all the fundamental laws of Newtonian mechanics*, summarized in Chapter 12. Recent advances in mathematical and computational techniques have expanded our insight into the many complex types of behavior that solutions to the equations of motion may have. But the principles and conservation laws that we use to establish the differential equations are unchanged. Mechanical systems remain governed by laws and concepts laid down by Isaac Newton more than 300 years ago, and extended by Albert Einstein early in the 20th century.

16.2 Linear and Nonlinear Differential Equations

The mathematical description of mechanical systems that you have seen in the preceding chapters, has been in the form of ordinary differential equations. In mathematics courses you will learn about the many interesting properties of this class of differential equations. It is important to distinguish between linear and nonlinear differential equations. An ordinary differential equation in the unknown function $x(t)$ is said to be *linear*, if the differential equation only contains terms linear in $x(t)$ and its derivatives, i.e., if it can be written in the following form:

$$a_n(t)\frac{\mathrm{d}^n x}{\mathrm{d}t^n} + \cdots + a_1(t)\frac{\mathrm{d}x}{\mathrm{d}t} + a_0(t)x + f(t) = 0 \,, \qquad (16.1)$$

where the functions $a_i(t)$ and the function $f(t)$ are arbitrary. Ordinary differential equations that are not linear, are called *nonlinear*. One can subsequently distinguish all mechanical systems (or, more precisely, all models of systems) into two classes: the linear and the nonlinear systems.

Example 16.1. Superposition. If the $x(t)$-independent term, $f(t)$ is absent, so that (16.1) takes the form

$$a_n(t)\frac{\mathrm{d}^n x}{\mathrm{d}t^n} + \cdots + a_1(t)\frac{\mathrm{d}x}{\mathrm{d}t} + a_0(t)x = 0\,, \tag{16.2}$$

the equation is said to be *homogeneous.* The single most important property of a homogeneous linear ordinary differential equation (sometimes taken to be the defining property) is the fact that any two solutions can be added to form a new solution. If $x(t)$ solves (16.2), and $y(t)$ solves (16.2), then any *linear combination* $z(t) = Ax(t) + By(t)$, where A and B are constants, will also solve (16.2) .

An $x(t)$-independent term, $f(t)$, is sometimes present on the right-hand side (making the equation *inhomogeneous*), and interpreted as a 'driving' or 'input' term. Consider now two linear equations that differ only in these driving terms:

$$\begin{aligned} a_n(t)\frac{\mathrm{d}^n x}{\mathrm{d}t^n} + \cdots + a_1(t)\frac{\mathrm{d}x}{\mathrm{d}t} + a_0(t)x &= f_1(t)\,, \\ a_n(t)\frac{\mathrm{d}^n x}{\mathrm{d}t^n} + \cdots + a_1(t)\frac{\mathrm{d}x}{\mathrm{d}t} + a_0(t)x &= f_2(t)\,. \end{aligned}$$

If $x_1(t)$ solves the equation with $f_1(t)$, and $x_2(t)$ solves the equation with $f_2(t)$, then

$$x_1(t) + x_2(t)$$

will solve the equation

$$a_n(t)\frac{\mathrm{d}^n x}{\mathrm{d}t^n} + \cdots + a_1(t)\frac{\mathrm{d}x}{\mathrm{d}t} + a_0(t)x = f_1(t) + f_2(t)\,.$$

This additive property is called the principle of *superposition.* If an arbitrary function $f(t)$ on the right-hand side can be written as a sum

$$f(t) = \sum_{n=1}^{N} f_n(t)\,,$$

where a solution of the equation with each of the f_n on the right-hand side is known, we can simply sum these, to form the solution to the equation with f on the right-hand side. For instance, as mentioned in Chapter 15, every periodic driving function f can be approximated very accurately by a linear combination of harmonic functions. Thus, if we can find the solution to the differential equation with *any* harmonic function as the driver, we can superpose these solutions to approximate the solution with the arbitrary periodic right-hand side function f. △

The class of *nonlinear* differential equations do *not* in general permit superposition of their solutions (indeed, a simple test for whether a given homogeneous differential equation is linear or not is to check if an arbitrary linear combination of two solutions satisfies the equation). It is among the class of nonlinear equations that we shall find examples of chaotic behavior. Geometry becomes an important tool in this study.

16.3 Phase Space

The form of Newton's second law suggests that many mechanical systems can be modeled by *second order* ordinary differential equations. For each coordinate q used to describe the configuration of the system, one can (formally) regard its time derivative $\dot{q} = w$ as an independent variable w. This allows us to write, say, an equation of motion

$$m\ddot{q} = F(q)$$

as a *system* of first-order equations

$$\begin{aligned} \frac{\mathrm{d}q}{\mathrm{d}t} &= w\,, \\ \frac{\mathrm{d}w}{\mathrm{d}t} &= \frac{1}{m}F(q)\,. \end{aligned} \tag{16.3}$$

The right-hand side of this system describes a vector field in $(q, \dot{q})$-space, and this vector field can, for each set of initial conditions, be integrated to the curve for which the tangent vector at each point is the specified vector field. Along such a curve within the $(q, \dot{q})$-space (often called the *phase plane* or

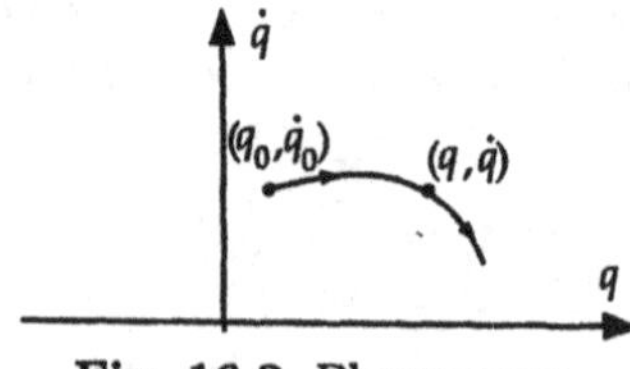

Fig. 16.2. Phase space

the *phase space*),[1] one can track simultaneously both the value of the variable

[1] Many texts refer to this as position-velocity space and reserve the term *phase space* for a more general construction, involving the so-called canonical momentum. In Selected References see, e.g., the book by Goldstein.

q and the value of its time derivative $\dot{q}$ for each solution to the equation of motion. The time evolution of all the phase-space points is called the *phase flow.* See Figure 16.2.

When the force F does not depend on time (as in the system described by equation 16.3), one can find the location of each phase space curve without integrating the system (16.3). For, as discussed in Chapter 8, the total mechanical energy E, which is the sum of kinetic and potential energy:

$$E(q,\dot{q}) = \frac{m}{2}\dot{q}^2 + U(q) ,$$

is constant along the curve in phase space. Consequently, each phase space orbit is a collection of points where the function $E(q,\dot{q})$ is constant, i.e., (part of) a *level curve* for the function $E(q,\dot{q})$. The particular value of the energy on each curve is set by the initial conditions.

Example 16.2. The Simple Harmonic Oscillator. For the simple harmonic oscillator,

$$m\ddot{q} + kq = 0 ,$$

the mechanical energy

$$E = \frac{1}{2}m\dot{q}^2 + \frac{1}{2}kq^2$$

is constant along each solution curve. The points in $(q,\dot{q})$-space, for which

$$\frac{k}{2E}q^2 + \frac{m}{2E}\dot{q}^2 = 1 ,$$

form an ellipse, with semi axis $\sqrt{2E/k}$ and $\sqrt{2E/m}$ respectively. For the harmonic oscillator, each solution curve in phase space is an ellipse. △

A solution curve can consist of a single point. Such a point is called an *equilibrium point,* and is necessarily located on the q-axis (why?).

Equilibrium points correspond to stationary points of the potential energy function U. If the potential energy function has a local maximum, the equilibrium point is unstable (neighboring phase space points are carried away by the phase flow), and if the potential energy function has a local minimum, the equilibrium point is stable (neighboring phase space points are not carried away by the phase flow). For an intuitive picture, consider the motion of a ball rolling on the 'hills' of the potential energy function. The ball will be in stable equilibrium at the bottom of an energy 'valley', and in unstable equilibrium if resting on a 'peak'. See Figure 16.3.

Note. For the equations (16.3), two different solution curves in phase space will never intersect or even touch. This is a simple consequence of the unique-

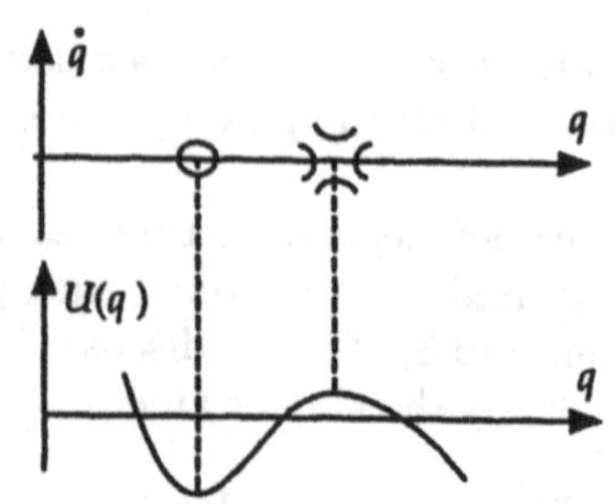

Fig. 16.3. Maxima of the potential function U give rise to unstable equilibrium points, and minima give rise to stable equilibrium points

ness property for the solutions. If two different solution curves did have one point in common, such a point could be seen as one initial condition evolving to two separate solutions of the differential equation, in violation of uniqueness.

Example 16.3. Phase Space of the Pendulum. Consider again the mathematical pendulum (see Section 3.4). We now assume that a mass m is fastened to the end of a stiff rod. We neglect the mass of the rod.

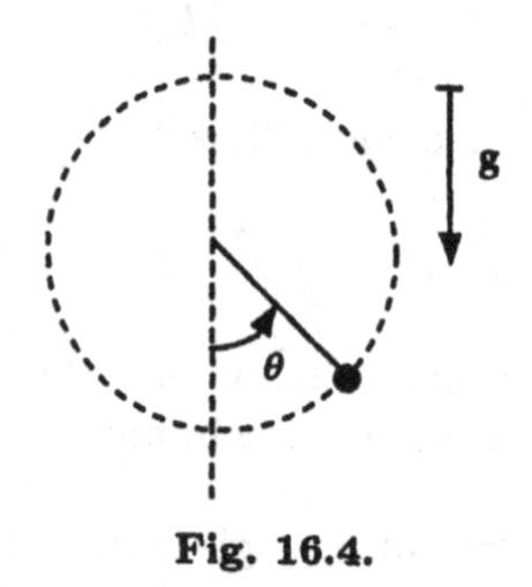

Fig. 16.4.

The configuration of the system is uniquely specified by the angle θ formed with the vertical. The equation of motion,

$$ml\ddot{\theta} = -mg\sin\theta$$

or,

$$l\,\ddot{\theta} + g\sin\theta = 0$$

may be re-written as a system:

$$\left.\begin{aligned} \dot{\theta} &= w \\ \dot{w} &= -g/l\sin\theta \end{aligned}\right\} . \tag{16.4}$$

We first consider the so-called linear (or simple) pendulum, where we assume that the amplitude in the oscillations are so small that $\sin\theta \approx \theta$ (small-oscillation approximation; see Section 3.4). With this simplification, the above equations take the form

$$\left.\begin{aligned}\dot{\theta} &= w \\ \dot{w} &= -g/l\,\theta\end{aligned}\right\} . \tag{16.5}$$

If we choose $\theta(0) = 0$ and $w(0) = w_0$, the equations can be integrated to

$$\begin{aligned}\theta(t) &= w_0\sqrt{l/g}\sin\left(\sqrt{\frac{g}{l}}t\right) ,\\ w(t) &= w_0\cos\left(\sqrt{\frac{g}{l}}t\right) .\end{aligned}$$

This may be immediately recognized as a parameter representation for an ellipse, with semi axes w_0 and $w_0\sqrt{l/g}$ respectively. For each set of initial conditions, the solution traces out an ellipse in the phase plane. The period of oscillation is

$$T = 2\pi\sqrt{\frac{l}{g}} .$$

Figure 16.5 shows a few of these ellipses, corresponding to different values of $w_0 = \dot{\theta}(0)$.

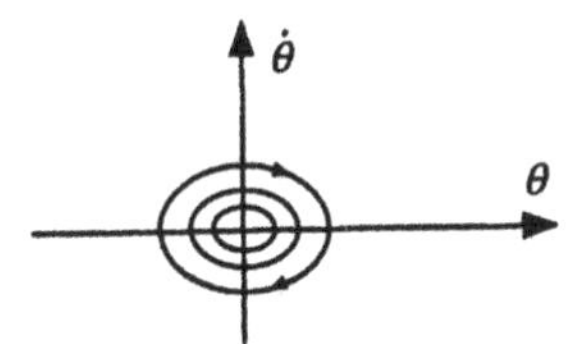

Fig. 16.5. Phase space for a pendulum performing small oscillations

Next we investigate what happens if we do *not* assume a small amplitude, and consider the more general set of equations (16.4).

The phase-space curves of the nonlinear pendulum fall into two classes. We first discuss the curves that resemble those of the linear pendulum and cross the θ axis. When a phase-space curve crosses the θ axis, the pendulum has no velocity at this instant; the pendulum has then swung as far as its initial energy allows, and is turning back (for this reason such points on the solution curves are called *turning points*). These curves are called *libration* curves. Suppose a particular solution has turning point at θ_0. The equation for the mechanical energy along this phase-space curve can then be written as

$$\frac{1}{2}m(l\dot{\theta})^2 + mg(l - l\cos\theta) = mg(l - l\cos\theta_0)$$

or,

$$\begin{aligned}\left(\frac{d\theta}{dt}\right)^2 &= 2\frac{g}{l}(\cos\theta - \cos\theta_0) \\ &= 4\frac{g}{l}\left(\sin^2\frac{\theta_0}{2} - \sin^2\frac{\theta}{2}\right).\end{aligned}$$

Thus,

$$2\sqrt{\frac{g}{l}}\ dt = \frac{d\theta}{\sqrt{\sin^2\frac{\theta_0}{2} - \sin^2\frac{\theta}{2}}}.$$

Placing the origin of time where $\theta = 0$, we integrate to

$$2\sqrt{\frac{g}{l}}\ t = \int_0^\theta \frac{d\theta}{\sqrt{\sin^2\frac{\theta_0}{2} - \sin^2\frac{\theta}{2}}}.$$

We now perform a change of variables. A new variable angle ϕ is defined through the equation

$$\sin\phi = \frac{\sin\frac{\theta}{2}}{\sin\frac{\theta_0}{2}}.$$

Changing to the ϕ variable in the integral, we get

$$\begin{aligned}\sqrt{\frac{g}{l}}\ t &= \int_0^\phi \frac{d\phi}{\sqrt{1 - \sin^2\frac{\theta_0}{2}\sin^2\phi}} \\ &= \mathrm{sn}^{-1}\left(\sin\phi, \sin\frac{\theta_0}{2}\right),\end{aligned}$$

where the function sn is a Jacobi elliptic function.[2] Inverting, we have that

$$\sin\phi(t) = \mathrm{sn}\left(\sqrt{\frac{g}{l}}\ t, \sin\frac{\theta_0}{2}\right),$$

so the $\theta(t)$ curve is given through the equation:

$$\sin\theta(t) = \sin\frac{\theta_0}{2}\,\mathrm{sn}\left(\sqrt{\frac{g}{l}}t, \sin\frac{\theta_0}{2}\right).$$

The elliptic function 'sn' oscillates between +1 and −1 in a manner similar to that of the elementary sine function, and consequently $\sin\theta(t)$ oscillates

[2] See, e.g., D.F. Lawden (1989): Elliptic Functions and Applications, Applied Mathematical Sciences (vol. 80), Springer, Berlin.

between $+\sin\theta_0$ and $-\sin\theta_0$. The period T is equal to four times the duration needed to reach the first turning point ($\phi = \pi/2$):

$$\begin{aligned} T &= \frac{4}{\sqrt{g/l}} \int_0^{\pi/2} \frac{d\phi}{\sqrt{1-\sin^2\frac{\theta_0}{2}\sin^2\phi}} \\ &= \frac{4}{\sqrt{g/l}} K(k) , \end{aligned}$$

with $k = \sin^2\frac{\theta_0}{2}$. The function $K(k)$ is the complete elliptic integral. As the angle $\theta_0 \to 0$, $K(\sin^2\frac{\theta_0}{2}) \to \pi/2$. Thus, for small values of θ_0 (i.e., small oscillations), we recover the familiar oscillation period. For arbitrary values of θ_0, the curves qualitatively resemble the ellipses of the linear pendulum, except that the oscillation periods are no longer independent of the amplitude, rather, *the period increases with increasing amplitude.* For $\theta_0 \to \pi$, $T \to \infty$ (see the discussion of the *separatrix*, below).

The other class of curves in phase space have no turning points and connect the ($\theta = -\pi$) line with the ($\theta = +\pi$) line. Since these two lines represent the same configuration of the pendulum, the curves that connect them with nonzero velocity correspond to rotational states of the pendulum, i.e., the mass moves around in a vertical circle. Such orbits are known as *rotation* orbits. In the positive $\dot{\theta}$ half plane the pendulum rotates counter-clockwise; in the negative $\dot{\theta}$ half plane the pendulum rotates clockwise.

For these solutions, the kinetic energy never vanishes. Writing the conservation of mechanical energy as:

$$\frac{l}{2g}\dot{\theta}^2 - \cos\theta = \lambda ,$$

where λ is a constant, we see (since $\dot{\theta}^2$ is never zero) that the value of λ must be greater than 1 for these curves. Defining a parameter k by $k = \sqrt{2/(\lambda+1)}$, we have that k must vary between 0 and 1; the greater the energy, i.e., the larger the value of λ, the closer k is to 0. Introducing a new angle variable ψ by $\psi = \theta/2$, we write the energy equation as

$$\dot{\psi}^2 = \frac{g}{k^2 l}\left(1 - k^2\sin\psi\right) .$$

Separating variables and integrating, we get, in analogy with the libration orbits,

$$\sqrt{\frac{g}{k^2 l}}\, t = \mathrm{sn}^{-1}(\sin\psi, k)$$

or,

$$\sin\frac{\theta(t)}{2} = \mathrm{sn}\left(\sqrt{\frac{g}{k^2 l}}\, t, k\right) .$$

This leads to a period of

$$T = 2\,k\,\sqrt{\frac{l}{g}}\;K(k)\,,$$

where again $K(k)$ denotes the elliptic integral. For $k \to 0$, the period $T \to 0$; these solutions have large kinetic energies and correspond to the pendulum rotating with large angular velocities.

Separating the two classes of solutions is a pair of curves (called jointly the *separatrix*), which at first sight seems to violate the rule that distinct phase-space curves never touch or intersect. The top half of the separatrix represents the solution for which the pendulum starts out in an upright (inverted) position, then infinitely slowly begins to swing counter-clockwise, completes a full circle, comes back to the upright position with zero velocity and comes to rest there. The lower-half solution is similar, except that the motion takes place in a clockwise direction.

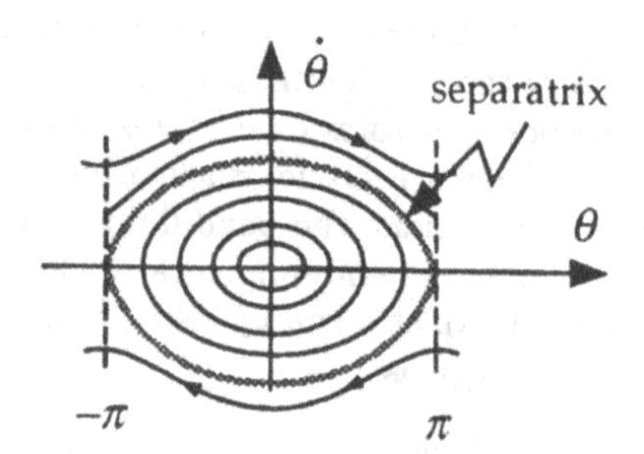

Fig. 16.6. Phase space for the nonlinear pendulum

Both solutions take an infinitely long time to complete the full revolution; most of the time being spent near the upright position where the angular velocity drops to zero. The curves do not touch at any finite time value, but rather originate and terminate at the unstable equilibrium point $\theta = \pm\pi$ as $t \to \pm\infty$, and uniqueness is not violated.

The pendulum is nonlinear, but it is not chaotic. All solutions can be given explicitly in terms of analytic functions. All orbits (except the two separatrix solutions) are periodic, and initial conditions that are sufficiently 'close' will stay 'close' for many oscillation periods.

Note also that near the origin the phase-space plot for the nonlinear pendulum coincides with the phase-space plot for the linear pendulum model. This is as it should be. Nevertheless, the phase portrait for the nonlinear pendulum contains, as we have seen, features that could never be guessed from the linear model. △

Example 16.4. Bifurcation in a Nonlinear Model. Consider a particle having mass m and being able to move without friction on a circular hoop of radius R. The hoop is rotating about a diameter, with an impressed angular velocity ω. This angular frequency is a parameter that can be changed at will. A homogeneous gravitational field $\mathbf{g}$, parallel to the axis of rotation, acts on the system. See Figure 16.7.

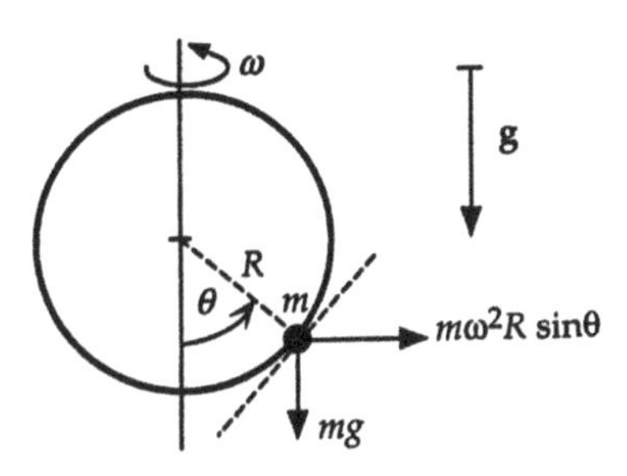

Fig. 16.7. Particle on a rotating hoop

The position of the particle is specified by the angle θ. Three forces act on the particle: the gravitational force, the centrifugal force, and the reaction force from the hoop. The reaction force from the hoop has no component in the tangential direction. Thus, the motion is governed by the competition between the tangential component, $-mg\sin\theta$, of the force of gravity (pulling the mass towards the bottom of the hoop), and the tangential component, $(m\omega^2 R\sin\theta)\cos\theta$, of the centrifugal force.

In the frame rotating with the hoop, Newton's second law for the tangential acceleration is

$$mR\ddot{\theta} \quad = \quad -mg\sin\theta + m\omega^2 R\sin\theta\cos\theta\ .$$

This is a nonlinear differential equation in θ. Note, that when $\omega = 0$, the system becomes the familiar mathematical pendulum, the hoop being simply another way to implement the constraint on the position of the mass.

An equilibrium position for the mass exists where $\ddot{\theta} = 0$, or

$$\omega^2 \sin\theta\left(\cos\theta - \frac{g}{R\omega^2}\right) = 0\ ,$$

i.e., where the two force components in the θ direction exactly balance, or both vanish. The latter only occurs when $\sin\theta = 0$, i.e., at the bottom or the top of the hoop.

For $\sin\theta \neq 0$, the angle $\theta = \theta_e$ will be an equilibrium position if

$$\cos\theta_e = \frac{g}{R\omega^2}\cdot$$

This equation has a solution for θ_e only if ω^2 is so large that

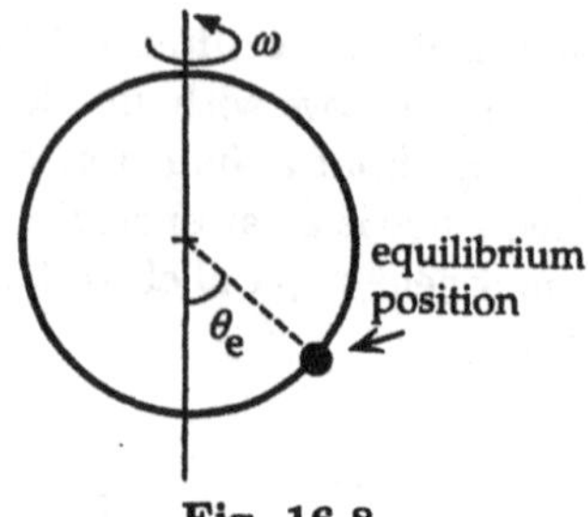

Fig. 16.8.

$$\frac{g}{R\omega^2} \leq 1,$$

or,

$$\omega \geq \sqrt{\frac{g}{R}} \, .$$

The critical value, $\omega = \sqrt{g/R}$, of the impressed angular frequency happens to coincide with the value of the frequency for small oscillations of the particle at about the bottom of the hoop.

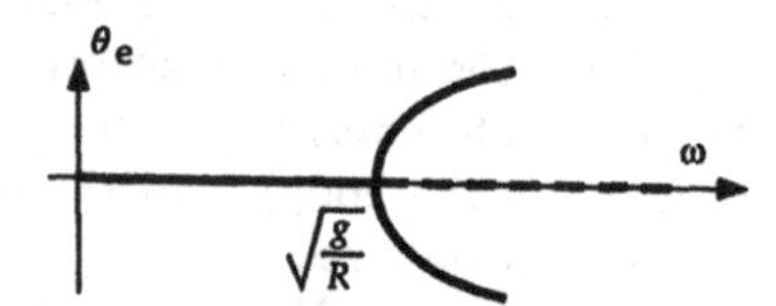

Fig. 16.9. Equilibrium positions as a function of the impressed rotation frequency ω

For values of ω greater than the critical value, the mass will be in equilibrium at the positions (symmetrical about $\theta = 0$):

$$\theta_e = \pm \arccos\left(\frac{g}{R\omega^2}\right) \, .$$

Figure 16.9 shows the θ_e equilibrium position(s) as a function of the impressed hoop rotation parameter ω.

For small values of ω, only the $\theta = 0$ position is an equilibrium. Then, at the critical value of ω, two new equilibrium positions appear, symmetrically around $\theta = 0$. For $\omega \to \infty$, the two new equilibrium values θ_e approach $\pm\pi/2$. Let us examine the motion around each of the equilibrium positions. For fixed ω, the system has, in the rotating frame, a potential-energy function

$$U(\theta) = \frac{1}{4} m\omega^2 R^2 \cos 2\theta - mgR\cos\theta \, .$$

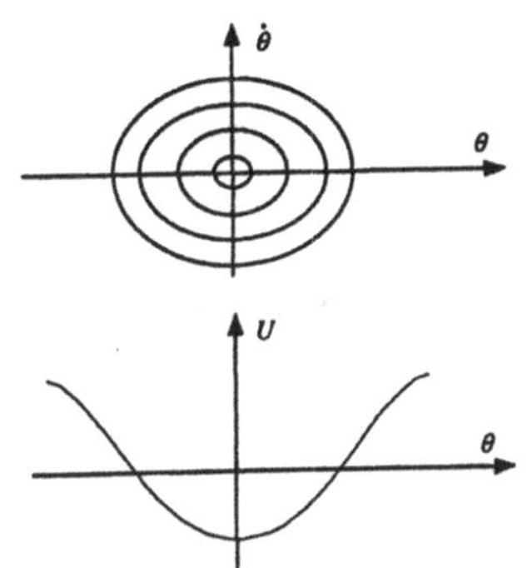

Fig. 16.10. The potential energy and the phase-space flow for ω less than the critical value

Consider first the phase-space plot for the system for small values of ω.

For ω exactly equal to 0, the system is, from a mechanical point of view, identical to the pendulum. For $\omega \ll \sqrt{g/R}$, the potential energy U still has a local minimum at $\theta = 0$, and the phase-space curves remain similar to that of the pendulum. For higher values of ω, the phase-space curves begin to

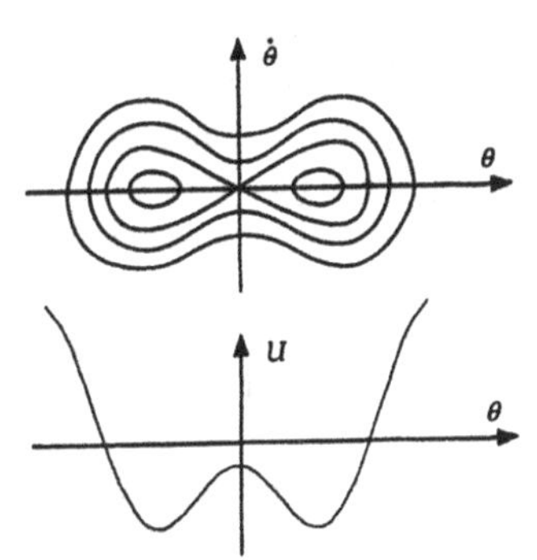

Fig. 16.11. The potential energy and the phase-space flow for ω greater than the critical value

deform, and when ω passes through the value $\sqrt{g/R}$, two new foci appear, symmetrically placed around the origin. The potential-energy function U now has a local maximum at $\theta = 0$, and local minima at $\pm \arccos\left(\frac{g}{\omega^2 R}\right)$.

The two new equilibrium points are consequently *stable*, whereas the origin has turned into an *unstable* equilibrium point. On Figure 16.9, this is indicated by the line at $\theta_e = 0$ being dashed to the right of the critical ω value. We say that the stable equilibrium point at the origin has undergone a *bifurcation* to two new stable equilibrium points.

The analysis and classification of such bifurcations is of immense value in the study of nonlinear differential equations. We refer the reader to the relevant texts given in the Selected References. △

16.4 A Forced, Damped Nonlinear Oscillator

In Chapter 15 we saw that the equation of motion for the periodically forced, damped harmonic oscillator is:

$$\ddot{x} + \gamma\dot{x} + \omega_0^2 x = \frac{F_0}{m}\cos\omega t \, .$$

Here, m is the mass of the particle, γ is the coefficient of friction, ω_0 is the oscillation frequency of the unforced and undamped oscillator, and F_0 is the amplitude of the impressed force. The solutions are well understood and can be expressed in terms of elementary functions, for all values of the parameters in the equation.

We shall now replace the *linear* restoring force $F = -m\omega_0^2 x$, with a different type of restoring force. This is a *nonlinear* restoring force of the form of $F = \alpha x - \beta x^3$, where α and β are positive constants. A restoring force with this dependence on x makes the origin $x = 0$ into an unstable equilibrium point, and the two symmetrically placed points $x = \pm\sqrt{\alpha/\beta}$ into stable equilibrium points, about which the unforced mass may oscillate. A mass moving on a rotating hoop (see Example 16.4) is subject to a force law of similar type when the impressed rotation frequency exceededs the critical value.

With this form for the restoring force, the forced, damped nonlinear oscillator has an equation of motion of the form of

$$\ddot{x} + \gamma\dot{x} - \eta x + \delta x^3 = \frac{F_0}{m}\cos\omega t \, , \tag{16.6}$$

where $\eta = \alpha/m$ and $\delta = \beta/m$. A nonlinear equation of this type is known as a *Duffing equation*, after the German engineer H. Duffing who early in this century studied oscillators of this type.

We do not attempt to solve the Duffing equation. We give only a brief, qualitative description of some of the solutions of the Duffing oscillator (16.6), in order to give a feeling for the problems encountered.

Let us first consider the *undamped* ($\gamma = 0$) and *unforced* ($F_0 = 0$) case. With no forcing or damping, the system is conservative, and the phase-space curves of the solutions have a fairly simple appearance; see Figure 16.12. All motions are bounded, and the motions that encircle both stable equilibrium points are separated from the motions that encircle only one equilibrium point by a pair of separatrix curves, which start from and terminate on the unstable equilibrium point at the origin.

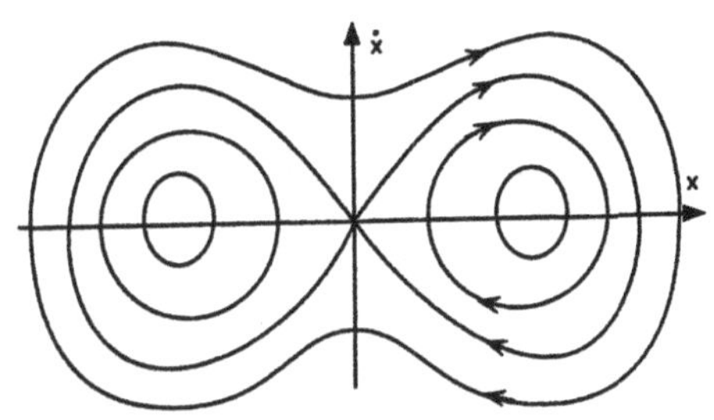

Fig. 16.12. Phase-space curves of the undamped, unforced, nonlinear oscillator

If we introduce damping (friction) into the system, the phase-space portrait changes character. With the exception of two solution curves that terminate at the unstable origin with zero energy, all phase-space curves will now eventually lose energy and spiral in towards one or the other of the stable equilibrium points. Two separatrix curves terminate on the unstable origin. See Figure 16.13. There is no chaos in this system.

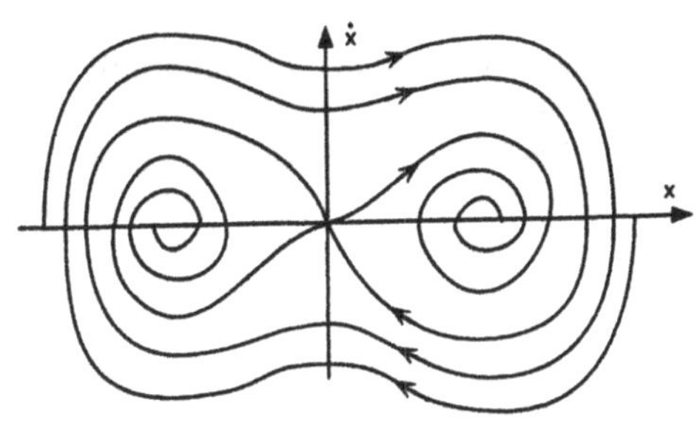

Fig. 16.13. Phase-space curves of the damped, unforced, nonlinear oscillator

Finally we shall briefly describe some of the solutions to the full (i.e., forced and damped) Duffing equation. The time-dependent forcing puts mechanical energy back into the system in competition with the friction force. For some initial conditions, this competition finds a balance, leading to a periodic orbit qualitatively like that of the unforced, undamped oscillator.

For other initial conditions, however, the solution curve winds about in a completely nonperiodic fashion (Figure 16.14).

The fact that such a solution curve has self-intersections in $(x, \dot{x})$-space is a consequence of the time dependence of the system (in $(x, \dot{x}, t)$-space the solution curves do not cross). Two solution curves that begin at nearly the same point in phase space (i.e., with nearly similar initial conditions) quickly evolve into quite different solutions. The system is deterministic, but unpredictable, due to the fact that there is a limit to the precision with which the initial values can be specified. What this limit is, may have some dependence

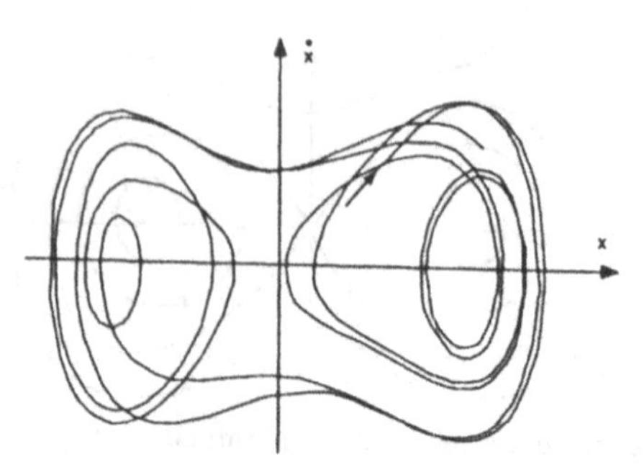

Fig. 16.14. A single phase-space curve of the driven, damped Duffing oscillator

on the exprimental procedure used to determine the initial conditions, but the existence of such a limit is unavoidable.

Clearly, not much can be gained from studying particular nonperiodic solution curves. Instead, we shall turn to a technique known as a *Poincaré section*, after the French mathematician H. Poincaré (1854–1912). In the present case, this amounts to marking, instead of the complete solution curve, only those points for which the time variable is an integer number of, say, the period $2\pi/\omega$ of the forcing (this is a natural time unit in the system).

Instead of a continuous curve, we get a sequence of dots in phase space. After many periods, one finds that the dots settle into a fairly narrow region of phase space (see Figure 16.15) and wander about only within this region.

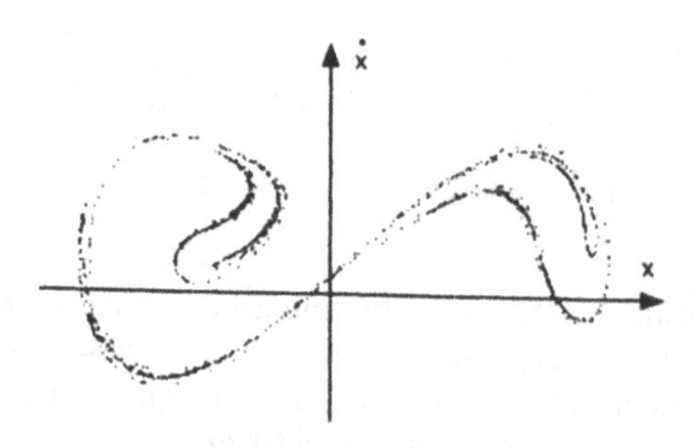

Fig. 16.15. A 'strange attractor' in the Duffing system

Remarkably, many different initial conditions eventually display the same Poincaré section, i.e., the dots on the solution curve are eventually attracted into the same region in phase space. For this reason, such a region is called an *attractor*. Moreover, computer studies indicate that this attracting region has unusual geometric properties; we shall not go into these here, but merely note that they have given rise to the term *strange attractors* being used about this and similar point sets. To get qualitative information about the long-term behavior of the system, one must study the motion on the attractor. Often, the *map* from one dot of the Poincaré section to the next is studied in its own right.

For many initial values, the final state of this forced, damped, nonlinear oscillator is chaotic: a nonperiodic motion on an attracting set, with sensitive dependence on initial conditions.

16.5 Liapunov Exponents

In the phase space for a system, we pick two points close to each other and let each of these points be the initial conditions for a solution curve. See Figure 16.16. In the figure, phase-space curves evolving from such a pair of initial conditions are drawn.

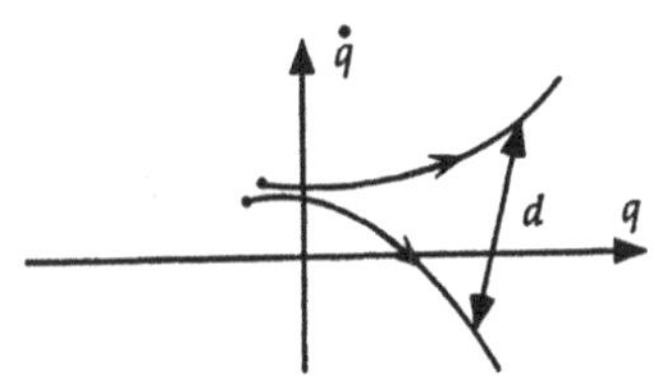

Fig. 16.16. Phase-space curves evolving from two initial conditions

The distance d between the phase-space points at time t tends to increases as time passes. If this distance d increases *exponentially* with time, i.e., if

$$d = d_0 \exp(\lambda t) ,$$

where d_0 and $\lambda > 0$ are constants, the system will evidently have a sensitive dependence on the initial conditions. The positive constant λ gives a quantitative measure of the rate at which solutions corresponding to two nearly identical initial conditions are separated by the phase-space flow. This is the idea behind the so-called *Liapunov exponents.*

Consider in phase space a flow that is generated by the set of equations:

$$\begin{aligned} \dot{q} &= f_1(q, w) , \\ \dot{w} &= f_2(q, w) . \end{aligned}$$

Let $(q_0(t), w_0(t))$ be a particular solution to this system. A measure of how rapidly nearby orbits separate from $(q_0(t), w_0(t))$ can be extracted from the corresponding *linearized system* (the linear approximation to the above flow, around $(q_0(t), w_0(t)$). This linearized system is given by:

$$\begin{bmatrix} \dot{q} \\ \dot{w} \end{bmatrix} = \begin{bmatrix} \frac{\partial f_1}{\partial q} & \frac{\partial f_1}{\partial w} \\ \frac{\partial f_2}{\partial q} & \frac{\partial f_2}{\partial w} \end{bmatrix} \begin{bmatrix} q \\ w \end{bmatrix} ,$$

where the elements of the matrix are to be evaluated along the solution $(q_0(t), w_0(t))$. Next, let $\mathbf{e}$ be a unit vector in phase space at $(q_0(0), w_0(0))$. See Figure 16.17.

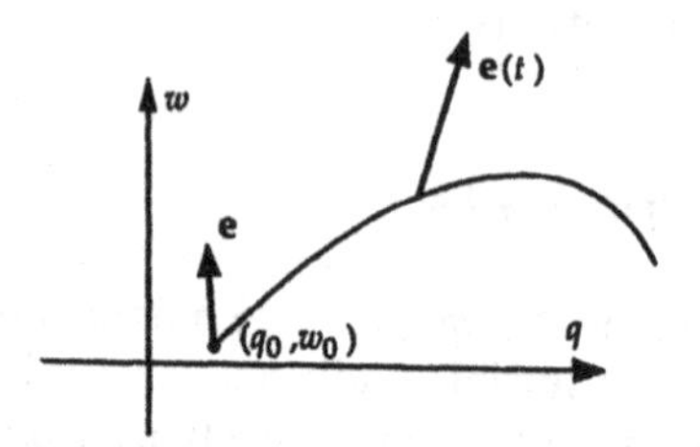

Fig. 16.17. The linearized flow evolves the vector $\mathbf{e}$ to the vector $\mathbf{e}(t)$

The linearized system will evolve the vector $\mathbf{e}$ forward to a new vector $\mathbf{e}(t)$. The Liapunov exponent is designed to extract the average exponential growth rate (if any) of the size of the vector $\mathbf{e}(t)$, as it evolves with the phase-space flow.

We therefore define the Liapunov exponent $\lambda_{(q_0,w_0)}(\mathbf{e})$, for the orbit $(q_0(t), w_0(t))$ and the direction $\mathbf{e}$, to be:

$$\lambda_{(q_0,w_0)}(\mathbf{e}) = \lim_{t\to\infty} \frac{1}{t} \ln \mid \mathbf{e}(t) \mid ,$$

where $\mid \mathbf{e}(t) \mid$ denotes the magnitude of the vector $\mathbf{e}(t)$.

Consider as an example the following (linear) system

$$\begin{bmatrix} \dot{q} \\ \dot{w} \end{bmatrix} = \begin{bmatrix} a & 0 \\ 0 & b \end{bmatrix} \begin{bmatrix} q \\ w \end{bmatrix} . \tag{16.7}$$

As this system is linear, we can immediately write down the solution $\mid \mathbf{e}(t) \mid$ for any initial vector $\mathbf{e} = (e_1, e_2)$:

$$\begin{bmatrix} e_1(t) \\ e_2(t) \end{bmatrix} = \begin{bmatrix} \exp(at) & 0 \\ 0 & \exp(bt) \end{bmatrix} \begin{bmatrix} e_1 \\ e_2 \end{bmatrix} = \begin{bmatrix} e_1 \exp(at) \\ e_2 \exp(bt) \end{bmatrix} .$$

The magnitude of the vector $\mathbf{e}(t)$ is consequently:

$$\mid \mathbf{e}(t) \mid = \sqrt{e_1^2 \exp(2at) + e_2^2 \exp(2bt)} .$$

If, say, $a > b$, and $e_1 \neq 0$, we have that

$$\begin{aligned} \lambda(\mathbf{e}) &= \lim_{t\to\infty} \frac{1}{t} \ln \left(\exp(at) \sqrt{e_1^2 + e_2^2 \exp(2(b-a)t)} \right) \\ &= a + \lim_{t\to\infty} \frac{\ln \left(e_1^2 + e_2^2 \exp(2(b-a)t) \right)}{2t} = a . \end{aligned}$$

Thus, for the flow (16.7), the above limit will extract the *dominant* exponential rate a among the directions for $\mathbf{e}$. This is true in general. For an n-dimensional system, one can find, by varying the vector $\mathbf{e}$, up to n distinct exponents.

For nonlinear systems, Liapunov exponents are mainly accessible through numerical studies,[3] and can be defined in a manner intrinsic to Poincaré maps. The existence of a *positive* Liapunov exponent is generally taken to be one of the indicators for chaos.

16.6 Chaos in the Solar System

With the realization that all the planets in the solar system interact with each other, slightly perturbing each other's orbits, the question arises whether or not these interactions might over time lead to significant changes in the orbital parameters for a given planet. One might imagine that many small perturbations might accumulate and cause a planet to fall into the Sun, or, in the other extreme, to escape from the solar system, drawing its energy to do so from all the other planets combined.

Between the orbits of Mars and Jupiter (see Figure 16.18), a vast number of masses ranging in size from large rocks to small planets orbit the Sun. Though some of the largest are known by individual names, they are all collectively called *asteroids*, and the most densely occupied region between Mars and Jupiter is known as the *asteroid belt*. Each of these small masses orbits the Sun in accordance with Kepler's laws. But each small mass is also subjected to a time-varying gravitational field, particularly from Jupiter (which is by far the most massive of the planets; see Section 14.8).

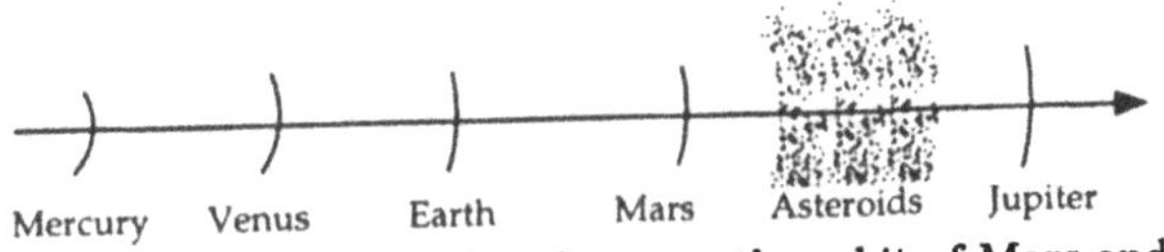

Fig. 16.18. Most asteroids have orbits between the orbit of Mars and the orbit of Jupiter

Responding to the effective potential (see Section 14.6), the radial motion of a planet or an asteroid is, disregarding the influence from other planets, an oscillation around a mean radius. This oscillation is practically undamped.

Consider now the time-varying gravitational tug on an asteroid from Jupiter. The radial equation of motion for the asteroid is now roughly like that of the weakly damped, forced oscillator (see Section 15.5). The magnitude of the gravitational disturbance from Jupiter is quite small compared to that from the Sun, but magnitude alone is not decisive. In Section 15.5 we

[3] see, e.g., Bennetin *et al*, Meccanica, **15** (1980).

saw that a weakly damped oscillator reacts very strongly to a time-varying force if that force oscillates with a frequency near the natural frequency of the oscillator. This is the phenomenon of *resonance*.

How can the asteroid be in resonance with Jupiter, when, as we know from Kepler's third law, they have different semi-major axes and subsequently different periods? The answer is that a *rational* relation between the orbital periods of the two bodies will lead to a similar phenomenon. If, say, Jupiter's period is 5/3 that of the asteroid, the two bodies will return to exactly the same relative position for every 5 orbital revolutions of the asteroid. This type of resonant interaction can, over many orbital periods, lead to a significant change of, e.g., the excentricity of the orbit of the asteroid, causing the asteroid to depart from its nearly circular orbit in the asteroid belt.

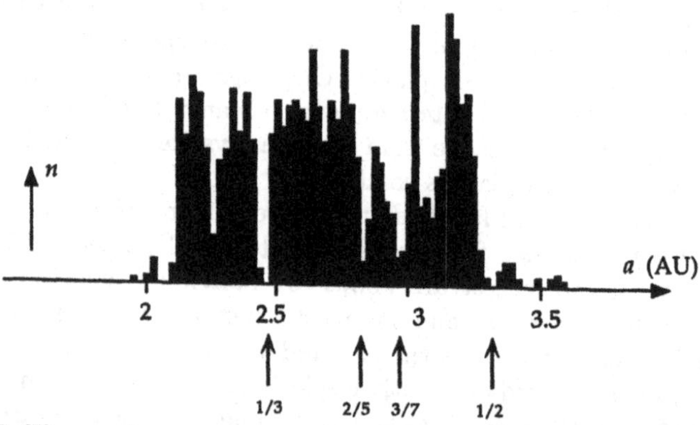

Fig. 16.19. The number n of asteroids as a function of the semi-major axis a of their orbits. The most important resonances (period of asteroid/period of Jupiter) are indicated

The so-called Kirkwood gaps (named after the astronomer who discovered them) are regions in the asteroid belt where the density of asteroids is remarkably low compared to other regions.

A plot of the number of asteroids as a function of the radius (Figure 16.19) clearly shows the Kirkwood gaps. On the figure some of the values of the semi-major axis that (through Kepler's 3rd law) correspond to a rational ratio between the period for an asteroid at that location and the period of Jupiter are indicated.

How will the *planets* affect each other? About 100 years after Newton, the French mathematician Lagrange (1736–1813) demonstrated that it would take at least several millions of years for a dramatic perturbation of, e.g., the Earth's orbit to occur. Thus, from the time perspective of the human civilization, the solar system may be considered as being stable.

Nevertheless, the mathematical question of determining the stability of the planetary orbits *for all time*, still remained open. In 1885, the Swedish

King Oscar II announced a prize contest for a mathematical proof that the planetary orbits would remain stable forever. The prize was awarded to Poincaré who submitted, not the requested proof, but a remarkable investigation into the properties of the simplest nontrivial problem, namely, the problem of three masses interacting gravitationally. Poincaré's pioneering work laid the basis for most of the later studies into what is now known as the theory of deterministic chaos.

Even today, the issue about the long-term stability of the solar system is not settled permanently. Recent extensive numerical calculations[4] strongly suggest that the orbital parameters of most planets vary slowly, but chaotically. The axes of rotation for some planets (e.g., Mars) also seem to be undergoing chaotic fluctuations in their directions.

16.7 Problems

Problem 16.1. Which of the following (systems of) differential equations are linear?

(a) $\exp(t)\ddot{x} - \dot{x} + \sin t \; x = 0 \,,$

(b) $\ddot{x} - 2x\dot{x} + t \; x = 0 \,,$

(c) $\dot{x} = w \,, \quad \dot{w} = \cos\omega t \; x + 2w \,,$

(d) $\dot{x} = w \,, \quad \dot{w} = \sin x \,,$

(e) $\frac{\mathrm{d}^2\psi(x)}{\mathrm{d}x^2} + (E - V(x))\,\psi(x) = 0 \,.$

Problem 16.2. Sketch qualitatively the phase-space curves corresponding to the potential functions below:

(a) $U(x) = 5\sqrt{x^2+5} - 0.5x^2 \,,$

(b) $U(r) = -\frac{1}{r} + \frac{1}{r^2} \,, \quad r > 0 \,,$

(c) $U(r) = -\frac{1}{R}$ for $0 < r < R$ and $U(r) = -\frac{1}{r}$ for $0 < R < r < \infty$.

Problem 16.3. For the nonlinear pendulum (Example 16.3), show that, if we choose $\theta = 0$ for $t = 0$, motion on the separatrix is given by

[4] see, e.g., Laskar and Robutel, Nature **361**, 608, (1993), or J. Wisdom, p. 109 in *Dynamical Chaos*, M. Berry *et al.* (eds.) Royal Society of London (1987).

$$\theta_{\mathrm{sep}}(t) = 4 \arctan \left[\exp \left(\sqrt{\frac{g}{l}}\, t \right) \right] - \pi \quad \text{for} \quad -\infty < t < \infty .$$

Problem 16.4. Show that the Liapunov exponents for the harmonic oscillator

$$\begin{bmatrix} \dot{x} \\ \dot{y} \end{bmatrix} = \begin{bmatrix} 0 & 1 \\ -\omega^2 & 0 \end{bmatrix} \begin{bmatrix} x \\ y \end{bmatrix}$$

all vanish.

Problem 16.5. How many days would the Earth year be shorter, if the Earth were to be in exact 1/12 resonance with Jupiter ?

Appendix. Vectors and Vector Calculus

Addition and Subtraction of Vectors

Physical quantities possessing both magnitude and direction can be described mathematically as *vectors*. A vector can be visualized as an oriented line segment (an arrow). Vectors can be added (and subtracted) to form other vectors, and they can be multiplied by real numbers (see the figure below).

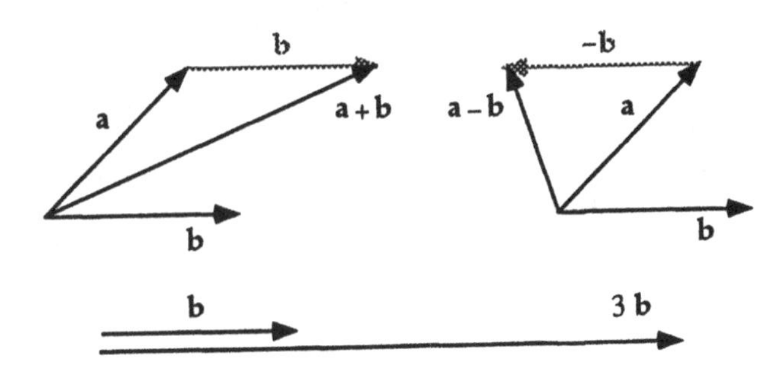

The Dot Product, $\mathbf{a} \cdot \mathbf{b}$

Let $\mathbf{a}$ be a vector with length a, and $\mathbf{b}$ be a vector with length b. The dot product $\mathbf{a} \cdot \mathbf{b}$, also known as the scalar product, is defined by

$$\mathbf{a} \cdot \mathbf{b} \equiv ab \cos(\mathbf{a}, \mathbf{b}),$$

where $\cos(\mathbf{a}, \mathbf{b})$ denotes cosine of the angle between the two vectors. Note that no coordinate system is involved in the definition of the scalar product.

The value of the scalar product does not depend on the orientation of the angle $(\mathbf{a}, \mathbf{b})$ (why?). From this, we see that

$$\mathbf{a} \cdot \mathbf{b} = \mathbf{b} \cdot \mathbf{a}.$$

The scalar product is *commutative*. If $\mathbf{a} \cdot \mathbf{b} = 0$, we say that the vectors are orthogonal. And, since $\cos 0 = 1$, we have that $\mathbf{a} \cdot \mathbf{a} = a^2$; we often use the

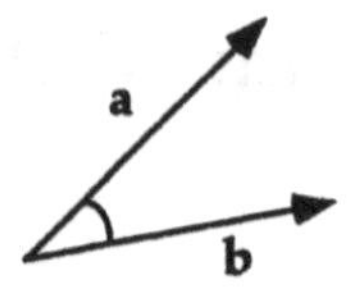

notation $\mathbf{a} \cdot \mathbf{a} = |\,\mathbf{a}\,|^2$. The vector $\mathbf{e_a} \equiv \mathbf{a}/a$ is a unit vector, i.e., a vector of length 1, in the direction of **a**. We can divide vectors with a number, but not vice versa – the scalar product has no inverse. The equation $\mathbf{a} \cdot \mathbf{x} = c$ thus has no unique solution for the vector **x**.

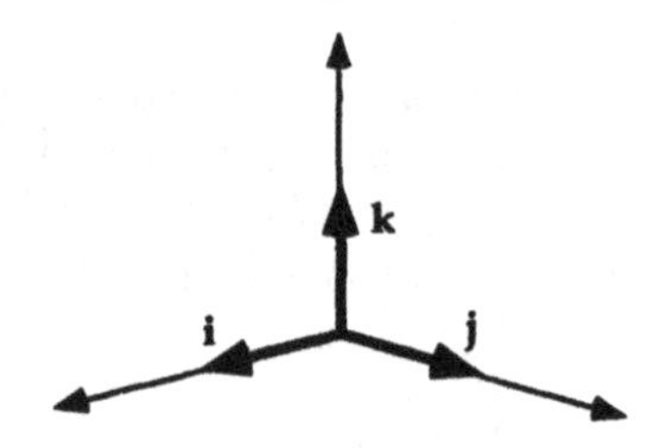

We often employ the set $(\mathbf{i}, \mathbf{j}, \mathbf{k})$ of three unit vectors. These are unit vectors, $\mathbf{i} \cdot \mathbf{i} = \mathbf{j} \cdot \mathbf{j} = \mathbf{k} \cdot \mathbf{k} = 1$, directed along the axis of a cartesian coordinate system, and mutually orthogonal: $\mathbf{i} \cdot \mathbf{j} = \mathbf{j} \cdot \mathbf{k} = \mathbf{k} \cdot \mathbf{i} = 0$.

An arbitrary vector **a** can be written as a combination of the three unit vectors:

$$\mathbf{a} = a_x \mathbf{i} + a_y \mathbf{j} + a_z \mathbf{k}\,.$$

The three numbers, a_x, a_y and a_z are called respectively the x, y, and the z coordinates of the vector **a**. We have that $a_x = \mathbf{a} \cdot \mathbf{i}, a_y = \mathbf{a} \cdot \mathbf{j}$ and $a_z = \mathbf{a} \cdot \mathbf{k}$. We identify a vector with its coordinates, writing $\mathbf{a} = (a_x, a_y, a_z)$. Expressed in terms of coordinates, $\mathbf{a} \pm \mathbf{b} = (a_x \pm b_x, a_x \pm b_x, a_x \pm b_x)$, and $k\mathbf{a} = (ka_x, ka_y, ka_z)$, where k is a constant. Expressed in terms of its coordinates, the length of vector **a** is given by

$$a = \sqrt{\mathbf{a} \cdot \mathbf{a}} = \sqrt{a_x^2 + a_y^2 + a_z^2}\,,$$

and the unit vector in the direction of the vector **a** thus has coordinates

$$\mathbf{e_a} = \left(\frac{a_x}{\sqrt{a_x^2 + a_y^2 + a_z^2}}, \frac{a_y}{\sqrt{a_x^2 + a_y^2 + a_z^2}}, \frac{a_z}{\sqrt{a_x^2 + a_y^2 + a_z^2}} \right).$$

Furthermore, $\mathbf{a} \cdot \mathbf{b} = a_x b_x + a_y b_y + a_z b_z$.

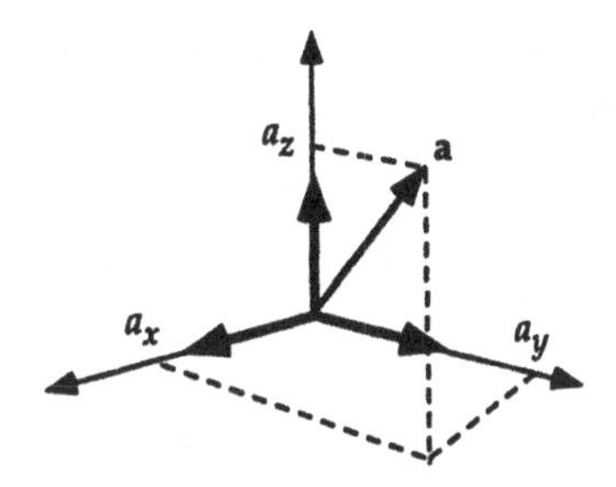

The Cross Product, a × b

Let **a** be a vector with length a, and **b** be a vector with length b. The cross product, also known as the vector product, is defined as

$$\mathbf{a} \times \mathbf{b} \equiv ab \sin(\mathbf{a}, \mathbf{b}) \mathbf{n}(\mathbf{a}, \mathbf{b})$$

where $\mathbf{n}(\mathbf{a}, \mathbf{b})$, is a unit vector giving the direction of $\mathbf{a} \times \mathbf{b}$ according to the following *right-hand rule*:

To find the direction of the cross product, place your right hand *such that the fingers are in the direction of the first vector* **a** *and the palm of your hand is towards the second vector* **b**. *The direction of* $\mathbf{a} \times \mathbf{b}$ *is then perpendicular to the plane spanned by* **a** *and* **b**, *and in the direction of your thumb.*

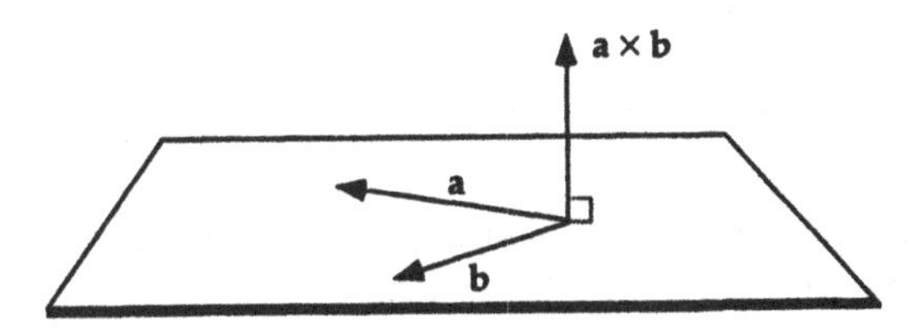

Note that the result of the multiplication this time is a new vector, not a number, and the result does depend on the relative orientation between the vectors **a** and **b**. We employ no coordinate system, but an orientation convention (right handed in the present case). The angle between the vectors **a** and **b** is now the *smallest* angle through which one can turn the vector **a** until it is aligned with **b**. That angle is always between 0 and 180°. If we interchange **a** and **b**, the angle is unchanged, but the right hand rule makes the vector $\mathbf{n}(\mathbf{b}, \mathbf{a})$ point opposite the vector $\mathbf{n}(\mathbf{a}, \mathbf{b})$. Consequently,

$$\mathbf{b} \times \mathbf{a} = -\mathbf{a} \times \mathbf{b} \,,$$

so the vector product is *anti-commutative*. If **b** is along the direction of **a**, $\sin(\mathbf{b}, \mathbf{a}) = 0$, so that, in particular, $\mathbf{a} \times \mathbf{a} = 0$.

In a coordinate system with orthonormal vectors $(\mathbf{i}, \mathbf{j}, \mathbf{k})$, we can find the coordinates for $\mathbf{a} \times \mathbf{b}$ in terms of the coordinates of $\mathbf{a}$ and the coordinates of $\mathbf{b}$.

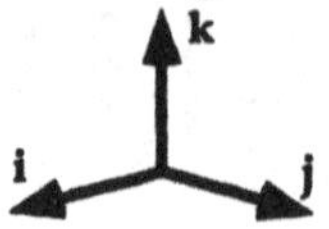

$$\mathbf{i} \times \mathbf{j} = \mathbf{k}$$

$$\mathbf{j} \times \mathbf{k} = \mathbf{i}$$

$$\mathbf{k} \times \mathbf{i} = \mathbf{j}$$

Note: the vectors $\mathbf{i}$, $\mathbf{j}$, and $\mathbf{k}$ are oriented in such a way that they form the basis of a right-handed coordinate system.

Using the above formulae, and the anti-commutative law, one obtains the cartesian coordinate expression for the cross product:

$$\begin{aligned} \mathbf{a} \times \mathbf{b} &= (a_x\mathbf{i} + a_y\mathbf{j} + a_z\mathbf{k}) \times (b_x\mathbf{i} + b_y\mathbf{j} + b_z\mathbf{k}) \\ &= (a_y b_z - b_y a_z)\mathbf{i} + (a_z b_x - b_z a_x)\mathbf{j} + (a_x b_y - b_x a_y)\mathbf{k}. \end{aligned}$$

Briefly,

$$\mathbf{a} \times \mathbf{b} = (a_y b_z - b_y a_z, a_z b_x - b_z a_x, a_x b_y - b_x a_y) .$$

An important formula which we shall use several times, and which can be verified by direct (and repeated) use of the coordinate expression for the vector product, is the so called "bac-cab" formula:

$$\mathbf{a} \times (\mathbf{b} \times \mathbf{c}) = \mathbf{b}(\mathbf{a} \cdot \mathbf{c}) - \mathbf{c}(\mathbf{a} \cdot \mathbf{b})$$

The scalar and vector products give rise to a number of similar identities that can be verified by checking the coordinate expressions of both sides of the equation. Below we list just a couple of these:

$$\mathbf{a} \cdot (\mathbf{b} \times \mathbf{c}) = \mathbf{b} \cdot (\mathbf{c} \times \mathbf{a}) = \mathbf{c} \cdot (\mathbf{a} \times \mathbf{b}),$$

$$(\mathbf{a} \times \mathbf{b}) \cdot (\mathbf{c} \times \mathbf{d}) = (\mathbf{a} \cdot \mathbf{c})(\mathbf{b} \cdot \mathbf{d}) - (\mathbf{a} \cdot \mathbf{d})(\mathbf{b} \cdot \mathbf{c}).$$

Vector Calculus

Consider a vector $\mathbf{r}$ which is the position vector for a particle P (see the figure below). Suppose that $\mathbf{r}$ has coordinates (x, y, z) in some cartesian coordinate system:

$$\mathbf{r} = x\mathbf{i} + y\mathbf{j} + z\mathbf{k}.$$

If the particle P is in motion relative to the coordinate system, the vector $\mathbf{r}$ varies with time in such a way that the coordinate values $(x(t), y(t), z(t))$ are differentiable functions of time. We define the time derivative of the vector $\mathbf{r}$ to be

$$\frac{\mathrm{d}\mathbf{r}}{\mathrm{d}t} \equiv \left(\frac{\mathrm{d}x}{\mathrm{d}t}\right)\mathbf{i} + \left(\frac{\mathrm{d}y}{\mathrm{d}t}\right)\mathbf{j} + \left(\frac{\mathrm{d}z}{\mathrm{d}t}\right)\mathbf{k}.$$

The vector $\mathbf{v} \equiv \mathrm{d}\mathbf{r}/\mathrm{d}t$ is called the velocity of the particle P relative to the chosen coordinate frame. The vector $\mathbf{a} \equiv \mathrm{d}\mathbf{v}/\mathrm{d}t = \mathrm{d}^2\mathbf{r}/\mathrm{d}t^2$ is called the acceleration of the particle P relative to the chosen coordinate frame.

$$\mathbf{a} \equiv \frac{\mathrm{d}\mathbf{v}}{\mathrm{d}t} = \frac{\mathrm{d}^2\mathbf{r}}{\mathrm{d}t^2} = \left(\frac{\mathrm{d}^2x}{\mathrm{d}t^2}\right)\mathbf{i} + \left(\frac{\mathrm{d}^2y}{\mathrm{d}t^2}\right)\mathbf{j} + \left(\frac{\mathrm{d}^2z}{\mathrm{d}t^2}\right)\mathbf{k}.$$

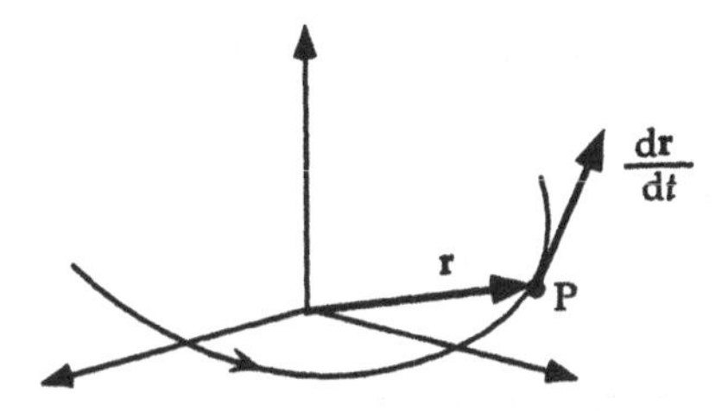

We now consider the particular case of motion confined to a plane.

Suppose the vector $\mathbf{r}$ is the position vector for a particle moving within the plane. The set (x, y) of cartesian coordinates specify the vector $\mathbf{r}$ in the sense that $\mathbf{r} = x\mathbf{i} + y\mathbf{j}$. We shall now introduce another set of mutually orthogonal unit vectors, $(\mathbf{e}_r, \mathbf{e}_\theta)$.

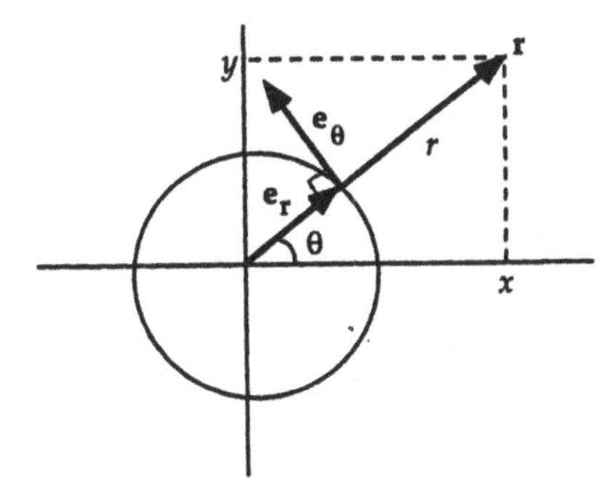

The vector $\mathbf{e}_r$ is the unit vector in the direction of $\mathbf{r}$, and $\mathbf{e}_\theta$ is perpendicular to $\mathbf{e}_r$ and in the direction of increasing θ (see the figure). The two direction vectors are known when θ is known.

If the position of the particle changes with time, $\dot{\mathbf{r}} = ((\mathrm{d}x/\mathrm{d}t), (\mathrm{d}y/\mathrm{d}t))$ and $\ddot{\mathbf{r}} = ((\mathrm{d}^2x/\mathrm{d}t^2), (\mathrm{d}^2y/\mathrm{d}t^2))$ are the cartesian coordinates of the velocity vector $\mathbf{v}$ and acceleration vector $\mathbf{a}$, respectively.

What are the coordinates of the vectors $\mathbf{v}$ and $\mathbf{a}$ expressed by $\mathbf{e_r}$ and $\mathbf{e}_\theta$? We need to know how the vectors $\mathbf{e_r}$ and $\mathbf{e}_\theta$ change with time. In cartesian coordinates these unit vectors are given by

$$\begin{aligned} \mathbf{e_r} &= (\cos\theta, \sin\theta)\,, \\ \mathbf{e}_\theta &= (-\sin\theta, \cos\theta)\,. \end{aligned}$$

By differentiation with respect to time we get

$$\begin{aligned} \frac{\mathrm{d}\mathbf{e_r}}{\mathrm{d}t} = \dot{\mathbf{e}}_\mathbf{r} &= \dot{\theta}(-\sin\theta, \cos\theta)\,, \\ \frac{\mathrm{d}\mathbf{e}_\theta}{\mathrm{d}t} = \dot{\mathbf{e}}_\theta &= \dot{\theta}(-\cos\theta, -\sin\theta)\,. \end{aligned}$$

The direction vectors thus change with time according to

$$\begin{aligned} \dot{\mathbf{e}}_\mathbf{r} &= \dot{\theta}\mathbf{e}_\theta\,, \\ \dot{\mathbf{e}}_\theta &= -\dot{\theta}\mathbf{e_r}\,. \end{aligned}$$

This result can also be seen geometrically by inspection of the figure. The unit vectors change in direction, but not in magnitude.

For the position, velocity and acceleration we obtain by differentiation:

$$\begin{aligned} \mathbf{r} &= r(t)\mathbf{e_r}\,, \\ \dot{\mathbf{r}} &= \dot{r}(t)\mathbf{e_r} + r(t)\dot{\mathbf{e}}_\mathbf{r}\,, \\ \ddot{\mathbf{r}} &= \ddot{r}(t)\mathbf{e_r} + 2\dot{r}(t)\dot{\mathbf{e}}_\mathbf{r} + r(t)\ddot{\mathbf{e}}_\mathbf{r}\,. \end{aligned}$$

By using the formulae for $\dot{\mathbf{e}}_\mathbf{r}$ and $\dot{\mathbf{e}}_\theta$, and rearranging, we finally get

$$\boxed{\begin{aligned} \mathbf{v} &= \dot{\mathbf{r}} = \frac{\mathrm{d}r}{\mathrm{d}t}\mathbf{e_r} + r\frac{\mathrm{d}\theta}{\mathrm{d}t}\mathbf{e}_\theta\,, \\ \mathbf{a} &= \ddot{\mathbf{r}} = \left[\frac{\mathrm{d}^2r}{\mathrm{d}t^2} - r\left(\frac{\mathrm{d}\theta}{\mathrm{d}t}\right)^2\right]\mathbf{e_r} + \left[\frac{1}{r}\frac{\mathrm{d}}{\mathrm{d}t}\left(r^2\frac{\mathrm{d}\theta}{\mathrm{d}t}\right)\right]\mathbf{e}_\theta\,. \end{aligned}}$$

The *polar* coordinates of the velocity and the acceleration vectors are consequently

$$\begin{aligned} (v_r, v_\theta) &= \left(\dot{r}, r\dot{\theta}\right) = \left(\frac{\mathrm{d}r}{\mathrm{d}t}, r\frac{\mathrm{d}\theta}{\mathrm{d}t}\right) \\ (a_r, a_\theta) &= \left(\ddot{r} - r\dot{\theta}^2, 2\dot{r}\dot{\theta} + r\ddot{\theta}\right) = \left(\frac{\mathrm{d}^2r}{\mathrm{d}t^2} - r\left(\frac{\mathrm{d}\theta}{\mathrm{d}t}\right)^2, \frac{1}{r}\frac{\mathrm{d}}{\mathrm{d}t}\left(r^2\frac{\mathrm{d}\theta}{\mathrm{d}t}\right)\right)\,. \end{aligned}$$

Selected References

[1.] Born M. (1962): Einstein's Theory of Relativity, Dover Publications, New York
[2.] Born M. (1949): Natural Philosophy of Cause and Chance, The Clarendon Press, Oxford
[3.] Feynman R. P., Leighton R. B. and Sands M. (1963): The Feynman Lectures on Physics (vol.1), Addison-Wesley, Massachusetts
[4.] French A. P. (1971): Newtonian Mechanics, W. W. Norton, New York
[5.] Goldstein H. (1980): Classical Mechanics (2nd ed.), Addison-Wesley, Massachusetts
[6.] Kittel C., Knight W. D. and Ruderman M. A. (1973): Berkeley Physics Course (vol. 1), McGraw-Hill, San Francisco
[7.] Newton I. (1686): Philosophiæ Naturalis Principia Mathematica, (English translation by Motte A. 1729), University of California Press, Berkeley and Los Angeles (1962)
[8.] Guckenheimer J. and Holmes P. (1986): Nonlinear Oscillations, Dynamical Systems and Bifurcations of Vector Fields, Springer, Berlin
[9.] Ott E. (1993): Chaos in Dynamical Systems, Cambridge University Press, Cambridge
[10.] Wiggins S. (1990): Introduction to Applied Nonlinear Dynamical Systems and Chaos, Springer, Berlin

Answers to Problems

Chapter 1

1.1 (1) $\omega = 33.02\ \mathrm{s}^{-1}$. Between the hand and the first mass
(2) No
1.2 (1) $M = 6.48 \times 10^{23}$ kg
(2) $T = 684$ days
1.3 $S = Mg$
1.4 (1) $S_1 = \frac{m(l\omega^2+\sqrt{2}g)}{2}$, $S_2 = \frac{m(l\omega^2-\sqrt{2}g)}{2}$
(2) $\omega^2 \geq \sqrt{2}g/l$
1.5 $v = \sqrt{Mgr/m}$
1.6 (1) $S = ml\omega^2$
(2) $\cos\theta = g/l\omega^2$
(3) $\sin\theta = 0$
1.7 (1) $S = \frac{m}{2}\left(\frac{l\omega^2}{2} + \sqrt{3}g\right)$, $R = \frac{m}{2}\left(g - \frac{\sqrt{3}}{2}l\omega^2\right)$
(2) $\omega^2 = 2g/\sqrt{3}l$
1.8 (1) $\mathbf{v} = (\sqrt{6}, 2, \sqrt{2})$
(2) $v = 2\sqrt{3}$
(3) $\beta = 54.7°$
(4) $\gamma = 65.9°$
1.9 70.5°
1.10 $t = \frac{3}{2}$ s, $\mathbf{d} = (\frac{1}{2}, -\frac{1}{2}, 0)$ m
1.12 (1) $a_T = 4.9\ \mathrm{m\,s^{-2}}$, $a_N = 0$
(2) $\omega = \dot{\theta} = 0$, $\ddot{\theta} = \dot{\omega} = 7\ \mathrm{s}^{-2}$
(3) $S = 4.24$ N

Chapter 2

2.1 $v_e = 5.03 \times 10^3\ \mathrm{m\,s^{-1}}$, $v_e = 2.38 \times 10^3\ \mathrm{m\,s^{-1}}$
2.2 (1) $h = 22.5$ m
(2) 42.9 m
2.3 $a = g/9$, $S = 2.18$ N
2.4 $f = \frac{M}{m+M}F$

2.5 $\tan\theta > \mu$
2.6 (1) dry $\approx 174.6\ \mathrm{km\,h^{-1}}$ icy $\approx 87.1\ \mathrm{km\,h^{-1}}$
(2) $r_{\mathrm{max}} = 4.4$ cm
2.7 (1) $\mu = \tan\varphi_0$
(2) $F = \frac{\mu M g}{\sqrt{1+\mu^2}}$
2.8 (1) $d_{\mathrm{max}} = 4.9$ cm
(2) $y_0 = 4.9$ cm
(3) $a = g$, upwards
2.9 Hint : $K(\Delta x_1 + \Delta x_2) = k_2 \Delta x_2$; $k_1 \Delta x_1 = k_2 \Delta x_2$
$T = 2\pi\sqrt{\frac{M(k_1+k_2)}{k_1 k_2}}$
2.10 (1) $F = \frac{Mg}{L} z$
(2) $F(r) = \frac{M\omega^2}{2L}(L^2 - r^2)$
2.11 (1) $v_{\mathrm{Moon}}/v_{\mathrm{esc}} = \sqrt{1 - 1/60} = 0.9916$
(2) $v_{\mathrm{max}} \approx 42\ \mathrm{km\,s^{-1}} + 30\ \mathrm{km\,s^{-1}} = 72\ \mathrm{km\,s^{-1}}$
2.12 $F = (m_2/m_1)(M + m_1 + m_2)g$
2.13 (1) $a = (1/3)g$
(2) $S = 70$ kg $= 686$ N
2.14 $v = 331.9\ \mathrm{m\,s^{-1}}$
2.15 $\theta = 45°$
2.16 (1) $f = 3(Mg/L)S$
(2) $f = 3Mg$
2.17 (1) $u = \frac{m}{M+m} v_0$
(2) $T = \frac{v_0}{\mu g(1+m/M)}$
(3) $D = \frac{1}{2}\frac{v_0^2}{\mu g(1+m/M)}$
2.18 (1) $F = \rho a u^2$
(2) $F = 9000$ N
(3) 27 m
(4) $F = \rho a(u - v)^2$
2.20 (1) $u_1 = \frac{m-M}{m+M} v$
(2) $u_2 = \frac{2m}{m+M} v$
(3) $f = \frac{4mM}{(m+M)^2}$, $r = \frac{2}{1+m/M}$
(4) $f = \frac{1}{52}$
2.21 $v = \frac{2M}{M-m} V_0$
2.22 (1) $\mu = 2mg/v_{\mathrm{rel}} \approx 26\ \mathrm{kg\,s^{-1}}$
(2) $b = g/v_{\mathrm{rel}} \approx 3.3 \times 10^{-3}\ \mathrm{s^{-1}}$
2.23 $F = \mu v$

Chapter 3

3.1 (1) 11.4 kg* = 111.7 N

(2) 26.4 kg* = 258.7 N
(3) 84.3 kg*
(4) 55.7 kg*

Chapter 4

4.1 $x'(t') = 5t'^2 + 4$
$v'_x(t') = 10t'\ \mathrm{m\,s^{-1}}$
$a'_x = a_x = 10\ \mathrm{m\,s^{-2}}$

Chapter 5

5.1 (1) $\mathbf{v} = \boldsymbol{\omega} \times \mathbf{r} = 2\pi(2, -4\sqrt{2}, 2\sqrt{2})\ \mathrm{m\,s^{-1}}$
(2) $\mathbf{a} = \boldsymbol{\omega} \times (\boldsymbol{\omega} \times \mathbf{r}) = 4\pi^2(12\sqrt{2}, -4, -20)\ \mathrm{m\,s^{-2}}$
5.2 $a = R\omega^2 = 3.3 \times 10^{-2}\ \mathrm{m\,s^{-2}}$

Chapter 6

6.1 Cork (floats) to the left, lead (sinks) to the right.
6.2 $a = g/\mu$
6.3 (2) $T = 2\pi\sqrt{\frac{l}{a}}$
6.4 $T = 84.4$ min
6.5 $T = 84.4$ min
6.6 $d = 2.66$ mm, east
6.7 (1) $\omega = \sqrt{g/\mu R}$
(2) $v = 8.6\ \mathrm{m\,s^{-1}}$
6.8 (1) Centrifugal force: $m\omega^2\boldsymbol{\rho}$. Coriolis force: $2m(\mathbf{v} \times \boldsymbol{\omega})$
(2) $x_0 = \frac{1}{\omega^2}\sqrt{\mu^2 g^2 - 4v^2\omega^2}$
6.9 $\theta = 30°$
6.10 (1) $\omega_0 = \sqrt{(g/r)\tan\theta}$
(2) $\omega = 2\pi\nu$, $\omega^2 \geq \frac{g}{r}\frac{\sin\theta + \mu\cos\theta}{\cos\theta - \mu\sin\theta}$
(3) $\mu^{-1} = \tan\theta_{\max}$
6.11 (1) $\boldsymbol{\omega} = (0, \omega\cos\Phi, -\omega\sin\Phi)$
(2) $m\ddot{x} = -2m\omega\dot{y}\sin\Phi$
$m\ddot{y} = 2m\omega\dot{x}\sin\Phi$
(3) $y = 10.4$ m
6.13 $\Delta t = \frac{nv}{2(M+m)\mu g}$
6.14 $T = \frac{\pi}{\omega\sin\phi}$

Chapter 8

8.1 31.9 m

8.2 $59.6 \times 10^3 \text{ m s}^{-1} \approx 60 \text{ km s}^{-1}$

8.3 $\cos\theta = \frac{2}{3}$

8.4 (1) $v_0 = 7.9 \text{ km s}^{-1}$
(2) $t_0 = \pi\sqrt{R/g} = 42.2$ min

8.5 (1) $R = \frac{mv_0^2}{r} - mg(2 - 3\cos\theta)$
(2) $\theta = 60°$
(3) $F = 4mg$
(4) $\theta = 45°$

8.6 (1) $h = R/2$
(2) $F = 4mg$, horizontal

8.7 (1) $T_0 = (1/2)mgL$
(2) $u = \frac{2}{n+1}\sqrt{gL}, \quad v = \frac{n-1}{n+1}\sqrt{gL}, \quad q = \frac{4n}{(n+1)^2}$
(3) $n \to \infty, \quad q \to 0$, reflection from infinite mass.
(4) $\cos\theta_0 = 1 - \frac{1}{2}\left(\frac{n-1}{n+1}\right)^2$
(5) $n \to \infty \quad \cos\theta_0 \to 1/2$

8.8 $h_{\max} = a + l^2/(2\pi^2 a)$

Chapter 9

9.1 (1) $\Delta l = v\sqrt{m/k}$
(2) $\Delta l = v\sqrt{\frac{Mm}{(M+m)k}}$

9.2 $x = D\frac{M}{M+m}$

9.3 (1) $d = \mu v\sqrt{\frac{M}{(m+\mu)(M+m+\mu)k}}$
(2) $\tau = T/4 = (\pi/2)\sqrt{n/k}, \quad n = \frac{(m+\mu)M}{m+\mu+M}$
(3) interchange M and m.

9.4 (1) $S = \frac{1}{2\mu g}\left(\frac{m}{m+M}\right)^2 v^2$
(2) $\Delta p \approx 1 \text{ kg m s}^{-1}, \quad \Delta P \approx 720 \text{ kg m s}^{-1}$

9.5 Highest when the people jump off one by one, as described in (b)

9.6 $d = 18.1$ m

9.7 (1) $u = \frac{m}{M+m}v \approx \frac{m}{M}v$
(2) $P = mv$
(3) $v_{\text{CM}} = \frac{m}{2M+m}v \approx \frac{m}{2M}v$
(4) $K = \frac{1}{2}(2M+m)v_{\text{CM}}^2$
(5) $E = \frac{1}{2}\frac{m^2v^2}{M+m} \approx \frac{1}{2}\frac{m^2}{M}v^2$
(6) $\Delta l = \frac{mv}{\sqrt{2kM}}$
(7) $\tau = 2\pi\sqrt{M/2k}$

9.8 $v = \sqrt{2gh}\frac{M+m}{m}$
9.9 (1) $y_{\rm CM} = \frac{1}{2}gt^2$
(2) $a(1) = 2g$
$a(2) = 0$
(3) $z = (L/2) + (mg/2k)\cos\sqrt{2k/m}\,t$

Chapter 11

11.1 $T = 2\pi\sqrt{\frac{3r}{2g}} \approx 1.1$ s
11.2 (1) $T = 2\pi\sqrt{\frac{R^2/2+L^2}{gL}}$
(2) $T = 2\pi\sqrt{\frac{R^2/4+L^2}{gL}}$
11.3 (1) $T_1 = \frac{1}{2}(2M+m)l^2\omega_1^2$
(2) $\omega_2 = \frac{2M+m}{2M}\omega_1\,, \quad T_2 = T_1\left(1 + \frac{m}{2M}\right)$
(3) $\omega_3 = \omega_1\,, \quad T_3 = T_1$
(4) $v = l\omega_1\sqrt{1+\frac{m}{2M}}$
11.4 (1) CM falls along the vertical.
(2) $T = \frac{1}{8}ML^2\dot{\theta}^2\left(\sin^2\theta + \frac{1}{3}\right)$
$U = -Mg\frac{L}{2}(1-\cos\theta)$
(3) $\dot{y} = \sin\theta\sqrt{\frac{3gL(1-\cos\theta)}{4-3\cos^2\theta}}$
11.5 (1) $\mathbf{v}_{\rm CM} = \frac{\mathbf{P}}{M}$
(2) $T = \frac{P^2}{2M}\left(1 + 12\frac{d^2}{L^2}\right)$
(3) OC $= L^2/12d$, below O
$d \geq L/6$
11.6 (1) $u = \frac{m\sqrt{2gh}}{2M+m+I/r^2}$
(2) $Q = mgh\frac{2M+I/r^2}{2M+m+I/r^2}$
11.7 (1) $v = \frac{mgt}{3m+M}$
(2) $S = (1/2)l\,, \quad d = (3/2)l$

Chapter 12

12.1 (1) $v_{\rm CM} = \frac{3\sqrt{2}}{2}\frac{mv}{4M+3m}$
(2) $v_{\rm m} = \sqrt{\frac{2(M+m)ga(\sqrt{2}-1)(4M+3m)}{3m^2}}$
12.2 (1) $v = 5.4\,{\rm m\,s^{-1}}$
(2) $v = 4.4\,{\rm m\,s^{-1}}$
12.3 $\alpha = (mg/Mr)\sin\theta_0$
12.4 $h = (3/2)r$

12.5 $t_0 = \pi\sqrt{\frac{7}{5}\frac{R}{g}} \approx 50$ min.

12.6 (1) $t_r = \frac{2}{7}\frac{v_0}{\mu g}$

(2) $D = \frac{12}{49}\frac{v_0^2}{\mu g}$

(3) $W = \frac{1}{7}MV_0^2$

12.7 $\mu_{\min} = 2a/7g$

12.8 $\omega = \frac{mv_0}{MR+2mR}$

12.9 $F = \frac{1}{4}Mg$, upwards.

12.10 (1) $\omega_0 = \frac{3}{2}\frac{mv}{ML}$

(2) $v = (2M/m)\sqrt{gL/3}$

(3) $F_1 = \frac{5}{2}Mg$

(4) $F_2 = \frac{1}{2}\frac{M}{\Delta t}\sqrt{\frac{gL}{3}}$

(5) $\Delta t \ll T,\ T = 2\pi\sqrt{\frac{2L}{3g}}$

12.11 (1) $v_{\mathrm{CM}} = v/2$

(2) $\omega = 6v/5L$

(3) $\Delta T = (1/10)Mv^2$

(4) $d = (2/3)L$

12.12 (1) $\dot{\theta}^2 = (3g/L)(\sin\theta_0 - \sin\theta)$

(2) $\sin\theta_1 = (2/3)\sin\theta_0$

(3) $\dot{x}_{\mathrm{CM}} = (1/3)\sqrt{gL}$

12.13 (1) $F_1 = (1/28)Mg$

(2) $F_2 = (41/14)Mg$

(3) $h = \left(\frac{7\pi^2}{72} - \frac{1}{2}\right)L$

Chapter 13

13.1 $R = \frac{Mr^3\omega_0^2}{2l^2} + Mg$

13.2 (1) $\Delta\theta = \frac{2ml\sqrt{2gh}}{Mr^2\omega_0} \approx 0.04$ rad $\approx 2°$

(2) $\Delta t \ll \sqrt{2h/g} \approx 1/3$ s

13.3 (1) $\Omega = \sqrt{\frac{2gl}{l^2+r^2/4}}$

(2) $F_{\mathrm{B}} = \frac{1}{2}m\left[g + l\Omega^2 + \frac{r^2}{2a}\omega_0\Omega\right]$

$F_{\mathrm{C}} = \frac{1}{2}m\left[g + l\Omega^2 - \frac{r^2}{2a}\omega_0\Omega\right]$

13.4 $\nu = \omega_0/2\pi = 10.8\ \mathrm{s}^{-1}$

Chapter 14

14.1 (1) $v_0 = 11.2\ \mathrm{km\,s^{-1}}$ (independent of θ)

(2) $v \approx 42\ \mathrm{km\,s^{-1}}$
14.3 (1) $\omega = 6.83 \times 10^{-10}\mathrm{s}^{-1}$
(2) $L_{\mathrm{CM}} = \frac{Mm}{M+m}R^2\omega \approx 1.17 \times 10^{47}\ \mathrm{kg\,m^2\,s^{-1}}$
14.4 $r_{\mathrm{CM}} = 4660\ \mathrm{km}$
$T = 2.35 \times 10^6\ \mathrm{s}, \quad v \approx 1\ \mathrm{km\,s^{-1}}$
14.5 (1) $T = 75.6$ year ≈ 76 year
(2) $V_{\mathrm{P}} = \sqrt{\frac{GM(1+e)}{a(1-e)}} \approx 5.5 \times 10^4\ \mathrm{m\,s^{-1}} \approx 55\ \mathrm{km\,s^{-1}}$
$V_{\mathrm{A}} = \sqrt{\frac{GM(1-e)}{a(1+e)}} \approx 0.09 \times 10^4\ \mathrm{m\,s^{-1}} \approx 0.9\ \mathrm{km\,s^{-1}}$
14.6 (1) $h_{\mathrm{a}} = 1047\ \mathrm{km}$
(2) $T = 1.62 \ \approx 1\ \mathrm{h}\ 37\ \mathrm{min}$
14.7 (1) $v_0 = 27.2\ \mathrm{km\,s^{-1}}$
$v_{\mathrm{E}} = 29.8\ \mathrm{km\,s^{-1}}, \quad \delta v = -2.6\ \mathrm{km\,s^{-1}}$
(2) $\tau = \frac{T}{2} = 145.6$ days
(3) $v_1 = 37.8\ \mathrm{km\,s^{-1}} \quad v_v = 35.0\,,\ \mathrm{km\,s^{-1}}$
(4) $\theta = 54°$

Chapter 15

15.1 (1) $T = 1.64\ \mathrm{s}$
(2) $T_{\mathrm{min}} = 1.52\ \mathrm{s}$
15.2 (1) $T_{\mathrm{d}}/T_{\mathrm{f}} = 2/\sqrt{3}$
(2) $x_2/x_1 = \exp\left(-2\pi/\sqrt{3}\right)$
15.3 (1) $x(0) = \frac{F_0}{m}\frac{1}{\omega_0^2-\omega^2}$
$v(0) = 0$
(2) $x_0 = \frac{F_0}{m}\frac{1}{\omega_0^2-\omega^2}$
$\theta = 0$
15.4 (1) $W = \frac{1}{2}\frac{F_0^2}{m\gamma}$
(2) $b = 1/2\pi\ \mathrm{N\,s\,m^{-1}}$
15.5 $Q \approx 26240$

Chapter 16

16.1 (a), (c), (e)
16.5 About 4 days

Index

Springer-Verlag and the Environment

We at Springer-Verlag firmly believe that an international science publisher has a special obligation to the environment, and our corporate policies consistently reflect this conviction.

We also expect our business partners – paper mills, printers, packaging manufacturers, etc. – to commit themselves to using environmentally friendly materials and production processes.

The paper in this book is made from low- or no-chlorine pulp and is acid free, in conformance with international standards for paper permanency.

Fundamental Physical Constants

Quantity	Value	Unit
Gravitational constant G	6.668×10^{-11}	$\mathrm{Nm^2kg^{-2}}$
Acceleration of gravity g (near the surface of the Earth)	9.8	$\mathrm{m\,s^{-2}}$
Mass of:		
Sun	1.99×10^{30}	kg
Earth	5.98×10^{24}	kg
Mars	6.42×10^{23}	kg
Venus	4.89×10^{24}	kg
Jupiter	1.90×10^{27}	kg
Saturn	5.69×10^{26}	kg
Moon (of the Earth)	7.35×10^{22}	kg
Radius (mean radius) of:		
Earth	6.37×10^{6}	m
Moon	1.73×10^{6}	m
Mars	3.39×10^{6}	m
Jupiter	7.13×10^{7}	m
Sun	6.96×10^{8}	m
Orbital Radius (mean radius) of:		
Earth (1 AU = 1 astronomical unit)	1.495×10^{11}	m
Mars	1.52	AU
Venus	0.72	AU
Jupiter	5.20	AU
Moon (around the Earth)	3.84×10^{8}	m
Orbital Periods (sidereal periods):		
Earth (1 year)	3.16×10^{7}	s
Mars (687 days)	5.94×10^{7}	s
Moon (27.32 days around the Earth)	2.36×10^{6}	s
Rotation Period (relative to the Heliocentric Reference Frame) of		
Earth $\approx$ 23 hours 56 min	8.62×10^{4}	s
Atomic Units		
Mass of electron	9.11×10^{-31}	kg
Mass of proton	1.67×10^{-27}	kg
Charge of proton	1.60×10^{-19}	C
Velocity of light	2.998×10^{8}	$\mathrm{m\,s^{-1}}$

Printed in the United States
By Bookmasters